Remote Sensing of Precipitation

Remote Sensing of Precipitation

Volume 2

Special Issue Editor

Silas Michaelides

MDPI • Basel • Beijing • Wuhan • Barcelona • Belgrade

MDPI

Special Issue Editor
Silas Michaelides
The Cyprus Institute
Cyprus

Editorial Office
MDPI
St. Alban-Anlage 66
4052 Basel, Switzerland

This is a reprint of articles from the Special Issue published online in the open access journal *Remote Sensing* (ISSN 2072-4292) from 2018 to 2019 (available at: https://www.mdpi.com/journal/remotesensing/special_issues/precipitation_rs)

For citation purposes, cite each article independently as indicated on the article page online and as indicated below:

LastName, A.A.; LastName, B.B.; LastName, C.C. Article Title. *Journal Name* **Year**, *Article Number*, Page Range.

Volume 2
ISBN 978-3-03921-287-3 (Pbk)
ISBN 978-3-03921-288-0 (PDF)

Volume 1-2
ISBN 978-3-03921-289-7 (Pbk)
ISBN 978-3-03921-290-3 (PDF)

Contents

About the Special Issue Editor

Silas Michaelides is currently an Adjunct Professor at the Cyprus Institute. He is a former Director of the Department of Meteorology of Cyprus, a position that he reached having served through all the scientific ranks of this governmental organization for more than 40 years. He holds a Ph.D. in Meteorology, an M.Sc. in Agricultural Meteorology, a Master's degree in Public Sector Management, and a B.Sc. in Mathematics. He has published 115 papers in peer-reviewed scientific journals, many of which address precipitation and related issues. He has also published a book on precipitation. He was awarded the International Research Award for Young Scientist by the World Meteorological Organization. He is a member of the European Geosciences Union, a member of the American Meteorological Society, a Fellow of the Royal Meteorological Society and a member of the Hellenic Meteorological Society. He is currently the Vice Chairman of the Cyprus Remote Sensing Society.

remote sensing

MDPI

Article

The Passive Microwave Neural Network Precipitation Retrieval (PNPR) Algorithm for the CONICAL Scanning Global Microwave Imager (GMI) Radiometer

Paolo Sanò [1,*], Giulia Panegrossi [1], Daniele Casella [2], Anna C. Marra [1], Leo P. D'Adderio [1], Jean F. Rysman [1] and Stefano Dietrich [1]

[1] Institute of Atmospheric Sciences and Climate (ISAC), Italian National Research Council (CNR), 00133 Rome, Italy; giulia.panegrossi@artov.isac.cnr.it (G.P.); a.marra@isac.cnr.it (A.C.M.); leopio.dadderio@artov.isac.cnr.it (L.P.D.); jeanfrancois.rysman@artov.isac.cnr.it (J.F.R.); S.Dietrich@isac.cnr.it (S.D.)
[2] SERCO SpA, 00044 Frascati, Italy; daniele.casella@serco.com
* Correspondence: paolo.sano@artov.isac.cnr.it; Tel.: +39-064-993-4267

Received: 22 June 2018; Accepted: 14 July 2018; Published: 16 July 2018

Abstract: This paper describes a new rainfall rate retrieval algorithm, developed within the EUMETSAT H SAF program, based on the Passive microwave Neural network Precipitation Retrieval approach (PNPR v3), designed to work with the conically scanning Global Precipitation Measurement (GPM) Microwave Imager (GMI). A new rain/no-rain classification scheme, also based on the NN approach, which provides different rainfall masks for different minimum thresholds and degree of reliability, is also described. The algorithm is trained on an extremely large observational database, built from GPM global observations between 2014 and 2016, where the NASA 2B-CMB (V04) rainfall rate product is used as reference. In order to assess the performance of PNPR v3 over the globe, an independent part of the observational database is used in a verification study. The good results found over all surface types (CC > 0.90, ME < −0.22 mm h^{-1}, RMSE < 2.75 mm h^{-1} and FSE% < 100% for rainfall rates lower than 1 mm h^{-1} and around 30–50% for moderate to high rainfall rates), demonstrate the good outcome of the input selection procedure, as well as of the training and design phase of the neural network. For further verification, two case studies over Italy are also analysed and a good consistency of PNPR v3 retrievals with simultaneous ground radar observations and with the GMI GPROF V05 estimates is found. PNPR v3 is a global rainfall retrieval algorithm, able to optimally exploit the GMI multi-channel response to different surface types and precipitation structures, that provide global rainfall retrieval in a computationally very efficient way, making the product suitable for near-real time operational applications.

Keywords: satellite precipitation retrieval; neural networks; GPM; GMI; remote sensing

---

1. Introduction

Precipitation is an essential element of the global hydrological and energy cycles and its measurements are of great importance in a variety of research areas, such as climate studies, management of water resources, natural hazards and hydrology. In spite of its crucial role in many aspects of economic and social life on Earth, precipitation estimation still has many problems to overcome to meet the needs of hydrological and climate research and of operational applications.

Basically, precipitation is one of the most difficult atmospheric parameters to measure accurately as its estimate (from satellite and from the ground) is complicated by several factors: its large spatial and temporal variability; its phase (liquid and solid, mixed phase) and its composition in terms of different hydrometeor types, densities and sizes; problems in the conversion of radiometric measurements into quantitative precipitation estimates [1–4].

Rain gauge measurement are the only available direct measurement of precipitation, however they are affected by errors (e.g., effects of wind and evaporation) and suffer from their spatial distribution often too sparse over land to resolve rainfall intensity variability and virtually non-existent over the oceans [5]. Ground-based weather radars, on the other hand, provide high-resolution indirect measurements of precipitation with uncertainties related to the conversion of reflectivity to rain intensity (calibration, range effects, beam-blocking, clutter, etc.) and their geographic coverage is inadequate for global precipitation monitoring.

These problems have highlighted the need to rely on satellite-based observations, which currently represent the most promising way for obtaining long-term global precipitation records. Although significant developments have been made, since the first measurements in the 1970s, quantitative precipitation estimation still poses several problems as the relation between surface precipitation rate and satellite-based observations is very complex. Remote sensing of precipitation from satellites is largely based on estimates made by observing some cloud top properties (e.g., cloud cover and cloud-top temperatures) in visible or infrared (IR) images, or by analysing the effects (absorption and scattering) of rain drops or large ice particles on microwave (MW) radiation. Since the early 1990s several studies have demonstrated that spaceborne passive microwave (PMW) observations have great potential for quantitative precipitation estimates because of the ability of MW radiation to penetrate the precipitating cloud (e.g., [2,6]). However, MW-based precipitation retrieval has to overcome some difficulties, such as: the inability to resolve the extreme variability of precipitation both temporally and spatially; the complex emission and scattering effects by hydrometeors on the upwelling radiation, associated to the extremely variable background surface conditions; the "ambiguity" (or "nonuniqueness") in the relationships between the satellite observed spectral signatures and the surface precipitation [7–9].

The launch of the Tropical Rainfall Measuring Mission (TRMM) (1997) has provided an important opportunity for the development of MW techniques, for the presence onboard of both a Ku-band precipitation radar (PR) and a PMW multi-frequency imaging radiometer (TMI). This has allowed the development of retrieval algorithms that have exploited the combined use of the information of the two sensors [10–12]. Moreover, during its operational period the TRMM-PR radar has provided accurate estimates of instantaneous rain rate, as well as calibration for other precipitation-relevant sensors in sun-synchronous orbits [13–15].

An important step forward towards the improvement of global precipitation monitoring has been achieved in 2014 with the advent of the Global Precipitation Measurement (GPM) mission, thanks to the availability of the NASA/JAXA GPM Core Observatory (GPM-CO) (equipped with the GPM conical-scanning Microwave Imager (GMI) and the Dual-frequency Precipitation Radar (DPR) (Ku and Ka band). The GPM mission contributes to the constellation of pre-existing and future radiometers onboard Low Earth Orbit (LEO) satellites and equipped with precipitation-sensing channels, ensuring 3-hourly global coverage between 65°S and 65°N [16]. One of the GPM mission goals is to provide global precipitation products [16–18] by harmonizing the different products obtained from the heterogeneous constellation of MW sensors. The Goddard Profiling Algorithm (GPROF), that is the NASA official PMW precipitation retrieval algorithm, is a well-known physically based (Bayesian) algorithm, aimed at this goal. Since its first release by Kummerow and Giglio [19], it has continuously improved and evolved: Its current parametric approach allows its use in order to provide operational products for all PMW sensors available in TRMM and GPM era (1997 to present) [7,20]. In this same direction the EUMETSAT H SAF (Satellite Application Facility on Support to Operational Hydrology and Water Management) program [21] has evolved to ensure optimal temporal and spatial monitoring of precipitation. This program, established in 2005, was designed to deliver satellite products (precipitation, soil moisture and snow cover parameters) mainly for operational hydrological applications and precipitation monitoring. The scientific collaboration established in 2014 between

H SAF and GPM, endorsed by EUMETSAT and approved by the NASA PMM Research Program, has officially promoted a joint research activity towards development and refinement of retrieval techniques and validation strategies. In this context PMW instantaneous precipitation rate products exploiting all radiometers in the GPM constellation are being released within H SAF. These products are designed to be readily available and distributed in near-real time, also to be merged with GEO IR observations to provide high spatial and temporal resolution MW/IR precipitation products for operational applications.

The H SAF PMW precipitation products are based on two different precipitation retrieval approaches [22]: the physically based Bayesian Cloud Dynamics and Radiation Database (CDRD) algorithm [23–25], originally designed for conically scanning radiometers and the Passive microwave Neural network Precipitation Retrieval (PNPR) algorithm primarily developed for cross-track scanning radiometers, AMSU/MHS and ATMS [26,27]. In this work, a new algorithm based on the PNPR approach designed for the conically scanning GMI radiometer to provide global rainfall rate estimates for near-real time applications, is presented.

Artificial neural networks (NNs) represent a highly flexible tool alternative to regression and classification techniques, widely applied in an increasing field of environmental sciences for their capability to approximate complex nonlinear and imperfectly known functions to an arbitrary degree of accuracy (e.g., [28–30]). The opportunities offered by their ability to learn and generalize and to be quite robust to noise, have encouraged their use in precipitation estimation from satellite and ground measurements, precipitation being, as mentioned, one of the most difficult of all atmospheric variables to measure. NN techniques have proven to be effective in this area of research and have been successfully used in many rainfall estimation and monitoring applications (e.g., [31–34]).

While previous versions of PNPR algorithms, developed for cross-track scanning radiometers AMSU/MHS (PNPR v1) and ATMS (PNPR v2) (H SAF precipitation products H02 and H18 respectively), were based on a training database obtained from cloud resolving model (CRM) simulations coupled to a radiative transfer equation (RTE) model, the new PNPR for GMI (hereafter referred to as PNPR v3) is based on an observational database. This database is built from global GMI-DPR observations during a period of 27 months between 2014 and 2016. The purpose was to develop a computationally efficient global precipitation retrieval algorithm able to handle the extremely large and rich observational database available from GPM-CO observations, meeting, at the same time, the H SAF requirement of delivering products useful to near-real time operations. It is worth noting that only liquid precipitation is considered in this work, while a separate module dedicated to snowfall retrieval from GMI measurements has been recently developed [35] and will be soon incorporated in PNPR v3.

In this paper the design methodology of PNPR v3 algorithm for GMI is described in detail and the results of a verification study using the NASA DPR-GMI combined product [36] are presented. Two case studies over Italy with a comparison of the PNPR v3 retrieval with co-located ground-based radar observations and with the GPROF (V5) instantaneous precipitation rate estimates, are also analysed. Section 2 presents a brief description of the GMI radiometer characteristics. Section 3 presents the relevant features of the observational database. In Section 4 the design of the NN with the selection of the inputs is described and in Section 5 the rain-no rain classification methodology is presented. In Sections 6 and 7 the results of the verification study and of the two case studies are presented. Finally, Section 8 provides the concluding remarks.

2. The GMI Instrument

The GMI aboard the GPM-CO is a multichannel, conical-scanning, total power MW radiometer equipped with 10 dual-polarization (V and H) window channels at 5 frequencies (10.65, 18.70, 36.5, 89.0 and 166.0 GHz) and three single-polarization (V) channels, one at 23.8 GHz and two in the water vapour absorption band at 183.31 GHz (V polarization).

All these frequencies are actually considered as the most appropriate for detecting the wide spectrum (heavy, moderate and light) of precipitation intensities [16]. The four high frequency, millimetre-wave channels at 166 GHz and 183.31 GHz, can be exploited for light precipitation and snowfall retrieval at higher latitudes (e.g., [37]). Having a low (407 km) orbital altitude and a 1.22 m diameter antenna, GMI can provide, on a 904 km wide swath, a better spatial resolution than most of the previous radiometers (up to roughly 4 km × 7 km at the high frequency channels and around 11 km × 18 km at 19 GHz). Moreover, compared to other radiometers, GMI has significant improvements in the calibration system [38]. Because of these features, GMI represents a significant advancement in satellite PMW imagery.

The central portion of the GMI swath overlaps the Ku-band and Ka-band DPR swaths, 245 km and 125 km wide, respectively. The dual-frequency radar observations are beam-matched over the Ka-band swath, with a horizontal resolution of approximately 5 km and a vertical resolution of 250 m in standard observing mode. The measurements within the overlapped swaths of DPR and GMI are used to provide combined DPR-GMI products, besides DPR (dual-frequency and single frequency) and GMI (GPROF) products. These products serve as a precipitation/radiometric standard for the other GPM constellation members and are used to build a priori databases to support MW precipitation retrieval algorithms [16,17,23,39,40].

3. The Observational Database

The approach based on NNs requires a "training phase", that uses a large sample of data representative of the input and output variables of the retrieval process (in this case the TBs with ancillary parameters and the surface precipitation rate, respectively). The performance of the NN is largely dependent on the completeness and representativeness of the database and on its consistency with the actual observations. In this research, an observational database was used to train the NN. It is worth noting that a database derived from observations has various advantages over one provided by simulations of a CRM coupled to a RTE model. There are, in fact, some limitations associated with the use of CRMs, such as uncertainties in surface property characterization (e.g., surface emissivity), single scattering properties of ice or mixed phase hydrometeors, cloud microphysics parameterizations (particle size distributions, bulk densities, conversion processes), vertical and horizontal distribution of solid and liquid hydrometeors [7,8,11,23].

On the other hand, the use of observational databases is subject to other limitations such as calibration and stability, sensitivity of the instruments and accuracy of the precipitation products used as reference [41–44].

Table 1 presents the main characteristics of the observational database used in this research. The period analysed consisted of 27 months between 2014 and 2016, with about 1120 million observations (precipitating and non-precipitating). The GPM 2B-CMB product (version 04D), which combines DPR and GMI measurements, is used as reference [36,39]. In particular, the precipitation estimates used in the observational database are provided on the Ku-band radar swath and obtained from DPR Ku-band reflectivity and GMI brightness temperatures (1C-R product). Since only liquid precipitation is considered in this study, only pixels where 2B-CMB liquid fraction is larger than 0.8 are selected. The observational database is made of co-located vectors of GMI brightness temperatures (TBs) and 2B-CMB surface rainfall rate matching the GMI high frequency channel nominal resolution (4.4 × 7.2 km), with over 170 million precipitating and over 945 million non-precipitating elements.

Table 1. Description of the observational database features.

Period	1 April 2014 to 30 June 2016
Geographical area	65°S–65°N, 180°W–180°E
Number of precipitating observations	173,901,578
Number of non-precipitating observations	945,897,262
Horizontal resolution	4.4 km × 7.2 km
Reference precipitation product	2B-CMB level-2 GMI/DPR combined V04D on Ku-band radar swath
GMI TBs	1C-R GMI V04C

It is important to add that, during the network design, the complete available database is normally divided into three parts: the ground truth (training) database to be used for the actual training (i.e., for estimating the parameters of the model), the validation database to be used for the verification of the performance of the trained NN and the test database, used to check the validity and usefulness of different NN models during the selection of the optimal NN (O-NN) (see Section 4.2). The use of an independent test sample is generally quite effective in overcoming the overfitting tendency of NN models and for obtaining an unbiased estimate of the NN performance. These three databases are used, in different ways, in the NN design phase. In our study, we have reserved an additional (fourth) part of the database to be used in a verification study once the NN design has been defined.

It should be noted that all the selected databases need to be representative of all the precipitation events within the original complete database. The choice of the size and of the specific members of each database is thus crucial in obtaining an effective evaluation of the final NN performance. Bearing in mind that there are no general guidelines on how to split the complete database into smaller (four, in our case) parts, in our research the sizes of the four obtained databases were defined considering on one side the need to have sufficiently large and representative samples and on the other side the need to reduce the computation time. Considering that in the period chosen for the analysis (27 months), the coincident GMI-DPR pixels are about 170 million globally, in the design phase the sizes of the databases have been set at about 60, 30 and 30 million data for the training, validation and test databases, respectively. The size of the fourth database, reserved for a verification study on the final NN (PNPR v3) performance (Section 6), has been set at about 50 million data. In order to ensure a homogeneous seasonal and spatial distribution of the GMI/ 2B-CMB pixels among the four datasets, an orbit by orbit selection with different steps depending on the relative size of the various databases was carried out. Moreover, a statistical analysis was made by dividing each dataset into four bins of precipitation rate (0.0001–1, 1–10, 10–30, >30 mm h^{-1}) and verifying that the percentage of occurrences in each bin was comparable for all the datasets.

4. PNPR v3

The new PNPR algorithm for GMI (PNPR v3) has been developed following the approach used for PNPR v1 and v2, (for AMSU/MHS and ATMS) in the NN design but with some important innovations (besides the use of the observational database): a new selection procedure of channels to be used as input to the NN, to fully exploit GMI capabilities for global precipitation retrieval (as opposed to the sounding capabilities of AMSU/MHS and ATMS [45]); a reduction of the number of inputs, based on a new principal component analysis (PCA) approach; a new precipitation detection (screening) procedure based on a NN approach. Finally, unlike the previous PNPR algorithms designed for the Meteosat Second Generation (MSG) full disk area (60°S–75°N, 60°W–60°E), PNPR v3 provides precipitation retrieval over the whole globe (between 65°S–65°N).

However, some results of the previous PNPR approaches are retained in the new algorithm. The goal of using a single NN for different types of background surfaces has been preserved in the new version. Usually, the number of NNs used in the precipitation retrieval algorithms is defined so as to optimize the algorithm performance under different operating conditions. For example, distinct NN algorithms are proposed to deal separately with stratiform and convective precipitation (e.g., [46]), or with different surface types (i.e., land or sea [31]). However, the use of different networks can often lead to discontinuity of the estimates in correspondence with transitions between different conditions. The approach of a single NN-based algorithm requires a greater effort in the NN design phase and in the input selection but prevents discontinuities or inconsistencies in the retrieved precipitation patterns.

Another heritage of the previous PNPR algorithms is the use of the TB differences in the water vapour absorption band channels at 183.31 GHz as input to the NN. Opaque channels around 183 GHz were originally designed to retrieve water vapour profiles due to their different sensitivity to specific layers of the atmosphere (e.g., [29,47]). However, these channels have shown great potential for precipitating cloud characterization and for precipitation retrieval [48–52].

4.1. The Neural Network

A detailed description of the neural network design process used in this work can be found in [26,27]. Here only a short summary is given, for the sake of clarity. A NN consists of a number of neurons (also called perceptrons) that exchange information with each other. The NN scheme, shown in Figure 2 in Sanò et al. [26], is characterized by three blocks: (1) the input layer, that receives the input signals, (2) the hidden layer(s) and (3) the output layer, which provides the network response. Each layer holds a number of neurons determined, along with the number of hidden layers, during the design of the network. Each node has its own transfer function and receives, as input, a weighted sum of the outputs of the previous layer. The output of the transfer function corresponds to the output of each node.

The estimation of the weights of each neuron-neuron connection is performed in the NN training phase, during which a training database, providing the network with the inputs (e.g., TBs) and the expected output (e.g., 2B-CMB rainfall rate) is used and the value of each weight is modified to reduce the error between the network and the expected outputs. The training continues in order to minimize the error. At the end of the training, the final values of the weights connecting the neurons of the different layers, store the knowledge of the NN [53]. The design of the network architecture is normally quite complex. Model selection in neural networks aims at finding as few hidden units and neuron-neuron connections, as necessary for a good approximation of the true function. Unfortunately, this is not a simple problem.

4.2. Input Selection and Design of Optimal Neural Network

An important step in the PNPR v3 development was the identification of the optimal set of inputs to the NN, and, the definition of one optimal NN (O-NN) able to provide global rainfall estimates (for all surface types). To this purpose, different typologies of TB-derived variables have been considered and several test networks were designed. In addition to the use of measured TBs in the different GMI channels, also their differences and their linear combinations to maximize the correlation with rain rate (following the CCA approach) [23,27], have been examined. Other TBs derived variables (e.g., Scattering Index (SI) [54,55], Polarization Corrected Temperature (PCT) [32,46,56,57] and TB-space transformation based on Principal Component Analysis (PCA) [12,31] were also analysed. For each set of inputs, the optimal NN (O-NN) was designed (in terms of number of hidden layer and perceptrons) following the cross validation procedure [58,59]. For a detailed description of the procedure see Sanò et al. [26,27].

In the cross validation strategy, the comparison between two models (two different NN for a given set of inputs in this case) is based on the mean square prediction errors (*MSPE*) which is obtained applying the two models to different databases. For this purpose, a test database (see Section 3) is used, divided into M subsets containing n observations each. The model is repeatedly applied to different datasets made of $n(M-1)$ observations, leaving out a different subset each time. The average *MSPE* defines the cross-validation error, *CV*:

$$CV = \frac{1}{M} \sum_{m=1}^{M} MSPE_m$$

In this procedure, the first step consists in determining the number of hidden layers of the NN for a given set of inputs. Starting from a simple architecture, two NNs are compared, one of which contains an additional hidden layer. For both NNs the *CV* is evaluated and, if the more complex unit shows a smaller *CV* error, the additional hidden layer is accepted. The procedure stops when no further hidden layer is able to reduce the *CV* error. At this point, with a similar procedure, the optimal number of perceptrons in each layer (for the same set of inputs) are found. Considering that there is a trade-off between these steps (the number of layers and the number of perceptrons in each layer are interdependent), the design procedure requires alternately tuning the number of layers, the number of perceptrons for each set of inputs. In order to obtain the O-NN about 100 NNs for each set of inputs were tested. Then the procedure was repeated modifying progressively the number or type of inputs and comparing the *CV* errors for the different O-NNs. The optimal set of inputs, associated to the O-NN with minimum *CV* error, was selected.

A preliminary analysis showed that the use of Principal Components (PCs) lead to low *CV* error compared to other tested input variables. Moreover, by estimating the variance of the signal associated with each component, it is possible to recognize the signal related to precipitation with respect to that of the background surface. Furthermore, the PCs selection procedure allows to reduce the size of the training database, simplifying the NN learning process.

The PCA approach was applied to all the GMI channels and to subsets of channels with similar characteristics, that is, water vapour absorption channels and low or high frequency window channels in double polarization. For each subset and for each surface type (vegetated land, arid land, ocean, coast), the PCs with correlation coefficient greater or equal to 0.7 were selected. Then the *CV* approach was used to further reduce the number of PCs and to define four O-NNs (one for each surface type). Following Sanò et al. [27], additional inputs were also considered in this phase: the TB difference in the 183.31 GHz channels (TB183 ± 3–TB183 ± 7 GHz), the surface height (altitude), the total precipitable water (TPW) derived from ECMWF Era Interim forecast (0.125° × 0.125° resolution). Two subsets, respectively composed of channels 10.6 GHz H/V, 18.7 GHz H/V (subset #1, 4 channels) and channels 36.5 GHz H/V, 89.0 GHz H/V, 166.0 H/V GHz (subset #2, 6 channels), produced the best results.

Table 2 lists the PCs selected for each surface type from subset #1 and subset #2. The different PCs are labelled as PCx.y where x is the subset index and y is the order of the PC in that subset (i.e., PC1.3 is the third order PC corresponding to subset #1).

Table 2. List of the first three neural network (NN) inputs, selected for the four surface typologies, based on principal component analysis (PCA).

Vegetated Land	Arid Land	Ocean	Coast
PC 1.2	PC 1.3	PC 1.1	PC 1.2
PC 2.1	PC 2.2	PC 2.1	PC 2.2
PC 2.3	PC 2.3	PC 2.2	PC 2.3

The seven selected inputs are: the PCs listed in Table 2 (three different PCs for each surface), the 183 GHz TB difference, the three ancillary variables (surface height, TPW and the surface type flag). The last step consisted in the design and development of one final O-NN able to provide global surface rainfall rate estimates over all the surfaces. As mentioned in Section 4.1 the NN architecture is composed by three blocks: (1) the input layer, that receives the selected inputs, (2) the hidden layer(s), that contributes to the processing phase and (3) the output layer, that provides the surface rain rate estimate. The resulting NN architecture consists of one input layer with number of perceptrons equal to the number of inputs (seven) and two hidden layers with 23 and 10 perceptrons in the first and in the second layer, respectively. The tan-sigmoid transfer function is used for the input and the hidden layers, while a linear transfer function is used for the output node [26,27].

Table 3 shows the performance of the final O-NN compared with those of the O-NNs trained for different surfaces. The correlation coefficient computed on the test database (RCV) in the *CV* procedure and the *CV*, are indicated. From the table, it turns out that the final O-NN has performance that is on average better than the other NNs optimized for each surface.

Table 3. Performance of the final optimal neural network (O-NN) compared with those of the O-NNs trained for different surfaces.

	O-NN Veg. Land	O-NN Arid Land	O-NN Ocean	O-NN Coast	O-NN Final
R_{CV}	0.81	0.76	0.84	0.78	0.83
CV	0.40	0.56	0.38	0.49	0.39

It is worth adding that the resulting O-NN is extremely fast, allowing a processing time of about 2 min to retrieve the surface precipitation over a complete orbit using a computer equipped with an Intel-I7 processor and 16 Gb of ram.

5. The Rain/No-Rain Classification

In general, the identification of rain areas, or Rain/No-rain Classification (RNC) of pixels, represents a preliminary step to the MW precipitation retrieval and is considered crucial to obtain good performances of a PMW retrieval algorithm [54,60–63].

In our study, the RNC has been based on the NN approach, designed following the *CV* methodology already illustrated in Section 4 and using the same observational database (Section 3). However, some changes were made to the database in order to make it suitable for the discrimination between rain and no-rain areas. First, to the database used for the rainfall rate retrieval NN, consisting only of precipitating pixels, an equal number of non-precipitating pixels was added (about 350 million of elements). Secondly, the precipitation values were grouped into three categories labelled with three RNC flag values: 0 in the absence of precipitation (precipitation rate < 0.1 mm h^{-1}), 1 for precipitation rate between 0.1 mm h^{-1} and 1 mm h^{-1} and 2 for precipitation rate > 1 mm h^{-1}.

The RNC algorithm is based on two different NNs used jointly, each based on different input parameters. The use of two separate networks, was more convenient than the use of multiple parameters in a single more complex network, due to the extremely large size of the training database. This approach allowed us to fully exploit the potential offered by the 13 GMI channels. The first NN input selection was based on the PCA, as for the precipitation retrieval, with the addition of two TB-based indexes widely used for the RNC: the SI [54,55,64] and PCT [56,65]. For the first resulting NN (SNN1), the inputs selected are: PCs listed in Table 2, the 183 GHz TB difference, the SI and PCT (at 89 GHz), the surface type (vegetated and arid land, sea, coast) and the 2-m temperature (T2m), provided by the ECMWF forecast (0.125° × 0.125° resolution).

A second NN (SNN2), supporting the previous one (SNN1), was based on the use of polarization signal as input (TB differences between channels with same frequency and different polarization V and H) that can be very useful not only for the RNC itself but also for the surface characterization [54,66]. The inputs of SNN2 are: the surface type (vegetated and arid land, sea, coast), the ECMWF T2m and the polarization signal (TBv-TBh) of the 5 window channels at 10.65, 18.70, 36.5, 89.0 and 166.0 GHz.

According to the categorization of the training database (associated to three RNC flag values), the two networks return the outputs in terms of the same flag values (0, 1 and 2). A rain/no-rain classification index (RNCI) was therefore built by combining the outputs from the two NNs as shown in Table 4. The value of RNCI is 0 when none of the two NNs detects precipitation (output flags equal to 0 for both SSN1 and SSN2). The value is 1 when only one NN detects precipitation (output flags 0 and 1–2 or 1–2 and 0, for SSN1 and SSN2 respectively). The value is 2 when both NNs indicate the presence of very light precipitation (output flags 1 for both SSN1 and SSN2). When the two NNs find precipitation but in different categories (output flags 1 and 2 or 2 and 1), RNCI is equal to 3. Finally, RNCI equals 4 when both NNs indicate precipitation rate greater than 1 mm h^{-1} (output flags equal to 2 for both SSN1 and SSN2). With this criterion, in addition to exploiting the information of both networks, the RNC returns a parameter (RNCI) that can be used as a preliminary identification of areas with different rainfall intensity.

Table 4. Rain/no-rain classification index (RNCI) values corresponding to the NN outputs.

RNCI	SNN1	SNN2
0	0	0
1	0	1–2
1	1–2	0
2	1	1
3	1	2
3	2	1
4	2	2

Some tests have been done and dichotomous statistical indices have been calculated in order to evaluate the performance of the screening. For this purpose, the new verification database (100 million data not included in the training/validation/test databases) was used and the 2B-CMB rainfall rate (grouped according to the RNC categories) was used as truth. Four cases corresponding to the four RNCI values were analysed and the POD (probability of detection; range 0 to 1, perfect score 1), FAR (false alarm rate; range 0 to 1, perfect score 0), CSI (critical success index; range 0 to 1, perfect score 1) and HSS (Heidke skill score; range $-\infty$ to 1, perfect score 1) were calculated.

Table 5 shows the results obtained. As expected, the POD decreases and FAR decreases as the RNCI increases. CSI and HSS have a maximum value at RNCI equal to 2. The low values of CSI (and HSS) for RNCI ≥ 1, in spite of the POD = 0.96, is due to the high FAR found in this case (FAR = 0.65). Therefore, the best scores are obtained for RNCI ≥ 2 (i.e., when both SSN1 and SSN2 indicate a minimum of 0.1 mm h^{-1} rainfall rate) and this threshold has been selected as the optimal rain-no rain flag indication for PNPR v3.

Table 5. Dichotomous statistical indexes corresponding to RNCI values.

RNCI	POD	FAR	CSI	HSS
≥ 1	0.96	0.65	0.25	0.15
≥ 2	0.72	0.15	0.63	0.67
≥ 3	0.64	0.12	0.59	0.65
4	0.57	0.11	0.54	0.62

In order to illustrate an example of the effect of the different RNCI flags on the rain/no-rain pattern, the results of the screening scheme applied to one GMI overpass of Hurricane Maria near Puerto Rico on 20 September 2017 at 02:03 UTC are shown. Maria, a category 5 hurricane, was moving toward the west-northwest, over the extreme north-eastern Caribbean Sea. On 20 September, the centre of the hurricane was located southeast of Puerto Rico.

Figure 1 shows the measured TBs in the GMI channels at 10.7 GHz (H-pol) and 166.0 GHz (V-pol). The image clearly shows the structure and the eye of the hurricane, with the eye well defined at 166 GHz thanks to its high spatial resolution (around 5 km). Around the eye, regions of higher TBs at 10.7 GHz correspond to regions with lower TB at 166.0 GHz, likely representing convective precipitation embedded in the eyewall.

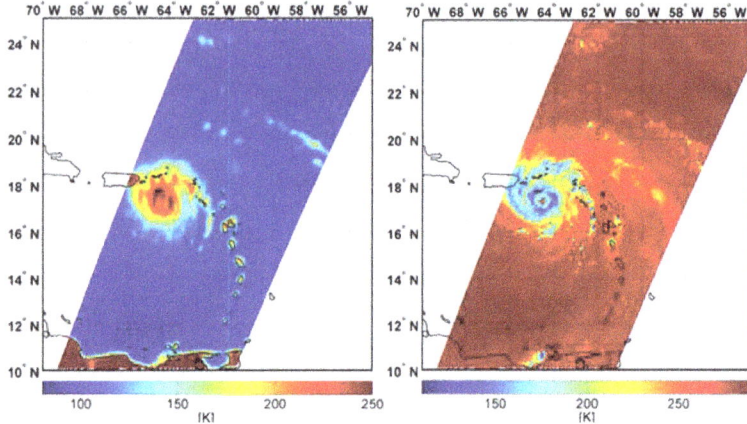

Figure 1. Hurricane Maria southeast of Puerto Rico, on 20 September 2017 at 02:03 UTC. TB (K) at 10.7 H-pol GHz (**left panel**) at 166.0 V-pol GHz (**right panel**) from GMI overpass.

Figure 2 shows the precipitation patterns obtained by the RNC for different RNCI values and the pattern obtained by GPROF V05 with a precipitation threshold of 0.1 mm h^{-1}.

As RNCI increases, the rainfall region gets smaller and for RNCI = 4 it is more or less coincident with the area with a more distinct rainfall emission signal at 10 GHz (higher TBs). The GPROF V05 pattern obtained for a minimum threshold of 0.1 mm h^{-1}, results more similar to the PNPR v3 rainfall pattern obtained for RNCI $\geq$ 1. This example shows that, even if for RNC the best statistical scores globally are obtained for RNCI $\geq$ 2 (Table 5), by providing all the RNCI values (0 to 4), PNPR v3 allows to adjust the rain/no rain pattern for specific cases and to perform analysis of the effect of RNC on the results.

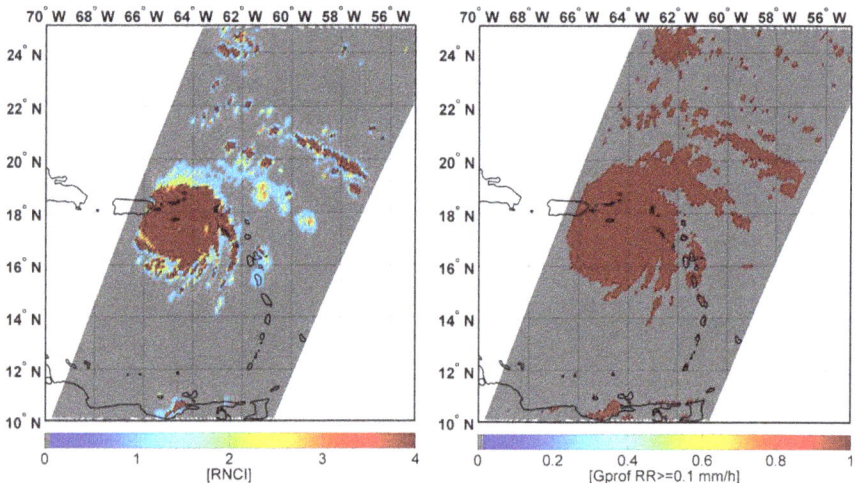

Figure 2. Hurricane Maria southeast of Puerto Rico, on 20 September 2017 at 02:03 UTC. **Left panel** shows the rainfall patterns obtained from Passive microwave Neural network Precipitation Retrieval (PNPR) v3 with the different RNCI values (0 to 4). The **right panel** shows the pattern obtained from the Goddard Profiling Algorithm (GPROF) V05 with a minimum rainfall rate threshold of 0.1 mm h^{-1}.

6. Verification Study

An assessment of the final PNPR v3 retrieval has been carried out using the verification database (i.e., an independent part of the observational database, based on 2B-CMB V04 product, not used in the training and design phase of the algorithm, see Section 3). Figure 3 shows a pixel-based comparison of the surface rain rate estimates by PNPR v3 and the 2B-CMB, for vegetated and arid land, ocean and coast. Only pixels where both products provide rainfall rate ≥ 0.1 mm h^{-1} (hits) were considered. Pixels with 2B-CMB liquid fraction <0.8, or with surface height greater than 2000 m, or with likely presence of ice or snow on the ground (according to the Snow Depth and Sea Ice Cover available from the ECMWF Era Interim re-analysis (0.125° × 0.125° resolution) have been also eliminated. Therefore, in the comparison liquid precipitation only is considered and cases that might likely be affected by larger uncertainty (in both products) are excluded. The figure shows a general good consistency between the two products, with a quite homogeneous trend in all the panels. Most of the points are close to the main diagonal for both vegetated land and ocean, with slight overestimation of very low precipitation (rain rates less than 1 mm h^{-1}), extending to moderate precipitation (rain rate up to 10 mm h^{-1}) mostly over ocean. A similar but less marked, trend is also found for arid land and coast.

The agreement between PNPR v3 and 2B-CMB points out the good outcome of the network training, as well as of its optimization and complex input selection procedure, allowing the use of one unique NN over the different surface types. PNPR v3, applied to an independent global dataset, with precipitation rates extending over a wide range of values (up to 80 mm h^{-1}), shows a good ability to retrieve global precipitation without anomalous inhomogeneities in the estimates.

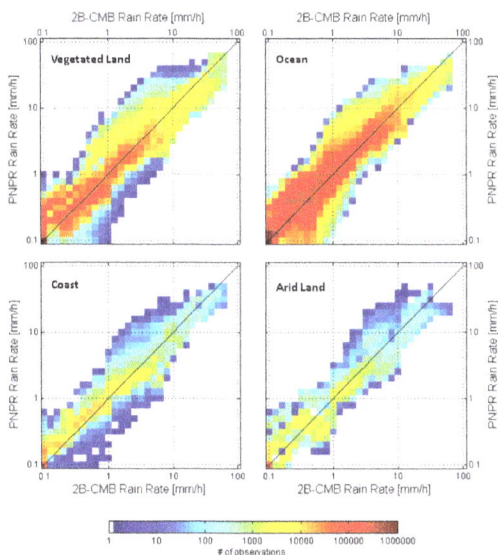

Figure 3. Density scatter plots of pixel based comparison of surface rain rate estimates from PNPR v3 and 2B-CMB (see Table 1) over different surfaces: vegetated land (**top left**), ocean (**top right**), coast (**bottom left**) and arid land (**bottom right**). Only pixels where both PNPR v3 and 2B-CMB provide liquid precipitation (hits) are shown.

Table 6 shows the contingency table of PNPR v3 versus 2B-CMB. Each column represents the rain rate class for 2B-CMB, while each row represents the rain rate class for the PNPR v3. Four rainfall rate intervals were selected in this comparison, 0.1–1 mm h^{-1}, 1–10 mm h^{-1}, 10–30 mm h^{-1} and ≥ 30 mm h^{-1}. The percentages shown in a given column, provided for the four surface types, represent how PNPR v3 classifies the precipitation assigned to the corresponding 2B-CMB class. There is a general consistency between PNPR v3 and 2B-CMB estimates, as shown by the largest percentages found on the main diagonal for each surface type. In the first column, corresponding to rain rate lower than 1 mm h^{-1}, where about 84–89% of hits are found (depending on the background surface), the agreement is particularly evident. Also for the second column the percentages of hits are fairly large, between 80% and 89%. For the third and fourth columns, corresponding to higher values of precipitation, the percentages on the main diagonal are lower, highlighting an underestimation (up to 44.5% of cases for vegetated land) and a maximum overestimation of 7.9% (arid land).

Table 6. Contingency table of PNPR v3 retrievals relative to 2B-CMB combined product estimates (pixel based).

		2B-CMB	Rain Rate (mm h^{-1})	
Vegetated Land	0.1 < 2B-CMB ≤ 1	1 < 2B-CMB ≤ 10	10 < 2B-CMB ≤ 30	2B-CMB > 30
0.1 < PNPR v3 ≤ 1	88.4%	9.2%	0.0%	0.0%
1 < PNPR v3 ≤ 10	11.6%	87.7%	36.3%	0.0%
10 < PNPR v3 ≤ 30	0.0%	3.1%	63.7%	44.5%
PNPR v3 > 30	0.0%	0.0%	0.0%	55.5%
Ocean	0.1 < 2B-CMB ≤ 1	1 < 2B-CMB ≤ 10	10 < 2B-CMB ≤ 30	2B-CMB > 30
0.1 < PNPR v3 ≤ 1	89.1%	9.2%	0.0%	0.0%
1 < PNPR v3 ≤ 10	10.9%	88.9%	38.4%	0.0%
10 < PNPR v3 ≤ 30	0.0%	1.9%	58.3%	40.5%
PNPR v3 > 30	0.0%	0.0%	3.3%	59.5%
Coast	0.1 < 2B-CMB ≤ 1	1 < 2B-CMB ≤ 10	10 < 2B-CMB ≤ 30	2B-CMB > 30
0.1 < PNPR v3 ≤ 1	86.6%	3.1%	0.0%	0.0%
1 < PNPR v3 ≤ 10	13.4%	80.1%	36.3%	0.0%
10 < PNPR v3 ≤ 30	0.0%	16.8%	57.1%	29.8%
PNPR v3 > 30	0.0%	0.0%	6.6%	70.2%
Arid Land	0.1 < 2B-CMB ≤ 1	1 < 2B-CMB ≤ 10	10 < 2B-CMB ≤ 30	2B-CMB > 30
0.1 < PNPR v3 ≤ 1	84.2%	9.6%	0.0%	0.0%
1 < PNPR v3 ≤ 10	15.8%	80.7%	31.2%	0.0%
10 < PNPR v3 ≤ 30	0.0%	9.7%	60.9%	19.9%
PNPR v3 > 30	0.0%	0.00%	7.9%	80.1%

Table 7 shows the continuous statistical indexes calculated for the different surfaces (mean error (ME), mean absolute error (MAE), root mean square error (RMSE), standard deviation (SD), adjusted root mean square error (ARMSE) and correlation (CC)) obtained for PNPR v3 in the comparison with 2B-CMB estimates. The adjusted root mean square error is the RMSE corrected removing the bias [67].

Table 7. Statistical indexes obtained in the comparison of PNPR v3 and 2B-CMB estimates.

	Vegetated Land	Arid Land	Ocean	Coast
N. PIXELS	7,525,522	346,799	36,745,126	696,054
ME (mm h^{-1})	−0.21	−0.22	0.10	−0.20
MAE (mm h^{-1})	0.94	0.90	0.60	0.64
RMSE (mm h^{-1})	2.75	2.60	1.62	1.86
SD (mm h^{-1})	2.78	2.63	1.65	1.90
ARMSE (mm h^{-1})	2.74	2.59	1.63	1.85
CC	0.90	0.92	0.91	0.93

The table confirms the good agreement between the two products, with very low values of the errors (ME ranging from -0.22 to 0.10 mm h^{-1} and MAE from 0.60 to 0.94 mm h^{-1}), low RMSE (less than 2.75 mm h^{-1}) and a good correlation (≥ 0.90).

A further test on the performance of PNPR v3 was carried out by analysing the relative fractional standard error percentage (FSE%), that is, the ratio between RMSE and the mean "true" value, as a function of the 2B-CMB mean rainfall rate value computed for different rain rate intervals (0.5 mm h^{-1} bins):

Figure 4 shows a general agreement in the trends over the four surfaces. For precipitation rates lower than 1 mm h^{-1}, FSE% drops from 200 to 250% to values below 100% for all surface (with high FSE% for arid land and vegetated land). For rain rates between 1 and 10 mm h^{-1}, the FSE% varies for the different surface types between 80% and 40% (with overall better scores over ocean), while for higher rain rates FSE% does not vary much across the different surface types, ranging between 30% and 50%.

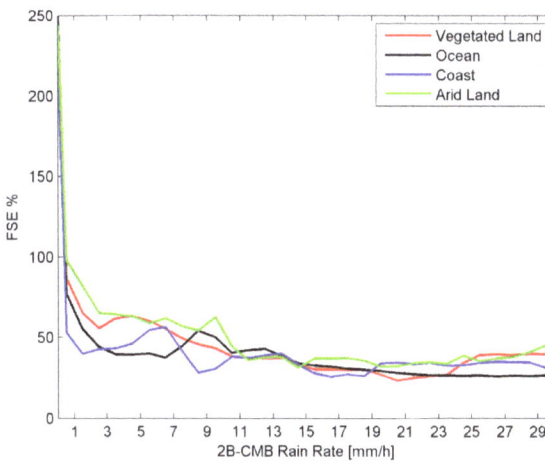

Figure 4. Standard error percentage (FSE%) of PNPR v3 retrieval with respect to the mean 2B-CMB rainfall rate value computed within 0.5 mm h^{-1} bins.

The results in the verification study were obtained using the same product (2B-CMB NS V04) used in the training phase of the algorithm, although for a completely independent dataset. It should be noted that, given the homogeneity of the (independent) verification database with the databases used in the training and design phase of the algorithm, good results on the performance of PNPR v3 were reasonably expected. In the next section two case studies over Italy are analysed, where ground-based radar precipitation estimates are used for comparison.

7. Case Studies

The first case study is a heavy rainfall storm that struck the Tuscany region, particularly the city of Livorno (Lat 43°32'N, Lon 10°19'E), between 9 and 10 September 2017. The convergence of a flow of warm, moist air of North Atlantic origin, from the Channel of Sicily towards the Tuscan archipelago with cooler winds from the west, gave rise to a condition of great instability, with heavy rainfall enhanced by the orography of the area.

Figure 5 shows the PNPR v3 and GPROF V05 precipitation estimates for the GMI overpass on 10 September 2017 at 01:17 UTC, compared with quality-controlled rainfall rate estimates of national radar network mosaic (at 01:20 UTC), provided by the Department of Civil Protection of Italy, used as reference [68].

In the figure, the PNPR v3 (left panel) and GPROF V05 (middle panel) precipitation estimates on the area covered by the Italian radar network are shown. Simultaneous radar observations (right panel) were averaged to match the nominal resolution of the corresponding satellite retrievals using a two-dimensional Gaussian function with an elliptic horizontal section (oriented considering the orientation of the satellite orbit with respect to the surface at the time of the overpass, see [69] for details).

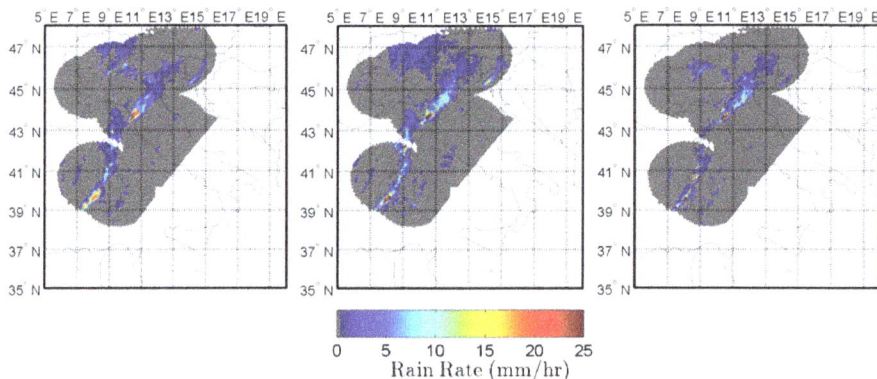

Figure 5. Heavy storm over Livorno (Tuscany region, Italy) on 10 September 2017, at 01:17 UTC. Precipitation estimates from PNPR v3 (**left panel**), GPROF V05 (**middle panel**) and ground radar observations, at 01:20 UTC (**right panel**) are shown. The GMI products are shown only on the area covered by the radar network.

The minimum threshold for the GPROF rainfall rate has been set to 0.1 mm h^{-1}. Some differences in the rainfall patterns identified by the two GMI products can be noticed, with a larger extension of the GPROF light precipitation area. A good agreement between the two products and with respect to the radar is evident in the heavy precipitation areas (also around Livorno). Table 8 shows the continuous statistical indexes evaluated for PNPR v3 and GPROF V05 compared with radar estimates (hits only). Both products tend to overestimate the precipitation (ME equal to 1.00 and 0.99 mm h^{-1} for PNPR v3 and GPROF V05 respectively). The statistical scores are comparable but with differences in favour of the GPROF product.

Table 8. Continuous statistical indexes (ME, MAE, RMSE, SD, ARMSE, CC) obtained, for PNPR v3 and GPROF V05 with respect to ground-based radar estimates, for the Livorno flash-flood case (GMI overpass at 01:17 UTC on 10 September 2017).

	PNPR v3	GPROF V05
N. PIXELS (RR > 0 mm h^{-1})	1447	2455
ME (mm h^{-1})	1.00	0.99
MAE (mm h^{-1})	2.16	1.54
RMSE (mm h^{-1})	3.77	3.58
SD (mm h^{-1})	4.16	3.97
ARMSE (mm h^{-1})	3.63	3.44
CC	0.61	0.60

Table 9 shows the dichotomous statistical scores for both PNPR v3 and GPROF V05 algorithms, with respect to the radar estimates. For PNPR v3 the statistical scores obtained considering the four different RNCI thresholds are shown. The results obtained confirm that the selection of the rain/no-rain pixels using RNCI $\geq$ 2 provides the best performance (according to the results showed in

Table 5). These results also confirm a good agreement in the pattern of the precipitation obtained by the two products, with a slightly better performance of PNPR v3 especially in terms of FAR.

Table 9. Dichotomous statistical scores (POD, FAR, CSI and HSS) for PNPR v3 (for different RNCI thresholds) and GPROF with respect to ground-based radar estimates for the Livorno flash-flood case (GMI overpass at 01:17 UTC on 10 September 2017).

	RNCI ≥ 1	RNCI ≥ 2	RNCI ≥ 3	RNCI ≥ 4	
	PNPR v3	PNPR v3	PNPR v3	PNPR v3	GPROF V05
POD	0.78	0.70	0.34	0.31	0.86
FAR	0.55	0.37	0.20	0.19	0.53
CSI	0.40	0.48	0.31	0.29	0.43
HSS	0.49	0.59	0.40	0.38	0.51

The second case study is a precipitation event that affected a large part of Italy, on 9 August 2015 and that produced some intense and complex rainfall pattern, with scattered limited areas of heavy rainfall (e.g., in Tuscany and Liguria region) and extended areas of light to moderate rainfall. This situation was entirely captured by a GMI overpass at 16:35 UTC (Figure 6).

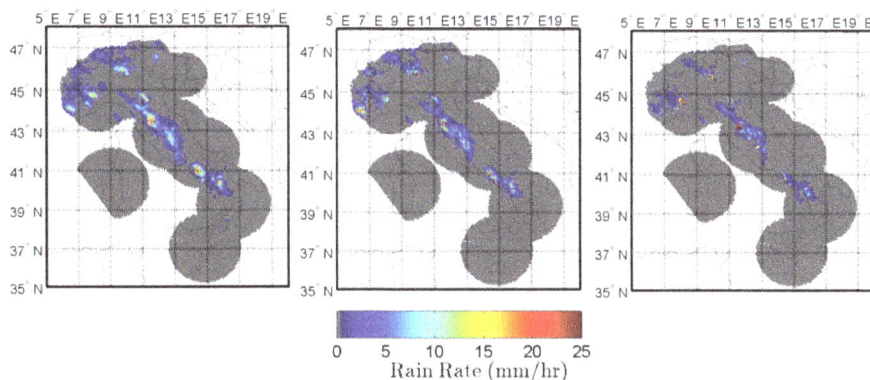

Figure 6. Precipitation event over Italy on 9 August 2015, at 16:35 UTC. Precipitation estimates from PNPR v3 (**left panel**), GPROF V05 (**middle panel**) and ground radar observations, at 16:40 UTC (**right panel**) are shown. The GMI products are shown only on the area covered by the radar network.

Also in this case there is a good agreement between the two products, showing very similar patterns and with respect to the radars, although PNPR v3 shows larger precipitation areas and overestimation of moderate precipitation. Table 10 shows worse scores than the Livorno case, both for PNPR v3 and for GPROF V05, with larger overestimation (ME equal to 1.48 and 1.17 for PNPR v3 and GPROF V05 respectively). For this case, PNPR v3 shows higher correlation and lower RMSE than GPROF. Table 11 confirms the good agreement in the pattern of the precipitation obtained by the two products, with a slightly better performance of PNPR v3 also for this case.

It has been evidenced by several authors [70] that in a validation exercise spatial colocation and horizontal averaging of measurements (both ground-based and satellite-based) are key elements critically affecting the results. We want here to highlight that all the statistical scores in this study have been obtained using pixel-based comparison of instantaneous precipitation rate estimates. The good agreement with ground-based radars in terms of spatial pattern and statistical scores for both PNPRv3 and GPROF confirms the high quality of the two GMI-based algorithms.

Table 10. Same as Table 8 but, for the GMI overpass at 16:35 UTC on 9 August 2015.

	PNPR v3	GPROF V05
N. PIXELS (RR > 0 mm h^{-1})	1605	1309
ME (mm h^{-1})	1.48	1.17
MAE (mm h^{-1})	2.43	2.34
RMSE (mm h^{-1})	4.33	5.50
SD (mm h^{-1})	5.04	5.90
ARMSE (mm h^{-1})	4.07	5.40
CC	0.56	0.51

Table 11. Same as Table 9 but for the GMI overpass at 16:35 UTC on 9 August 2015.

	PNPR v3	GPROF V05
POD	0.83	0.80
FAR	0.36	0.33
CSI	0.58	0.57
HSS	0.70	0.68

8. Summary and Conclusions

In this work, a new algorithm based on the Passive microwave Neural network Precipitation Retrieval approach (PNPR v3), designed to work with the conically scanning GMI radiometer, is thoroughly described and results about its performance are presented and discussed. This algorithm, developed within the EUMETSAT H SAF program, follows the two previous PNPR (v1 and v2) products designed to work with cross-track scanning radiometers, AMSU/MHS and ATMS [26] and [27]. Although it benefits from the experience acquired with the development of the previous algorithms, this new version of PNPR presents several innovations with respect to the others.

A fundamental new aspect is that PNPR v3 is a global rainfall rate retrieval algorithm (between 65°N–S), while PNPR v1 and v2 algorithms are designed to work over the MSG full disk area, [60°S–75°N, 60°W–60°E]). As opposed to previous PNPR algorithms, based on a training database obtained from cloud-radiation model simulations, the new PNPR v3 for GMI is completely based on an extremely large observational database (with over 150 million elements with rainfall) built from coincident GMI/DPR global observations during a period of 27 months between 2014 and 2016. In the database, the NASA 2B-CMB V04 rainfall rate product (Normal Scan swath, liquid fraction >0.8) is used as reference. In addition, a new PCA-based input selection procedure has been designed to fully exploit the potentials of GMI channels compared to AMSU/MHS and ATMS. Another difference is a new rain/no-rain classification scheme (RNC), also based on the NN approach, which provides different rainfall masks for different minimum thresholds and degree of reliability. The procedure to design one unique NN for all surface types has been also described.

In order to assess the performance of PNPR v3 over the globe, an independent part of the observational database, composed of about 50 million data distributed globally in 27 months period, not used in the algorithm design phase, is used in a verification study. A good agreement is found between PNPR v3 and 2B-CMB products, with a good CC (greater than 0.90), very low values of the ME (between 0.10 and −0.22 mm h^{-1}), MAE (between 0.60 and 0.94 mm h^{-1}) and RMSE (between 1.62 and 2.75 mm h^{-1}). Also, the FSE% drops below 100% for rainfall rates lower than 1 mm h^{-1}, ranging around 30–50% for moderate to high rainfall rates, over all surface types. These good results, also compared to previous versions of PNPR v1 and v2 [26,27], or other studies [67], besides demonstrating the good outcome of the input selection procedure, as well as of the training and design phase of the NN, also show how PNPR v3 is able exploit the great capabilities of GMI for rainfall retrieval over the different surface types. It is worth noting, however, that, since the algorithm has been trained for (mostly) liquid precipitation, ancillary variables indicating the possible occurrence of (mostly) frozen precipitation have been used in the PNPR v3 assessment to limit the verification to rainfall cases.

The further verification carried out on two case studies over Italy, showed a good consistency of PNPR v3 retrievals with GPROF V05 estimates and with simultaneous ground-based radar estimates. The dichotomous statistical indexes indicated good rainfall detection skills by PNPR v3 (FAR = 0.37 and 0.36, POD = 0.70 and 0.83 and HSS = 0.59 and 0.70 for the two cases), with consistently better scores than GPROF (especially the lower FAR), except for the higher POD for GPROF in one case. For the rainfall rate retrieval, the continuous statistical scores indicated slight overestimation by the two GMI products (with higher ME and MAE for PNPR v3 than for GPROF), while the other scores indicated better performance of PNPR v3 in one case and of GPROF for the other case. It is worth noticing the ability of PNPR v3 to provide rainfall rate estimates in agreement with the ground-based radar and with the well consolidated GPROF v05 product for these two cases, characterized by extremely variable precipitation intensity across different surface typologies and complex orography.

The new PNPR v3 is a global rainfall retrieval algorithm, able to optimally exploit the GMI multi-channel response to different surface types and precipitation structures found around the globe. The PCA-based input selection procedure, where the precipitation signal is separated from the signal related to the different surfaces types, with minimal use of model-derived ancillary parameters, as well as the training and optimization of one unique NN allow to: (1) handle the extremely rich database based on the GPM-CO measurements and products; (2) provide global rainfall retrieval in a computationally very efficient way, making the product suitable for near-real time operational applications; (3) avoid discontinuities across different background surfaces.

The approach can be easily adapted to new versions of GPM products as they become available (including high quality snowfall rate estimates), or to background surface conditions that have not been considered in this study (e.g., frozen soil, snow cover, sea ice). However, due to the limited sensitivity of GPM DPR to snowfall especially at higher latitudes, separate modules for snowfall detection and retrieval recently developed by Rysman et al. [35] will be coupled to PNPR v3. These modules are based on the use of observational databases built from GMI and Cloudsat snowfall coincident observations [37].

In the future, a more extensive independent validation of PNPR v3 precipitation rate product using ground-based radars and rain gauges will be carried out over Europe within the EUMETSAT HSAF program. Moreover, in order to evaluate the PNPR v3 globally and compare it to GPROF (the GPM official PMW product) a more exhaustive analysis of different case studies around the globe will be carried out within the ongoing scientific collaboration between the EUMETSAT HSAF and GPM, on precipitation algorithm development and validation.

Author Contributions: P.S. designed and implemented the PNPR v3 algorithm. All co-authors have contributed to the group discussions on the development of the algorithm and on the results and to the final draft.

Funding: This research was funded by EUMETSAT through the project "Satellite Application Facility on Support to Operational Hydrology and Water Management" (H SAF), Third Continuous Development and Operational Phase (CDOP-3)".

Acknowledgments: This research was supported by EUMETSAT through the project "Satellite Application Facility on Support to Operational Hydrology and Water Management" (H SAF) and by PRIN 2015 4WX5NA. The Precipitation Measurement Mission (PMM) Research Program and EUMETSAT are warmly acknowledged for supporting the H-SAF and GPM collaboration through the approval of the no-cost proposal "H-SAF and GPM: precipitation algorithm development and validation activity." The authors want to gratefully thank the PMM Science Team for the interactions and discussion on critical aspects related to PMW precipitation retrieval algorithm development and in particular Dave Randel and Chris Kummerow for sharing the research database of coincident GMI/DPR observations which is the basis for this study. The GMI measurements have been collected from the public NASA PPS data archive ftp://arthurhou.pps.eosdis.nasa.gov. Finally, we would thank the Italian Civil Protection Department for providing the radar data utilized for the case studies over Italy.

Conflicts of Interest: The authors declare no conflict of interest. The funding sponsors had no role in the design of the study; in the collection, analyses, or interpretation of data; in the writing of the manuscript and in the decision to publish the results.

References

1. Mugnai, A.; Smith, E.A.; Tripoli, G.J. Foundations for statistical physical precipitation retrieval from passive microwave satellite measurement. Part II: Emission-source and generalized weighting-function properties of a time-dependent cloud-radiation model. *J. Appl. Meteorol.* **1993**, *32*, 17–39. [CrossRef]
2. Bennartz, R.; Petty, G.W. The Sensitivity of Microwave Remote Sensing Observations of Precipitation to Ice Particle Size Distributions. *J. Appl. Meteorol.* **2001**, *40*, 345–364. [CrossRef]
3. Tian, Y.; Peters-Lidard, C.D.; Eylander, J.B.; Joyce, R.J.; Huffman, G.J.; Adler, R.F.; Hsu, K.; Turk, F.J.; Garcia, M.; Zeng, J. Component analysis of errors in satellite-based precipitation estimates. *J. Geophys. Res.* **2009**, *114*. [CrossRef]
4. Kirstetter, P.-E.; Hong, Y.; Gourley, J.J.; Chen, S.; Flamig, Z.; Zhang, J.; Schwaller, M.; Petersen, W.; Amitai, E. Toward a Framework for Systematic Error Modeling of Spaceborne Precipitation Radar with NOAA/NSSL Ground Radar–Based National Mosaic QPE. *J. Hydrometeorol.* **2012**, *13*, 1285–1300. [CrossRef]
5. Kidd, C.; Becker, A.; Huffman, G.J.; Muller, C.L.; Joe, P.; Skofronick-Jackson, G.; Kirschbaum, D.B. So, How Much of the Earth's Surface *Is* Covered by Rain Gauges? *Bull. Am. Meteorol. Soc.* **2017**, *98*, 69–78. [CrossRef]
6. Wilheit, T.; Adler, R.; Avery, S.; Barrett, E.; Bauer, P.; Berg, W.; Chang, A.; Ferriday, J.; Grody, N.; Goodman, S.; et al. Algorithms for the retrieval of rainfall from passive microwave measurements. *Remote Sens. Rev.* **1994**, *11*, 163–194. [CrossRef]
7. Kummerow, C.D.; Ringerud, S.; Crook, J.; Randel, D.; Berg, W. An Observationally Generated A Priori Database for Microwave Rainfall Retrievals. *J. Atmos. Ocean. Technol.* **2011**, *28*, 113–130. [CrossRef]
8. Panegrossi, G.; Dietrich, S.; Marzano, F.S.; Mugnai, A.; Smith, E.A.; Xiang, X.; Tripoli, G.J.; Wang, P.K.; Poiares Baptista, J.P.V. Use of Cloud Model Microphysics for Passive Microwave-Based Precipitation Retrieval: Significance of Consistency between Model and Measurement Manifolds. *J. Atmos. Sci.* **1998**, *55*, 1644–1673. [CrossRef]
9. You, Y.; Liu, G. The relationship between surface rainrate and water paths and its implications to satellite rainrate retrieval. *J. Geophys. Res. Atmos.* **2012**, *117*. [CrossRef]
10. Grecu, M.; Olson, W.S.; Anagnostou, E.N. Retrieval of Precipitation Profiles from Multiresolution, Multifrequency Active and Passive Microwave Observations. *J. Appl. Meteorol.* **2004**, *43*, 562–575. [CrossRef]
11. Grecu, M.; Olson, W.S. Bayesian Estimation of Precipitation from Satellite Passive Microwave Observations Using Combined Radar–Radiometer Retrievals. *J. Appl. Meteorol. Climatol.* **2006**, *45*, 416–433. [CrossRef]
12. Petty, G.W.; Li, K. Improved Passive Microwave Retrievals of Rain Rate over Land and Ocean. Part II: Validation and Intercomparison. *J. Atmos. Ocean. Technol.* **2013**, *30*, 2509–2526. [CrossRef]
13. Bellerby, T.; Todd, M.; Kniveton, D.; Kidd, C. Rainfall Estimation from a Combination of TRMM Precipitation Radar and GOES Multispectral Satellite Imagery through the Use of an Artificial Neural Network. *J. Appl. Meteorol.* **2000**, *39*, 2115–2128. [CrossRef]
14. Schumacher, C.; Houze, R.A. Stratiform Rain in the Tropics as Seen by the TRMM Precipitation Radar. *J. Clim.* **2003**, *16*, 1739–1756. [CrossRef]
15. Lin, X.; Hou, A.Y. Evaluation of Coincident Passive Microwave Rainfall Estimates Using TRMM PR and Ground Measurements as References. *J. Appl. Meteorol. Climatol.* **2008**, *47*, 3170–3187. [CrossRef]
16. Hou, A.Y.; Kakar, R.K.; Neeck, S.; Azarbarzin, A.A.; Kummerow, C.D.; Kojima, M.; Oki, R.; Nakamura, K.; Iguchi, T. The Global Precipitation Measurement Mission. *Bull. Am. Meteorol. Soc.* **2014**, *95*, 701–722. [CrossRef]
17. Skofronick-Jackson, G.; Petersen, W.A.; Berg, W.; Kidd, C.; Stocker, E.F.; Kirschbaum, D.B.; Kakar, R.; Braun, S.A.; Huffman, G.J.; Iguchi, T.; et al. The Global Precipitation Measurement (GPM) Mission for Science and Society. *Bull. Am. Meteorol. Soc.* **2017**, *98*, 1679–1695. [CrossRef]
18. Panegrossi, G.; Casella, D.; Dietrich, S.; Marra, A.C.; Sano, P.; Mugnai, A.; Baldini, L.; Roberto, N.; Adirosi, E.; Cremonini, R.; et al. Use of the GPM Constellation for Monitoring Heavy Precipitation Events over the Mediterranean Region. *IEEE J. Sel. Top. Appl. Earth Obs. Remote Sens.* **2016**, *9*, 2733–2753. [CrossRef]
19. Kummerow, C.; Giglio, L. A Passive Microwave Technique for Estimating Rainfall and Vertical Structure Information from Space. Part I: Algorithm Description. *J. Appl. Meteorol.* **1994**, *33*, 3–18. [CrossRef]

20. Kummerow, C.D.; Randel, D.L.; Kulie, M.; Wang, N.-Y.; Ferraro, R.; Joseph Munchak, S.; Petkovic, V. The Evolution of the Goddard Profiling Algorithm to a Fully Parametric Scheme. *J. Atmos. Ocean. Technol.* **2015**, *32*, 2265–2280. [CrossRef]

21. Mugnai, A.; Casella, D.; Cattani, E.; Dietrich, S.; Laviola, S.; Levizzani, V.; Panegrossi, G.; Petracca, M.; Sanò, P.; Di Paola, F.; et al. Precipitation products from the hydrology SAF. *Nat. Hazards Earth Syst. Sci.* **2013**, *13*, 1959–1981. [CrossRef]

22. Mugnai, A.; Smith, E.A.; Tripoli, G.J.; Bizzarri, B.; Casella, D.; Dietrich, S.; Di Paola, F.; Panegrossi, G.; Sanò, P. CDRD and PNPR satellite passive microwave precipitation retrieval algorithms: EuroTRMM/EURAINSAT origins and H-SAF operations. *Nat. Hazards Earth Syst. Sci.* **2013**, *13*, 887–912. [CrossRef]

23. Casella, D.; do Amaral, L.M.C.; Dietrich, S.; Marra, A.C.; Sanò, P.; Panegrossi, G. The Cloud Dynamics and Radiation Database Algorithm for AMSR2: Exploitation of the GPM Observational Dataset for Operational Applications. *IEEE J. Sel. Top. Appl. Earth Obs. Remote Sens.* **2017**, *10*, 3985–4001. [CrossRef]

24. Casella, D.; Panegrossi, G.; Sanò, P.; Dietrich, S.; Mugnai, A.; Smith, E.A.; Tripoli, G.J.; Formenton, M.; Di Paola, F.; Leung, W.-Y.H.; et al. Transitioning from CRD to CDRD in Bayesian retrieval of rainfall from satellite passive microwave measurements: Part 2. Overcoming database profile selection ambiguity by consideration of meteorological control on microphysics. *IEEE Trans. Geosci. Remote Sens.* **2013**, *51*, 4650–4671. [CrossRef]

25. Sanò, P.; Casella, D.; Mugnai, A.; Schiavon, G.; Smith, E.A.; Tripoli, G.J. Transitioning from CRD to CDRD in Bayesian retrieval of rainfall from satellite passive microwave measurements: Part 1. Algorithm description and testing. *IEEE Trans. Geosci. Remote Sens.* **2013**, *51*, 4119–4143. [CrossRef]

26. Sanò, P.; Panegrossi, G.; Casella, D.; Di Paola, F.; Milani, L.; Mugnai, A.; Petracca, M.; Dietrich, S. The Passive microwave Neural network Precipitation Retrieval (PNPR) algorithm for AMSU/MHS observations: Description and application to European case studies. *Atmos. Meas. Tech.* **2015**, *8*, 837–857. [CrossRef]

27. Sanò, P.; Panegrossi, G.; Casella, D.; Marra, A.C.; Di Paola, F.; Dietrich, S. The new Passive microwave Neural network Precipitation Retrieval (PNPR) algorithm for the cross-track scanning ATMS radiometer: Description and verification study over Europe and Africa using GPM and TRMM spaceborne radars. *Atmos. Meas. Tech.* **2016**, *9*, 5441–5460. [CrossRef]

28. Liou, Y.-A.; Tzeng, Y.C.; Chen, K.S. A neural-network approach to radiometric sensing of land-surface parameters. *IEEE Trans. Geosci. Remote Sens.* **1999**, *37*, 2718–2724. [CrossRef]

29. Blakwell, W.J.; Chen, F.W. Neural Network Applications in High Resolution Atmospheric Remote Sensing. *Linc. Lab. J.* **2005**, *15*, 299–322.

30. Aires, F.; Prigent, C.; Rossow, W.B.; Rothstein, M. A new neural network approach including first guess for retrieval of atmospheric water vapor, cloud liquid water path, surface temperature, and emissivities over land from satellite microwave observations. *J. Geophys. Res. Atmos.* **2001**, *106*, 14887–14907. [CrossRef]

31. Surussavadee, C.; Staelin, D.H. Global millimeter-wave precipitation retrievals trained with a cloud-resolving numerical weather prediction model, Part I: Retrieval design. *IEEE Trans. Geosci. Remote Sens.* **2008**, *46*, 99–108. [CrossRef]

32. Mahesh, C.; Prakash, S.; Sathiyamoorthy, V.; Gairola, R.M. Artificial neural network based microwave precipitation estimation using scattering index and polarization corrected temperature. *Atmos. Res.* **2011**, *102*, 358–364. [CrossRef]

33. Hong, Y.; Hsu, K.-L.; Sorooshian, S.; Gao, X. Precipitation estimation from remotely sensed imagery using an artificial neural network cloud classification system. *J. Appl. Meteorol.* **2004**, *43*, 1834–1853. [CrossRef]

34. Tapiador, F.J.; Navarro, A.; Levizzani, V.; García-Ortega, E.; Huffman, G.J.; Kidd, C.; Kucera, P.A.; Kummerow, C.D.; Masunaga, H.; Petersen, W.A.; et al. Global precipitation measurements for validating climate models. *Atmos. Res.* **2017**, *197*, 1–20. [CrossRef]

35. Rysman, J.-F.; Panegrossi, G.; Sanò, P.; Marra, A.C.; Dietrich, S.; Milani, L.; Kulie, M.S. SLALOM: An all-surface snow water path retrieval algorithm for the GPM Microwave Imager. *Remote Sens.* **2018**, for peer review.

36. Grecu, M.; Olson, W.S.; Munchak, S.J.; Ringerud, S.; Liao, L.; Haddad, Z.; Kelley, B.L.; McLaughlin, S.F. The GPM combined algorithm. *J. Atmos. Ocean. Technol.* **2016**, *33*, 2225–2245. [CrossRef]

37. Panegrossi, G.; Rysman, J.-F.; Casella, D.; Marra, A.; Sanò, P.; Kulie, M. CloudSat-Based Assessment of GPM Microwave Imager Snowfall Observation Capabilities. *Remote Sens.* **2017**, *9*, 1263. [CrossRef]

38. Draper, D.W.; Newell, D.A.; Wentz, F.J.; Krimchansky, S.; Skofronick-Jackson, G.M. The global precipitation measurement (GPM) microwave imager (GMI): Instrument overview and early on-orbit performance. *IEEE J. Sel. Top. Appl. Earth Obs. Remote Sens.* **2015**, *8*, 3452–3462. [CrossRef]

39. Olson, W.S.; Masunaga, H.; GPM Combined Radar-Radiometer Algorithm Team. *GPM Combined Radar-Radiometer Precipitation Algorithm Theoretical Basis Document (Version 4)*; NASA: Washington, DC, USA, 2016.

40. Wilheit, T.; Berg, W.; Ebrahimi, H.; Kroodsma, R.; McKague, D.; Payne, V.; Wang, J. Intercalibrating the GPM constellation using the GPM Microwave Imager (GMI). In Proceedings of the 2015 IEEE International Geoscience and Remote Sensing Symposium (IGARSS), Milan, Italy, 26–31 July 2015; pp. 5162–5165.

41. Shimozuma, T.; Seto, S. Evaluation of KUPR algorithm in matchup cases of GPM and TRMM. In Proceedings of the 2015 IEEE International Geoscience and Remote Sensing Symposium (IGARSS), Milan, Italy, 26–31 July 2015; pp. 5134–5137.

42. Toyoshima, K.; Masunaga, H.; Furuzawa, F.A. Early evaluation of Ku- and Ka-band sensitivities for the global precipitation measurement (GPM) dual-frequency precipitation radar (DPR). *SOLA* **2015**, *11*, 14–17. [CrossRef]

43. Kubota, T.; Iguchi, T.; Kojima, M.; Liao, L.; Masaki, T.; Hanado, H.; Meneghini, R.; Oki, R. A Statistical Method for Reducing Sidelobe Clutter for the Ku-Band Precipitation Radar on board the GPM *Core Observatory*. *J. Atmos. Ocean. Technol.* **2016**, *33*, 1413–1428. [CrossRef]

44. Wentz, F.J.; Draper, D. On-orbit absolute calibration of the global precipitation measurement microwave imager. *J. Atmos. Ocean. Technol.* **2016**, *33*, 1393–1412. [CrossRef]

45. Boukabara, S.-A.; Garret, K.; Blackwell, B. ATMS Description & Expected Performances. In Proceedings of the 3rd Post-EPS User Consultation Workshop, Darmstadt, Germany, 29–30 September 2011.

46. Sarma, D.K.; Konwar, M.; Sharma, S.; Pal, S.; Das, J.; De, U.K.; Viswanathan, G. An artificial-neural-network-based integrated regional model for rain retrieval over land and ocean. *IEEE Trans. Geosci. Remote Sens.* **2008**, *46*, 1689–1696. [CrossRef]

47. Wang, J.R.; Zhan, J.; Racette, P. Storm-associated microwave radiometric signatures in the frequency range of 90–220 GHz. *J. Atmos. Ocean. Technol.* **1997**, *14*, 13–31. [CrossRef]

48. Ferraro, R.R. *The Status of the NOAA/NESDIS Operational AMSU Precipitation Algorithm*, 2nd ed.; Workshop of the International Precipitation Working Group: Monterey, CA, USA, 2004; p. 9.

49. Hong, G.; Heygster, G.; Miao, J.; Kunzi, K. Detection of tropical deep convective clouds from AMSU-B water vapor channels measurements. *J. Geophys. Res.* **2005**, *110*. [CrossRef]

50. Qiu, S.; Pellegrino, P.; Ferraro, R.; Zhao, L. The improved AMSU rain-rate algorithm and its evaluation for a cool season event in the western United States. *Weather Forecast.* **2005**, *20*, 761–774. [CrossRef]

51. Funatsu, B.M.; Claud, C.; Chaboureau, J.-P. Potential of Advanced Microwave Sounding Unit to identify precipitating systems and associated upper-level features in the Mediterranean region: Case studies. *J. Geophys. Res.* **2007**, *112*. [CrossRef]

52. Funatsu, B.M.; Claud, C.; Chaboureau, J.-P. Comparison between the large-scale environments of moderate and intense precipitating systems in the Mediterranean region. *Mon. Weather Rev.* **2009**, *137*, 3933–3959. [CrossRef]

53. McCann, D.W. A neural network short-term forecast of significant thunderstorms. *Weather Forecast.* **1992**, *7*, 525–534. [CrossRef]

54. Ferraro, R.R.; Smith, E.A.; Berg, W.; Huffman, G.J. A screening methodology for passive microwave precipitation retrieval algorithms. *J. Atmos. Sci.* **1998**, *55*, 1583–1600. [CrossRef]

55. Grody, N.C. Classification of snow cover and precipitation using the special sensor microwave imager. *J. Geophys. Res.* **1991**, *96*, 7423. [CrossRef]

56. Kidd, C. On rainfall retrieval using polarization-corrected temperatures. *Int. J. Remote Sens.* **1998**, *19*, 981–996. [CrossRef]

57. Spencer, R.W. A satellite passive 37-GHz scattering-based method for measuring oceanic rain rates. *J. Clim. Appl. Meteorol.* **1986**, *25*, 754–766. [CrossRef]

58. Anders, U.; Korn, O. Model selection in neural networks. *Neural Netw.* **1999**, *12*, 309–323. [CrossRef]

59. Marzban, C. Neural networks for postprocessing model output: ARPS. *Mon. Weather Rev.* **2003**, *131*, 1103–1111. [CrossRef]

60. Seto, S.; Kubota, T.; Takahashi, N.; Iguchi, T.; Oki, T. Advanced rain/no-rain classification methods for microwave radiometer observations over land. *J. Appl. Meteorol. Climatol.* **2008**, *47*, 3016–3029. [CrossRef]

61. Sudradjat, A.; Wang, N.-Y.; Gopalan, K.; Ferraro, R.R. Prototyping a generic, unified land surface classification and screening methodology for GPM-era microwave land precipitation retrieval algorithms. *J. Appl. Meteorol. Climatol.* **2011**, *50*, 1200–1211. [CrossRef]

62. Kirstetter, P.-E.; Viltard, N.; Gosset, M. An error model for instantaneous satellite rainfall estimates: Evaluation of BRAIN-TMI over West Africa. *Q. J. R. Meteorol. Soc.* **2013**, *139*, 894–911. [CrossRef]

63. Kacimi, S.; Viltard, N.; Kirstetter, P.-E. A new methodology for rain identification from passive microwave data in the Tropics using neural networks. *Q. J. R. Meteorol. Soc.* **2013**, *139*, 912–922. [CrossRef]

64. Ferraro, R.; Grody, N.; Kogut, J. Classification of geophysical parameters using passive microwave satellite measurements. *IEEE Trans. Geosci. Remote Sens.* **1986**, *GE-24*, 1008–1013. [CrossRef]

65. Spencer, R.W.; Goodman, H.M.; Hood, R.E. Precipitation retrieval over land and ocean with the SSM/I: Identification and characteristics of the scattering signal. *J. Atmos. Ocean. Technol.* **1989**, *6*, 254–273. [CrossRef]

66. Islam, T.; Srivastava, P.K.; Dai, Q.; Gupta, M.; Zhuo, L. Rain Rate Retrieval Algorithm for Conical-Scanning Microwave Imagers Aided by Random Forest, RReliefF, and Multivariate Adaptive Regression Splines (RAMARS). *IEEE Sens. J.* **2015**, *15*, 2186–2193. [CrossRef]

67. Tang, L.; Tian, Y.; Lin, X. Validation of precipitation retrievals over land from satellite-based passive microwave sensors. *J. Geophys. Res. Atmos.* **2014**, *119*, 4546–4567. [CrossRef]

68. Rinollo, A.; Vulpiani, G.; Puca, S.; Pagliara, P.; Kaňák, J.; Lábó, E.; Okon, L.; Roulin, E.; Baguis, P.; Cattani, E.; et al. Definition and impact of a quality index for radar-based reference measurements in the H-SAF precipitation product validation. *Nat. Hazards Earth Syst. Sci.* **2013**, *13*, 2695–2705. [CrossRef]

69. Derin, Y.; Anagnostou, E.; Anagnostou, M.N.; Kalogiros, J.; Casella, D.; Marra, A.C.; Panegrossi, G.; Sano, P. Passive microwave rainfall error analysis using high-resolution x-band dual-polarization radar observations in complex terrain. *IEEE Trans. Geosci. Remote Sens.* **2018**, *56*, 2565–2586. [CrossRef]

70. Omranian, E.; Sharif, H.O. Evaluation of the Global Precipitation Measurement (GPM) Satellite Rainfall Products over the Lower Colorado River Basin, Texas. *JAWRA J. Am. Water Resour. Assoc.* **2018**. [CrossRef]

remote sensing

MDPI

Article

Benefits of the Successive GPM Based Satellite Precipitation Estimates IMERG–V03, –V04, –V05 and GSMaP–V06, –V07 Over Diverse Geomorphic and Meteorological Regions of Pakistan

Frédéric Satgé [1,*], Yawar Hussain [2], Marie-Paule Bonnet [3], Babar M. Hussain [4], Hernan Martinez-Carvajal [5], Gulraiz Akhter [6] and Rogério Uagoda [7]

[1] CNES, UMR HydroSciences, University of Montpellier, Place E. Bataillon, 34395 Montpellier, France
[2] Departamento de Engenharia Civil e Ambiental, Universidade de Brasilia, 70910-900 Brasília-DF, Brazil; yawar.pgn@gmail.com
[3] IRD, UMR Espace–Dev, Maison de la télédétection, 500 rue JF Breton, 34093 Montpellier, France; marie-paule.bonnet@ird.fr
[4] Department of Physics, The University of Lahore, Gujrat Campus, Main GT Road, Gujrat 50700, Pakistan; gentlemanscientist2012@gmail.com
[5] Faculty of Mines, National University of Colombia at Medellín, Cl. 59a, No. 63-20, Medellín, Colombia; hmartinezc30@gmail.com
[6] Department of Earth Science, Quaid-i-Azam University, Islamabad 45320, Pakistan; agulraiz@qau.edu.pk
[7] Post-Graduation Program in Geography, University of Brasilia, ICC Norte, 70910-900 Brasilia-DF, Brazil; rogeriouagoda@unb.br
* Correspondence: frederic.satge@gmail.com; Tel.: +33-4671-49060

Received: 4 July 2018; Accepted: 13 August 2018; Published: 30 August 2018

Abstract: Launched in 2014, the Global Precipitation Measurement (GPM) mission aimed at ensuring the continuity with the Tropical Rainfall Measuring Mission (TRMM) launched in 1997 that has provided unprecedented accuracy in Satellite Precipitation Estimates (SPEs) on the near-global scale. Since then, various SPE versions have been successively made available from the GPM mission. The present study assesses the potential benefits of the successive GPM based SPEs product versions that include the Integrated Multi–Satellite Retrievals for GPM (IMERG) version 3 to 5 (–v03, –v04, –v05) and the Global Satellite Mapping of Precipitation (GSMaP) version 6 to 7 (–v06, –v07). Additionally, the most effective TRMM based SPEs products are also considered to provide a first insight into the GPM effectiveness in ensuring TRMM continuity. The analysis is conducted over different geomorphic and meteorological regions of Pakistan while using 88 precipitations gauges as the reference. Results show a clear enhancement in precipitation estimates that were derived from the very last IMERG–v05 in comparison to its two previous versions IMERG–v03 and –v04. Interestingly, based on the considered statistical metrics, IMERG–v03 provides more consistent precipitation estimate than IMERG–v04, which should be considered as a transition IMERG version. As expected, GSMaP–v07 precipitation estimates are more accurate than the previous GSMaP–v06. However, the enhancement from the old to the new version is very low. More generally, the transition from TRMM to GPM is successful with an overall better performance of GPM based SPEs than TRMM ones. Finally, all of the considered SPEs have presented a strong spatial variability in terms of accuracy with none of them outperforming the others, for all of the gauges locations over the considered regions.

Keywords: Satellite Precipitation Estimates; GPM; TRMM; IMERG; GSMaP; TMPA; CMORPH; assessment; Pakistan

1. Introduction

Precipitation is a key component of the water cycle, which is facing unprecedented pressure due to the combined effects of population growth and climate change. Precipitation estimates are therefore crucial to adapt and anticipate ongoing changes. However, precipitations are generally retrieved from sparse and unevenly distributed gauges network introducing large uncertainty over the remote regions, such as tropical forests, mountainous, and desert regions. In this context, Satellite Precipitation Estimates (SPEs) offer the possibility to monitor precipitation on regular grids at the near-global scale representing an unprecedented measurement opportunity.

Launched in 1997 by NASA (National Aeronautics and Space Administration) and the Japan Aerospace Exploration Agency (JAXA), the Tropical Rainfall Measuring Mission (TRMM) was the first mission that was dedicated to SPEs production. From the TRMM mission, numerous SPEs were made available to follow precipitation from space on regular grid at near global coverage. TRMM based SPEs includes the TRMM Multisatellite Precipitation Analysis (TMPA) [1], the Climate prediction centre MORPHing (CMORPH) [2], the Precipitation Estimation from remotely Sensed Information using Artificial Neural Networks (PERSIANN) [3], and the Global Satellite Mapping Precipitation (GSMaP) [4]. TRMM based SPEs have proved effective in precipitation estimation over the world, as reviewed by [5,6]. From the success of TRMM based SPEs, a new generation of SPEs took advantage of previous SPEs to estimate precipitation over larger time window. This is the case with PERSIANN–Climate Data Record (PERSIANN–CDR) [7], Multi–Source Weighted–Ensemble Precipitation (MSWEP) [8], and Climate Hazards Group InfraRed Precipitation (CHIRP) with Station data (CHIRPS) [9]. These SPEs have proven effective to follow regional drought processes [10–12] and long term hydrological survey [12,13].

Recently, the Global Precipitation Measurement (GPM) mission was launched on 27 February 2014 to ensure the continuity of the TRMM mission. From the GPM mission, two new SPEs were made available: the Integrated Multi–Satellite Retrievals for GPM (IMERG) [14] and a new version of GSMaP product. Available at a 0.1° and half–hourly and hourly temporal scales, respectively, they offer the opportunity of capturing finer local precipitation variations in space and time [15]. Due to their recent release, few studies report on GPM based SPEs.

A first attempt was made by, [16] while comparing IMERG to its predecessor TMPA at the monthly timescale. The study highlighted the differences in both precipitation datasets, which vary according to surfaces and precipitations rates. Since then, numerous studies were dedicated to provide more insights into this discrepancy and highlighted the potential IMERG benefits over its predecessor TMPA at a more local scale. For example, in India, IMERG was found more accurate in the estimation of mean monsoon precipitation than TMPA [17]. In China, IMERG precipitation estimates were compared with the gauges observation at the national level considering daily [18] and monthly temporal scales [19], and a local level study reported on IMERG precipitation estimates over the Chinese Beijang river basin [20]. All of these studies confirmed the benefits that are brought by IMERG over its predecessor TMPA with equivalent to higher performance according to the considered protocol and region. In Korea and Japan, despite of some inconsistencies induced by orographic convection and coastal effects, IMERG performed about 8% better than TMPA [21]. In Singapore, while considering different temporal scales, IMERG appeared as a slight improved version in comparison to TMPA [22]. In Iran, [23] compared to gauges observation, IMERG potential was found superior to TMPA at both daily and monthly temporal scales. In Bolivia, IMERG potential assessment in comparison to gauges observations for different spatiotemporal scales reported the IMERG benefits over TMPA [15]. Despite the temporal continuity objective from TRMM (TMPA) to GPM (IMERG) some authors focused on IMERG datasets without inter–comparing GPM to TRMM based SPEs. All of these studies have highlighted the promising perspectives of IMERG precipitation retrieval algorithm [24–27].

However, all of the above-mentioned studies reported on the first IMERG released version (IMERG–v03) while two new versions were successively released since then (IMERG–v04 and –v05). According to our current knowledge, few studies have reported on the second IMERG version (IMERG–v04), while the latest released version (IMERG–v05) has remained unreported so far. The IMERG–v04 was assessed in Malaysia [28], Italy [29], and Pakistan [30,31] while using gauges and/or radar precipitation estimates as references. These studies brought relevant information about IMERG–v04 relative performance in comparison to the others considered SPEs versions. However, in these studies, the benefits of IMERG–v04 over its predecessor IMERG–v03 remained neglected. This is a crucial missing information towards the ongoing IMERG algorithm enhancement in order to ensure the best retrieval of precipitation estimates possible. In this context, two studies assessed IMERG–v03 to IMERG–v04 transition effectiveness in China [32,33]. According to these studies, IMERG–v04 did not exhibit the anticipated improvement when compared to IMERG–v03 with higher bias, lower Correlation Coefficient and Root Mean Square Error. This unexpected feature is very relevant for both IMERG algorithm developers and data users. Recently, the last IMERG version (IMERG–v05) was released, and yet, no study has reported on its potential.

In the above-described context it appears as a major concern to assess IMERG–v05 in comparison to (i) TRMM based SPEs to observe transition benefit from TRMM to GPM and to (ii) IMERG–v03 and –v04 to follow algorithm enhancement for the precipitation retrieval. Additionally, GSMaP–v06 and –v07 consist in the other GPM based SPEs group. However, few studies reported on their potential, while considering the first version (GSMaP–v06) only [15,17,33]. Therefore, as for IMERG datasets, GSMaP successive versions transition (GSMaP–v06 to GSMaP –v07) has to be reported to follow potential benefits of successive versions and to compare with IMERG and TRMM based SPEs.

To provide first feedback on the GPM successive SPEs versions performance, the present study assesses IMERG–v03, –v04, –v05 and GSMaP–v06, –v07 against gauges observation over Pakistan. Pakistan is selected as a study area because of its contrasted geomorphologic features and climatic divisions from the very wet Himalayan region including permanent snow–glacial region to the very dry regions which are known to interfere on SPEs potential spatial variability. To follow potential benefits from TRMM to GPM, TMPA, and CMORPH datasets are also considered in the present study.

2. Materials and Methods

2.1. Study Area

Pakistan is situated in the western zone of South Asia between 23.5–37.5°N latitude and 60–78°E longitude. Geographically, it is bounded at north by China, east by India, while Afghanistan and Iran cover the western side. Arabian Sea marks the southern border (Figure 1). The total area is 803,940 km^2 where elevation as varies from a maximum of 8011 m at K2 (second elevated peak on earth) to about 0 m at the Arabian Sea. Climatologically, the study area is very diverse in space and time, and it is divided into different climatic zones [34]. For the present study, four regions are considered to assess SPEs performance over variable geomorphic and meteorological contexts (Table 1).

The glacial region is located at the extreme northern edge of Pakistan having mean elevation of about 4158 m. It is mostly covered with glaciers and permanent snow. The summer snow melting contributes to the rivers flowing and is used as a source of agriculture in the country. Excessive snow melts and under ice streams can cause flooding in the lower elevated areas [35].

The humid region includes very high mountains of the Hindukush, Karakoram, and Himalaya (HKH) regions and all major rivers of Pakistan rivers (Indus, Kabul, Swat, Chitral, Gilgit, Hunza, Kurram, Jhelum, and Panjkora) originate from these regions. The region is located at a mean elevation of 1286 m and it counts with the highest precipitation amount estimated as 852 mm/year.

The arid region is characterized by low–lying plains consisting in the major agriculture regions (Punjab plain) that are drained by the Indus River and its tributaries [35]. The region counts with an average elevation and annual rainfall of 633 m and 322 mm/year, respectively.

The extreme arid region consists in barren lands located at the southern end of the country towards Arabian Sea and Iran. There are low dry mountain ranges with a mean average elevation of 444 m and very low amount of precipitation estimated as 133 mm/year

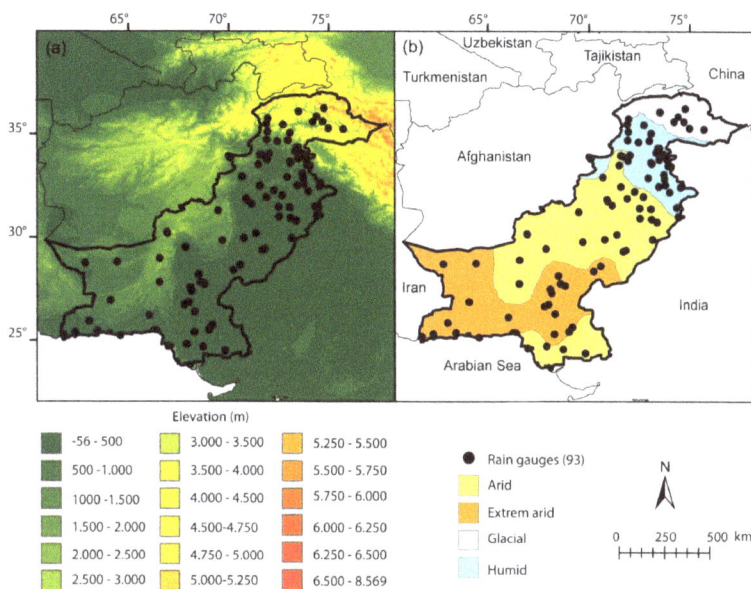

Figure 1. Study area with the location of precipitation gauges used as references. (**a**) Pakistan elevation derived from the Shuttle Radar Topography Model (SRTM). (**b**) Location of the four considered climatic zones.

Table 1. Main features of the considered regions. Annual average precipitations are derived from Tropical Rainfall Measuring Mission Multisatellite Precipitation Analysis (TMPA) over the 1998–2017 period and average elevation are derived from the Shuttle Radar Topography Model (SRTM).

Region	Pakistan	Humid	Glacial	Arid	Extreme Arid
Surface area (km^2)	878.400	115.810	82.070	396.320	286.511
Average elevation (m)	1044	1286	4158	633	444
Annual average precipitation (mm)	338	852	348	322	133
Number of stations	88	26	8	32	22

2.2. Datasets

2.2.1. GPM Based Satellite Precipitation Estimates

IMERG is a product of the National Aeronautics and Space Administration (NASA). It uses both passive microwave (PMW) and infra–red (IR) sensors available from Low Earth Orbital (LEO) and geostationary satellites, respectively. First, precipitation estimates that were derived from PMW datasets while using the Goddard Profiling algorithm (GPROF) are calibrated (i) to the GPM Combined Radar-Radiometer (CORRA) precipitation estimates that were derived from GPM Microwave Imager (GMI) and Dual-Frequency Precipitation Radar (DPR) and (ii) to the Precipitation Climatology Centre (GPCP) monthly precipitation estimates. Then, the PMW precipitation estimates are blended using the CMORPH–Kalman Filter (CMORPH–KF) Lagrangian time interpolation and the Precipitation Estimation from Remotely Sensed Information using Artificial Neural Networks–Cloud Classification System (PERSIANN–CSS). The blending process relies on motion vectors derived from the Climate Prediction Center IR (CPC–IR) cloud top temperature to produce half hourly precipitation estimates.

The IMERG system is run twice to produce the IMERG Early, Late and Final run product at 4 h, 12 h, and 2.5 month after observation time, respectively. Because of the availability at short latency, the 45-day

25

interval CORRA, the 30-day interval PERSIANN-CSS, and the three-month interval CMORPH-KF interpolation calibrations steps are necessarily trailing for the Early and Late runs, whereas a centered approach is used for the Final run. Ancillary products required on routine basis for IMERG algorithm also change according to the considered version (Early, Late, Final). Required data of surface temperature, relative humidity and surface pressure are provided by the Japanese Meteorological Agency (JMA), the GANAL gridded assimilation, and the European Centre for Medium-range Weather Forecasting (ECMWF) for the Early, Late, and Final run, respectively. Finally, the post processing adjustment step involved climatological based coefficient to produce the Early and Late run product while Global Precipitation Climatology Project (GPCC) monthly precipitation gauge analysis based coefficients are used to produce the Final run product. For more information on IMERG processing, please refer to [36].

From the year 2014 to 2018, three successive IMERG versions were made available. IMERG–v03 was the first version released in 2014, followed by IMERG–v04 in 2017, and then IMERG–v05 in early 2018. The main change among IMERG–v03, –v04, and –v05 rely on the use of successively different GPROF algorithm versions (GPROF–v03, –v04, and –v05). Differently to GPROF–v03, GPROF–v04 and –v05 used threshold precipitations rate to adjust fractional coverage. Others changes from IMERG–v03 to –v04 (and –v05) are: (1) the use of Global Precipitation Climatology Project (GPCP) v2.3 to compensate for the GPM combined instrument dataset (2BCMD) bias over non tropical oceans and land; (2) inclusion of the Advanced Technology microwave Sounder (ATMS) dataset; (3) dynamic calibration of PERSIANN–CSS by PMW derived precipitation estimates; (4) increase in HQ precipitation field spatial coverage from 60°N–S to 90°N–S; and, (5) removal of the GPCC grid box volume adjustment to eliminate blocky gauge adjustment (Final run only). The benefit of blocky gauge adjustment is clearly observable in Figure 2 with the total removed of blocky effects from IMERG–v03 to IMERG–v04 and –v05. Main changes in IMERG–v05 in comparison to previous –v04 and –v03 versions include some restrictions in Microwave Humidity Sounder (MHS) and ATMS swaths (for the five and eight footprints, respectively) and the no-consideration of TRMM Microwave Imager (TMI). For the present study, we used the IMERG–v03, –v04, and –v05 Final run.

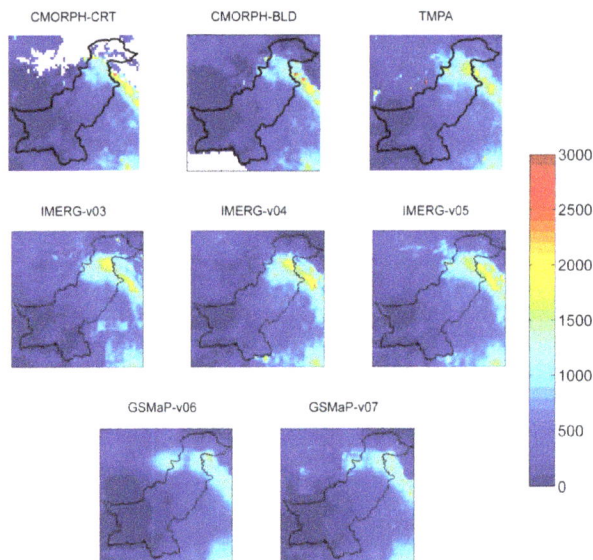

Figure 2. Annual precipitation maps for 2015 retrieved from all Satellite Precipitation Estimates (SPEs) at their original grid size. For each SPEs, only the pixels with more than 80% available daily data were considered.

GSMaP is a product of the Japan Science and Technology (JST) agency under the Core Research for Evolutional Science and Technology (CREST). Its precipitation estimates are based on PMW and IR datasets combination. In the first step, the precipitation estimates are derived from PMWs brightness temperature [37]. Then, a Kalman filter model is used to refine the precipitation rate propagated using the atmospheric moving vector derived from CPC–IR data [38], providing hourly precipitation estimates. The resultant product, GSMaP–MVK, is finally calibrated using the CPC unified gauge–based analysis of global daily precipitation. It is worth mentioning that GSMaP–MVK is available three days after observation time. To reduce data latency, a near real time version (GSMaP–NRT) using simplification in the processing is made available three hours after observation.

From 2014 to 2018, two successive versions of GSMaP were made available. The GSMaP–v06 was the first version released in 2015 and followed by GSMaP–v07 in 2017. The main modifications from GSMaP–v06 to GSMaP–v07 algorithm are: (1) the use of GPM/DPR observation as database; (2) implementation of snowfall estimation over high latitudes; and, (3) improvement of (i) gauge–calibration, (ii) orographic rain correction method, and (iii) weak precipitation detection over the ocean. For the present study, we only consider GSMaP–v06 and –v07 gauges adjusted versions (MVK).

2.2.2. TRMM Based Satellite Precipitation Estimates

Climate Prediction Center Morping (CMORPH) is a product of NOAA/CPC. Precipitation estimates are derived from PMW datasets and propagated in space and time while using a motion vector derived from CPC–IR data [2]. Recently, CMORPH was reprocessed using a fixed algorithm and homogeneous input to fix substantial inconsistencies in its first version between 2003 and 2006. The new version (CMORPH–v1.0) covers the entire period 1998 to the present, whereas the first version's (CMORPH–v0x) coverage started in 2002. There is an only satellite based version (CMORPH–RAW), a bias corrected version (CMORPH–CRT), and a satellite gauge blended version (CMORPH–BLD). CMORPH–CRT is derived from CMORPH–RAW adjustment while using the CPC unified gauge-based analysis over land and the pentad GPCP over the ocean [39]. CMORPH–CRT is then combined with the gauge analysis through the optimal interpolation technique to generate CMORPH–BLD [40]. For the present study, we consider CMORPH–CRT and –BLD. The choice is based on recent study in which CMORPH–CRT was found to outperform CMORPH–RAW [35]. As CMORPH–BLD is derived after an adjustment made to CMORPH–CRT, it is naturally selected to highlight potential benefit from the adjustment process.

TRMM Multisatellite Precipitation Analysis (TMPA) is a product of NASA in collaboration with the JAXA. Precipitation estimates are derived from PMW datasets and gaps are filled using CPC–IR, Meteorological Operational satellite program (MetOp), and GridSat–B1 datasets [1]. TMPA–v07 is the last available TMPA version since its predecessor (TMPA–v06) ended in July 2011. Enhancement form –v06 to –v07 version came from the consideration of a larger amount of PMW and IR data. It has a satellite based version (TMPA–RT) and a gauges adjusted version (TMPA–Adj). The TMPA–Adj version is derived from TMPA–RT adjustment using GPCP and Climate Assessment and Monitoring System (CAMS) datasets [1]. For the present study, we only consider the TMPA–Adj as it was found to outperform TMPA–RT version over Pakistan [35].

2.2.3. Gauges Precipitation Data

Meteorological data of Pakistan is owned by different organizations, such as Pakistan Meteorological Department (PMD), Water and Power Development Authority (WAPDA) of Pakistan, University of Boon under the Culture Areas Karakoram (CAK) program in the Bagrot valley and Yasin catchment of Gilgit basin during 1990–91, and Ev-K2-CNR (an Italian based organization). However, for the present, only the data handled by the PMD at 88 stations over one year period (1 January–31 December 2015) was made available and used. The data are manually collected and therefore subject to error introduce by personnel and instrumental errors. Additional errors for the gauges that are located in high elevation regions come from the wind effect, which might bias the precipitation collection by the gauges. In this context, the PMD follows the World Meteorological Organization (WMO) standard code WMO-N for the evaluation and correction of gauge-based precipitation data [41] to ensure as consistency as possible in the measurements. Therefore, the provided precipitation data by PMD were considered as ground truth for evaluation of the SPEs.

The dataset spread over the considered regions with 8, 26, 32, and 22 gauges for the glacial, humid, arid, and extreme arid regions, respectively (Table 1). Some of the stations located in the mountainous northern region (glacial and humid region) count with snowfall events. For these specific stations, the recorded snowfall is melted with hot water. Then, the hot water volume is removed from the total water (snowfall + hot water) to retrieve the snowfall water equivalent.

Finally, Figure 3 shows the Global Precipitation Climatology Project (GPCC) and Climate Prediction Center (CPC) gauges network used for the monthly and daily gauges based precipitation grid for the SPEs adjustment over the 2015 year. It is worth mentioning that both GPCP and CPC share common gauges with the gauges network that is used as reference in the present study. Therefore, the gauges network that was used for the SPEs assessment is not totally independent of the assessed SPEs and could influence SPEs performances conclusions. It is worth mentioning that Figure 3 aims at providing a general overview of potential overlapping between the reference gauge network with both CPC and GPCC one. Indeed, gauges from the other above-mentioned organizations might also be part of the CPC and GPCC gauges network.

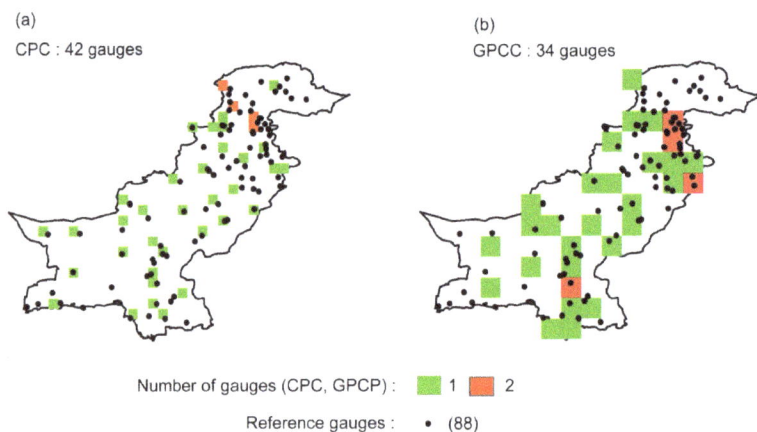

(a)
CPC : 42 gauges

(b)
GPCC : 34 gauges

Number of gauges (CPC, GPCP) : ▢ 1 ▢ 2

Reference gauges : • (88)

Figure 3. Gauges network used to produce (**a**) CPC (0.5°) and (**b**) GPCC (1°) gauges based precipitation grid for the year 2015 with gauges network used as reference data in the present study.

2.3. Method Used

2.3.1. SPEs and Gauges Pre-Processing

For the inter-comparison of CMORPH–CRT, –BLD, and TMPA with IMERG–v03, –v04, –v05 and GSMaP–v06 –v07, all IMERG and GSMaP datasets were first resampled from their original grid size (0.1°) to the CMORPHs and TMPA grid size (0.25°), according to the protocol that was proposed by [15]. To do so, IMERG and GSMaP were first resampled from 0.1° to 0.05° grid scale and the precipitation at the 0.25° grid is obtained by taking the mean precipitation value of the 25 pixels (0.05°) included into the 0.25° pixel.

For each of the grid box (0.1° and 0.25°), including gauges, daily temporal records are computed from all SPEs using 8 h to 8 h (local time) temporal windows to match with daily gauges observations. Then, monthly records are computed from daily records only for the month with more than 80% of available daily records for all SPEs and the gauges. Figure 4 shows the percentage of available daily and monthly values at each pixels location. To ensure robustness in the analysis, the database used for the SPEs assessment only considered stations with more than 90% of available data over the considered period (1 January–31 December 2015).

Finally, mean regional daily and monthly records are computed for the entire Pakistan, glacial, humid, arid, and extreme arid regions from the gauges (P_{ref}) and all SPEs by aggregating the records from all of the pixels (including gauges) included in the considered region.

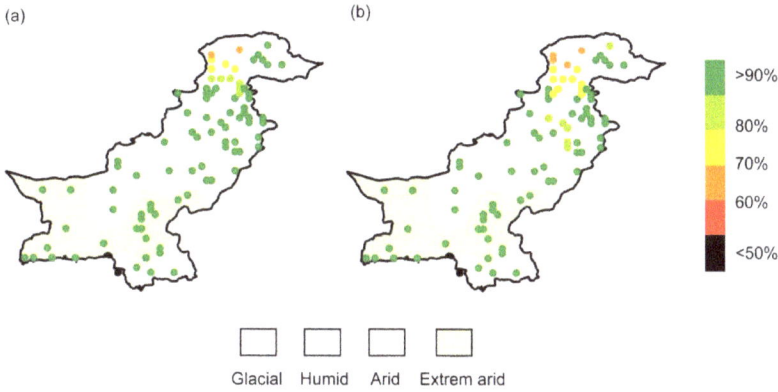

Figure 4. Available precipitations estimates from all SPEs and gauges for the 2015 year at (**a**) daily and (**b**) monthly time steps.

2.3.2. SPEs against Gauges at the Monthly Time Step

First, for each region (entire Pakistan, glacial, humid, arid, and extreme arid) SPEs mean monthly regional series were compared to the gauges in terms of Correlation Coefficient (*CC*), Standard Deviation (*STD*), Centered Root Mean Square Error (*CRMSE*), and percentage Bias (*%B*) (Equations (1)–(4)).

$$CC = \frac{Cov\left(SPE, P_{ref}\right)}{STD_{SPE} \times STD_{ref}} \tag{1}$$

where *CC* is the correlation coefficient, *SPE* and P_{ref} are the SPE and P_{ref} precipitation time series, and *Cov* is the covariance.

$$STD = \sqrt{\frac{1}{n}\sum_{i=1}^{n}\left(P_i - \overline{P}\right)^2} \tag{2}$$

where *STD* is the standard deviation in mm, *n* is the number of values, and *P* the precipitation value in mm (SPE or P_{ref}).

$$\%B = \frac{\frac{1}{n}\sum_{i=1}^{n}\left(SPE_i - P_{ref_i}\right)}{\frac{1}{n}\sum_{i=1}^{n}P_{ref_i}} \times 100 \tag{3}$$

where *%B* is the *SPE* Bias value in percentage, *n* is the number of values; *SPE* the precipitation estimate of the considered *SPE* value in mm, and P_{ref} the reference precipitation value in mm.

$$CRMSE = \sqrt{\frac{1}{n}\sum_{i=1}^{n}\left(\left(SPE_i - \overline{SPE}\right) - \left(P_{ref_i} - \overline{P}_{ref}\right)\right)^2} \tag{4}$$

where *CRMSE* is the centered root mean square error in mm, *n* is the number of values, *SPE* the precipitation estimate of the considered *SPE* value in mm, and P_{ref} the reference precipitation value in mm.

To facilitate the inter-comparison among all considered SPEs, the results are presented in the form of Taylor diagram [42]. Taylor diagram consists in an integrated graphical representation of CC and normalized STD and CRMSE values. STD and CRMSE are normalized by dividing SPEs STD and CRMSE by P_{ref} STD. In this context, SPEs optimum statistic scores are reached for CC, STD, and CRMSE values equal to 1, 1, and 0, respectively (P_{ref} score). Therefore, in the Taylor diagram, the closest is the SPEs to the reference dot the closest are the SPEs and P_{ref} estimates. Additionally, %B is considered to observe SPEs potential over/underestimation.

Secondly, CRMSE is computed at the pixel level (including gauges) for all SPEs and are plotted to observe very local SPEs performance at the pixel scale. Only CRMSE is considered as it consists in the most discriminating statistical scores. Indeed, in the Taylor diagram, CRMSE constrained SPEs relative position to the dot reference. %B is also computed and plotted to observe SPEs potential over/underestimation spatial variability.

2.3.3. SPEs against Gauges at the Daily Time Step

The daily performance assessment is based on categorical statistics aiming at measuring SPEs capacity for the detection of daily precipitation events. The statistics are based on a contingency table considering daily precipitation as a discrete value with two possible cases: day with or without precipitation (Table 2).

Table 2. Contingency table used to define daily categorical scores for the verification of SPEs against gauge data.

		Gauges (P_{ref})	
		Precipitation	No precipitation
SPE	Precipitation	a	b
	No Precipitation	c	d

Four indexes are considered: the Probability of Detection (*POD*), the False Alarm Ratio (*FAR*), Critical Success Ratio (*CSI*) the Bias, and the Heidke Skill Score (*HSS*) (Equations (5)–(9)).

$$POD = \frac{a}{(a+c)} \tag{5}$$

$$FAR = \frac{b}{(a+b)} \tag{6}$$

$$CSI = \frac{a}{(a+b+c)} \tag{7}$$

$$Bias = \frac{(a+b)}{(a+c)} \tag{8}$$

$$HSS = \frac{2*(a*d-b*c)}{[(a+c)*(c+d)+(a+b)*(b+d)]} \tag{9}$$

POD aims at representing SPE's ability to correctly forecast precipitation events. Values vary from 0 to 1, with 1 as a perfect score.

FAR aims at representing how often SPE detect precipitation event, when, actually, it does not occur. Values vary from 0 to 1, with 0 as a perfect score. FAR is also represented in form of the Success Ratio (SR = 1 − FAR).

CSI is the ratio between the number of precipitation events that are correctly detected by the SPE and the number of all precipitation events, as registered by the gauge and the SPE. Values vary between 0 and 1 with a perfect score of 1.

Bias aims at representing SPEs tendency to under forecast (Bias < 1) or over forecast (Bias > 1) precipitation events with a perfect score of 1.

HSS consists in a generalized skill score on precipitation events forecasting compared to a random based prediction. Values range from $-\infty$ to 1 with a perfect score of 1 and negative values indicating that the random based prediction outperforms the SPE one.

POD, SR, CSI and Bias are computed for all SPEs and considered regions (entire Pakistan, glacial, humid, arid, and extreme arid) from the respective mean regional daily precipitation series while using a threshold value of 1 mm/day to differentiate precipitation to no precipitation events [15,24]. It is worth mentioning that some authors used an increasing threshold to assess SPEs performance for different precipitation intensities (e.g., [43–45]). However, for the present study, most of the assessed pixels count with only one gauge. Therefore, the representativeness between areal (SPE pixel) and point (gauges) measurement should decrease with increase in threshold value.

To facilitate SPEs inter-comparison, results are presented in form of a performance diagram [45]. The performance diagram integrates POD, SR, CSI, and Bias in a geometric way in which the relative position of the reference dot located in the upper left corner permits the SPEs performance inter-comparison. The reference dot corresponds to POD, SR, CSI, and Bias values of 1, 0, 1, and 1, respectively. Closer is the SPE to the reference dot, higher are the SPEs forecasting ability.

Finally, HSS is computed at the pixel level (including gauges) for all SPEs and are plotted to observe very local SPEs potential in daily precipitation event forecasting ability.

2.3.4. Benefits of GPM Based SPEs Successive Versions

The objective of this section is to have first insight into the potential benefits of the successive GSMaP and IMERG versions at the monthly scale.

For each pixel including gauges, the GSMaP versions (–v06 or –v07) with the lowest CRMSE being plotted to localize the most suited version in space. Additionally, the absolute CRMSE difference between the two versions are also plotted to quantify the spatial enhancement between both versions.

Similarly, for each pixel including gauges, the IMERG versions (–v03, –v04, –v05) with the lowest CRMSE are also plotted. Because three versions are available for IMERG estimates, the absolute CRMSE difference is based on the versions with the highest and lowest CRMSE values.

2.3.5. Benefits of GPM over TRMM Based SPEs

This section aims at presenting which SPEs between GPM and TRMM considered in this study provides the most accurate precipitation estimates over Pakistan. At each gauge location, the SPEs with the lowest CRMSE and %B are plotted to highlight the SPE that should be used for the specific pixel location.

3. Results

3.1. SPEs Monthly Performance at the Regional Scale

Figure 5 shows the SPEs performance for each considered regions in the form of Taylor diagram. On a general way, all SPEs provide relatively accurate precipitation estimates over the entire Pakistan, humid, arid, and extreme arid regions with CRMSE lower than 0.5 and CC higher than 0.9. On the contrary, all the SPEs performances are very low over the glacial region with CRMSE higher than 0.5 and CC lower than 0.5.

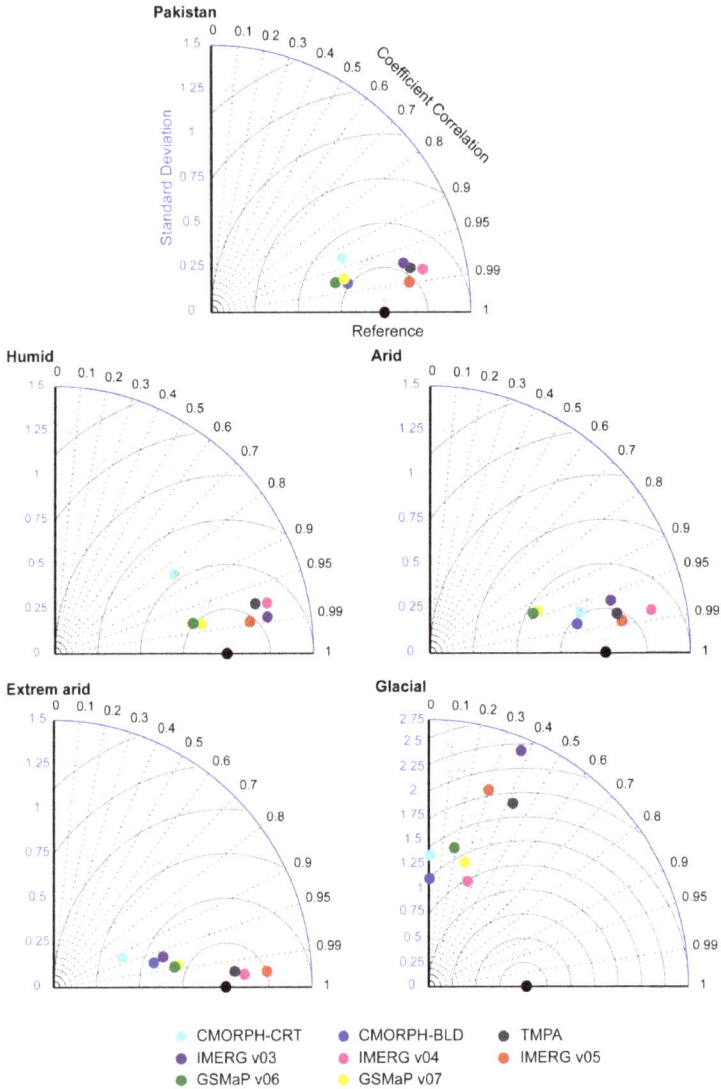

Figure 5. SPEs monthly performance for all the considered regions expressed in form of Taylor diagram. The reference black dot represents the perfect statistical scores (CC = 1, STD = 1, CRMSE = 0). Centered Root Mean Square Error (CRMSE) is represented by curved solid grey lines.

When considering the GSMaP versions, the new released GSMaP–v07 always provide closer estimates to the reference in comparisons to the previous GSMaP–v06.

Regarding to IMERG versions, the enhancement from successive versions are more contrasted. Indeed, IMERG–v05 precipitation estimates are closer to the reference for the entire Pakistan, humid and arid regions with CRMSE value of approximately 0.2 for all of the regions. However, over the extreme arid region, IMERG–v04 is found better than IMERG–v05. A similar inconsistency is observable over the humid and arid regions with IMERG–v03 closer to the reference than IMERG–v04.

When compared to the TRMM based SPEs versions (CMORPH–CRT, CMORPH–BLD, and TMPA), the benefit of the new GPM based SPEs (IMERG and GSMaP) is not evident. When considering the entire Pakistan, CMORPH–BLD estimates performance (CRMSE = 0.26) is closer to GSMaP–v07 (CRMSE = 0.29) and IMERG–v05 (CRMSE = 0.21). The same is true for the humid region. Over the arid region, GSMaP–v06 and –v07 are outperformed by TRMM based SPEs, while IMERG–v05 still provides more accurate precipitation estimates (CRMSE = 0.19). Finally, for the extreme arid region, TMPA (CRMSE = 0.1) outperformed all GPM based SPEs.

Figure 6 shows the %B observed for all SPEs over the different considered regions. Regarding GSMaP–v06 and –v07, they both underestimate precipitation for the entire of Pakistan, humid, arid, and extreme arid regions with %B around −20% and highly overestimate precipitation over the glacial region. Overall, GSMaP–v07 is slightly less biased than GSMaP–v06.

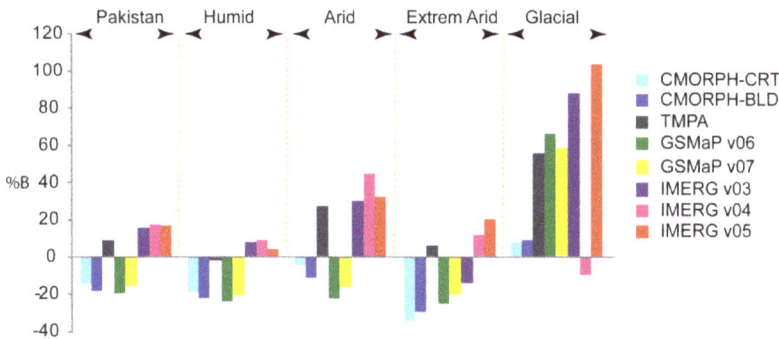

Figure 6. %B for all considered SPEs and all the considered regions.

Regarding IMERG–v03, –v04, and –v05, the opposite trend is observed for the entire Pakistan, humid, arid, and extreme arid as they all overestimate the precipitation. For all IMERG versions, the %B is around 5% for the humid region and superior to 30% for the arid region. For the glacial region, IMERG–v03 and IMERG–v05 highly overestimate the precipitation estimates with %B superior to 80% and 100%, respectively, while IMERG–v04 is the less biased (absolute %B < 15%) product among GPM based SPEs.

TMPA %B is very low for the entire Pakistan, humid and extreme arid regions with respective values of 9%, −2%, and 6%, while it considerably overestimates precipitation over the arid and glacial regions with %B of 27% and 56%.

Similar to GSMaP–v06 and –v07, CMORPH–CRT and CMORPH–BLD underestimate precipitation over the entire Pakistan, humid, arid, and extreme arid regions. The bias is very low and very high for the arid and extreme arid regions, respectively. Interestingly, unlike the other SPEs, CMORPH–CRT, and BLD present %B lower than 10% over the glacial region.

3.2. SPEs Monthly Performance at the Gauge scale

Figure 7 shows the CRMSE obtained for each pixel including gauges at the monthly time step. CRMSE value is selected as it controls SPEs relative position in the Taylor diagram, and therefore can be used easily for the evaluation of inter-comparative performance of SPEs. All SPEs are better able to represent precipitation spatial variability in the eastern than western part in the arid and extremely arid regions. All SPEs performance is very low over the northern glacial region with CRMSE value higher than 1.

Among TRMM based SPEs, the enhancement from CMORPH–CRT to CMORPH–BLD is observable over the extreme arid and humid regions with an overall increase in the number of pixel with CRMSE value lower than 0.5 (almost three times as many). TMPA presents a quite similar

CRMSE pattern than CMORPH–BLD, except over the arid region where CMORPH–BLD has better captured the precipitation spatial variability with higher proportion of pixels with CRMSE values lower than 0.5.

When considering the IMERG successive versions (–v03, –v04, –v05), a high discrepancy is observed in their respective ability to represent the precipitation spatial variability. The most suitable version (in term of CRMSE values) changes along the considered regions. IMERG–v04 has the highest number (13) of pixel with CRMSE lower than 0.5 over the extreme arid region in comparison to IMERG–v03 (7) and –v05 (9). Over the arid and humid regions, the lastest released IMERG–v05 outperformed its previous versions (IMERG–v03, –v04) with many pixels having a CRMSE value lower than 0.5. However, over the glacial region, all IMERG versions perform very poorly with unsatisfactory CRMSE values systematically higher than 0.5.

When considering the GSMaP SPEs successive versions (–v06, –v07), the error spatial distribution in the form of CRMSE values are very close over the entire Pakistan, irrespective of the considered region. Both versions performed very poorly over the glacial region with a CRMSE value higher than 1.

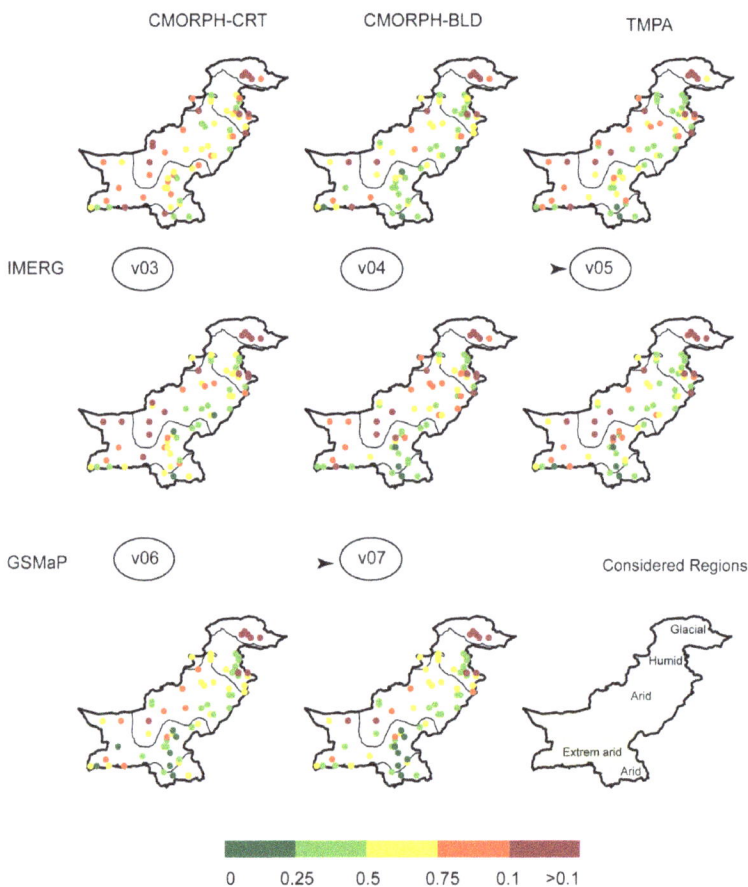

Figure 7. SPEs performance at the gauges point measurement expressed in form of CRMSE.

Figure 8 shows the %B that was obtained for each pixel, including gauges, at the monthly time step. CMORPH–CRT and CMORPH–BLD present very similar %B distribution with relatively low %B value and many pixels in the −15% to 15% ranges over the humid and northern arid regions.

Besides, at the regional scale they both tend to underestimate monthly precipitation with a dominant proportion of pixels with negative %B values. Over the glacial region, CMORPH products strongly overestimate (underestimate) precipitation for the western (eastern) located pixels. On the contrary, TMPA tends to overestimate monthly precipitation over the entire of Pakistan, especially over the humid and northern arid regions. The pixels with −15% to 15% (%B) ranges are mainly located in the central Pakistan over the arid and extreme arid regions. Over the glacial region, TMPA strongly overestimates precipitation for the western located pixels, whereas reasonable %B values are found for the two eastern located pixels.

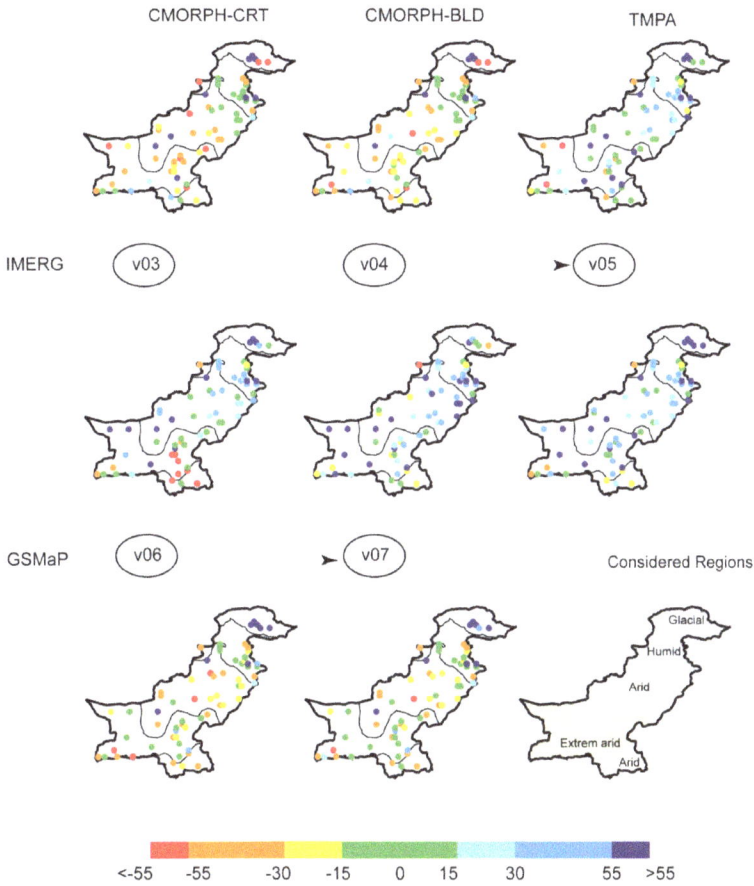

Figure 8. SPEs performance at the gauges point measurement expressed in form of %B.

Regarding IMERG successive versions (–v03, –v04, –v05), they all tend to overestimate monthly precipitation at the regional scale. The %B pattern is very close for all IMERG versions over the humid region. Differences account over the arid region, where IMERG–v03 presents the highest number of pixel (7), with %B values ranging from −15% to 15%. Interestingly, for the pixels that are located southeast, the strong positive %B for the first IMERG version (IMERG–v03) was consistently corrected for the following IMERG versions (–v04, –v05). Over the glacial region, IMERG–v04 presents the lowest %B values in comparison to its previous (–v03) and following (–v05) versions.

Similar to the CRMSE values, GSMaP–v06 and –v07 present very close %B spatial distribution with an overall trend of underestimation of the monthly precipitation. Pixels, with %B ranging from

−15% to 15%, are mainly located over the extreme arid and humid regions. As observed for TRMM based SPEs and IMERG datasets, GSMaP–v06 and –v07 %B are very high for the pixels that are located in the glacial region.

3.3. SPEs Daily Potential at the Regional Scale

Figure 9 shows the SPEs daily precipitation events forecasting ability for the entire Pakistan and each considered regions in the form of performance diagram. On a general way, SPEs performance at the daily time step is higher for humid and arid regions than for the extreme arid and glacial regions. SPEs performances are closer for the humid and arid region with a close relative position in the performance diagram than for the extreme arid and glacial regions.

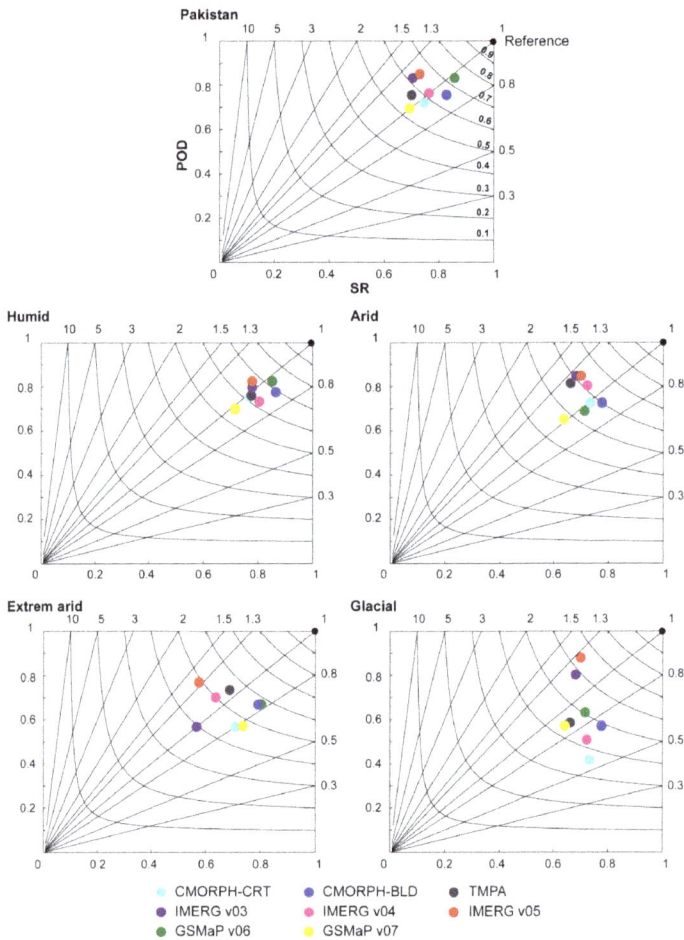

Figure 9. SPEs daily performance expressed in form of performance diagram. Curved lines and straight lines represent the CSI and Bias values, respectively. The reference black dot represents the perfect statistical scores (POD = 1, SR = 1, Bias = 1 and CSI = 1).

When considering the TRMM based SPEs, CMORPH–BLD performance is highest than its non-gauges adjusted version CMORPH–CRT. Indeed for all regions, CMORPH–BLD is found to be closest to the reference dot. This shift is induced by an increase of SR over the entire Pakistan, humid and arid regions, whereas an increase of POD also contributes to CMORPH–BLD enhancement over the extreme arid and glacial regions. In comparison to TMPA, CMORPH–BLD presents the highest ability for daily precipitation events forecasting for all considered regions.

When considering the three successive IMERG versions, their performances for the entire Pakistan are close with IMERG–v05 slightly closer to the reference dot. Similar observation is true for the humid and arid regions where IMERG–v05 outperformed its two previously released versions (–v03 and –v04). Much variability in IMERG relative performance is observable over the extreme arid and glacial regions with IMERG–v04 (–v05) performing better over the extreme arid (glacial) regions.

Interestingly, for all considered regions, GSMaP–v07 is less able to forecast daily precipitation event than it previous version (GSMaP–v06). Indeed, from GSMaP–v06 to –v07 a decrease is observed in all considered indexes (POD, SR, CSI and Bias). GSMaP–v06 outperformed all considered SPEs at the daily time step for the entire Pakistan, humid and extreme arid regions.

3.4. SPEs Daily Potential at the Gauges Scale

Figure 10 shows the SPEs spatial consistency at the daily time step expressed in the form of HSS. From all the considered TRMM based SPEs, CMORPH–BLD is far from the SPEs with the highest ability for daily precipitation events forecasting. It has the highest number of pixels with HSS that is higher than 0.6 homogeneously distributed all over the Pakistan. However, its performance is particularly bad over the glacial regions as illustrated by the top northern located pixel with a negative HSS value.

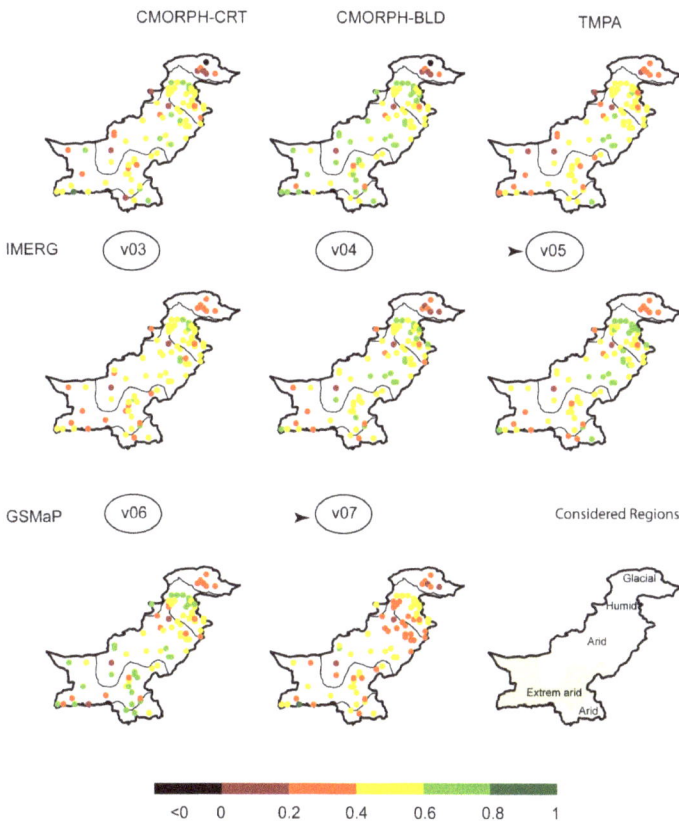

Figure 10. SPEs performance at the daily time step expressed in form of Heidke Skill Score (HSS) index.

Contrary to the observation made for the monthly time step, the successive IMERG versions consist in a continuously improvement in daily precipitation event forecasting. Indeed, the number of pixels with HSS higher than 0.6 considerably increased from IMERG–v03 to –v04 (especially over the arid region) and slightly increased from IMERG–v04 to –v05. However, over the glacial regions, the HSS is low (HSS < 0.4) for all pixels and considered IMERG versions.

When considering the GSMaP–v06 and –v07 versions, the first GSMaP version (–v06) presents the highest number of pixels with HSS higher than 0.6, confirming the observation that is presented in Figure 9. It is clearly observable over the humid and extreme arid regions. As observed for TRMM based SPEs and IMERG datasets, GSMaP–v06 and –07 performance are very low over the glacial region (HSS < 0.4).

3.5. Benefits of GPM Based SPEs Successive Versions

Figure 11 shows GSMaP and IMERG most suited version at each gauges location, according to the CRMSE values. Additionally, the absolute CRMSE difference between the most and least suited version is given to quantify the potential enhancement of successive versions.

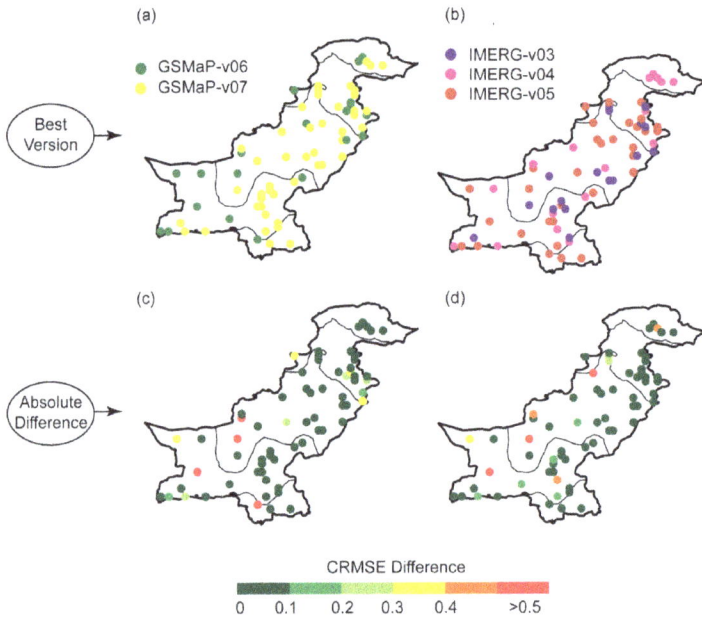

Figure 11. Improvement on monthly precipitation estimates from successive Global Precipitation Measurement (GPM) based SPEs: best SPEs version from successive Global Satellite Mapping of Precipitation (GSMaP) (**a**) and Integrated Multi–Satellite Retrievals for GPM (IMERG) (**b**) versions based on CRMSE values; absolute difference of CRMSE value between the most and least accurate GSMaP versions (**c**) and IMERG versions (**d**).

Regarding GSMaP successive versions, GSMaP–v07 is the most accurate version with lowest CRMSE value obtained at most of the gauges location. It is worth mentioning that for all pixels where GSMaP–v07 performed better than GSMaP–v06, the difference in CRMSE value is very low (<0.1), meaning that both versions performed quite similarly at those pixels location. However, for the pixels where GSMaP–v06 outperforms GSMaP–v07, CRMSE difference is much higher with differences that are higher than 0.5 in the arid and extreme arid regions. Therefore, the enhancement of GSMap–v07 is not very efficient and brought local inconsistency, especially over the arid and extreme arid regions.

When considering IMERG successive versions, the last IMERG–v05 is the most efficient with the highest numbers of pixels (32), where it outperformed its previous versions IMERG–v03 and IMERG–v04. Both IMERG–v03 and –v04 count with 17 pixels where they provide the best monthly precipitation estimates. IMERG–v04 (–v05) is the most efficient to capture precipitation variability over the glacial (arid) region, confirming previous results at the regional scale (Figure 5). However, the absolute difference in CRMSE values is very low overall (<0.1), meaning that all versions performed quite similarly over the Pakistan. Interestingly, the pixels with the highest absolute difference in CRMSE value (>0.5) account for the pixels where the last IMERG–v05 provides the best monthly precipitation estimates. This shows the benefits that are brought by the last IMERG–v05 over the previously released IMERG–v03 and –v04.

3.6. Benefits of GPM over TRMM Based SPEs

Figure 12 shows the SPEs with the highest performance at the monthly time step in terms of CRMSE. A strong heterogeneity is observed with all SPEs that are represented with at least one pixel over all considered regions except over the glacial region. Therefore, no SPEs consistently outperformed the others. At the regional scale, with 14 and 12 pixels, GSMaP–v07 and IMERG–v05, respectively, are the most represented SPEs (Figure 12b). It highlights the global enhancement from TRMM based SPEs to the lastest GPM based SPEs versions.

More generally, GPM based SPEs (GSMaP and IMERG) are significantly more represented than TRMM based SPEs (CMORPH and TMPA) (Figure 12c). It highlights the GPM benefit on the ongoing SPEs generation. However, for some pixels that are mainly located in the central arid Pakistan region, CMORPH and TMPA still outperformed the GSMaP and IMERG datasets.

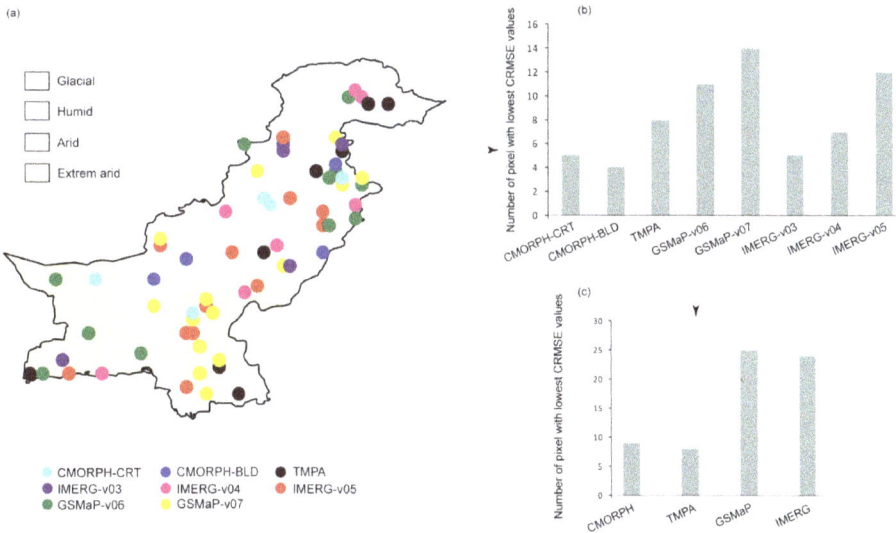

Figure 12. SPEs with the highest performance among all considered SPEs in term of CRMSE. (**a**) Spatial distribution of SPEs ranking in term of CRMSE, (**b**) numbers of pixels in which the considered SPEs outperformed the others, and (**c**) same as (**b**) but aggregated to the SPEs group.

4. Discussion

Numerous studies reported on GPM based SPEs accuracy in different regions. However, yet, none study reported on the potential benefit brought by the successive IMERG and GSMaP available versions. Therefore, it should be reminded that this is the first study reporting on IMERG–v05 performance and considering the potential benefit that is brought by IMERG–v05 over its two previously released versions (IMERG–v03, –v04). Regarding GSMaP, this is the first study reporting on both GSMaP–v06 and –v07.

Interestingly, at the regional scale and monthly time step, IMERG–v03 outperformed IMERG–v04, which is in line with previous observations that were made at the daily time step over China [32,33]. This unexpected feature is also observed over the humid and arid regions. However, the very last IMERG–v05 version consists in a real enhancement, as it provides more accurate precipitation estimates than its two previous versions at the regional scale and over both humid and arid regions. Interestingly, the enhancement in precipitation estimates that is brought by the successive IMERG versions (–v03, –v04, –v05) is much marked at the daily than monthly timescale (Figures 7 and 9). All the considered IMERG versions use monthly gauge adjustment (GPCC), so the precipitation estimates at the monthly timescale is strongly dependent on the gauge adjustment. Actually, IMERG similar performance at the monthly timescale means that this component of the algorithms has improved little from IMERG–v03 to –v05 (Figure 7). On the other hand, IMERG estimates at daily timescale are more dependent on other components of the algorithms. Therefore, IMERG precipitation improvement at the daily timescale is a reflection of IMERG algorithms enhancement from IMERG–v03 to –v05 (Figure 10).

Regarding GSMaP datasets, as expected, GSMaP–v07 systematically provides a more realistic precipitation estimate than its previous GSMaP–v06 version at the monthly scale. However, the enhancement is very small, bringing inconsistencies locally. Currently, no studies are available to compare our findings regarding to IMERG–v05 and GSMaP–v07, and therefore other regions should be considered to reinforce the results of the present study.

When comparing the last GPM based SPE versions (IMERG–v05, GSMaP–v07) performance, IMERG–v05 and GSMaP–v07 performed better at the monthly and daily timescale, respectively (Figures 5 and 9). This difference might be the consequence of the gauges adjustment used for IMERG–v05 and GSMaP–v07 processing. Actually, IMERG–v05 relies on a monthly gauge adjustment (GPCC), whereas GSMaP–v07 relies on a daily gauge adjustment (CPC). On a similar way, regarding TRMM based SPEs (CMORPH–BLD and TMPA), CMORPH–BLD (CPC daily gauge adjustment) systematically performed better than TMPA (GPCC monthly gauge adjustment) at the daily timescale. Therefore, the gauges adjustment step appeared to be a considerably influencing factor in SPEs performance.

On a general way, all the SPEs are unsuitable over the glacial region with very low statistical scores at both monthly and daily time steps. This is related to PMW difficulties to retrieve accurate precipitation estimates over the frozen area, which appear to be similar to ice precipitation aloft in the scattering signal in microwave channels on satellites [46–48]. Therefore, SPEs algorithms have difficulties in distinguishing between the emission from frozen surfaces and the scattering from ice precipitation aloft. As a result, all of the tested SPEs globally overestimate precipitation over Pakistani glacial region (Figure 8), which is in line with the previous study results focused on TRMM based SPEs [35].

Overall, SPEs relative performance ranking highly varies according to the considered region and the considered pixels. Actually, despite an overall better performance for GPM than TRMM based SPEs, none of the SPEs systematically outperformed the others (Figure 12). In this line, SPEs merging should be considered to take advantages from all available SPEs. Previous studies have already reported on the benefit of such an approach to retrieve the most realistic satellite based precipitation estimates. For example, the blended MSWEP precipitation estimates [8,49] which merge precipitation estimate from (i) different SPEs, (ii) gauges observation and (iii) atmospheric model was tested over the Lake Titicaca basin. In comparison to 11 SPEs, MSWEP provided the most realistic precipitation estimate at the gauge level and the most realistic streamflow and snow–cover simulation if used as a forcing data in the models [50]. In this line, [51] merged TMPA, CMORPH, and PERSIANN datasets to retrieve precipitation estimates over the Tibetan plateau. The merge product was found to outperform MSWEP estimates at both gauges and for streamflow simulation. However, those merged product did not include GPM based SPEs. Therefore according to the present study, future SPE merging attempt should be considered using IMERG and GSMaP datasets as they provide the most accurate precipitation estimates over most regions of the Pakistan.

It is worth mentioning that this study only considers gauge based assessment. Indeed, SPEs potential conclusions may differ when using different indicators than precipitation gauges and especially when comparing SPEs relative performance [50]. Therefore, future study should test GPM based SPEs as forcing data in hydrological modelling to complete the present study. However, such efforts are still complicated, as only years data are available for IMERG and GSMaP, which compromise models calibration/validation steps.

Finally, the SPEs analysis at the daily time step should be taken with caution as spatial representability between areal (SPE pixel) and point (gauges) measurement may suffer some inconsistency. Indeed, precipitation event detected by the SPE may not be detected by the gauges as it might be raining in other point location of the pixel area [15,35,44]. In this line, a recent study shows that SPEs potential conclusions should be influenced by the reference gauges density used for the assessment [52]. Therefore, similar protocol should be reiterated while using a denser gauges precipitation network to provide robustness results and conclusions on the considered SPEs performance.

5. Conclusions

The present study is a first attempt to highlight the potential benefits that are brought from the successive released versions of GPM based SPEs. IMERG–v03, –v04, –v05 and GSMaP–06, v07 were successively assessed at the daily and monthly times scales over the contrasted geomorphologic and climatic divisions of Pakistan. An inter-comparison between SPE products from GPM with its predecessor TRMM is also included in this study that has increased the merits of this study. Despite the limited coverage and scarcity of the ground reference points, some consistent features emerged from the analysis:

- SPEs accuracy is region dependent with variable ranking in SPEs performance according to the considered region. All SPEs have presented a strong deficiency over the glacial regions that will remain a major challenge for the future SPEs algorithm development. Additionally, for the same region, the SPEs ranking changed at the very local pixel scale.
- When considering IMERG datasets, IMERG–v04 should be taken as a well named transitional version between IMERG–v03 and –v05. Indeed, with the exception of the extreme arid region, it has provided globally worst precipitation estimates than its predecessor IMERG–v03. IMERG–v05 fulfilled the expected improvement in precipitation estimates with more realistic precipitation estimates than its predecessor IMERG–v03 and –v04 at both monthly and daily timescale, except over the extreme arid region where IMERG–v04 appeared as a most suitable IMERG version.
- Considering GSMaP datasets, at the monthly timescale, the two successive versions (–v06, –v07) have performed quite similarly with an overall light enhancement from GSMaP–v06 to –v07 for all the considered regions. A contradiction is observed at the daily timescale at which GSMaP–v06 become more sensitive to precipitation event detection.
- When comparing IMERG and GSMaP datasets performance, IMERG monthly precipitation estimates are more realistic than GSMaP ones over the arid and extreme arid regions.
- When considering TRMM based SPEs, CMORPH–BLD highly outperformed CMORPH–CRT. On a general way, CMORPH–BLD outperformed TMPA, except over the extreme arid region and at the monthly timescale.
- Overall, the transition from TRMM to GPM constitutes a clear enhancement of precipitation estimates over Pakistan with GPM based SPEs provided more realistic monthly precipitation estimates than TRMM based SPEs.

- No SPE is found to outperform the others promoting the development of SPEs merging approach to improve the precipitation representation over Pakistan.

Author Contributions: F.S., conceived the experiments and F.S. and Y.H. analysed the data. M.-P.B., Y.H., B.M.H., H.M.-C., G.A., R.U. provided reviews and suggestions. F.S. and Y.H. wrote the paper.

Funding: This research received no external funding.

Acknowledgments: This work is part of a postdoctoral fellowship funded by the CNES (Centre National d'Etudes Spatiales, France). The authors are grateful to SPE datasets providers and to the Pakistan Meteorological Department (PMD) for the in situ precipitation observations.

Conflicts of Interest: The authors declare no conflict of interest.

References

1. TRMM and Other Data Precipitation Data Set Documentation. Available online: https://www.researchgate.net/profile/George_Huffman/publication/228892338_TRMM_and_Other_Data_Precipitation_Data_Set_Documentation/links/575f0bde08ae9a9c955fac32/TRMM-and-Other-Data-Precipitation-Data-Set-Documentation.pdf (accessed on 13 August 2018).
2. Joyce, R.J.; Janowiak, J.E.; Arkin, P.A.; Xie, P. CMORPH: A Method that Produces Global Precipitation Estimates from Passive Microwave and Infrared Data at High Spatial and Temporal Resolution. *J. Hydrometeorol.* **2004**, *5*, 487–803. [CrossRef]
3. Sorooshian, S.; Hsu, K.-L.; Gao, X.; Gupta, H.V.; Imam, B.; Braithwaite, D. Evaluation of PERSIANN System Satellite–Based Estimates of Tropical Rainfall. *Bull. Am. Meteorol. Soc.* **2000**, *81*, 2035–2046. [CrossRef]
4. GSMaP. User's Guide for Global Satellite Mapping of Precipitation Microwave-IR Combined Product (GSMaP_MVK), Version 5. Available online: http://sharaku.eorc.jaxa.jp/GSMaP/document/DataFormatDescription_MVK&RNL_v6.5133A.pdf (accessed on 13 August 2018).
5. Maggioni, V.; Meyers, P.C.; Robinson, M.D. A Review of Merged High-Resolution Satellite Precipitation Product Accuracy during the Tropical Rainfall Measuring Mission (TRMM) Era. *J. Hydrometeorol.* **2016**, *17*, 1101–1117. [CrossRef]
6. Sun, Q.; Miao, C.; Duan, Q.; Ashouri, H.; Sorooshian, S.; Hsu, K.-L. A review of global precipitation datasets: data sources, estimation, and intercomparisons. *Rev. Geograp.* **2018**, *56*, 79–107. [CrossRef]
7. Ashouri, H.; Hsu, K.L.; Sorooshian, S.; Braithwaite, D.K.; Knapp, K.R.; Cecil, L.D.; Nelson, B.R.; Prat, O.P. PERSIANN-CDR: Daily precipitation climate data record from multisatellite observations for hydrological and climate studies. *Bull. Am. Meteorol. Soc.* **2015**, *96*, 69–83. [CrossRef]
8. Beck, H.E.; Vergopolan, N.; Pan, M.; Levizzani, V.; van Dijk, A.I.J.M.; Weedon, G.; Brocca, L.; Pappenberger, F.; Huffman, G.J.; Wood, E.F. Global-scale evaluation of 23 precipitation datasets using gauge observations and hydrological modeling. *Hydrol. Earth Syst. Sci. Discuss.* **2017**. [CrossRef]
9. Funk, C.; Peterson, P.; Landsfeld, M.; Pedreros, D.; Verdin, J.; Shukla, S.; Husak, G.; Rowland, J.; Harrison, L.; Hoell, A.; et al. The climate hazards infrared precipitation with stations—A new environmental record for monitoring extremes. *Sci. Data* 2015. [CrossRef] [PubMed]
10. Agutu, N.O.; Awange, J.L.; Zerihun, A.; Ndehedehe, C.E.; Kuhn, M.; Fukuda, Y. Assessing multi-satellite remote sensing, reanalysis, and land surface models' products in characterizing agricultural drought in East Africa. *Remote Sens. Environ.* **2017**, *194*, 287–302. [CrossRef]
11. Bayissa, Y.; Tadesse, T.; Demisse, G.; Shiferaw, A. Evaluation of satellite-based rainfall estimates and application to monitor meteorological drought for the Upper Blue Nile Basin, Ethiopia. *Remote Sens.* **2017**, *9*, 669. [CrossRef]
12. Satgé, F.; Espinoza, R.; Zolá, R.; Roig, H.; Timouk, F.; Molina, J.; Garnier, J.; Calmant, S.; Seyler, F.; Bonnet, M.-P. Role of Climate Variability and Human Activity on Poopó Lake Droughts between 1990 and 2015 Assessed Using Remote Sensing Data. *Remote Sens.* **2017**, *9*, 218. [CrossRef]
13. Casse, C.; Gosset, M. Analysis of hydrological changes and flood increase in Niamey based on the PERSIANN-CDR satellite rainfall estimate and hydrological simulations over the 1983–2013 period. *IAHS-AISH Proc. Rep.* **2015**, *370*, 117–123. [CrossRef]

14. Integrated Multi-satellitE Retrievals for GPM (IMERG) Technical Documentation. Available online: https://docserver.gesdisc.eosdis.nasa.gov/public/project/GPM/IMERG_doc.05.pdf (accessed on 13 August 2018).

15. Satgé, F.; Xavier, A.; Zolá, R.; Hussain, Y.; Timouk, F.; Garnier, J.; Bonnet, M.-P. Comparative Assessments of the Latest GPM Mission's Spatially Enhanced Satellite Rainfall Products over the Main Bolivian Watersheds. *Remote Sens.* **2017**, *9*, 369. [CrossRef]

16. Liu, Z. Comparison of Integrated Multisatellite Retrievals for GPM (IMERG) and TRMM Multisatellite Precipitation Analysis (TMPA) Monthly Precipitation Products: Initial Results. *J. Hydrometeorol.* **2016**, *17*, 777–790. [CrossRef]

17. Prakash, S.; Mitra, A.K.; AghaKouchak, A.; Liu, Z.; Norouzi, H.; Pai, D.S. A preliminary assessment of GPM-based multi-satellite precipitation estimates over a monsoon dominated region. *J. Hydrol.* **2016**, *556*, 865–876. [CrossRef]

18. Tang, G.; Ma, Y.; Long, D.; Zhong, L.; Hong, Y. Evaluation of GPM Day-1 IMERG and TMPA Version-7 legacy products over Mainland China at multiple spatiotemporal scales. *J. Hydrol.* **2016**, *533*, 152–167. [CrossRef]

19. Chen, F.; Li, X. Evaluation of IMERG and TRMM 3B43 Monthly Precipitation Products over Mainland China. *Remote Sens.* **2016**, *8*, 472. [CrossRef]

20. Wang, Z.; Zhong, R.; Lai, C.; Chen, J. Evaluation of the GPM IMERG satellite-based precipitation products and the hydrological utility. *Atmos. Res.* **2017**, *196*, 151–163. [CrossRef]

21. Kim, K.; Park, J.; Baik, J.; Choi, M. Evaluation of topographical and seasonal feature using GPM IMERG and TRMM 3B42 over Far-East Asia. *Atmos. Res.* **2017**, *187*, 95–105. [CrossRef]

22. Tan, M.L.; Duan, Z. Assessment of GPM and TRMM precipitation products over Singapore. *Remote Sens.* **2017**, *9*, 720. [CrossRef]

23. Sharifi, E.; Steinacker, R.; Saghafian, B. Assessment of GPM-IMERG and Other Precipitation Products against Gauge Data under Different Topographic and Climatic Conditions in Iran: Preliminary Results. *Remote Sens.* **2016**, *8*, 135. [CrossRef]

24. Oliveira, R.; Maggioni, V.; Vila, D.; Morales, C. Characteristics and Diurnal Cycle of GPM Rainfall Estimates over the Central Amazon Region. *Remote Sens.* **2016**, *8*, 544. [CrossRef]

25. Mahmoud, M.T.; Al-Zahrani, M.A.; Sharif, H.O. Assessment of Global Precipitation Measurement Satellite Products over Saudi Arabia. *J. Hydrol.* **2018**, *559*, 1–12. [CrossRef]

26. Sharifi, E.; Steinacker, R.; Saghafian, B. Multi time-scale evaluation of high-resolution satellite-based precipitation products over northeast of Austria. *Atmos. Res.* **2018**, *206*, 46–63. [CrossRef]

27. Sungmin, O.; Foelsche, U.; Kirchengast, G.; Fuchsberger, J.; Tan, J.; Petersen, W.A. Evaluation of GPM IMERG Early, Late, and Final rainfall estimates using WegenerNet gauge data in southeastern Austria. *Hydrol. Earth Syst. Sci.* **2017**, *21*, 6559–6572. [CrossRef]

28. Tan, M.L.; Santo, H. Comparison of GPM IMERG, TMPA 3B42 and PERSIANN-CDR satellite precipitation products over Malaysia. *Atmos. Res.* **2018**, *202*, 63–76. [CrossRef]

29. Chiaravalloti, F.; Brocca, L.; Procopio, A.; Massari, C.; Gabriele, S. Assessment of GPM and SM2RAIN-ASCAT rainfall products over complex terrain in southern Italy. *Atmos. Res.* **2018**, *206*, 64–74. [CrossRef]

30. Anjum, M.N.; Ding, Y.; Shangguan, D.; Ahmad, I.; Ijaz, M.W.; Farid, H.U.; Yagoub, Y.E.; Zaman, M.; Adnan, M. Performance evaluation of latest integrated multi-satellite retrievals for Global Precipitation Measurement (IMERG) over the northern highlands of Pakistan. *Atmos. Res.* **2018**, *205*, 134–146. [CrossRef]

31. Muhammad, W.; Yang, H.; Lei, H.; Muhammad, A.; Yang, D. Improving the regional applicability of satellite precipitation products by ensemble algorithm. *Remote Sens.* **2018**, *10*, 577. [CrossRef]

32. Wei, G.; Lü, H.; Crow, W.T.; Zhu, Y.; Wang, J.; Su, J. Evaluation of satellite-based precipitation products from IMERG V04A and V03D, CMORPH and TMPA with gauged rainfall in three climatologic zones in China. *Remote Sens.* **2018**, *10*, 30. [CrossRef]

33. Zhao, H.; Yang, S.; You, S.; Huang, Y.; Wang, Q.; Zhou, Q. Comprehensive evaluation of two successive V3 and V4 IMERG final run precipitation products over Mainland China. *Remote Sens.* **2018**, *10*, 34. [CrossRef]

34. Adnan, S.; Ullah, K.; Gao, S.; Khosa, A.H.; Wang, Z. Shifting of agro-climatic zones, their drought vulnerability, and precipitation and temperature trends in Pakistan. *Int. J. Climatol.* **2017**, *37*, 529–543. [CrossRef]

35. Hussain, Y.; Satgé, F.; Hussain, M.B.; Martinez-Caravajal, H.; Bonnet, M.-P.; Cardenas-Soto, M.; Llacer Roig, H.; Akhter, G. Performance of CMORPH, TMPA and PERSIANN rainfall datasets over plain, mountainous and glacial regions of Pakistan. *Theor. Appl. Climatol.* **2017**. [CrossRef]

36. Huffman, G.; Bolvin, D.; Braithwaite, D.; Hsu, K.; Joyce, R. Algorithm Theoretical Basis Document (ATBD) NASA Global Precipitation Measurement (GPM) Integrated Multi-satellitE Retrievals for GPM (IMERG). Available online: https://pmm.nasa.gov/sites/default/files/document_files/IMERG_ATBD_V5.2_0.pdf (accessed on 13 August 2018).

37. Shige, S.; Yamamoto, T.; Tsukiyama, T.; Kida, S.; Ashiwake, H.; Kubota, T.; Seto, S.; Aonashi, K.; Okamoto, K. The GSMaP precipitation retrieval algorithm for microwave sounders part I: Over-ocean algorithm. *IEEE Trans. Geosci. Remote Sens.* **2009**, *47*, 3084–3097. [CrossRef]

38. Ushio, T.; Sasashige, K.; Kubota, T.; Shige, S.; Okamoto, K.; Aonashi, K.; Inoue, T.; Takahashi, N.; Iguchi, T.; Kachi, M.; et al. A Kalman Filter Approach to the Global Satellite Mapping of Precipitation (GSMaP) from Combined Passive Microwave and Infrared Radiometric Data. *J. Meteorol. Soc. Jpn.* **2009**, *87A*, 137–151. [CrossRef]

39. Bias-Corrected CMORPH: A 13-Year Analysis of High-Resolution Global Precipitation Objective. Available online: https://meetingorganizer.copernicus.org/EGU2011/EGU2011-1809.pdf (accessed on 13 August 2018).

40. Xie, P.; Xiong, A.Y. A conceptual model for constructing high-resolution gauge-satellite merged precipitation analyses. *J. Geophys. Res. Atmos.* **2011**, *116*, 1–14. [CrossRef]

41. World Meteorological Organization Guide to Hydrological Practices: Data Acquisition and Processing, Analysis, Forecasting And Other Applications; 1994. Available online: http://www.innovativehydrology.com/WMO-No.168-1994.pdf (accessed on 13 August 2018).

42. Taylor, K.E. Summarizing multiple aspects of model performance in a single diagram. *J. Geophys. Res.* **2001**, *106*, 7183–7192. [CrossRef]

43. Ochoa, A.; Pineda, L.; Crespo, P.; Willems, P. Evaluation of TRMM 3B42 precipitation estimates and WRF retrospective precipitation simulation over the Pacific–Andean region of Ecuador and Peru. *Hydrol. Earth Syst. Sci.* **2014**, *18*, 3179–3193. [CrossRef]

44. Satgé, F.; Bonnet, M.-P.; Gosset, M.; Molina, J.; Hernan Yuque Lima, W.; Pillco Zolá, R.; Timouk, F.; Garnier, J. Assessment of satellite rainfall products over the Andean plateau. *Atmos. Res.* **2016**, *167*, 1–14. [CrossRef]

45. Roebber, P.J. Visualizing Multiple Measures of Forecast Quality. *Weather Forecast.* **2009**, *24*, 601–608. [CrossRef]

46. Mourre, L.; Condom, T.; Junquas, C.; Lebel, T.; Sicart, J.E.; Figueroa, R.; Cochachin, A. Spatio-temporal assessment of WRF, TRMM and in situ precipitation data in a tropical mountain environment (Cordillera Blanca, Peru). *Hydrol. Earth Syst. Sci.* **2016**, *20*, 125–141. [CrossRef]

47. Levizzani, V.; Amorati, R.; Meneguzzo, F. A Review of Satellite-Based Rainfall Estimation Methods. Available online: http://satmet.isac.cnr.it/papers/MUSIC-Rep-Sat-Precip-6.1.pdf (accessed on 13 August 2018).

48. Ferraro, R.R.; Smith, E.A.; Berg, W.; Huffman, G.J. A Screening Methodology for Passive Microwave Precipitation Retrieval Algorithms. *J. Atmos. Sci.* **1998**, *55*, 1583–1600. [CrossRef]

49. Beck, H.E.; van Dijk, A.I.J.M.; Levizzani, V.; Schellekens, J.; Miralles, D.G.; Martens, B.; de Roo, A. MSWEP: 3-hourly 0.25° global gridded precipitation (1979–2015) by merging gauge, satellite, and reanalysis data. *Hydrol. Earth Syst. Sci. Discuss.* **2016**, *21*, 589–615. [CrossRef]

50. Satgé, F.; Ruelland, D.; Bonnet, M.-P.; Molina, J.; Pillco, R. Consistency of satellite precipitation estimates in space and over time compared with gauge observations and snow-hydrological modelling in the lake Titicaca region. *Hydrol. Earth Syst. Sci.* **2018**. submitted.

51. Ma, Y.; Yang, Y.; Han, Z.; Tang, G.; Maguire, L.; Chu, Z.; Hong, Y. Comprehensive evaluation of Ensemble Multi-Satellite Precipitation Dataset using the Dynamic Bayesian Model Averaging scheme over the Tibetan Plateau. *J. Hydrol.* **2017**, *556*, 634–644. [CrossRef]

52. Tang, G.; Behrangi, A.; Long, D.; Li, C.; Hong, Y. Accounting for spatiotemporal errors of gauges: A critical step to evaluate gridded precipitation products. *J. Hydrol.* **2018**, *559*, 294–306. [CrossRef]

remote sensing

MDPI

Article

Using Multiple Monthly Water Balance Models to Evaluate Gridded Precipitation Products over Peninsular Spain

Javier Senent-Aparicio [1,*], Adrián López-Ballesteros [1], Julio Pérez-Sánchez [1,*], Francisco José Segura-Méndez [1] and David Pulido-Velazquez [1,2]

[1] Department of Civil Engineering, Catholic University of San Antonio, Campus de Los Jerónimos s/n, 30107 Guadalupe, Murcia, Spain; ballesteross_93@hotmail.com (A.L.-B.); fjsegura7@gmail.com (F.J.S.-M.); d.pulido@igme.es (D.P.-V.)

[2] Geological Survey of Spain (IGME), Granada Unit, Urb. Alcázar del Genil, 4, Edificio Zulema, 18006 Granada, Spain

* Correspondence: jsenent@ucam.edu (J.S.-A.); jperez058@ucam.edu (J.P.-S.); Tel.: +34-968-278-818 (J.S.-A.)

Received: 17 May 2018; Accepted: 10 June 2018; Published: 11 June 2018

Abstract: The availability of precipitation data is the key driver in the application of hydrological models when simulating streamflow. Ground weather stations are regularly used to measure precipitation. However, spatial coverage is often limited in low-population areas and mountain areas. To overcome this limitation, gridded datasets from remote sensing have been widely used. This study evaluates four widely used global precipitation datasets (GPDs): The Tropical Rainfall Measuring Mission (TRMM) 3B43, the Climate Forecast System Reanalysis (CFSR), the Precipitation Estimation from Remotely Sensed Information using Artificial Neural Networks (PERSIANN), and the Multi-Source Weighted-Ensemble Precipitation (MSWEP), against point gauge and gridded dataset observations using multiple monthly water balance models (MWBMs) in four different meso-scale basins that cover the main climatic zones of Peninsular Spain. The volumes of precipitation obtained from the GPDs tend to be smaller than those from the gauged data. Results underscore the superiority of the national gridded dataset, although the TRMM provides satisfactory results in simulating streamflow, reaching similar Nash-Sutcliffe values, between 0.70 and 0.95, and an average total volume error of 12% when using the GR2M model. The performance of GPDs highly depends on the climate, so that the more humid the watershed is, the better results can be achieved. The procedures used can be applied in regions with similar case studies to more accurately assess the resources within a system in which there is scarcity of recorded data available.

Keywords: TRMM; CFSR; PERSIANN; MSWEP; streamflow simulation; lumped models; Peninsular Spain

1. Introduction

Precipitation is one of the most important drivers for hydrological modelling because it has a strong impact on the accuracy of hydrological models [1]. Although the amount, intensity, and distribution of precipitation are clearly linked to various processes in the hydrological cycle, this relation is nonlinear. Nevertheless, the accurate assessment of precipitation is of the utmost importance for hydrological modelling, as it provides meteorological input for hydrological studies. Therefore, reliable and accurate precipitation information at sufficient spatial and temporal resolution is essential not only for the study of climate trends, but also for water resource management [2]. Traditionally, hydrologic simulations are usually based on historical gauge observations that may not be available for a specific basin due to the malfunctioning of the equipment installed or the low density of stations [3]. Moreover, there can be important deviations between point-scale gauge information and true areal precipitation [4–7]; thus, the use of a grid dataset rather than a single rain gauge is advisable.

In recent years, and to overcome the above limitations, global precipitation datasets (GPDs) have been widely used in the hydrology field. Besides being generally used as input data, GPDs are also employed for estimating input parameters for hydrological modelling [8]. Furthermore, reliable precipitation data are essential for hydrological modelling because their errors could lead to an inappropriate model setup, resulting in the wrong simulations and subsequent decisions [9]. Easy access, long-term series, and quality and homogeneity of data have encouraged the use of GPDs in hydrology [10]. These gridded datasets are very useful for hydrological modelling and provide potential alternative data sources for data-sparse and ungauged areas. The improvement of sensor technology has provided worldwide satellite observation data that are more spatially homogenous [1]. Some of the most commonly used products from satellite-derived data are the Tropical Rainfall Measuring Mission (TRMM) [11] and the Precipitation Estimation from Remotely Sensed Information using Artificial Neural Networks (PERSIANN) [12]. Moreover, it is becoming increasingly frequent to combine data from satellites with gauge measurements, resulting in more accurate tools in water balance models, for example, Multi-Source Weighted-Ensemble Precipitation (MSWEP) [13]. Beck et al. [14] validated MSWEP on a global scale using worldwide observations from more than 75,000 gauges, and gauge-corrected datasets were also evaluated using hydrological modelling for nearly 9000 catchments. The Climate Forecast System Reanalysis (CFSR) [15] is a third-generation reanalysis product [16]. It was designed and executed as a global, high-resolution, coupled atmosphere-ocean-land surface-sea ice system to provide the best estimate of the state of these coupled domains over the 1979–2014 period. The current CFSR will be extended as an operational, real-time product in the future. Gridded precipitation product errors may cause additional inconsistency in hydrologic simulations [17], and, owing to the fact that GPDs are integrated systems, the uncertainty related to internal processing of observations (missing data, homogenization, atmospheric biases, etc.) can become difficult to evaluate [10]. Therefore, the study of hydrological outputs using various GPDs requires further investigation. Although some studies have been reported comparing global gridded precipitation datasets and their performance in driving hydrological models [18], most of them were carried out over large river basins. There is a need to improve our understanding of satellite precipitation products' performance over data-sparse and ungauged small watersheds [19]. To our knowledge, no studies have been carried out to investigate the efficiency of GPDs in driving hydrological models over Peninsular Spain.

Furthermore, an appropriate hydrological model in a watershed is essential for providing accurate model predictions, and GPDs can be used for a better understanding of these processes [20,21], leading to improved model simulations. The development of monthly water balance models (MWBMs) is a complex task in a water resource system [22]. The appropriate analysis of their management is essential, especially in arid and semi-arid regions, where precipitation is very unevenly distributed with high evapotranspiration (ETP) rates. The spatial structure of a MWBM can be divided into three categories: Lumped, semi-distributed, and fully distributed [23]. In a lumped water balance model, catchment parameters and variables are averaged in space, so hydrological processes are approached through conceptual solutions formulated by using semi-empirical equations, while semi-distributed and fully distributed models process spatial variability by homogeneous zones or grid cells, respectively. However, it is not only spatial discretization that determines the quality of the simulation. The choice of model is dictated by the modelling purpose. When flow at the catchment outlet is the main required goal in water resource management, as in the present paper, lumped models may be the best choice [24]. Currently, the ABCD model was found to have satisfactory results in Greece [25]. Wriedt and Bouraoui [26] used GR2M in nearly 500 catchments in Germany, France, Spain, and Portugal and obtained good results both in the centre and north of Spain, and in Central European basins. The Australian water balance model (AWBM) is one of the most widely used rainfall/run-off models in Australia [27], but, nowadays, it is being used worldwide [28,29], for both humid and dry basins. The Thornthwaite and Mather water balance model has been successfully used in different water balance research studies in Spain [30]. Guo-5p [31,32] is an adaptation of Thornthwaite and Mather's model with five parameters, so it was chosen in order to compare with the latter. Finally, the Témez model has been widely used in Spanish catchments [33,34] and by the Spanish government in water management [35].

In this study, in order to include the main climate zones in Peninsular Spain, six MWBMs were constructed for four meso-scale basins (ranging in area from 70 to 414 km^2): Oceanic climate (Esva river basin at Trevías, TRE), Galicia variant of the oceanic climate (Tea river basin at Puenteareas, PUE), Mediterranean climate (Gargüera river basin at Gargüera, GAR), and semi-arid basin (Vallehermoso river basin at Camarenilla, RVA). Therefore, the four basins cover a wide range of climatic and physiographic conditions.

Thus, the goals of this research can be divided into three stages: (1) To compare and evaluate the robustness and the accuracy of the GPDs with gauged precipitation data in different climatic zones over Peninsular Spain, (2) to assess the performance of four different satellite precipitation products and rain gauge historical information as input into MWBMs for streamflow simulation over Peninsular Spain, and (3) to evaluate the performance of the simulated streamflow of four different GPDs in previously fitted MWBMs with rain gauge datasets. The contents of the paper are structured as follows: The study area and datasets used in this study are introduced in Section 2; the methodology is described in Section 3; Section 4 presents the results and discussion; and Section 5 highlights the main conclusions.

2. Materials and Methods

2.1. Study Area

Peninsular Spain features a wide range of climates, due to its position between the subtropical zone and the European temperate zone. It also includes some of the driest areas in the southeast, with a marked summer drought, and the rainiest areas in Europe in the northeast [24]. Four basins distributed over Peninsular Spain were used as study areas. The basins were selected based on the wide diversity of climate conditions representative of the types of weather found in Peninsular Spain. In addition, they are located in areas in which withdrawals are negligible and located upstream from reservoirs. As can be seen in Figure 1, the basins studied are well distributed over Peninsular Spain and represent four different climatic zones, according to the Köppen-Geiger classification system [36]. Table 1 shows basin sizes ranging from 70 km^2 to 414 km^2 and elevations ranging from 400 to 690 m. The average precipitation shown in Table 1 is from gauge measurements.

Table 1. Summary of the main characteristics of the selected basins (1998–2009).

Code	Area (km^2)	Altitude (m.a.s.l)	Average Slope (%)	Köppen Classification	Average Precipitation (mm/year)	Average ETP (mm)	Average Flow (hm^3/year)
RVA	86	608	18.33	Bsk	422	1034	2.2
GAR	70	690	32.44	Csa	1081	1040	18.9
PUE	264	400	18.43	Csb	1624	762	435.2
TRE	414	527	7.51	Cfb	1241	674	300.9

2.2. Datasets

In the following section, the gridded datasets used in this study are briefly introduced. All of the simulations were performed with monthly precipitation data, as well as monthly potential ETP and discharge time series for the common period (1998–2009) for all datasets (Table 2). The horizontal resolution of the gridded datasets used varied from 1 km grid to 0.3° × 0.3° spacing. The areal precipitation was estimated using Thiessen polygons. Even if this method does not take into account orography influences, it has been considered adequate in the present study due to the density of the spatial resolution of the datasets used. All of the MWBMs use rainfall and ETP time series as input data. The ETP data series in each basin were obtained from the official monthly series provided by the Centre of Studies and Experimentation of Civil Works (CEDEX) [37].

2.2.1. TRMM Dataset

The TRMM Multisatellite Precipitation Analysis (TMPA) provides a calibration-based sequential scheme for combining precipitation estimates from multiple satellites, as well as gauge analysis where feasible, at fine scales ($0.25° \times 0.25°$ and 3-hourly). The dataset covers the latitude band $50°$N–S for the period from 1998 to the delayed present. The monthly product TRMM 3B43 was used in this study. More information about this dataset can be found in [11,14].

2.2.2. CFSR Dataset

The Climate Forecast System Reanalysis (CFSR) is a global coupled atmosphere-ocean-land surface-sea ice system and forecast model. The CFSR is based on hourly forecasts generated using information from satellite-derived products and the global weather station network, covering any location in the world [38]. The CFSR data have a spatial resolution of approximately 38 km, and the data are available from 1979 to present on the SWAT website. More information about this dataset can be found in [15].

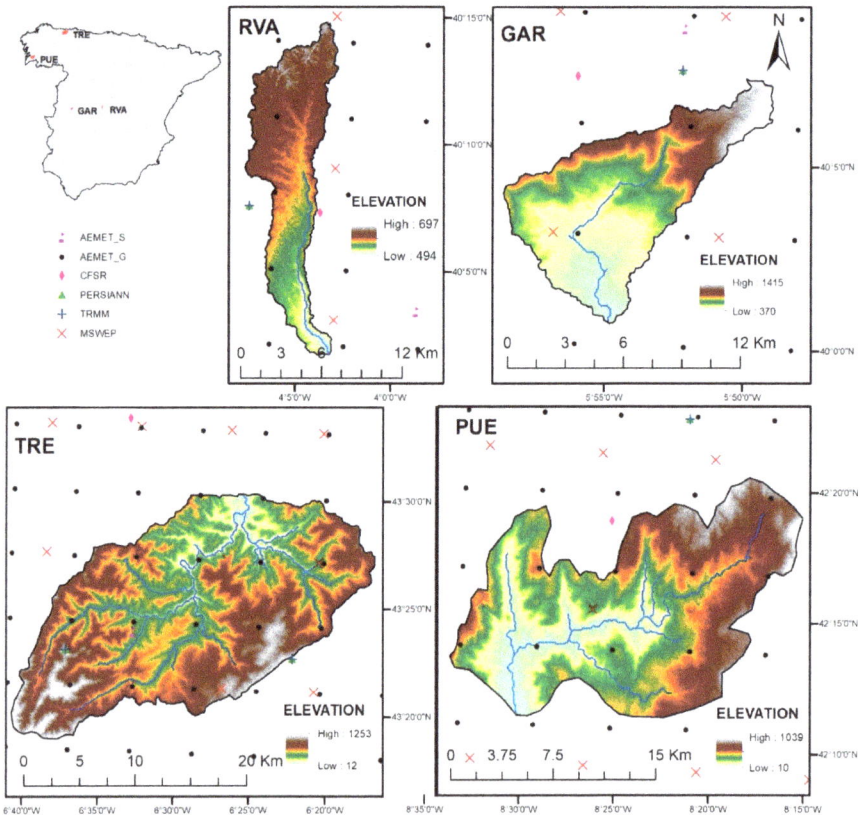

Figure 1. Location, elevation, and gridded precipitation datasets of the selected basins.

2.2.3. PERSIANN Dataset

The PERSIANN-Climate Data Record (CDR) provides daily rainfall estimates at 0.25° spatial resolution for the latitude band 60°N–60°S over the period of 1983 to the delayed present. PERSIANN-CDR is generated from the PERSIANN algorithm, using infrared brightness temperature data from geostationary satellites to estimate rainfall rate and updating its parameters using passive/active microwave observations from low-orbital satellites. More information about this dataset can be found in [12].

2.2.4. MSWEP Dataset

The MSWEP version 2.1 is a fully global precipitation dataset for the period 1979 to 2016 with a 3-hourly temporal and 0.1° spatial resolution, specifically designed for hydrological modelling. MSWEP uses the complementary strengths of gauge-based, satellite-based, and reanalysis-based data to provide precipitation estimates over the entire globe. More information can be found in [13].

Table 2. List of datasets used in this study and coverage periods.

Dataset	Version	Spatial Resolution	Areal Coverage	Temporal Resolution	Temporal Coverage	Data Sources	Source
CFSR	DS093.1	0.3° × 0. 3° (≈38 km)	Global	Daily	1979–present	Reanalysis	[15]
TRMM	3B43	0.25° × 0.25° (≈30 km)	Latitude band 50°N–S	3-hourly	1998–present	Gauge, satellite	[11]
PERSIANN	CDR	0.25° × 0.25° (≈30 km)	Latitude band 60°N–S	Daily	1983–present	Gauge, satellite	[12]
MSWEP	V2.1	0.1° × 0.1° (≈12 km)	Global	3-hourly	1979–2016	Gauge, satellite, reanalysis	[13]
AEMET_G	V1.0	5 km	Spain	Daily	1951–2017	Gauge	[39]

2.2.5. AEMET Dataset

The Spanish National Meteorological Agency (AEMET) grid, version 1.0, provides daily rainfall for the period of 1951 to the delayed present over Spain, with a spatial resolution of 5 km (AEMET_G). The method used is gauge analysis via Optimal Interpolation from the series of observations of the National Weather Data Bank of AEMET. More information about this dataset can be found in [39].

2.2.6. Rain Gauge Data

The gauged precipitation dataset consists of the nearest rain gauge records (AEMET_S) to each studied watershed (Figure 1), provided by the AEMET. The main characteristics of gauged stations are shown in Table 3.

Table 3. Rain gauge station characteristics used in the study.

Basin Code	Station Name	Latitude	Longitude	Altitude (m.a.s.l)	Source
RVA	Recas	39.96	−4.01	609	SIAR [1]
GAR	Valdastillas	40.14	−5.87	495	REDAREX [2]
PUE	Vigo/Peinador	42.24	−8.62	261	AEMET [3]
TRE	Zardain	43.39	−6.55	410	AEMET

[1] SIAR: Castilla-La Mancha Irrigation Consultancy Service; [2] REDAREX: Extremadura Irrigation Consultancy Service; [3] AEMET: Spanish National Meteorology Agency.

3. Materials and Methods

3.1. Monthly Water Balance Models

Six well-known and documented MWBMs were used in this study: ABCD, AWBM, GR2M, Guo 5P, Thornthwaite, Mather, and Témez models. All these models are lumped and use a low number of parameters (from 2 to 5). The water balance in these models is represented by different storages, the moisture content of which varies depending on physical or empirical relationships [40]. More information about these MWBMs can be found in [24]. A brief description of the models is given below:

- Thornthwaite-Mather (THM): It was developed in the early 1940s for the Delaware River, and several MWBMs are based on it. Based on the study of the model done by Alley [41], this model has two parameters (storage constant and soil moisture capacity) and two storages.
- ABCD: It is composed of two storages. It is characterised by allowing streamflow to occur even under conditions of moisture deficit [42]. It has four parameters and emerges as a tool for assessment of water resources in the United States.
- AWBM: It uses three storages to simulate partial surface run-off areas. The water balance in each of these storages is determined separately, using a total of six parameters [27]. It was developed in the 90s and today is one of the most widely used in Australia.
- GR2M: It is an evolution of the GR2 model that provides a simplified representation of the rainfall/runoff process. It is characterised by a small number of parameters, developed with empirical criteria, which do not correspond to specific physical attributes. This model is composed of four parameters and two storages. The model has been tested in numerous French stations. The description of this MWBM can be found in the work of Makhlouf and Michel [43].
- Guo: Described in [31,32], this MWBM is an adaptation of the model of Thornthwaite and Mather [44], increasing the number of parameters up to five. It has been applied in different sub-basins of the Dongjiang, in southern China, with good results. Xiong and Guo [31] compare it with the two-parameter model, concluding similar behaviour in practice.
- Témez (TEM): It is a purely empirical model that has been widely used in many Spanish basins, especially for assessment of water resources developed by the Hydrographical Study Centre. The model considers the land to be divided into two zones: Upper unsaturated, or soil moisture, and lower saturated, or aquifer, which functions as an underground reservoir that drains into the network of channels [45]. This model uses four parameters. It is a lumped model that has been applied in a distributed way in order to obtain an evaluation of the Spanish water resources [46].

3.2. MWBM Calibration and Validation Strategy

The calibration of the MWBM parameters was carried out by comparing predicted data with observed data for a period of seven years (2003–2009). This period of time was chosen because it includes dry, average, and wet years, which is desirable to reach a good model calibration [47]. Monthly streamflow data were collected from the national water agency of Spain [37]. The value that minimizes the differences between both flow series and the objective function that minimizes the sum of square of deviations (SSQ) were considered the optimal values for each parameter. The optimization algorithm is the generalized reduced gradient (GRG2) [48], which searches for the extreme values of the functions by the GRG2 algorithm method [49]. During the calibration process of MWBMs, it was necessary to consider some initial conditions, such as the value of the initial soil moisture. These initial conditions tend to influence the final results; therefore, an initial period of two years (1998–1999) was used for model warm-up. After calibration, MWBMs were validated using the monthly discharge data of three years (2000–2002).

3.3. Performance of GPDs in Simulating Streamflow

There are two different strategies to assess the performance of the GPDs in simulating streamflow [50]. The first approach is calibration and validation of MWBMs using rain gauge data or gridded rainfall datasets with monthly observed streamflow. The second approach is calibration and validation of MWBMs with rain gauge data followed by the best fitted parameters found in MWBMs being used to simulate streamflow with GPDs. Artan et al. [51] and Zeweldi et al. [52] indicated that an MWBM can be improved when using satellite-based data [51–53]. Nevertheless,

Habib et al. [54] considered that calibration achieved with GPDs could result in unrealistic parameter values in MWBMs to compensate for the large errors in input datasets. In this study, both approaches will be carried out (Figure 2).

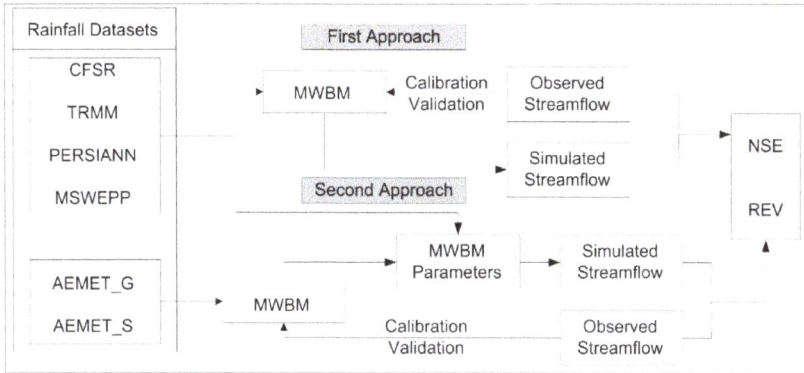

Figure 2. Flowchart for comparative study.

3.4. Statistical Analysis

To quantitatively compare GPDs with rain gauge observations, widely used validation statistical indices are used in this study. The correlation coefficient (R) reflects the degree of linear correlation between GPDs and gauge observations, the relative bias (BIAS) is used to measure the systematic bias of the GPDs and the root mean square error (RMSE) quantifies the average error magnitude, which is slightly biased towards larger errors. R values vary from -1 to 1, with values closer to 1 indicating a positive correlation and high model performance. BIAS and RMSE values of 0 indicate a perfect fit.

$$R = \frac{\sum_{i=1}^{n}\left(G_i - \overline{G}\right)\left(S_i - \overline{S}\right)}{\sqrt{\sum_{i=1}^{n}\left(G_i - \overline{G}\right)^2}\sqrt{\sum_{i=1}^{n}\left(S_i - \overline{S}\right)^2}} \tag{1}$$

$$BIAS = \frac{\sum_{i=1}^{n}\left(S_i - G_i\right)}{\sum_{i=1}^{n} G_i} \times 100\% \tag{2}$$

$$RMSE = \sqrt{\frac{1}{n}\sum_{i=1}^{n}\left(S_i - G_i\right)^2} \tag{3}$$

where S_i is the precipitation from the rain gauge grid (AEMET_G), $\overline{S}$ is the average precipitation from the rain gauge grid, G_i is the precipitation from GPDs, and $\overline{G}$ is the average precipitation from GPDs. In order to make a comparison among various MWBMs, some quantitative information is also required to measure model performance. In this study, the streamflow data measured at the outlet of the catchment was used to assess the model performance. Statistical performance indices, such as the Nash-Sutcliffe efficiency (NSE) [55] or the percentage difference between the total observed and modelled runoff (REV), have been calculated. NSE can range from $-\infty$ to 1, with NSE = 1 being the optimal value. The REV optimal value is 0.

$$NSE = 1 - \frac{\sum_{i=1}^{n}\left(O_i - M_i\right)^2}{\sum_{i=1}^{n}\left(O_i - \overline{O}\right)^2} \tag{4}$$

$$REV = \frac{M_T - O_T}{O_T} \times 100\% \tag{5}$$

where O_i is the observed discharge, M_i is the modelled discharge, $\overline{O}$ is the mean of observed discharge, M_T is the total modelled run-off, and O_T is the total observed run-off.

4. Results

4.1. Comparison of Areal Mean Rainfalls

The first study performed consisted of analysing and comparing the areal mean rainfalls in the studied watersheds for the five different datasets considered. Figure 3 shows the accumulated precipitation from 1998 to 2009. As expected, both values of the national grid (AEMET_G) and the nearest rain gauge (AEMET_S) are quite similar, and their disparities in the last third of the period varies depending on the difference between rain gauge elevation, the watershed's average altitude, and their proximity to the studied watershed. Thus, for example, TRE's nearest record station is located in this watershed and within the average range of its altitude, so the accumulated precipitation only differs slightly. However, RVA's rain gauge, which is located out of the boundaries of the watershed, is at a lower height than the whole study area and shows higher differences than the AEMET grid.

The long-term monthly areal rainfall for all the datasets considered and the four watersheds under study are shown in Figure 4. Although the monthly tendencies in all GPDs are similar to gauged data, the differences are mostly concentrated in the rainy seasons, while they are reduced in summer months. TRMM estimates are close to those from the rain gauge data and are lower than recorded precipitation, except in RVA, where MSWEP appears to be the dataset that fits the best with national grid or nearest rain gauges. Moreover, as in the accumulated precipitation analysis, GAR and RVA show the highest differences with GPDs, even wider in RVA, where TRMM and PERSIANN precipitation data are around 50% higher than the data from the AEMET dataset.

Previous findings are also confirmed with statistical analysis shown in Table 4, comparing the AEMET grid with the rest of the datasets. *R* values are no lower than 0.83, and, in most cases, higher than 0.95, which indicates a good lineal correlation, as can be seen in Figures 2 and 3.

Concerning RMSE results, TRMM appears to be the best GPD in two out of the four watersheds (TRE and GAR), while MSWEP shows better results in PUE and RVA. However, the difference of RMSE values in PUE for TRMM and MSWEP only varies by 0.3 mm, so both GPDs could be used in similar regions to the studied ones according this goodness-of-fit test. Furthermore, PERSIANN reaches the worst values for nearly all the watersheds, until it doubles the lowest result.

BIAS results are less clear than RMSE ones, especially for the higher values. In all the watersheds, rain gauges show percentages lower than 10%, in some cases the lowest of the analysed values. TRMM also results in a good dataset for all watersheds, except the semi-arid one (RVA), where MSWEP presents, along with the nearest rain gauge, the best data according to BIAS values. However, PERSIANN was not found to be the worst dataset, as RMSE showed; it showed the best BIAS value (-0.27% in TRE) for all the possible combinations of watershed datasets studied. Notwithstanding, the drier the region is, the worse the value for PBIAS shows when using the PERSIANN grid.

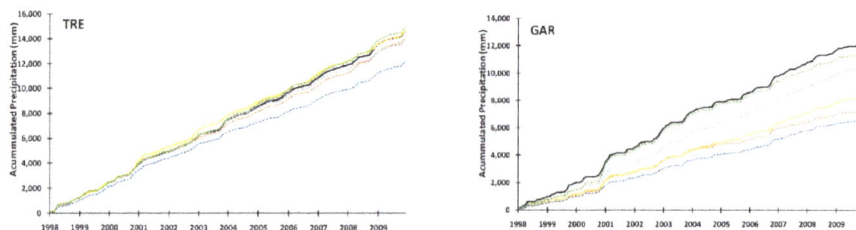

Figure 3. *Cont.*

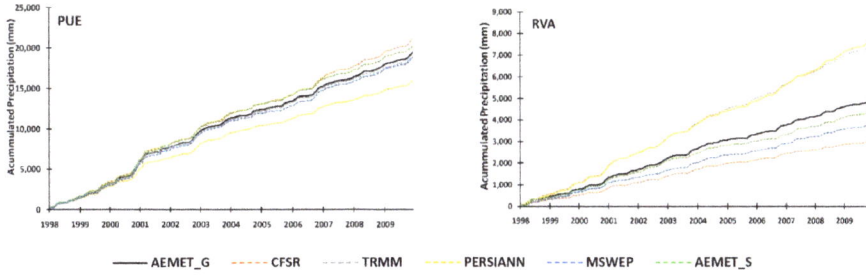

Figure 3. Accumulated monthly precipitation over selected basins for ground observations and gridded precipitation datasets in the 1998–2009 period.

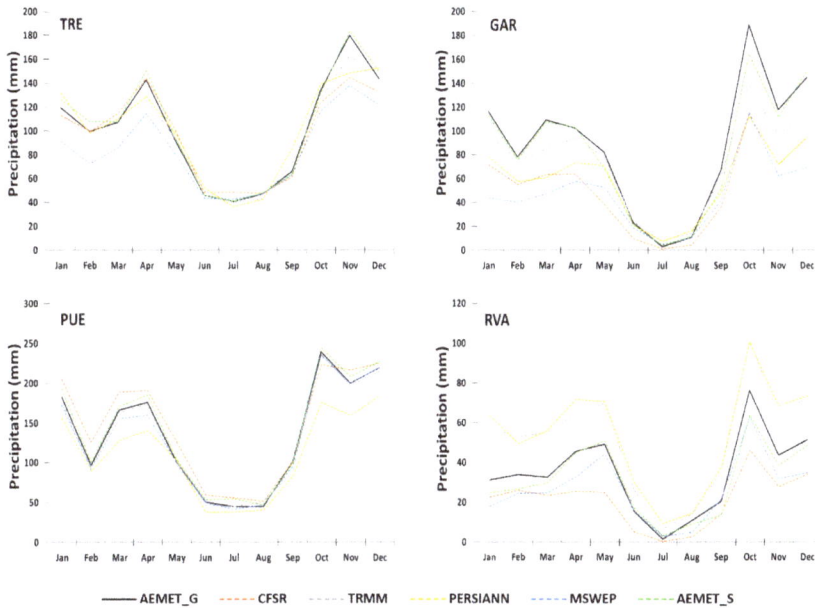

Figure 4. Long-term monthly areal precipitation of ground observations and gridded precipitation products in the 1998–2009 period.

Table 4. Statistical indices used to quantify the accuracy of precipitation estimates against precipitation from AEMET_G (best results in bold).

Basins	Dataset	R	RMSE (mm)	BIAS (%)
	AEMET_S	0.99	18.05	−3.82
	TRMM	**0.99**	18.58	**2.53**
PUE	PERSIANN	0.97	48.01	17.94
	CFSR	0.98	30.77	−8.90
	MSWEP	**0.99**	**18.28**	3.54
	AEMET_S	0.99	13.02	−2.46
	TRMM	**0.96**	**20.68**	4.19
TRE	PERSIANN	0.83	39.34	**−0.27**
	CFSR	0.94	23.15	3.43
	MSWEP	0.94	30.05	16.56

Table 4. *Cont.*

Basins	Dataset	R	RMSE (mm)	BIAS (%)
	AEMET_S	0.96	25.71	6.02
	TRMM	**0.95**	**41.19**	**14.88**
GAR	PERSIANN	0.95	55.73	32.00
	CFSR	0.95	53.65	40.39
	MSWEP	0.94	63.43	46.04
	AEMET_S	0.93	12.51	10.06
	TRMM	0.95	22.28	−50.29
RVA	PERSIANN	0.91	25.97	−56.53
	CFSR	0.88	21.34	38.88
	MSWEP	**0.94**	**13.98**	**22.75**

4.2. Evaluation of the Simulated Streamflow Using MWBM

The second phase of the research assessed the performance of the four GPDs and the two rain gauges' banks of data from the AEMET (nearest one and grid) as input into the MWBMs considered for streamflow simulation in the four studied watersheds. Table 5 lists the NSE values when comparing observed and modelled streamflow for the best performance of each of the MWBMs, with every dataset taken into account. Furthermore, means have been calculated in each watershed, both for MWBMs and GPDs. The AEMET grid showed the best results, both mean and individual, for all the watersheds, while the nearest rain gauge only performed similarly to the grid in PUE and TRE; however, the NSE in GAR and RVA was not as satisfactory as the AEMET grid, although similar values to GPDs were achieved in most cases. The NSE mean in GR2M reached a value over 0.75 for the four watersheds, achieving its best value (0.95) in PUE for the AEMET_S, AEMET_G, and TRMM data. The rest of the models showed different values depending on the watershed. Thus, in PUE, all models except GR2M reached a similar NSE, around 0.64, 30% lower than GR2M. TRE did not vary more than 5%, regardless of the model used. These differences are slightly higher in GAR and reach more than 50% in RVA when comparing GR2M with ABCD or THM.

Table 5. Nash-Sutcliffe efficiency (NSE) of simulated streamflow for the different datasets using monthly water balance models (MWBMs) (best results in bold).

Basin	Dataset	MWBM						
		ABCD	AWBM	GR2M	GUO	TEM	THM	Mean
	AEMET_G	0.65	0.65	0.95	0.70	0.69	0.65	0.71
	AEMET_S	0.69	0.71	0.95	0.75	0.73	0.70	**0.76**
	TRMM	0.63	0.62	0.95	0.67	0.67	0.62	0.69
PUE	PERSIANN	0.51	0.46	0.87	0.56	0.44	0.62	0.58
	CFSR	0.61	0.62	0.84	0.66	0.65	0.62	0.67
	MSWEP	0.64	0.63	0.95	0.69	0.67	0.64	0.70
	Mean	0.62	0.62	**0.92**	0.67	0.64	0.64	
	AEMET_G	0.86	0.79	0.86	0.83	0.82	0.82	**0.83**
	AEMET_S	0.86	0.80	0.84	0.82	0.80	0.80	0.82
	TRMM	0.78	0.71	0.76	0.71	0.73	0.77	0.74
TRE	PERSIANN	0.59	0.56	0.50	0.61	0.46	0.59	0.55
	CFSR	0.84	0.76	0.87	0.78	0.81	0.80	0.81
	MSWEP	0.78	0.64	0.78	0.71	0.73	0.77	0.73
	Mean	**0.79**	0.71	0.77	0.74	0.72	0.76	
	AEMET_G	0.89	0.98	0.92	0.89	0.90	0.86	**0.91**
	AEMET_S	0.82	0.79	0.93	0.70	0.78	0.58	0.77
	TRMM	0.74	0.90	0.79	0.88	0.72	0.89	0.82
GAR	PERSIANN	0.74	0.73	0.75	0.77	0.69	0.72	0.74
	CFSR	0.82	0.78	0.95	0.90	0.81	0.84	0.85
	MSWEP	0.63	0.62	0.39	0.54	0.63	0.90	0.60
	Mean	0.77	0.80	**0.86**	0.81	0.75	0.68	

Table 5. *Cont.*

Basin	Dataset	MWBM						
		ABCD	AWBM	GR2M	GUO	TEM	THM	Mean
	AEMET_G	0.71	0.81	0.81	0.57	0.39	0.56	**0.64**
	AEMET_S	0.49	0.68	0.78	0.66	0.53	0.15	0.55
	TRMM	0.22	0.64	0.69	0.65	0.48	0.24	0.49
RVA	PERSIANN	0.05	0.45	0.68	0.45	0.61	0.24	0.41
	CFSR	0.59	0.69	0.83	0.59	0.67	0.24	0.60
	MSWEP	−0.26	0.68	0.78	0.50	0.68	0.09	0.41
	Mean	0.30	0.66	**0.76**	0.57	0.56	0.25	

When comparing total volumes observed and modelled (Table 6) with REV, previous findings were confirmed. GR2M appeared to be the best model for all the watersheds [24], and, although AEMET_G did not give the lowest percentage in all cases, the difference was never over 5% compared with the rest of the datasets. TRMM and MSWEP results for GR2M were below 15% in all watersheds, except in GAR, where CFSR volume error was −1.59%, 20% lower than the others. CFSR also performed very satisfactorily in TRE and GAR, achieving a total streamflow error, compared to the observed, lower than 3.5% in GR2M, even better than in the AEMET datasets. PERSIANN appears to be the worst GPD in most cases, although greater value is shown with CFSR in RVA for THM, doubling the observed streamflow in this watershed. No trend was found related to over- or underestimating total streamflow when using this goodness-of-fit measure.

Table 6. Percentage difference between the total observed and modelled runoff (REV) of simulated streamflow for the different datasets using MWBMs (best results in bold).

Basin	Dataset	MWBM					
		ABCD	AWBM	GR2M	GUO	TEM	THM
	AEMET_G	−14.56	−36.70	**+1.43**	−17.37	−36.14	−36.23
	AEMET_S	−8.07	−31.58	+6.60	−12.03	−30.30	−31.46
PUE	TRMM	−15.72	−38.10	+3.26	−17.49	−37.30	−38.26
	PERSIANN	−31.91	−52.47	−14.53	−28.59	−51.57	−29.89
	CFSR	−22.95	−34.91	−12.30	−23.09	−35.41	−35.20
	MSWEP	−16.03	−38.20	+1.46	−16.81	−37.42	−37.29
	AEMET_G	−7.52	−14.66	−6.88	−6.01	−8.40	−14.49
	AEMET_S	−5.74	−13.97	−9.56	−8.97	−9.01	−3.39
TRE	TRMM	+3.31	−4.70	+5.86	+9.17	+0.62	−2.94
	PERSIANN	+6.00	+10.43	+12.14	+8.76	+12.14	+11.26
	CFSR	−3.03	−11.37	**−3.08**	−3.02	−4.85	−10.88
	MSWEP	+9.43	−21.52	+15.41	+14.22	−14.53	−15.58
	AEMET_G	−12.20	−7.77	−7.96	−9.15	−3.60	+8.51
	AEMET_S	−3.39	+2.89	+7.89	+13.39	+19.53	+26.08
GAR	TRMM	−33.31	−26.30	−28.18	−28.04	−22.48	−17.09
	PERSIANN	−34.02	−46.62	−35.89	−43.06	−39.56	−42.96
	CFSR	−28.17	−26.42	**−1.59**	−21.08	−21.60	−28.02
	MSWEP	−56.23	−54.13	−19.70	−44.45	−46.73	−58.91
	AEMET_G	−25.67	+19.43	**−1.04**	+64.41	+23.95	+62.01
	AEMET_S	−29.22	−9.19	+12.97	+44.64	+38.43	−22.71
RVA	TRMM	−25.58	−34.69	−10.59	−22.91	−16.43	−26.10
	PERSIANN	−44.34	−47.75	−16.30	+36.21	−19.19	−30.01
	CFSR	−21.18	+54.91	+20.33	+59.41	+45.42	+119.72
	MSWEP	−48.37	+25.48	−11.18	+52.98	+5.20	−44.89

4.3. Evaluation of GPDs Using the AEMET_G-Calibrated GR2M

Given the large number of parameters that were forced, in calibration analysis, to the extreme values range of MWBMs with satellite precipitation datasets, a 'second approach' was applied. Thus, once it was demonstrated that the AEMET grid is the best dataset for all watersheds and that

GR2M shows, on average, the best performance in the different climate regions in Peninsular Spain, experiments based on the well-calibrated model were conducted to evaluate streamflow predictions with input from rain gauge data and the four gridded rainfall datasets over the four watersheds. All watersheds reached the best result with the nearest rain gauge for both NSE and REV (Table 7). TRMM showed the best performance in three out of the four watersheds among the GPDs studied, because RVA does not exceed 0.33 NSE with MSWEP, and the lowest REV is over 60%, although the rest of the GPDs showed worse errors. There is, once again, a clear trend towards worse performance the drier the watershed is, substantially decreasing in NSE value from an average 0.82 in PUE to negative values in RVA. REV results indicated higher variation depending on the GPD, but this also followed same tendency of increasing errors the drier the watershed is. Despite these wide ranges in REV, an underestimation of streamflow in GPDs was confirmed, becoming near −80% in GAR, even if in RVA this trend was reversed with TRMM and PERSIANN. The latter reached the worst results for the four watersheds, except for GAR.

Table 7. NSE and REV of global precipitation datasets (GPDs) using the AEMET_G-calibrated GR2M (best results in bold).

Basins	Dataset	NSE	REV (%)
	AEMET_S	0.87	8.18
	TRMM	0.88	−3.70
PUE	PERSIANN	0.70	−29.26
	CFSR	0.80	14.76
	MSWEP	**0.89**	**−3.45**
	AEMET_S	0.82	0.99
	TRMM	**0.77**	**−9.66**
TRE	PERSIANN	0.57	−0.27
	CFSR	0.72	−11.27
	MSWEP	0.58	−30.67
	AEMET_S	0.90	−0.91
	TRMM	**0.57**	**−28.78**
GAR	PERSIANN	0.32	−60.02
	CFSR	0.48	−62.21
	MSWEP	0.17	−77.13
	AEMET_S	0.61	−31.72
	TRMM	−2.70	200.16
RVA	PERSIANN	−3.71	228.23
	CFSR	0.17	−74.86
	MSWEP	**0.33**	**−60.47**

5. Discussion

The use of point-scale gauge records can lead to important deviations in areal precipitations, as suggested by Tang et al. [4], especially the drier the watershed is. With regard to GPDs, as has been reported by other authors [56,57], the volumes of precipitation of the satellite precipitation products tended to be smaller than those of the gauged data in most cases, and these differences are greater the drier the watershed is, as in the cases of GAR and RVA. MSWEP and PERSIANN show the highest differences in accumulated precipitation, normally lower, except for the semi-arid watershed (RVA). Furthermore, even in RVA, PERSIANN and TRMM datasets show volumes higher than gauged records.

Concerning the goodness-of-fit tests, R does not seem to be a good measure to assess the validity of the studied datasets, because the results did not differ much from the others [58]. In fact, when using other criteria, both graphical and metrics, the performance of the datasets are clearly different.

The results confirmed that TRMM and CFSR are, on average, the models that reached the best performance in all the watersheds, with slight differences that are higher the drier the watershed is. The good performance of TRMM is also shown in other recent studies that used TRMM datasets worldwide [59,60]. However, the good results achieved with PERSIANN grid in [8] are not found in the present study, due to various and extreme climatic conditions which characterize Spain, as in the case of complex topography in Chile shown by Zambrano-Bigiarini et al. [61].

Regarding the MWBMs, GR2M gave the best results in all the watersheds for most of the rainfall products used, as previously reported by Pérez-Sánchez et al. [24] in Peninsular Spain. On the contrary, ABCD or THM may be considered the ones with the worst performance in a semi-arid watershed, but did not show such unsatisfactory results in the rest of the climate watersheds studied. The commonly used model in Spain, TEM, gave poor results (25% on average). In general, worse performance of MWBMs was observed the drier the watershed climate was. Because the gauge network density is similar in both the wet and dry regions, the uneven distribution of precipitation in semi-arid watersheds seems to be the reason why the performance of the datasets and models is very different.

Figure 5 shows the comparison between the observed and simulated streamflow when using GR2M, which resulted, as seen, generally, in the best MWBM in the four watersheds. The peaks of simulated flow are poorly modelled due to the underestimation in precipitation driven by the GPDs. These differences tend to be higher the drier the watershed is, as shown before, and regardless of the area. In fact, PUE and TRE observed peaks in the rainiest years (2001, 2003, and 2007) are nearer to simulated ones with most GPDs (especially TRMM and MSWEP) than in GAR and RVA watersheds, where simulated peak flows in 2001, 2003, and 2004 are lower than observed ones [62–64]. Nevertheless, there are other peak flows (though lesser in number) where simulated peak flows exceed observed ones, such as PERSIANN in TRE in 2001 and MSWEP and PERSIANN in RVA in 2007 and 2008, highlighting the higher precipitation volumes in these GPDs. However, base flows are well-modelled in all watersheds, reflecting the good results in NSE with GR2M (Table 4), despite differences in peaks, which are reflected in REV (Table 6).

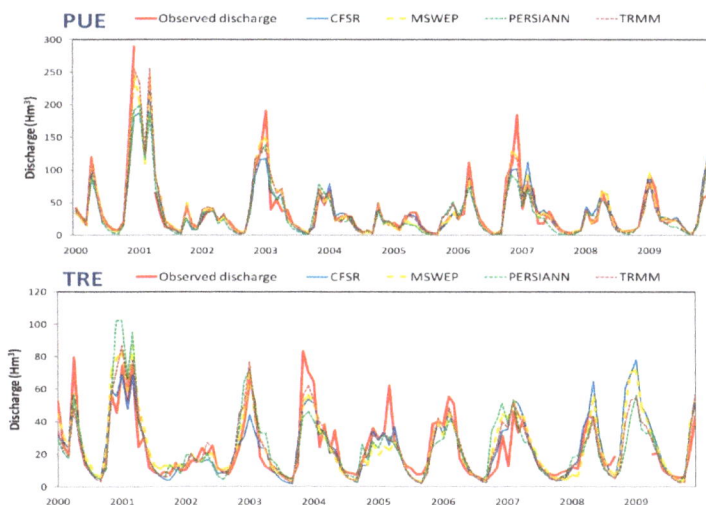

Figure 5. *Cont.*

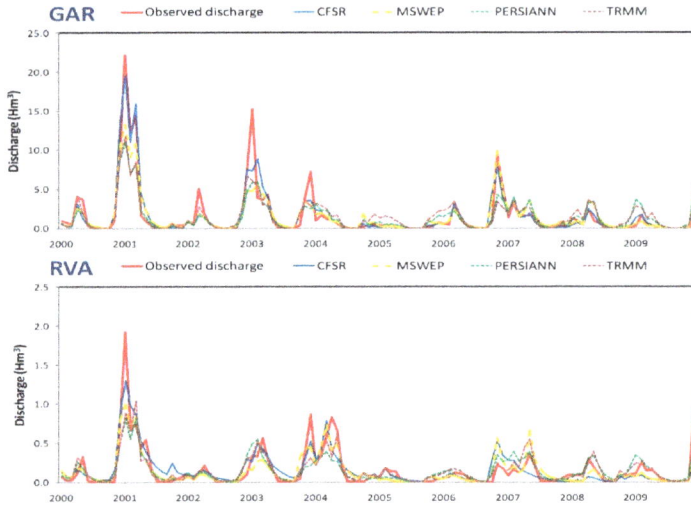

Figure 5. Observed and simulated monthly flow hydrographs for the different datasets using GR2M during the calibration and validation periods (2000–2009).

When using the second approach (Figure 6), whilst TRMM and MSWEP performed similarly both in base and peak flows in the more humid watershed (PUE), errors in peak flows were higher the more arid the watershed was. Although TRMM-simulated peak flows in the 2004–2006 period are lower than observed ones in TRE, total volume difference in the study period represents less than 10%. The CFSR dataset exhibited similar behaviour to TRMM, except in the last three years, where peak flows were higher than observed ones, as with PERSIANN in 2001. On the contrary, differences with MSWEP became even larger from 2002 in TRE, and streamflow was underestimated both in GAR and RVA, as with CFSR. None of the GPDs gave satisfactory results in this approach in the semi-arid watershed (RVA), overestimating total volume by up to 200% with TRMM and PERSIANN, and underestimating it by around 65% with CFSR and MSWEP.

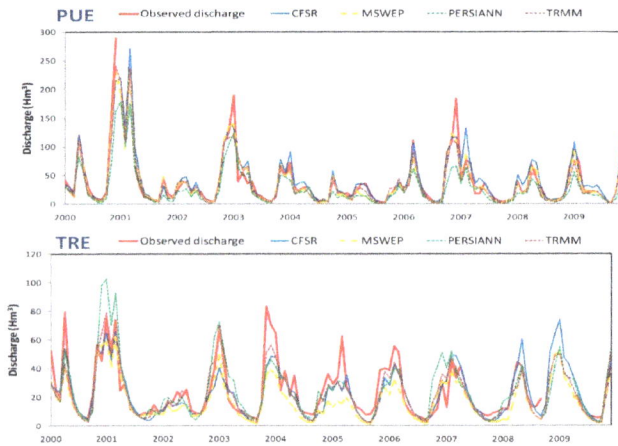

Figure 6. *Cont.*

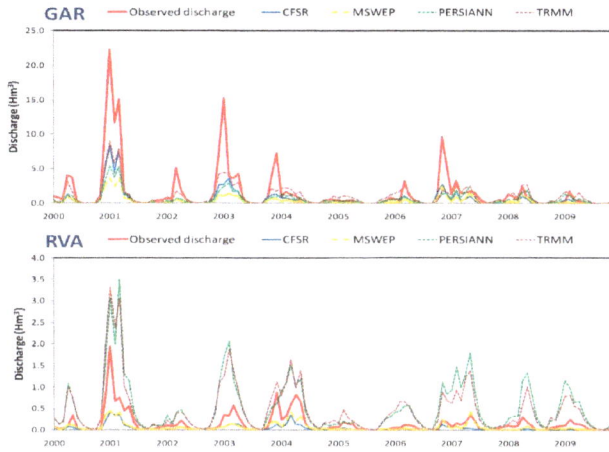

Figure 6. Observed and simulated monthly flow hydrographs for the different datasets using the AEMET_G-calibrated GR2M during the calibration and validation periods (2000–2009).

6. Conclusions

In this study, satellite rainfall products represented by PERSIANN, TRMM, CSFR, and MSWEP were assessed for the quality of their rainfall estimates on a monthly scale based on data from ground observations over four basins located in Peninsular Spain, which cover different climatic zones, for the period of January 2000 to December 2009. Due to the uniform coverage and no missing data, gridded datasets are much easier to use than station data. The following conclusions can be drawn from the results of this study:

1. The results underscore the superiority of the national gridded dataset over the other rainfall remote sensing products examined in this study.
2. The use of point-scale gauge records can lead to important deviations in areal precipitations, especially the drier the watershed is.
3. The better estimation of volumes of precipitation by using MSWEP would possibly be due to its finer resolution. However, that is not altogether necessary for success in better streamflow forecast.
4. The precipitation volumes of the GPDs tend to be smaller than those of the gauged data. However, PERSIANN and TRMM datasets show volumes higher than gauged records in semi-arid watersheds.
5. The lumped GR2M model provides a better streamflow forecast than the other MWBMs in Peninsular Spain watersheds. Notwithstanding, the performance of GPDs and MWBMs highly depends on the climate: The more humid the watershed is, the better results can be achieved.
6. When using GPDs in MWBM parameter calibration, TRMM rainfall data provides the best performance in simulating streamflow, with satisfactory precision in all watersheds according to NSE. However, CFSR achieves better results with regard to total volume recorded in sub-humid watersheds.
7. Calibration achieved directly with GPDs could result in unrealistic parameter values in MWBMs to compensate for the large errors in input datasets. Thus, an assessment of previously fitted value parameters should be taken in account. Likewise, a study of MWBM performance and best fitted parameters with rain gauge data should be used with GPDs, in order to avoid invalid or extreme parameter values in MWBMs.

8. When using rain gauge grid dataset-fitted parameters in MWBMs, TRMM was also the best GPD in humid and sub-humid watersheds, but its performance loses effectiveness the more arid the watershed is, as the rest of the GPDs showed, especially in peak flows, due to both the underestimation and overestimation of the extreme gauge precipitation in semi-arid watersheds.

9. The uneven distribution of precipitation in semi-arid watersheds seems to be the reason why the performance of datasets and models is worse than in humid and sub-humid regions.

10. Because semi-arid watersheds do not seem to provide very good results with the MWBMs and GPDs used, and because satellite rainfall datasets continue to improve, further analysis with other satellite data products and the joint use of (semi-) distributed models and downscaling datasets [65] are recommended for future studies, according to the methodology followed in developing this study. Likewise, sequential data assimilation techniques [66] may improve current hydrology model outputs using real-time observations.

The procedures used can be applied in regions with similar case studies to more accurately assess the resources within a system in which there is scarcity of recorded data available.

Author Contributions: J.S.-A. and J.P.-S. conceived and designed the experiments; A.L.-B. and F.J.S.-M. performed the experiments and analysed the data; D.P.-V. provided reviews and suggestions; J.P.-S. and J.S.-A. prepared the manuscript with contributions from all co-authors.

Acknowledgments: This research has been partially supported by the Euro-Mediterranean Water Institute (Grant No. 57/15). In addition, the authors acknowledge Papercheck Proofreading & Editing Services.

Conflicts of Interest: The authors declare no conflicts of interest.

References

1. Sun, Q.; Miao, C.; Duan, Q.; Ashouri, H.; Sorooshian, S.; Hsu, K.-L. A review of global precipitation data sets: Data sources, estimation, and intercomparisons. *Rev. Geophys.* **2018**, *56*. [CrossRef]

2. Liu, X.; Yang, T.; Hsu, K.; Liu, C.; Sorooshian, S. Evaluating the streamflow simulation capability of PERSIANN-CDR daily rainfall products in two river basins on the Tibetan Plateau. *Hydrol. Earth Syst. Sci.* **2017**, *21*, 169–181. [CrossRef]

3. Infante-Corona, J.A.; Lakhankar, T.; Pradhanang, S.; Khanbilvardi, R. Remote Sensing and Ground-Based Weather Forcing Data Analysis for Streamflow Simulation. *Hydrology* **2014**, *1*, 89–111. [CrossRef]

4. Tang, G.; Behrangi, A.; Long, D.; Li, C.; Hong, Y. Accounting for spatiotemporal errors of gauges: A critical step to evaluate gridded precipitation products. *J. Hydrol.* **2018**, *559*. [CrossRef]

5. Villarini, G.; Mandapaka, P.; Krajewski, V.; Witold, F.; Moore, R. Rainfall and sampling uncertainties: A rain gauge perspective. *J. Geophys. Res. Atmos.* **2008**. [CrossRef]

6. Jensen, N.E.; Pedersen, L. Spatial Variability of Rainfall. Variations within a Single Radar Pixel. *J. Atmos. Res.* **2005**, *77*, 269–277. [CrossRef]

7. Wood, S.J.; Jones, D.A.; Moore, R. Accuracy of rainfall measurement for scales of hydrological interest. *Hydrol. Earth Syst. Sci.* **2000**, *4*, 531–543. [CrossRef]

8. Tan, M.L.; Gassman, P.W.; Cracknell, A.P. Assessment of Three Long-Term Gridded Climate Products for Hydro-Climatic Simulations in Tropical River Basins. *Water* **2017**, *9*, 229. [CrossRef]

9. Faramarzi, M.; Srinivasan, R.; Iravani, M.; Bladon, K.D.; Abbaspour, K.C.; Zehnder, A.J.B.; Goss, G.G. Setting up a hydrological model of Alberta: Data discrimination analyses prior to calibration. *Environ. Model. Softw.* **2015**, *74*, 48–65. [CrossRef]

10. Raimonet, M.; Oudin, L.; Thieu, V.; Silvestre, M.; Vautard, R.; Rabouille, C.; Le Moigne, P. Evaluation of Gridded Meteorological Datasets for Hydrological Modeling. *J. Hydrometeorol.* **2017**, *18*, 3027–3041. [CrossRef]

11. Huffman, G.J.; Bolvin, D.T.; Nelkin, E.J.; Wolff, D.B.; Adler, R.F.; Gu, G.; Stocker, E.F. The TRMM Multisatellite Precipitation Analysis (TMPA): Quasi-global, multiyear, combined-sensor precipitation estimates at fine scales. *J. Hydrometeorol.* **2007**, *8*, 38–55. [CrossRef]

12. Ashouri, H.; Hsu, K.-L.; Sorooshian, S.; Braithwaite, D.K.; Knapp, K.R.; Cecil, L.D.; Prat, O.P. PERSIANN-CDR: Daily precipitation climate data record from multi-satellite observations for hydrological and climate studies. *Bull. Am. Meteorol. Soc.* **2015**, *96*, 69–83. [CrossRef]

13. Beck, H.E.; van Dijk, A.I.J.M.; Levizzani, V.; Schellekens, J.; Miralles, D.G.; Martens, B.; de Roo, A. MSWEP: 3-hourly 0.25° global gridded precipitation (1979–2015) by merging gauge, satellite, and reanalysis data. *Hydrol. Earth Syst. Sci.* **2017**, *21*, 589–615. [CrossRef]

14. Beck, H.E.; Vergopolan, N.; Pan, M.; Levizzani, V.; van Dijk, A.I.J.M.; Weedon, G.P.; Brocca, L.; Pappenberger, F.; Huffman, G.J.; Wood, E.F. Global-scale evaluation of 22 precipitation datasets using gauge observations and hydrological modeling. *Hydrol. Earth Syst. Sci.* **2017**, *21*, 6201–6217. [CrossRef]

15. Saha, S.; Moorthi, S.; Wu, X.; Wang, J.; Nadiga, S.; Tripp, P.; Behringer, D.; Hou, Y.T.; Chuang, H.Y.; Iredell, M. The NCEP climate forecast system version 2. *J. Clim.* **2014**, *27*, 2185–2208. [CrossRef]

16. Hyun-Goo, K.; Jin-Young, K.; Yong-Heack, K. Comparative Evaluation of the Third-Generation Reanalysis Data for Wind Resource Assessment of the Southwestern Offshore in South Korea. *Atmosphere* **2018**, *9*, 73. [CrossRef]

17. Yang, Y.; Wang, G.; Wang, L.; Yu, J.; Xu, Z. Evaluation of Gridded Precipitation Data for Driving SWAT Model in Area Upstream of Three Gorges Reservoir. *PLoS ONE* **2014**, *9*, 11. [CrossRef] [PubMed]

18. Xu, H.; Xu, C.Y.; Sælthun, N.R.; Zhou, B.; Xu, Y. Evaluation of reanalysis and satellite-based precipitation datasets in driving hydrological models in a humid region of southern China. *Stoch. Environ. Res. Risk Assess.* **2015**, *29*, 2003–2020. [CrossRef]

19. Dos Reis, J.B.C.; Rennó, C.D.; Lopes, E.S.S. Validation of Satellite Rainfall Products over a Mountainous Watershed in a Humid Subtropical Climate Region of Brazil. *Remote Sens.* **2017**, *9*, 1240. [CrossRef]

20. Seneviratne, S.I.; Corti, T.; Davin, E.L.; Hirschi, M.; Jaeger, E.B.; Lehner, I.; Orlowsky, B.; Teuling, A.J. Investigating soil moisture–climate interactions in a changing climate: A review. *Earth Sci. Rev.* **2010**, *99*, 125–161. [CrossRef]

21. Hafeez, M.; van de Giesen, N.; Bardsley, E.; Seyler, F.; Pail, R.; Taniguchi, M. *GRACE, Remote Sensing and Ground-Based Methods in Multi-Scale Hydrology: Proceedings of Symposium JHO1 Held during IUGG2011*; IAHS Publications: Boulder, CO, USA, 2011; ISBN 978-1-907161-18-6.

22. Wurbs, R.A. Texas water availability modeling system. *J. Water Resour. Plan. Manag.* **2005**, *131*, 270–279. [CrossRef]

23. Sitterson, J.; Knightes, C.; Parmar, R.; Wolfe, K.; Muche, M.; Avant, B. An Overview of Rainfall-Runoff Model Types. EPA 2017 Office of Research and Development (8101R) Washington, DC 20460. Available online: https://cfpub.epa.gov/si/si_public_file_download.cfm?p_download_id=533906 (accessed on 5 June 2018).

24. Pérez-Sánchez, J.; Senent-Aparicio, J.; Segura-Méndez, F.; Pulido-Velázquez, D. Assessment of lumped hydrological balance models in Peninsular Spain. *Hydrol. Earth Syst. Sci. Discuss.* **2017**, *424*. [CrossRef]

25. Marinou, P.G.; Feloni, E.G.; Tzoraki, O.; Baltas, E.A. An implementation of a water balance model in the Evrotas basin. *EWRA* **2017**, *57*, 147–154.

26. Wriedt, G.; Bouraoui, F. *Towards a General Water Balance Assessment of Europe*; Joint Research Centre—Institute for Environment and Sustainability; Office for Official Publications of the European Communities: Luxembourg, 2009.

27. Boughton, W.C. The Australian water balance model. *Environ. Model. Softw.* **2004**, *19*, 943–956. [CrossRef]

28. Zhang, Q.; Liu, J.; Singh, V.P.; Gu, X.; Chen, X. Evaluation of impacts of climate change and human activities on streamflow in the Poyang Lake basin, China. *Hydrol. Process.* **2016**, *30*, 2562–2576. [CrossRef]

29. Sharifi, M.A.; Rodriguez, E. Design and development of a planning support system for policy formulation in water resources rehabilitation: The case of Alcázar De San Juan District in Aquifer 23, La Mancha, Spain. *J. Hydroinform.* **2002**, *4*, 157–175. [CrossRef]

30. Barros, R.; Isidoro, D.; Aragüés, R. Long-term water balances in La Violada irrigation district (Spain): I. Sequential assessment and minimization of closing errors. *Agric. Water Manag.* **2011**, *102*, 35–45. [CrossRef]

31. Xiong, L.; Guo, S. A two-parameter monthly water balance model and its application. *J. Hydrol.* **1999**, *216*, 111–123. [CrossRef]

32. Guo, S. *Impact of Climate Change on Hydrological Balance and Water Resource Systems in the Dongjiang Basin, China*; Modeling and Management of Sustainable Basin-Scale Water Resource Systems (Proceedings of a Boulder Symposium; ed. by S. P. Simonovic, Z. W. Kundzewicz, D. Rosbjerg & K. Takeuchi); IAHS Publications: Boulder, CO, USA, 1995.

33. Escriva-Bou, A.; Pulido-Velazquez, M.; Pulido-Velazquez, D. Economic value of climate change adaptation strategies for water management in Spain's Jucar basin. *J. Water Res. Plan. ASCE* **2017**, *143*. [CrossRef]

34. MIMAM. *Libro Blanco del Agua en España*; Centro de Publicaciones del Ministerio de Medio Ambiente: Madrid, Spain, 2000.

35. Estrela, T. *Modelos Matemáticos Para la Evaluación de Recursos Hídricos*; Centro de Estudios Hidrográficos y Experimentación de Obras Públicas: Madrid, Spain, 1992.

36. Kottek, M.; Grieser, J.; Beck, C.; Rudolf, B.; Rubel, F. World map of the Köppen-Geiger climate classification. *Meteorol. Z.* **2006**, *15*, 259–263. [CrossRef]

37. CEDEX. 2018. Available online: http://www.cedex.es (accessed on 20 April 2018).

38. Radcliffe, D.E.; Mukundan, R. PRISM vs. CFSR Precipitation Data Effects on Calibration and Validation of SWAT Models. *JAWRA J. Am. Water Resour. Assoc.* **2016**, 1–12. [CrossRef]

39. Peral García, C.; Navascués Fernández-Victorio, B.; Ramos Calzado, P. *Serie de Precipitación Diaria en Rejilla con Fines Climáticos. Nota Técnica 24 de AEMET*; Spanish Meteorological Agency (AEMET): Madrid, Spain, 2017.

40. Xu, C.Y.; Singh, V.P. A review on monthly water balance models for water resources investigations. *Hydrol. Process.* **1998**, *12*, 31–50.

41. Alley, W.M. On the treatment of Evapotranspiration, Soil Moisture Accounting and Aquifer Recharge in Monthly Water Balance Models. *Water Resour. Res.* **1984**, *20*, 1137–1149. [CrossRef]

42. Thomas, H.A. *Improved Methods for National Water Assessment*; Report, contract: WR 15249270; U.S. Water Resources Council: Washington, DC, USA, 1981.

43. Makhlouf, Z.; Michel, C. A two-parameter monthly water balance model for french watersheds. *J. Hydrol.* **1994**, *162*, 299–318. [CrossRef]

44. Thornthwaite, C.W.; Mather, J.R. *Instructions and Tables for Computing Potential Evapotranspiration and the Water Balance*; Publications in Climatology, 10.3; Drexel Institute: Centerton, NJ, USA, 1957.

45. Témez, J.R. *Modelo Matemático de Transformación "Precipitación-Aportación"*; Asociación de Investigación Industrial Eléctrica (ASINEL): Madrid, Spain, 1977.

46. Estrela, T.; Cabezas, F.; Estrada, F. La evaluación de los recursos hídricos en el Libro Blanco del Agua en España. *Ingeniería del Agua* **1999**, *6*, 125–138. [CrossRef]

47. Gan, T.Y.; Dlamini, E.M.; Biftu, G.F. Effects of model complexity and structure, data quality, and objective functions on hydrologic modeling. *J. Hydrol.* **1997**, *192*, 81–103. [CrossRef]

48. Fylstra, D.; Lasdon, L.; Watson, J.; Waren, A. Design and use of the Microsoft Excel Solver. *Interfaces* **1998**, *28*, 29–55. [CrossRef]

49. Lasdon, L.S.; Waren, A.D.; Jain, A.; Ratner, M. Design and testing of a generalized reduced gradient code for nonlinear programming. *ACM Trans. Math. Softw.* **1978**, *4*, 34–50. [CrossRef]

50. Zeweldi, D.A.; Gebremichale, M.; Downer, C.W. On CMORPH rainfall for streamflow simulation in a small, Hortonian watershed. *J. Hydrometeorol.* **2011**, *12*, 456–466. [CrossRef]

51. Artan, G.; Gadain, H.; Smith, J.L.; Asante, K.; Bandaragoda, C.J.; Verdin, J.P. Adequacy of satellite derived rainfall data for streamflow modelling. *Nat. Hazards* **2007**, *43*, 167–185. [CrossRef]

52. Yilmaz, K.K.; Hogue, T.S.; Hsu, K.L.; Sorooshian, S.; Gupta, H.V.; Wagener, T. Intercomparison of rain gauge, radar, and satellite-based precipitation estimates with emphasis on hydrologic forecasting. *J. Hydrometeorol.* **2005**, *6*, 497–517. [CrossRef]

53. Bitew, M.M.; Gebremichael, M.; Ghebremichael, L.T.; Bayissa, Y.A. Evaluation of high resolution satellite rainfall products through streamflow simulation in a hydrological modeling of a Small Mountainous Watershed in Ethiopia. *J. Hydrometeorol.* **2012**, *13*, 338–350. [CrossRef]

54. Habib, E.; Haile, A.T.; Sazib, N.; Zhang, Y.; Rientjes, T. Effect of bias correction of satellite-rainfall estimates on runoff simulations at the source of the Upper Blue Nile. *Remote Sens.* **2014**, *6*, 6688–6708. [CrossRef]

55. Nash, J.E.; Sutcliffe, J.V. River flow forecasting through conceptual models. Part I—A discussion of principles. *J. Hydrol.* **1970**, *10*, 282–290. [CrossRef]

56. Sidike, A.; Chen, X.; Liu, T.; Durdiev, K.; Huang, Y. Investigating alternative climate data sources for hydrological simulations in the upstream of the Amu Darya River. *Water* **2016**, *8*, 441. [CrossRef]

57. Yoshimoto, S.; Amarnath, G. Applications of Satellite-Based Rainfall Estimates in Flood Inundation Modeling—A Case Study in Mundeni Aru River Basin, Sri Lanka. *Remote Sens.* **2017**, *9*, 998. [CrossRef]

58. Pérez-Sánchez, J.; Senent-Aparicio, J.; Díaz-Palmero, J.M.; Cabezas-Cerezo, J.D. A comparative study of fire weather indices in a semiarid south-eastern Europe region. Case of study: Murcia (Spain). *Sci. Total Environ.* **2017**, *15*, 590–591, 761–774. [CrossRef] [PubMed]

59. Caparoci Nogueira, S.M.; Moreira, M.A.; Lordelo Volpato, M.M. Evaluating Precipitation Estimates from Eta, TRMM and CHRIPS Data in the South-Southeast Region of Minas Gerais State—Brazil. *Remote Sens.* **2018**, *10*, 313. [CrossRef]

60. Cao, Y.; Zhang, W.; Wang, W. Evaluation of TRMM 3B43 data over the Yangtze River Delta of China. *Sci. Rep.* **2018**, *8*, 5290. [CrossRef] [PubMed]

61. Zambrano-Bigiarini, M.; Nauditt, A.; Birkel, C.; Verbist, K.; Ribbe, L. Temporal and spatial evaluation of satellite-based rainfall estimates across the complex topographical and climatic gradients of Chile. *Hydrol. Earth Syst. Sci.* **2017**, *21*, 1295–1320. [CrossRef]

62. Dinku, T.; Ceccato, P.; Grover-Kopec, E.; Lemma, M.; Connor, S.J.; Ropelewski, C.F. Validation of satellite rainfall products over East Africa's complex topography. *Int. J. Remote Sens.* **2007**, *28*, 1503–1526. [CrossRef]

63. Karaseva, M.O.; Prakash, S.; Gairola, R.M. Validation of high-resolution TRMM-3B43 precipitation product using rain gauge measurements over Kyrgyzstan. *Theor. Appl. Climatol.* **2012**, *108*, 147–157. [CrossRef]

64. Guo, R.; Liu, Y. Evaluation of Satellite Precipitation Products with Rain Gauge Data at Different Scales: Implications for Hydrological Applications. *Water* **2016**, *8*, 281. [CrossRef]

65. Omranian, E.; Sharif, H.O. Evaluation of the Global Precipitation Measurement (GPM) Satellite Rainfall Products over the Lower Colorado River Basin, Texas. *J. Am. Water Resour. Assoc.* **2018**. [CrossRef]

66. Javaheri, A.; Nabatian, M.; Omranian, E.; Babbar-Sebens, M.; Noh, S.J. Merging Real-Time Channel Sensor Networks with Continental-Scale Hydrologic Models: A Data Assimilation Approach for Improving Accuracy in Flood Depth Predictions. *Hydrology* **2018**, *5*, 9. [CrossRef]

![remote sensing logo] *remote sensing*

MDPI

Article

Classification of Hydrometeors Using Measurements of the Ka-Band Cloud Radar Installed at the Milešovka Mountain (Central Europe)

Zbyněk Sokol [1,*], Jana Minářová [1] and Petr Novák [2]

[1] Institute of Atmospheric Physics, Czech Academy of Sciences, 141 31 Prague, Czech Republic; jana.minarova@ufa.cas.cz

[2] Czech Hydrometeorological Institute, 143 00 Praha-Komořany, Czech Republic; petr.novak@chmi.cz

*   Correspondence: sokol@ufa.cas.cz; Tel.: +420-272-016-037

Received: 5 September 2018; Accepted: 22 October 2018; Published: 23 October 2018

Abstract: In radar meteorology, greater interest is dedicated to weather radars and precipitation analyses. However, cloud radars provide us with detailed information on cloud particles from which the precipitation consists of. Motivated by research on the cloud particles, a vertical Ka-band cloud radar (35 GHz) was installed at the Milešovka observatory in Central Europe and was operationally measuring since June 2018. This study presents algorithms that we use to retrieve vertical air velocity (Vair) and hydrometeors. The algorithm calculating Vair is based on small-particle tracers, which considers the terminal velocity of small particles negligible and, thereby, Vair corresponds to the velocity of the small particles. The algorithm classifying hydrometeors consists of calculating the terminal velocity of hydrometeors and the vertical temperature profile. It identifies six hydrometeor types (cloud droplets, ice, and four precipitating particles: rain, graupel, snow, and hail) based on the calculated terminal velocity of hydrometeors, temperature, Vair, and Linear Depolarization Ratio. The results of both the Vair and the distribution of hydrometeors were found to be realistic for a thunderstorm associated with significant lightning activity on 1 June 2018.

Keywords: precipitating hydrometeor; hydrometeor classification; cloud radar; Ka-band; thunderstorm; thundercloud; vertical air velocity; terminal velocity; Milešovka observatory

1. Introduction

Measurements from a millimeter-wave Doppler radars are suitable for research on cloud microphysics at a high spatial and temporal resolution [1–3]. Therefore, a vertically pointing polarimetric Ka-band cloud radar (35 GHz) was installed at the Milešovka observatory (Czech Republic, Central Europe) as part of the running project Cosmic Rays and Radiation Events in the Atmosphere (CRREAT). CRREAT is focused on the relationships between cloud hydrometeors/precipitation particles and the electric field in the atmosphere. The Milešovka observatory is situated at a mountain top at an elevation of 837 m, which exceeds the surrounding landscape of more than 300 m and, thus, provides a 360° unobstructed view from the observatory. The observatory is equipped with a wide set of instruments (meteorological and non-meteorological) and its unique location and limited accessibility to the observatory counted among the reasons for selecting this type of cloud radar.

To the best of our knowledge, there are 17 Ka-band cloud radars operating in Europe (seven in Germany) including two mobile Ka-band cloud radars. The newly installed Ka-band cloud radar at the Milešovka observatory is the first of its kind operating in the Czech Republic. The installation of the radar at the observatory took place at the end of March 2018 and the radar started operating in June 2018.

The aim of this article is to describe two new functionalities that we added to the radar data processing to study the cloud structures, which is our research purpose. Specifically, we dealt with (i) the estimation of vertical air velocity and terminal velocity of hydrometeors and (ii) the classification of hydrometeors for which the vertical velocity and terminal velocity are the input parameters. Note that hydrometeors are any kind of liquid or solid water particles in the atmosphere that can result in precipitation, which may or may not reach the ground in the form of graupel, rain, snow, or hail.

Several pioneer studies that tried to retrieve the vertical air motion were often based on fixing an empirical relationship between the radar reflectivity and terminal velocity of hydrometeors depending on the diameter of hydrometeors [4,5]. However, a straightforward relationship among the variables is difficult to establish and is not known in the case of thunderclouds [6]. Kollias [7] applied a method for retrieving the vertical motion for W-band cloud radar that is valid for intense precipitation. The method was introduced by Lhermitte [8] and consists of retrieving the vertical air velocity from the signature of the observed Doppler spectra modulated by Mie scattering. However, Zheng et al. [6] pointed out that, by using measurements of millimeter (Ka/W) cloud radars, this method is not valid for the case of small particles including cloud droplets or light precipitation. In such a case, one can use the "small-particle-traced" method to retrieve the vertical air velocity [6]. The method assumes that small particles (i.e., tracers) have a negligible terminal velocity. Therefore, their velocity corresponds to that of the air [9–11]. This method was applied by Zheng et al. [6] during the TIPEX-III experiment over the Tibetan Plateau. Their retrieved air velocity was found to be reliable and in good agreement with other radar measurements and, as compared to retrievals based on disdrometer measurements, it provided more detailed information about the vertical air motion. Thus, we used this method in our study as well.

The classification of hydrometeors using cloud radar data has been discussed in many studies [12–15]. In general, hydrometeor classification algorithms using polarimetric measurements of any kind of Doppler radar can be based on the combination of radar reflectivity with a Linear Depolarization Ratio (LDR) and/or with differential reflectivity [16–18] or reflectivity difference [19]. In the past, the main target was the detection of hail from precipitation. In several studies [20,21], the hydrometeors were classified by using the decision tree method while, in others, by using fuzzy logic and neural networks [22]. Nowadays, most of the algorithms classifying hydrometeors using cloud radar data belong to the retrieving methods of Doppler spectra [12,13,23,24]. For instance, the retrieving methods for cloud properties were provided in Reference [25]. For upper tropospheric clouds, they have been compared in Reference [26] and, for stratospheric clouds, they were compared in Reference [27]. Other studies discussed the retrieving methods for cloud radar placed on satellites [28]. The algorithm that is used by the provider of our cloud radar was presented in Reference [29]. However, many classifying algorithms were either designed for weather radars (e.g., C-band) or limited to a specific kind of particle such as ice or precipitation.

In our study, we aim at classifying hydrometeors based mainly on the differences in terminal velocities of precipitating hydrometeors. The terminal velocity of hydrometeors suggests the occurrence of precipitation (the higher the terminal velocity of a hydrometeor, the higher the probability that the hydrometeor reaches the ground, i.e., precipitates). We illustrate the computational methods with an event that occurred on 1 June 2018. On 1 June 2018, a thunderstorm occurred near the Milešovka observatory producing many lightning strikes and precipitation at the observatory and its vicinity.

The article is organized as follows. After this introductory Section 1, Section 2 describes the Milešovka observatory and provides details about the Ka-band cloud radar installed at the observatory. Section 2 also depicts the thunderstorm on 1 June 2018 associated with strong lightning activity and it shows the radar data processing and the algorithms that we use to calculate the vertical air velocity and to classify the hydrometeors. Section 3 displays the resulting retrieved vertical air velocity and the distribution of hydrometeors during the thunderstorm on 1 June 2018 while Section 4 discusses the obtained results and compares it with those retrieved by the provided radar software. Conclusions are drawn in Section 5.

2. Materials and Methods

2.1. Milešovka Observatory

The Milešovka observatory is situated on the highest top of Central Bohemian Uplands in the Czech Republic in Central Europe (Figure 1) called the Milešovka Mountain (837 m a.s.l.; 50°33'18"N. and 13°55'54"E.). It is a meteorological and climatological observatory with continuous measurements since 1905. The location of the Milešovka observatory is suitable for atmospheric research due to a large 360° view and an absence of high obstacles in the surroundings, which makes it a unique meteorological observatory in the Czech Republic.

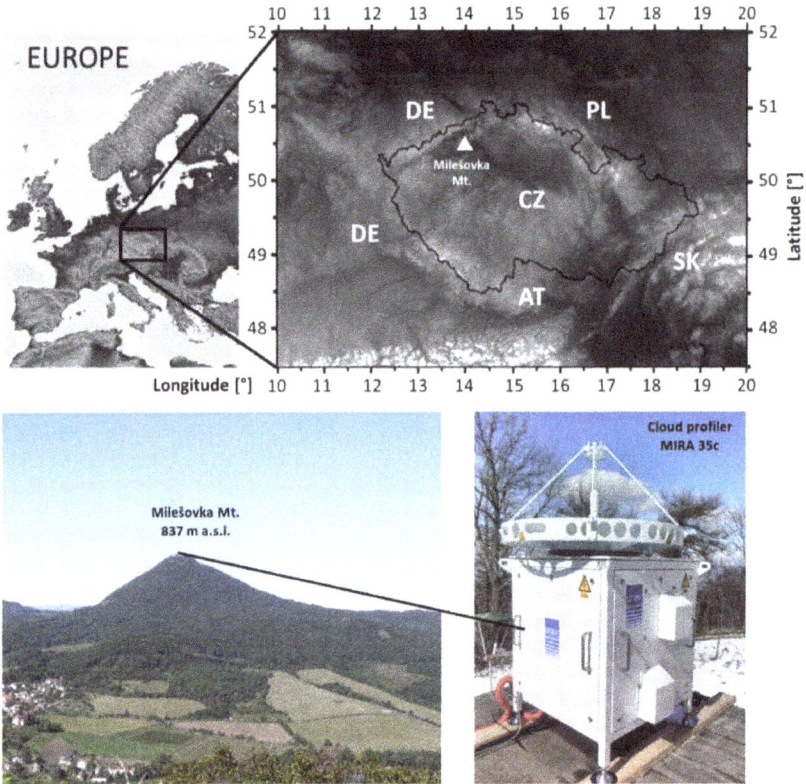

Figure 1. Geographical location of the Milešovka observatory at the Milešovka Mountain (837 m a.s.l.) where the cloud radar (profiler MIRA35c) was installed in March 2018.

The Milešovka observatory is operated by the Institute of Atmospheric Physics, Czech Academy of Sciences and controlled by an observer with a 24/7 service. The equipment includes instruments of a standard meteorological and climatological station providing e.g., measurements of temperature, precipitation, and wind. Moreover, it also includes two sonic anemometers, Vaisala ceilometer CL51, Thies Laser Precipitation Monitor etc. Besides various meteorological instruments, the Milešovka observatory is also equipped with instruments measuring the atmospheric electric field (Boltek Electric Field Monitor EFM-100), the magnetic field (SLAVIA sensors, Shielded Loop Antenna with a Versatile

Integrated Amplifier), and charged and neutral components of secondary cosmic rays (SEVAN) in order to investigate lightning in thunderstorms.

On 26 March 2018, a Ka-band vertically pointing cloud radar (profiler MIRA35c) was installed at the station (Figure 1) for detecting cloud particles in order to derive the distribution of hydrometeors in clouds. In clouds and thunderclouds, the hydrometeors might be responsible for precipitation and heavy rainfall, respectively.

2.2. Ka-Band Cloud Radar at the Milešovka Observatory

The Ka-band cloud radar (profiler MIRA35c) installed at the Milešovka observatory (Figure 1) was provided by METEK Gmbh (http://metek.de/). It is a Ka-band Doppler polarimetric radar with a center frequency of 35.12 +/−0.1 GHz. The cloud radar is vertically oriented and its technical specifications are listed in Table 1.

Table 1. Technical specifications of the cloud radar MIRA35c installed at the Milešovka observatory.

Radar Parameter	MIRA35c
Radar system	Doppler polarimetric
Radar band	Ka
Transmitter frequency [GHz]	35.12 +/−0.1
Radar core	Magnetron
Peak power [W]	2500
Antenna type	Casse grain
Antenna diameter [m]	1
Antenna gain [dB]	48.5
Antenna beam width [°]	0.6
Pulse repetition frequency [Hz]	2500–10,000
Pulse width [ns]	min. 100, max. 400
Detection unambiguous velocity range [m·s^{-1}]	±10.65
Original data measurements	Doppler spectra
Spectral moments	Reflectivity (Z)
	Doppler vertical velocity (DVV)
	spectrum width (σ)
Derived variables	Linear Depolarization Ratio (LDR)
	Signal to Noise Ratio (SNR)

The cloud radar is equipped with a software *"MIRA-3x IDL software for Data Processing and Visualization"* called IDLsoft hereafter (http://metek.de/product/mira-35c/). It performs the processing of radar data in several steps and the processed data are recorded in each step. Thus, it is possible to process the data by external algorithms at different levels of processing.

The processing of measured data is described in detail by Görsdorf et al. [1]. In this study, we point out that the calculation of Doppler spectra is preceded by incoherent averages (200 consecutive measurements are averaged), estimation of the noise floor of a spectrum and determination of the noise threshold S_{TH} [1]. Upon further processing, only values greater than S_{TH} are used. Other values are supposed to have no signal.

The cloud radar provides us with measurements of Doppler spectra from which three spectral moments (reflectivity, Doppler vertical velocity (DVV), and spectrum width), Linear Depolarization Ratio (LDR), and Signal-to-Noise Ratio are calculated. Using the measured quantities, cloud microphysical characteristics can be derived, e.g., type of hydrometeor and pure atmospheric vertical motion. After the installation of cloud radar at the Milešovka observatory in March 2018, the cloud radar was under testing for the first two months. Operational measurements are available since June 2018.

2.3. Thunderstorm on 1 June 2018

On 1 June 2018, a severe thunderstorm associated with precipitation and intense lightning occurred at the Milešovka observatory and its vicinity approximately between 12:00 and 12:30 UTC. Based on the information of the observer, the thunderstorm was related to a convective cell centered in the north of the observatory. Figure 2 shows the radar reflectivity that was measured by a C-band weather radar located 100 km southward from the Milešovka Mountain (Mt.) and operated by the Czech Hydrometeorological Institute.

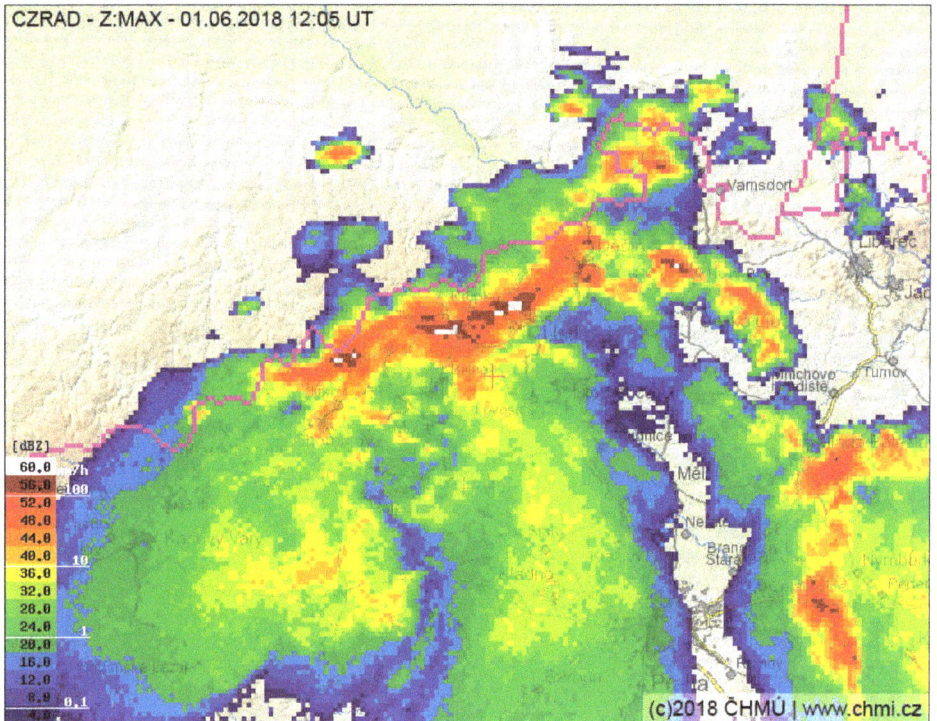

Figure 2. Radar reflectivity measured by the radar Brdy on 1 June 2018 at 12:05 UTC (source: Czech Hydrometeorological Institute). The location of the Milešovka Mt. is highlighted by the red cross.

The main precipitation cores (rain rates higher than 100 mm/h) were observed to be several kilometers in the North of the Milešovka Mt. (Figure 2). However, the one-minute precipitation maximum reached 1.5 mm at the Milešovka Mt., according to the rain gauge measurements (Figure 3).

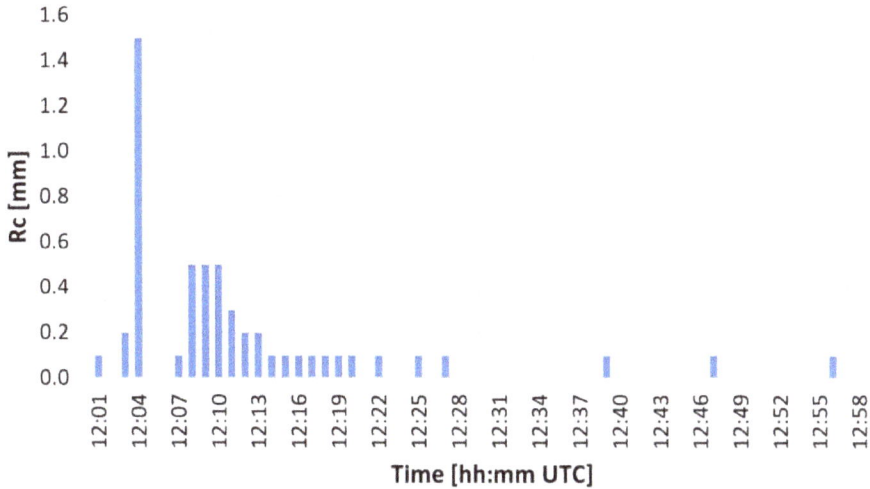

Figure 3. Cumulated one-minute precipitation from rain gauge measurements at the Milešovka observatory on 1 June 2018 from 12:00 to 13:00 UTC.

In addition, a strong lightning activity was observed during the event with most of the lightning detected several kilometers in the north of the Milešovka Mt. (not depicted). Nevertheless, a thunder struck straight at the Milešovka observatory at approximately 12:10 UTC, according to the observer. The thunder stroke caused a power failure at the station even though the cloud radar measured unceasingly due to an uninterruptible power supply. The lightning, which directly strikes the observatory, is registered only once per year on average, which is one of the reasons for studying this particular event. Moreover, we selected this event because, since 1 June 2018, there were very few precipitation cases that were observed close to the observatory due to an unusually dry and sunny summer in the Czech Republic. None of these cases were related to strong lightning activity (as on 1 June 2018). The selected storm on 1 June 2018 is also considered significant in the context of the northwestern part of the Czech Republic because events with similar manifestations (e.g., occurrence of intense lightning) generally occurs only 10 times per year on average.

Figure 4 displays the standard products of IDLsoft during the thunderstorm on 1 June 2018 at the Milešovka observatory: the time evolution of (i) equivalent radar reflectivity (Figure 4a) and (ii) DVV oriented upward (Figure 4b). It follows from Figure 4 that the storm started moving across the Milešovka Mt. around 12:00 UTC. Figure 4b shows an aliasing in the DVV short before 12:00 UTC (yellow to dark yellow colors). Large negative DVV around 12:00 UTC indicate intense downdrafts in the radar position.

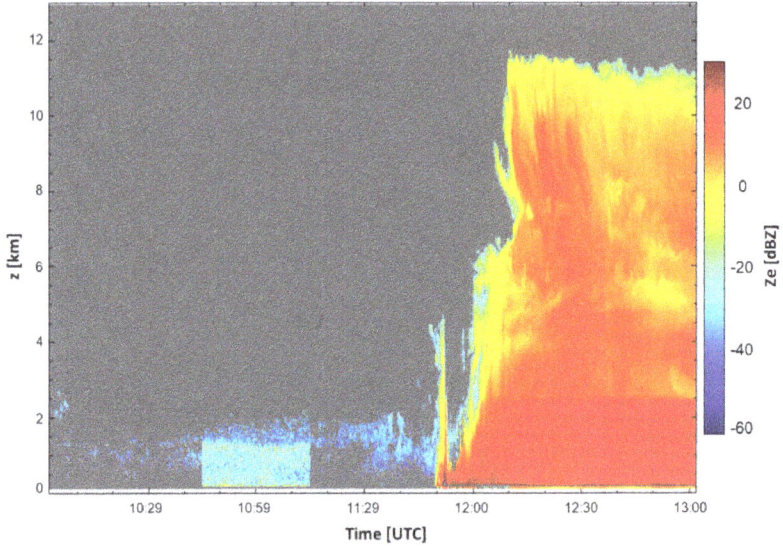

(a)

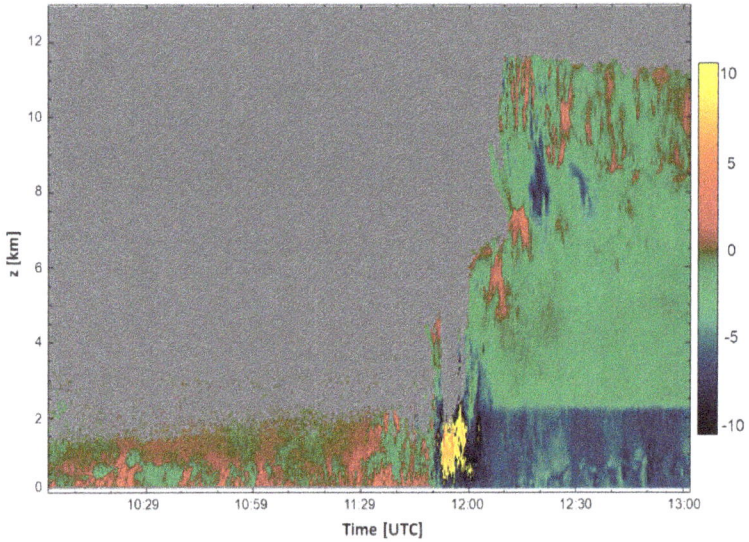

(b)

Figure 4. Standard products of IDLsoft based on the cloud radar measurements on 1 June 2018 from 10:02 to 13:01 UTC at the Milešovka observatory: (**a**) Equivalent radar reflectivity Ze [dBZ] and (**b**) Doppler vertical velocity DVV [m/s]. DVV is oriented upward from the radar position. Note that z [km] is the height in kilometers above the Milešovka Mt.

Table 2 shows the vertical temperature profile during the event based on the aerological sounding measurements at 12:00 UTC from Praha/Libuš station (No. 11520). The aerological station is situated approximately 60 km from the Milešovka observatory.

Table 2. Vertical temperature profile from sounding measurements at 12:00 UTC on 1 June 2018 based on the data from the Praha/Libuš station.

T [°C]	17.2	10	0	−10	−20
z [m]	0	842	2253	4278	5929

Note that z [m] corresponds to the height above the Milešovka Mt., which is situated at an elevation of 837 m a.s.l.

2.4. Radar Data Processing

The cloud radar at the Milešovka Mt. processes measured data using the IDLsoft. The IDLsoft analyses Doppler spectra for each gate and determines at most 15 peaks in the Doppler spectrum. It also determines discrete intervals of the Doppler spectrum (ID) that include one peak each (http://metek.de/product/mira-35c/). For each ID, quantities such as DVV are calculated.

The Doppler processing of the I-Q signal consists of the following steps:

1. Range gate decomposition
2. Phase correction
3. Calculation of the Fourier transform of the signal for each range gate
4. Non-coherent averaging of the Doppler spectra
5. Calculation of the noise level by applying the Hildebrand-Sekhon-Div algorithm based on Reference [30]
6. Estimation of the three first moments, i.e., radar reflectivity, DVV, and spectrum width
7. Estimation of derived quantities: Signal-to-Noise Ratio, equivalent radar reflectivity, and LDR

We used the measured data to retrieve the vertical air velocity (Section 2.5) and to classify the hydrometeors (Section 2.6).

2.5. Calculation of Vertical Air Velocity (Vair)

The calculation of vertical air velocity (Vair) is based on a known idea referred to a "small-particle-traced idea" in the literature. The small-particle-traced idea was described in References [9,10,31] and applied in Reference [6]. Our algorithm for retrieving the vertical air velocity from cloud radar measurements stems mostly from Reference [6]. Contrary to Reference [6], we developed a simple dealiasing algorithm for our vertically oriented radar. The dealiasing algorithm supposes that the velocity in the lowest gate is correct. The velocity in the next (upper) gate is checked and eventually corrected by using the condition that the difference in velocities in neighboring gates either does not exceed +10.65 m/s or is not lower than −10.65 m/s (see Table 1). An example of the dealiasing algorithm is given in Figure 5 where the vertical velocity is oriented downward towards the radar. The same (i.e., downward) orientation of the vertical velocity is used hereafter.

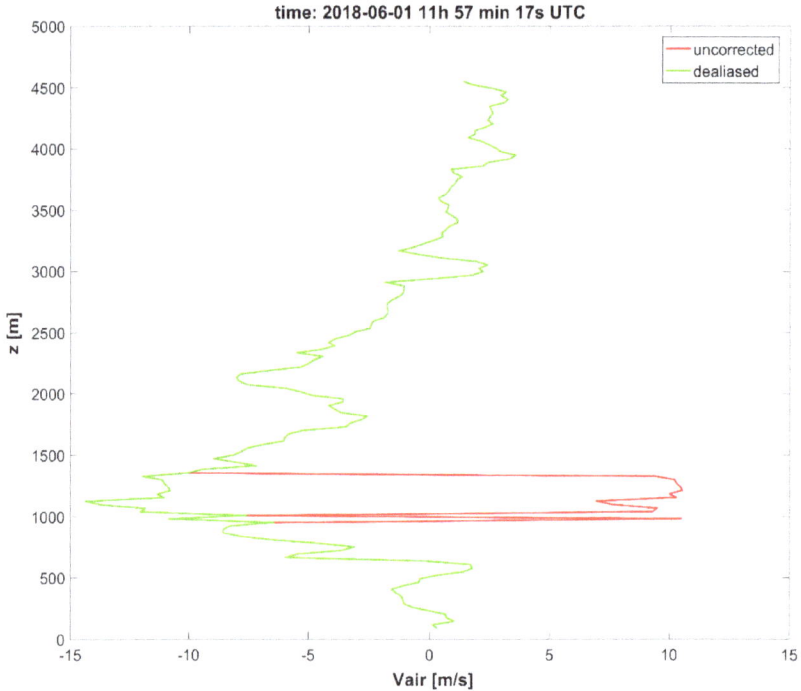

Figure 5. Vair [m/s] calculated and dealiased by our algorithm for 1 June 2018 at 11 h 57 min and 17 s UTC at the Milešovka Mt.: (i) uncorrected Vair (red) and (ii) dealiased Vair (green). Note that z [m] (vertical axis) is the height in meters above the Milešovka Mt.

The small-particle-traced idea is based on a supposition that the cloud droplets and small particles of ice or snow have negligible terminal velocity. Therefore, their vertical velocity corresponds to that of the air motion. If the orientation of vertical velocity is downward towards the radar (as in Figure 5 and hereafter), then the terminal velocity is always positive and the velocity of small particles (e.g., cloud droplets) can be estimated from the Doppler spectrum for vertically pointing radar based on known variances induced by (i) particle size distribution, (ii) turbulence, (iii) wind shear, and (iv) finite radar beam width [6].

In our study, we estimated Vair based on the procedure applied in Reference [6] with two modifications. The first modification that we required is that the amplitude of the left edge of the Doppler spectra is at least 0.1% of the maximum amplitude of Doppler spectra for any given measurement. This modification was motivated by the results of our tests, which revealed an insufficient removal of noise by S_{TH} (see Section 2.1). Specifically, Doppler spectra with a very low amplitude were not removed by S_{TH}. Therefore, the left edge of the Doppler spectra was giving unrealistic vertical velocities or velocities inconsistent with the velocities in neighboring gates.

The value of 0.1% is based on the testing of various S_{TH} ranging from 0.0001% to 1% of the maximum amplitude of Doppler spectra and on the evaluation of maximum differences in Vair between neighboring gates. The gates are above and below the evaluated gate. While the differences in Vair were large for S_{TH} = 0.05% and lower, the differences were significantly smaller for the S_{TH} ranging from 0.1% to 1%. Therefore, we considered 0.1% of the maximum amplitude of Doppler spectra as a suitable value of S_{TH} for our algorithm for any given measurement. It should be noted that we suppose the existence of "small-particles" even if they are not later recognized by the classification algorithm of hydrometeors (Section 2.6).

The second modification was related to horizontal wind shear calculations since our cloud radar is vertically oriented. We defined the horizontal wind shear df/dx by assuming that the horizontal wind does not change along the Lagrangian trajectories.

$$df/dx = -(1/u)(u(t) - u(t\text{-}dt))/dt, \tag{1}$$

where u is the horizontal wind velocity [m/s] measured by an aerological balloon at a time t while dt [s] is the time resolution of radar measurements (i.e., 2 s approximately).

2.6. Classification of Hydrometeors

The algorithm that we applied to classify hydrometeors stems from the assumption that the terminal velocity of various hydrometeors differs. Note that the terminal velocity of a hydrometeor might suggest whether the hydrometeor falls to the ground, i.e., becomes precipitation (rain, hail etc.), or it evaporates before reaching the ground depending on the air temperature. Thus, we also suppose in our algorithm that the occurrence of single hydrometeor depends on air temperature and partially on LDR, which indicates the shape of hydrometeors.

We used six types of hydrometeors: cloud, graupel, ice, snow, rain, and hail. Cloud droplets and small ice crystals are usually non-precipitating hydrometeors while rain, graupel, snow, and hail represent hydrometeors that can also be detected at the ground as precipitation. The interval of terminal velocity of a hydrometeor was derived from parameters of hydrometeors considered in the COSMO numerical weather prediction model. We selected the parameter values that belong to "standard" hydrometeors and that we use whenever we run the COSMO e.g., Reference [32]. Note that, in this study, we did not make any simulation in COSMO. We only took the parameter values of the six hydrometeors from COSMO. We slightly modified the original intervals of terminal velocity to avoid intersections within both the liquid and the solid hydrometeors (i.e., to get discrete intervals).

Table 3 shows the six types of hydrometeors that we retrieve. It displays minimum terminal velocity of individual hydrometeors (Vmin), maximum terminal velocity of individual hydrometeors (Vmax), and temperature intervals at which the individual hydrometeors can occur.

Table 3. Six types of hydrometeors and their minimum and maximum terminal velocity [m/s] (Vmin and Vmax, respectively) and minimum and maximum air temperature [°C] (Tmin and Tmax, respectively) within which the hydrometeors may occur.

Hydrometeor	Vmin [m/s]	Vmax [m/s]	Tmin [°C]	Tmax [°C]
Cloud	0.0001	0.1543	−20	40
Rain	0.1543	6.3384	−20	40
Snow	0.0290	1.2458	−70	0
Ice	1.2458	1.3133	−70	0
Graupel	1.3133	7.7747	−70	40
Hail	7.7747	10.0253	−70	40

The classification is performed for each gate on a condition that at least one ID is found. It should be noted that DVV (Figure 4b) is the weighted average of the spectrum of measured Doppler velocities (i.e., spectrum components) with the weight corresponding to the measured reflectivity of individual spectrum components. As a rule, the spectrum contains several peaks, which can significantly differ in corresponding speeds. Thus, they may correspond to different hydrometeors. The peaks and surrounding ID are determined by the IDLsoft during the basic processing of measured data.

The hydrometeor classification algorithm is performed for each peak and consists of two preliminary steps:

- Calculation of vertical temperature profile: We use aerological sounding measurements of temperature from station Praha/Libuš, which are linearly interpolated in time and height above the ground. The station is located 60 km southward from the Milešovka Mt. The measurements are regularly provided at 00, 06, and 12 UTC. Since we do not perform the classification of hydrometeors in real time, it is possible to interpolate the measurements in time.

- Computation of terminal velocity (Vter): terminal velocity is determined for each ID by subtracting Vair from the Doppler velocity for the given peak (DVP).

$$Vter(ID) = DVP - Vair. \tag{2}$$

Since it is well known that the occurrence of hydrometeors depends on air temperature T, we considered three temperature intervals in our study. We suppose that, below −20 °C, the water is fully frozen and therefore graupel, ice, snow, and hail can occur while cloud and rain droplets cannot occur. When the temperature is above 0 °C, we assume that cloud and rain droplets, graupel, and hail can appear while ice and snow cannot. In between (i.e., from −20 °C to 0 °C) in convective storms, it is assumed that solid hydrometeors occur while liquid hydrometeors can occur only in the case when the air rises up (Vair ≤ −0.01 m/s in the code) because, in this case, the cloud droplets are vertically advected. They do not freeze immediately and, instead, they become supercooled.

Based on Vter, Vmax, and Vmin (Table 3), Vair, and LDR, we define the hydrometeors in the three temperature intervals below.

1. T < −20 °C

    - If Vter < Vmin(ice), then the hydrometeor is classified as snow.
    - If Vmin(ice) ≤ Vter < Vmax(ice), then the hydrometeor is classified as ice.
    - If Vmin(graupel) ≤ Vter < Vmax(graupel), then the hydrometeor is classified as graupel.
    - If Vter ≥ Vmax(graupel), then the hydrometeor is classified as hail.

2. T > 0 °C

    - If Vter < Vmax(cloud), then the hydrometeor is classified as a cloud droplet.
    - If Vmin(rain) ≤ Vter < Vmin(graupel), then the hydrometeor is classified as rain.
    - If Vmin(graupel) ≤ Vter < Vmax(rain) and LDR < 0.05, then the hydrometeor is classified as rain. Otherwise, it is classified as graupel.
    - If Vmax(rain) ≤ Vter < Vmax(graupel), then the hydrometeor is classified as graupel.
    - If Vter ≥ Vmax(graupel), then the hydrometeor is classified as hail.

3. −20 °C ≤ T ≤ 0 °C

    - If Vair ≤ −0.01 m/s and Vter < Vmax(cloud), then two types of hydrometeors are supposed to occur: cloud droplet and snow.
    - If Vair > −0.01 m/s and Vter < Vmax(cloud), then the hydrometeor is classified as snow even if Vter < Vmin(snow).
    - If Vmax(cloud) ≤ Vter < Vmax(snow), then the hydrometeor is classified as snow.
    - If Vmin(ice) ≤ Vter < Vmax(ice), then the hydrometeor is classified as ice.
    - If Vmin(graupel) ≤ Vter < Vmax(graupel), then the hydrometeor is classified as graupel.
    - If Vter ≥ Vmax(graupel), then the hydrometeor is classified as hail.

Figure 6 shows an example of the hydrometeor classification on 1 June 2018 for a gate situated 4059 m above the radar. It shows the power spectrum corrected by Vair (Section 2.5). The peaks of the power spectrum and corresponding ID are determined by the IDLsoft and the hydrometeor classification is performed by using our algorithm following the above given rules.

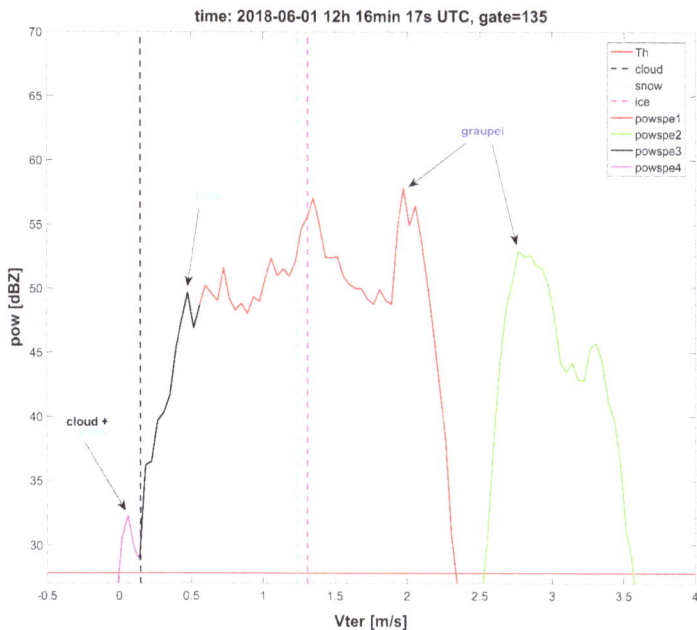

Figure 6. Hydrometeor classification on 1 June 2018 at 12 h 16 min and 17s UTC at the Milešovka Mt. for the gate 135, which corresponds to the height of 4059 m above the Milešovka Mt., gate temperature −9.5 °C, and Vair = −0.52 m/s. The solid lines (powspe1, ... , powspe4) represent the dependence of power spectrum on Vter corrected by Vair (Equation (2)). Single colors of the solid lines display ID determined by the IDLsoft. Vertical axis displays the power spectrum pow [dBZ] and the horizontal axis indicates Vter [m/s]. Vertical dashed lines depict maximum Vter corresponding to cloud, snow, and ice hydrometeors. The arrows indicate peaks where we determine hydrometeors. Th is the threshold, i.e., 0.1% of maximum power for the given power spectrum (Section 2.5).

3. Results

This section displays and comments the results of Vair that we computed during the studied thunderstorm on 1 June 2018 (Section 3.1). It also describes and evaluates the identified hydrometeor classes during the event (Section 3.2).

3.1. Vair during the Thunderstorm on 1 June 2018

Time development of Vair (calculated according to Section 2.5) during the thunderstorm on 1 June 2018 is displayed in Figure 7. Figure 7 also compares the calculated Vair to DVV during the thunderstorm. It is obvious from Figure 7 that the DVV is generally higher than Vair. The higher values of DVV are expected since DVV corresponds to the sum of Vair and terminal velocity (Vter) where Vter is always positive due to the downward orientation of the vertical velocity and the downward movement of hydrometeors.

We cannot explicitly verify the correctness of the derived values of Vair because we do not have any other available measurements to which we could compare our results. Therefore, we subjectively evaluated the results by using the general knowledge of the structure of vertical motion in storms. The obtained distribution of the derived vertical velocities mostly correspond to our expectations. Negative Vair is visible in the heights from 3 to almost 12 km prior to and during the observed maximum storm activity at the Milešovka Mt. (from 12:00 to 12:20 UTC). Regions with a significant negative Vair are obvious in heights from 6 to 8 km from 12:00 to 12:20 UTC, which corresponds to the layer of usually observed negative Vair minima in storms. Noticeable negative Vair values in the heights of around 10 km are less typical in storms even though they also appear in Figure 7.

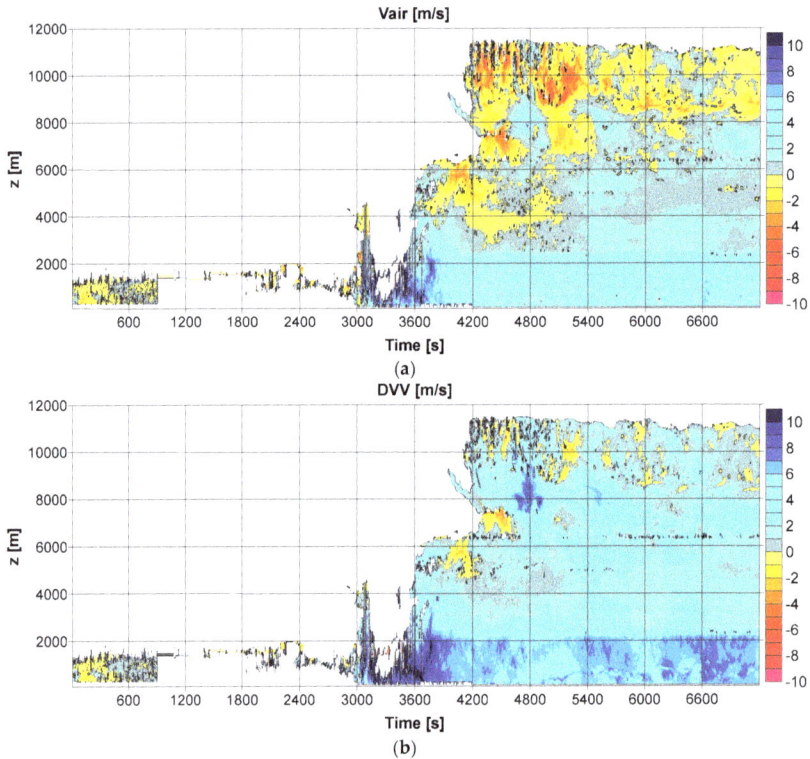

Figure 7. Time development of vertical velocity oriented downward on 1 June 2018 from 11:00 to 13:00 at the Milešovka observatory: (**a**) calculated Vair [m/s] and (**b**) calculated DVV [m/s] as a sum of Vair and mean Vter. Note that contours depict zero values and the white color depicts data where no target was detected and z [m] is the height in meters above the Milešovka Mt.

It should be noted that the minimum gate values of Vair were lower than −10 m/s in only several individual points in Figure 7a and, thus, they are not very noticeable in the Figure 7a. On the contrary, velocities lower than −5 m/s are clearly visible in quite large areas in Figure 7a. While at the beginning of the storm, i.e., around 12:00 UTC, negative values of Vair dominate the middle troposphere while after 12:30 UTC approximately Vair evinces mainly positive values in the low and middle troposphere. The positive values of Vair in the low and middle troposphere are typical for the mature and dissipation stage of storms.

One should be aware, while evaluating the values of Vair, that the vertically oriented cloud radar does not show any cross-section of the evolving storm, which the trajectory may significantly differ from a straight line. Thus, the results cannot be directly compared with drawings of conceptual models of storms. Moreover, the measured event consisted of several connected convective cores, which developed in time and moved in space (Figure 2). It means that, despite the uncertainty related to Vair, we do not have any means in our disposal to objectively verify the resulting Vair. However, we are convinced that the derived Vair has a physical justification and, therefore, can be used in further computations.

Concerning values of DVV given by the IDLsoft (Figure 4b), one can observe that our algorithm successfully applies the dealiasing algorithm, which removes the majority of unrealistic values of DVV that are most apparent before 12:00 UTC (i.e., 3600 s in Figure 7b). At that time, the IDLsoft shows a sharp difference of DVV from less than −10 m/s to more than +10 m/s (Figure 4b), which is very unlikely. Note that the DVV in Figure 7b differs from DVV in Figure 4b, produced by IDLsoft, by subtracting Vair, and the DVV in Figure 4b is of the opposite sign than DVV in Figure 7b due to the upward and downward orientation of vertical velocity, respectively.

3.2. Hydrometeors during the Thunderstorm on 1 June 2018

The classification of hydrometeors (Section 2.6) provides a distribution of hydrometeors in clouds. Figure 8 depicts the distribution of the six considered hydrometeors in the thundercloud during the thunderstorm on 1 June 2018 at the Milešovka Mt. Note that more than one hydrometeor can be detected at a point (as shown in Figure 6). The hydrometeors are allowed to overlap if they follow the conditions given in Section 2.6.

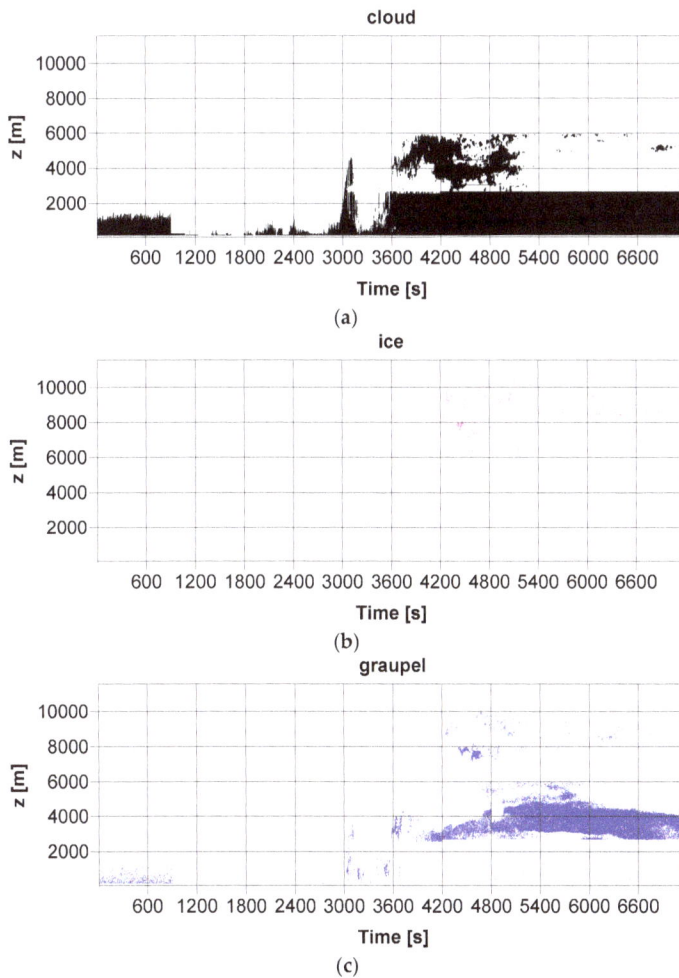

(a)

(b)

(c)

Figure 8. *Cont.*

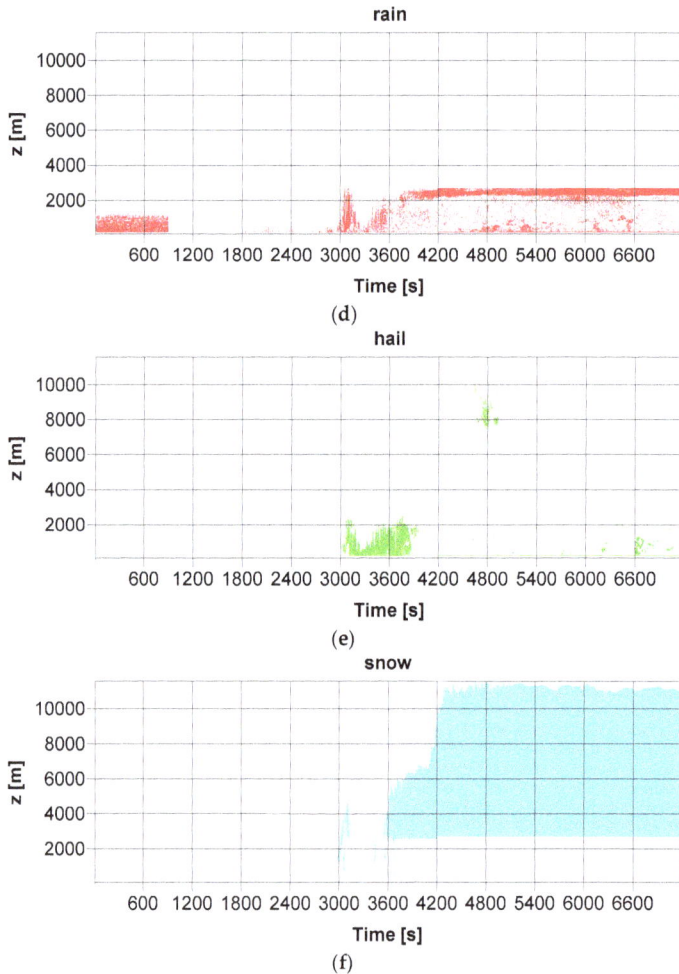

Figure 8. Distribution of hydrometeors using the classification given in Section 2.6 during the thunderstorm on 1 June 2018 at the Milešovka Mt.: (**a**) cloud droplets, (**b**) ice particles, (**c**) graupel, (**d**) rain, (**e**) hail, and (**f**) snow. The horizontal axis shows time in seconds from 11:00 to 13:00 UTC. Note that z [m] is the height in meters above the Milešovka Mt.

Figure 8 displays that precipitation in the form of rain, graupel, and hail are detected at the ground level by the algorithm. Figure 8a shows that cloud droplets are highly concentrated in lower layers up to approximately 2500 m above the Milešovka Mt., which is the altitude that roughly matches with the melting layer (i.e., bright band in Figure 4a, Table 2). The cloud droplets observed in higher altitudes correspond to the super-cooled droplets that can be observed above the melting layer in thunderclouds.

Contrary to cloud droplets, snow particles (Figure 8f) show the highest concentrations above the melting layer. Snow particles were the most dominant hydrometeor type during the study event on 1 June 2018. On the other hand, ice particles (Figure 8b) were identified the least often during the event by our algorithm due to the fact that the interval of terminal velocities is much smaller for ice than that for snow (Table 3). For our investigation, it is not crucial to distinguish between snow and ice.

Therefore, we consider a fusion of the two particles in one hydrometeor type in the future. However, we plan to test the current algorithm on more events first.

Graupel (Figure 8c) was mostly concentrated above the melting layer (2500–4500 m) while rain (Figure 8d) was mainly located in the melting layer or below. Our algorithm also identified hail in the layer of up to 2 km primarily around 12:00 UTC (Figure 8e). However, the occurrence of hail at approximately 12:00 UTC was not confirmed by the observer at the ground.

4. Discussion

We compared the results of the distribution of hydrometeors during the studied thunderstorm by our algorithm (Figure 8) with the results given by the provider of the cloud radar, i.e., the IDLsoft. The IDLsoft recognizes three types of hydrometeors based on Reference [29], which includes cloud, ice, and rain. Rain includes any kind of precipitation that has a significant fall velocity such as graupel or hail and can fall to the ground [29].

Figure 9 displays the distribution of the three hydrometeors identified by the IDLsoft during the thunderstorm on 1 June 2018. It shows that both ice and rain particles reached the ground level (i.e., precipitated), according to the IDLsoft. Moreover, the IDLsoft identified ice particles mainly in the melting layer and raindrops at the altitude of up to 12 km (Figure 9). As we mentioned above, in the IDLsoft rain is a precipitation that includes all particles (liquid and/or solid) with high terminal velocity while ice and cloud hydrometeors represent particles with small terminal velocities in the IDLsoft. However, a clear condition dividing particles between the cloud and the ice is not mentioned in Reference [29]. We expect that cloud particles represent solid particles in higher altitudes (i.e., snow and/or ice), which is similar to rain and includes solid particles in higher altitudes (graupel and/or hail). A personal communication with M. Bauer-Pfundstein pointed out that their algorithm might not be necessarily valid during thunderstorms.

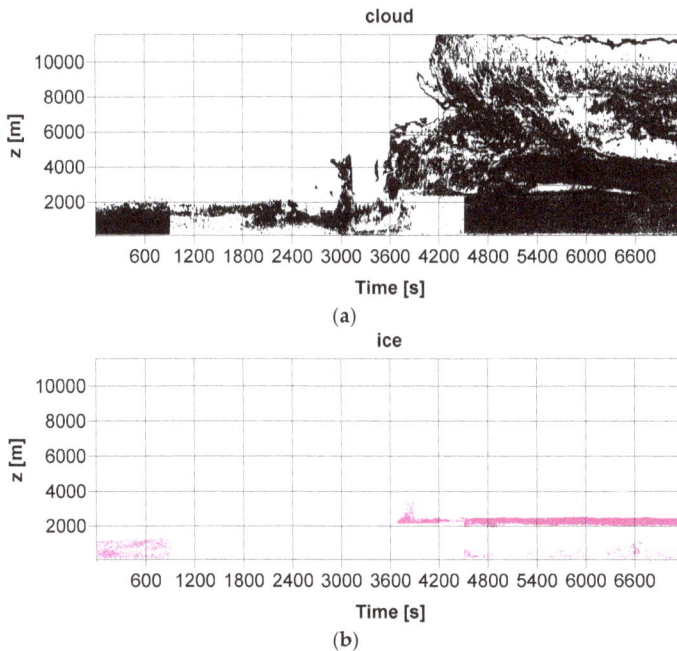

(a)

(b)

Figure 9. *Cont.*

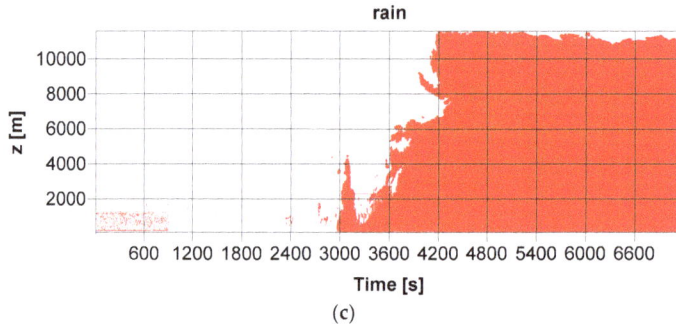

Figure 9. Distribution of classified hydrometeors by the IDLsoft during the thunderstorm on 1 June 2018 from 11:00 to 13:00 UTC at the Milešovka Mt.: (**a**) cloud droplets, (**b**) ice particles, and (**c**) rain. Note that z [m] is the height in meters above the Milešovka Mt.

As a result, we can assume that, although simple, our algorithm provides satisfying and plausible distribution of hydrometeors during thunderstorms (e.g., Figure 8) and the algorithm can be considered a suitable extension to the existing IDLsoft. However, further testing and verification of the algorithm are needed.

As far as LDR values are concerned, the use of it is rather marginal in our classification of hydrometeors. The analysis of the studied thunderstorm and of few clouds detected by the radar since June 2018 showed that the LDR values are usually available up to the height of the melting layer approximately. At higher altitudes, the LDR data are often unavailable due to strong attenuation of the signal. Therefore, we cannot base the hydrometeor classification on LDR values. In addition, for a vertically pointing radar, a slanted radar antenna (for instance 20°) might be more appropriate for the efficient use of LDR to classify the cloud particles. In any case, the use of LDR in the hydrometeor classification will be addressed in further research.

In the future, we will test the algorithm by using more thunderstorm events, which was not possible in this study due to an unusually dry and sunny summer in the Czech Republic (i.e., no similar thunderstorm has been detected at the station since the analyzed 1 June 2018). We will compare the results using six hydrometeor types to that using five hydrometeor types (ice and snow as one hydrometeor type). Moreover, we plan to compare the derived variables from the cloud radar with that of other instruments, e.g., disdrometer and ceilometer, situated at the Milešovka observatory.

5. Conclusions

The study presented two new functionalities that complement the provided software of a vertically pointing Ka-band cloud radar, which was located at the Milešovka observatory in Central Europe in 2018. The radar has been installed in order to study the cloud structure including thunderclouds. We improved the dealiasing algorithm and we applied a method for computing the vertical air velocity and the terminal velocity of hydrometeors by using the Doppler spectra. We developed an algorithm that enables one to classify the hydrometeors that can lead to precipitation. We illustrated the algorithms with a thunderstorm that crossed the Milešovka observatory on 1 June 2018 and was associated with significant lightning activity.

The method retrieving vertical air velocity, which is a variable needed for our classification of hydrometeors, was subjectively evaluated because we have no means to perform any objective verification. In our opinion, the obtained distribution of vertical air velocity in time and height is in good agreement with the structure of air velocity expected in storms and, therefore, we used the retrieved vertical air velocity in the algorithm classifying hydrometeors.

Remote Sens. **2018**, *10*, 1674

The algorithm classifying hydrometeors uses the information on vertical air velocity, temperature from sounding measurements from the nearest aerological station, terminal velocity, and LDR in non-precipitating/precipitating clouds. The resulting distribution of six considered hydrometeors (cloud droplets, ice and snow particles, rain, graupel, and hail) seems realistic for the thunderstorm on 1 June 2018 and more plausible than that obtained from the provided radar products for the thunderstorm. Nevertheless, the results pointed out that the method hardly distinguishes snow from ice and vice versa.

Author Contributions: Z.S. conceived the paper, conducted most of the analyses including the presented algorithms, and partly wrote the manuscript. J.M. conducted the literature review, tested the algorithms, processed results graphically, and wrote the majority of the manuscript. P.N. prepared and processed C-band weather radar data from the Czech Hydrometeorological Institute.

Funding: This research was funded by project CRREAT (reg. number: CZ.02.1.01/0.0/0.0/15_003/0000481) call number 02_15_003 of the Operational Programme Research, Development, and Education.

Acknowledgments: We owe thanks to Petr Pešice for his help in collecting and administrating data from the Milešovka observatory. We are also thankful to M.Phil. Syed Muntazir Abbas for his language corrections and to the three anonymous reviewers for their constructive and in-depth reviews.

Conflicts of Interest: The authors declare no conflict of interest. The funders had no role in the design of the study, in the collection, analyses, or interpretation of data, in the writing of the manuscript, or in the decision to publish the results.

References

1. Görsdorf, U.; Lehmann, V.; Bauer-Pfundstein, M.; Peters, G.; Vavriv, D.; Vinogradov, V.; Volkov, V. A 35-GHz Polarimetric Doppler Radar for Long-Term Observations of Cloud Parameters—Description of System and Data Processing. *J. Atmos. Ocean. Technol.* **2015**, *32*, 675–690. [CrossRef]

2. Kollias, P.; Clothiaux, E.E.; Miller, M.A.; Albrecht, B.A.; Stephens, G.L.; Ackerman, T.P. Millimeter-Wavelength Radars: New Frontier in Atmospheric Cloud and Precipitation Research. *Bull. Am. Meteorol. Soc.* **2007**, *88*, 1608–1624. [CrossRef]

3. Clothiaux, E.E.; Miller, M.A.; Albrecht, B.A.; Ackerman, T.P.; Verlinde, J.; Babb, D.M.; Peters, R.M.; Syrett, W.J. An Evaluation of a 94-GHz Radar for Remote Sensing of Cloud Properties. *J. Atmos. Ocean. Technol.* **1995**, *12*, 201–229. [CrossRef]

4. Rogers, R.R. An extension of the Z-R relation for Doppler radar. In Proceedings of the 11th Weather Radar Conference, Boulder, CO, USA, 14–18 September 1964; pp. 158–161.

5. Hauser, D.; Amayenc, P. A New Method for Deducing Hydrometeor-Size Distributions and Vertical Air Motions from Doppler Radar Measurements at Vertical Incidence. *J. Appl. Meteorol.* **1981**, *20*, 547–555. [CrossRef]

6. Zheng, J.; Liu, L.; Zhu, K.; Wu, J.; Wang, B. A Method for Retrieving Vertical Air Velocities in Convective Clouds over the Tibetan Plateau from TIPEX-III Cloud Radar Doppler Spectra. *Remote Sens.* **2017**, *9*, 964. [CrossRef]

7. Kollias, P. Cloud radar observations of vertical drafts and microphysics in convective rain. *J. Geophys. Res. Atmos.* **2003**, *108*. [CrossRef]

8. Lhermitte, R.M. Observation of rain at vertical incidence with a 94 GHz Doppler radar: An insight on Mie scattering. *Geophys. Res. Lett.* **1988**, *15*, 1125–1128. [CrossRef]

9. Gossard, E.E. Measurement of Cloud Droplet Size Spectra by Doppler Radar. *J. Atmos. Ocean. Technol.* **1994**, *11*, 712–726. [CrossRef]

10. Shupe, M.D.; Kollias, P.; Matrosov, S.Y.; Schneider, T.L. Deriving Mixed-Phase Cloud Properties from Doppler Radar Spectra. *J. Atmos. Ocean. Technol.* **2004**, *21*, 660–670. [CrossRef]

11. Shupe, M.D.; Kollias, P.; Poellot, M.; Eloranta, E. On Deriving Vertical Air Motions from Cloud Radar Doppler Spectra. *J. Atmos. Ocean. Technol.* **2008**, *25*, 547–557. [CrossRef]

12. Luke, E.P.; Kollias, P. Separating Cloud and Drizzle Radar Moments during Precipitation Onset Using Doppler Spectra. *J. Atmos. Ocean. Technol.* **2013**, *30*, 1656–1671. [CrossRef]

13. Kollias, P.; Luke, E.P. *A High Resolution Hydrometer Phase Classifier Based on Analysis of Cloud Radar Doppler Spectra*; Brookhaven National Laboratory: Washington, DC, USA, 2007.

14. Matrosov, S.; Schmitt, C.; Maahn, M.; de Boer, G. In Situ Validation of Cloud Radar-based Retrievals of Ice Hydrometeor Shapes. 2016; p. 11. Available online: https://asr.science.energy.gov/meetings/stm/2018/presentations/616.pdf (accessed on 23 October 2018).

15. Ge, J.; Zhu, Z.; Zheng, C.; Xie, H.; Zhou, T.; Huang, J.; Fu, Q. An improved hydrometeor detection method for millimeter-wavelength cloud radar. *Atmos. Chem. Phys.* **2017**, *17*, 9035–9047. [CrossRef]

16. Bringi, V.N.; Vivekanandan, J.; Tuttle, J.D. Multiparameter Radar Measurements in Colorado Convective Storms. Part II: Hail Detection Studies. *J. Atmos. Sci.* **1986**, *43*, 2564–2577. [CrossRef]

17. Hall, M.P.M.; Goddard, J.W.F.; Cherry, S.M. Identification of hydrometeors and other targets by dual-polarization radar. *Radio Sci.* **1984**, *19*, 132–140. [CrossRef]

18. Aydin, K.; Zhao, Y.; Seliga, T.A. A Differential Reflectivity Radar Hall Measurement Technique: Observations during the Denver Hailstorm of 13 June 1984. *J. Atmos. Ocean. Technol.* **1990**, *7*, 104–113. [CrossRef]

19. Tong, H.; Chandrasekar, V.; Knupp, K.R.; Stalker, J. Multiparameter Radar Observations of Time Evolution of Convective Storms: Evaluation of Water Budgets and Latent Heating Rates. *J. Atmos. Ocean. Technol.* **1998**, *15*, 13. [CrossRef]

20. Straka, J.M.; Dusan, S.Z. Algorithm to deduce hydrometeor types and contents from multi-parameter radar data. In Proceedings of the 26th International Conference on Radar Meteorology, Norman, OK, USA, 24–28 May 1993; pp. 513–515.

21. Höller, H. Radar-Derived Mass-Concentrations of Hydrometeors for Cloud Model Retrievals. In Proceedings of the 27th International Conference on Radar Meteorology, Vail, CO, USA, 9–13 October 1995; pp. 453–454.

22. Liu, H.; Chandrasekar, V. Classification of Hydrometeors Based on Polarimetric Radar Measurements: Development of Fuzzy Logic and Neuro-Fuzzy Systems, and In Situ Verification. *J. Atmos. Ocean. Technol.* **2000**, *17*, 140–164. [CrossRef]

23. Frisch, S.; Shupe, M.; Djalalova, I.; Feingold, G.; Poellot, M. The Retrieval of Stratus Cloud Droplet Effective Radius with Cloud Radars. *J. Atmos. Ocean. Technol.* **2002**, *19*, 8. [CrossRef]

24. Melchionna, S.; Bauer, M.; Peters, G. A new algorithm for the extraction of cloud parameters using multipeak analysis of cloud radar data First application and preliminary results. *Meteorol. Z.* **2008**, *17*, 613–620. [CrossRef] [PubMed]

25. Zhao, C.; Xie, S.; Klein, S.A.; Protat, A.; Shupe, M.D.; McFarlane, S.A.; Comstock, J.M.; Delanoë, J.; Deng, M.; Dunn, M.; et al. Toward understanding of differences in current cloud retrievals of ARM ground-based measurements. *J. Geophys. Res. Atmos.* **2012**, *117*. [CrossRef]

26. Comstock, J.M.; d'Entremont, R.; DeSlover, D.; Mace, G.G.; Matrosov, S.Y.; McFarlane, S.A.; Minnis, P.; Mitchell, D.; Sassen, K.; Shupe, M.D.; et al. An Intercomparison of Microphysical Retrieval Algorithms for Upper-Tropospheric Ice Clouds. *Bull. Am. Meteorol. Soc.* **2007**, *88*, 191–204. [CrossRef]

27. Austin, R.T.; Stephens, G.L. Retrieval of stratus cloud microphysical parameters using millimeter-wave radar and visible optical depth in preparation for CloudSat: 1. Algorithm formulation. *J. Geophys. Res. Atmos.* **2001**, *106*, 28233–28242. [CrossRef]

28. Marchand, R.; Mace, G.G.; Ackerman, T.; Stephens, G. Hydrometeor Detection Using *Cloudsat* —An Earth-Orbiting 94-GHz Cloud Radar. *J. Atmos. Ocean. Technol.* **2008**, *25*, 519–533. [CrossRef]

29. Bauer-Pfundstein, M.; Görsdorf, U. Target separation and classification using cloud radar Doppler-spectra. In Proceedings of the 33rd International Conference on Radar Meteorology, Cairns, Australia, 6–10 August 2007.

30. Hildebrand, P.H.; Sekhon, R.S. Objective Determination of the Noise Level in Doppler Spectra. *J. Appl. Meteorol.* **1974**, *13*, 808–811. [CrossRef]

31. Kollias, P.; Albrecht, B.A.; Lhermitte, R.; Savtchenko, A. Radar Observations of Updrafts, Downdrafts, and Turbulence in Fair-Weather Cumuli. *J. Atmos. Sci.* **2001**, *58*, 1750–1766. [CrossRef]

32. Sokol, Z.; Zacharov, P.; Skripniková, K. Simulation of the storm on 15 August, 2010, using a high resolution COSMO NWP model. *Atmos. Res.* **2014**, *137*, 100–111. [CrossRef]

remote sensing

MDPI

Article

Decorrelation of Satellite Precipitation Estimates in Space and Time

Francisco J. Tapiador [1,*], Cecilia Marcos [2], Andres Navarro [1], Alfonso Jiménez-Alcázar [1], Raul Moreno Galdón [1] and Julia Sanz [3]

[1] University of Castilla-La Mancha, Earth and Space Sciences Group (ESS), Institute of Environmental Sciences (ICAM), 45071 Toledo, Spain; Andres.Navarro@uclm.es (A.N.); Alfonso.JAlcazar@uclm.es (A.J.-A.); raulmorenogaldon@gmail.com (R.M.G.)

[2] National Meteorology Agency (AEMET), 28071 Madrid, Spain; cmarcosm@aemet.es

[3] Laboratory of Remote Sensing (LATUV), University of Valladolid, 47071 Valladolid, Spain; julia@latuv.uva.es

* Correspondence: Francisco.Tapiador@uclm.es; Tel.: +34-925-268-800 (ext. 5762)

Received: 25 April 2018; Accepted: 10 May 2018; Published: 14 May 2018

Abstract: Precise estimates of precipitation are required for many environmental tasks, including water resources management, improvement of numerical model outputs, nowcasting and evaluation of anthropogenic impacts on global climate. Nonetheless, the availability of such estimates is hindered by technical limitations. Rain gauge and ground radar measurements are limited to land, and the retrieval of quantitative precipitation estimates from satellite has several problems including the indirectness of infrared-based geostationary estimates, and the low orbit of those microwave instruments capable of providing a more precise measurement but suffering from poor temporal sampling. To overcome such problems, data fusion methods have been devised to take advantage of synergisms between available data, but these methods also present issues and limitations. Future improvements in satellite technology are likely to follow two strategies. One is to develop geostationary millimeter-submillimeter wave soundings, and the other is to deploy a constellation of improved polar microwave sensors. Here, we compare both strategies using a simulated precipitation field. Our results show that spatial correlation and RMSE would be little affected at the monthly scale in the constellation, but that the precise location of the maximum of precipitation could be compromised; depending on the application, this may be an issue.

Keywords: precipitation; geostationary microwave sensors; polar systems

1. Introduction

The importance of precise estimation of precipitation is apparent for assessing water availability for ecosystems and agriculture, and for other human activities. The usefulness of quality precipitation estimates is also evident for nowcasting and for data assimilation into numerical models. Thus, simulations have demonstrated that the assimilation of precipitation data leads to improved forecasting of a tropical cyclone in terms of its intensity and kinematical and precipitation structures [1,2]. The products after assimilating results in, for instance, significantly improved cyclone prediction, reflecting mostly in the cyclone's track, the associated frontal structure and the associated precipitation along the front [3]. Rainfall monitoring is also important to assess possible anthropogenic impacts on global climate [4–6], to monitor hydrometeorological natural disasters, such as flood and flash flood events [7–11], and to improve precipitation estimates in Earth System Models (ESMs) [12–14].

Satellites are the only means to provide homogeneous global estimates of precipitation. Gauges are limited to land areas, leaving oceans with little or no direct measurements; the same applies to ground radars. On the contrary, satellites cover the whole planet and have the potential to provide frequent estimates, not only of surface hydrometeors, but also of 3D precipitation profiles [15,16]. Unfortunately, measuring precipitation from space is a difficult task [17–19].

Satellite remote sensing of precipitation has evolved from the use of visible and/or infrared (IR) algorithms [20], to more direct strategies using passive microwave (PMW) radiometry [21] and orbiting radars, such as the Global Precipitation Measurement (GPM) mission [22] core observatory (GPM-Core), which is unique. PMW sensors measure the natural electromagnetic Earth emissions at microwave wavelengths, which are affected by rain drops in several ways. This allows for a more direct estimate of precipitation from space using radiative transfer modelling [23] (see References [24,25] for an update). Nonetheless, PMW sensors have poor temporal and spatial resolution due to their low orbits and the antenna diffraction limit at microwave wavelengths [26]. Infrared geostationary satellites, on the other hand, provide an indirect measure of the rainfall by establishing a relationship between cloud top temperature and surface precipitation [27], and have good temporal sampling and spatial resolution comparable with ground radars. The use of data fusion methods in rainfall estimation permits merging both datasets, aiming to create a high spatial and temporal resolution product [28].

Orbital radars are still scarce, and in spite of the effort in developing data fusion methods for IR and PMW sensors, the problem of precise remote sensing of precipitation is far to be solved in the near future. Routine comparisons between merged algorithms show large differences in performances between current methods depending on algorithm, season and location [29]. As space-borne sensors are the only means to homogeneously monitor land and ocean precipitation, the problem of reliably estimating global precipitation at appropriate spatial and temporal resolutions remains unsolved. Moreover, as precipitation estimates require a model error to be used when assimilated into NWP systems, algorithms and methods need to be physically based to be capable of estimating the covariance.

To further improve precipitation estimates from satellite, two major research directions are being followed. On the one hand, the GPM mission has increased the temporal resolution of the global estimates of precipitation by increasing the number of polar-orbiting microwave sensors, putting together a constellation of low-orbit satellites that reduce revisiting period. The contribution of all these satellites can generate improved (MW-based) and more frequent (more satellites involved) global precipitation estimates [30].

Another approach is the development of geostationary microwave sensors. While microwave antennas in the 6–90 GHz range would require antennas as large as 70 m for 10-km spatial resolution at 19 GHz [31], the exploitation of millimeter and submillimeter wavelengths would allow smaller (3-m) antennas, which are an affordable alternative for current engineering limits. Thus, a Geostationary Microwave Observatory (GEM) was proposed in 1998 [32] with a 2-m antenna yielding 15-km spatial resolution at nadir. In Europe, the Geostationary Observatory for Microwave Atmospheric Sounding (GOMAS) proposed a 3-m antenna aiming at 10-km spatial resolution [33]. Since then, other projects have been proposed and those are currently at different levels of maturity.

The rationale of millimeter and submillimeter estimation of precipitation differs from MW or PMW estimation. MW instruments such as the Precipitation Radar (PR) in TRMM measure the backscattered signal of a radar pulse, while PMW ones rely on the emission signature of cold hydrometeors over a warmer background (over the oceans), and on a variable relationship between the natural Earth PMW emission intercepted by hydrometeors and their emission signature over land. On the other hand, Geostationary Meteorological Satellite (GMS) estimate precipitation using absorption bands rather than windows to measure precipitation [34], which is an idea to measure how the atmospheric profiles are affected by the presence of hydrometeors. Preliminary studies show promising performances [31] in terms of not only precipitation identification but also raincell dynamics.

The aim of this paper is to analyze the theoretical differences in the spatial structure of precipitation between an ideal GMS and other alternatives such as a constellation of low-orbit PWM sensors or merged multi-satellite products. We follow a top-down approach by assuming an error-free GMS capable of providing precise measurements of precipitation. By degrading both the spatial resolution (to match the characteristics of hypothetical sensors spanning up to 250 km), and the revisiting periods (up to 6.0 h), we can compare the loss of performances with a perfect GMS estimate. As there is not an actual GMS sensor to compare with, we have used a simulation to generate a realistic rainfall field at 15 min/0.05 degrees (about 5 km) resolution. The advantage over a pure synthetic stochastic precipitation field is that the simulation using an observed cloud cover can account for rare events affecting precipitation such as landing hurricanes or land-surfaces processes that would be too complex or cumbersome to be used in a pure stochastic model using, for instance, Poisson statistics.

2. Data

We have selected the precipitation in October 2005 in Spain as our empirical basis (Figure 1). Climatologically, October is the rainiest month in the country, so a statistically-significant number of precipitation events both from Atlantics fronts and from Mediterranean convective systems can be expected. October 2005 is also interesting because of two major meteorological events: the 11th of October hurricane (Vince) that landed in Southwestern Spain. This was the first time ever a hurricane headed towards the Iberian Peninsula and landed in mainland Spain [35]. Cordoba airport (37.85N, −4.85W, 170 km inland) received an unusual 84 mm of rain in 4 h, with a maximum intensity of 88.8 mm in 10 min. The short period in which the hurricane landfall generated high rainfall rates makes this event an anomaly in the normal climatology of the area.

Monthly average Precipitation in Spain by Basin (2005)

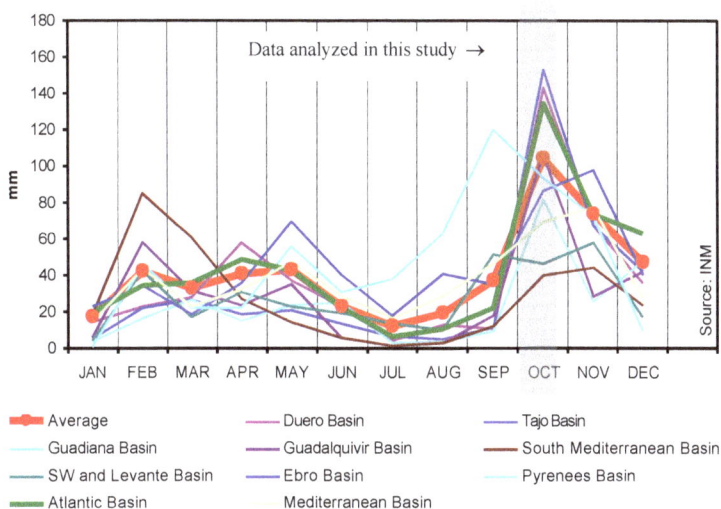

Figure 1. Rain gauge-derived monthly average precipitation in Spain during 2005 at basin level, indicating the data used for the present study. Data from the Spanish Meteorology Agency (AEMET) (formerly National Institute of Meteorology, INM).

Another event of interest is the major floods in Northeastern Spain (11–13 October) after the driest hydrological year on record. The driest year in Spain since 1947 was 2005, so monitoring high precipitation rates after such an event is relevant to erosion, urban drainage and agricultural analyses. Therefore, October 2005 presents a suitable benchmark for our study, as we have enough rain events in our comparison to be meaningful over the semi-arid environment of the Iberian Peninsula, and also we have the contribution of hurricane Vince and convective cells in the northeast. Satellite monitoring of the high precipitation rates associated with Vince require high temporal sampling, whereas convective cells need a high spatial resolution.

3. Methods

Precipitation rates in the Auto-Estimator [27] were based on the cloud top temperature using the following empirical relationship:

$$R = 1.1183 \cdot 10^{11} \cdot \exp(-3.6382 \cdot 10^{-2} \cdot T^{1.2})$$

where R is the rainfall rate in mm h^{-1} and T is the cloud top brightness temperature in Kelvin (K). The algorithm was calibrated for radar rainfall estimates from the US operational network of 5 and 10 cm radar (WSR-57S, WSR-74C, WSR-88D), indicated to provide rainfall estimates for fast-moving deep convective systems during summertime.

To analyze the spatial variability of the estimates we calculated a semivariogram [36]. For each precipitation estimate r_i, $i = 1, \ldots, M$ located at a d distance from the others r_j, $j = 1, \ldots, M$ the empirical semivariogram is given by:

$$\hat{\gamma}(d) \equiv \frac{1}{2N(d)} \sum_{(i,j) \in N(d)} |r_i - r_j|^2$$

where $N(d)$ denotes the set of estimates (i,j) located at d distance in every direction (omnidirectional semivariogram). The semivariogram provides an estimate of the spatial variance, thus characterizing the spatial variability of the precipitation.

Standard statistics, such as Pearson r^2, Root Mean Squared Error (RMSE) and bias, were used to compare many realizations. In addition, information entropy [37] was used to account for the informational content of the estimates. Entropy is defined, in this context, as:

$$S \equiv -\sum_i p(R_i = r) \log[p(R_i = r)]$$

where $p(R_i = r)$ indicates the probability of rainfall rate R_i being r.

4. Results and Discussion

To build our simulated precipitation field, we used the Auto-Estimator. As an IR source, we used Meteosat-8 (formerly Meteosat Second Generation) data from the EUMETSAT archive. In spite of the limitations of using an IR-based method, the Auto-Estimator is well suited to generating our simulated precipitation, as it can provide pixel-based estimates at a high temporal sampling using only geostationary imagery, while other more powerful methods are less suited to this purpose and require additional data or ancillary information. It is worth mentioning that further enhancements of this techniques give way to a new product, called the Hydro-estimator. Such improvements include cloud-top geometry, available atmospheric moisture, stability parameters, radar, and local topography.

Several strategies have been devised to merge IR and PMW data, including neural networks [28,38,39], histogram matching [40] and multivariate probability matching techniques [41]. The aim of these methods is to reduce the temporal gap between rainfall estimates without sacrificing the quality of the more direct PMW estimate. Other methods to merge IR and PMW data include advection techniques such as morphing techniques that advect PMW estimates through the IR, as in Joyce et al. [42]. Their CMORPH method uses a correlation window algorithm to find the IR trajectories on an almost global scale, then advecting PMW estimates on those trajectories to fill the gaps between PMW successive overpasses. It has been shown that this morphing procedure can outperform other methods [43], though seasonal and spatial variations exist.

A comparison between the Auto-Estimator monthly estimates and morphing techniques such as the CMORPH [42] and the UCLM algorithm [44] shows an overall agreement (Figure 2) for the purposes of the research in this paper.

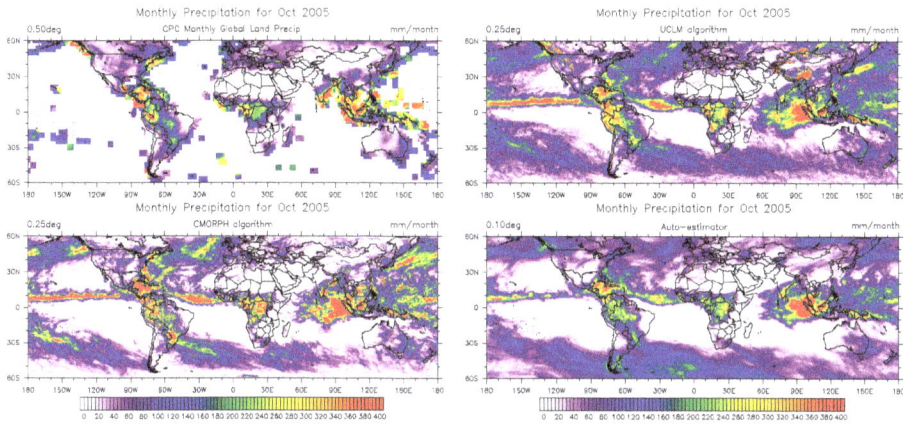

Figure 2. October 2005 precipitation estimates using a morphed microwave algorithm (top right, UCLM algorithm) and infrared-calibrated algorithms (bottom, CMORPH and AUTO-ESTIMATOR algorithms) compared to Climate Prediction Center (CPC) global land precipitation field (top left).

Daily comparisons (Figure 3) also show that the Auto-Estimator gives similar rainfall fields to those of other more complex algorithms in terms of detail and structure. Validation against the land-only, gauge-based Global Precipitation Climatology Project (GPCP) [45,46] (Figures 2 and 3, upper-left panel) also suggest that the Auto-Estimator is well-suited for use as a surrogate of a realistic global precipitation field.

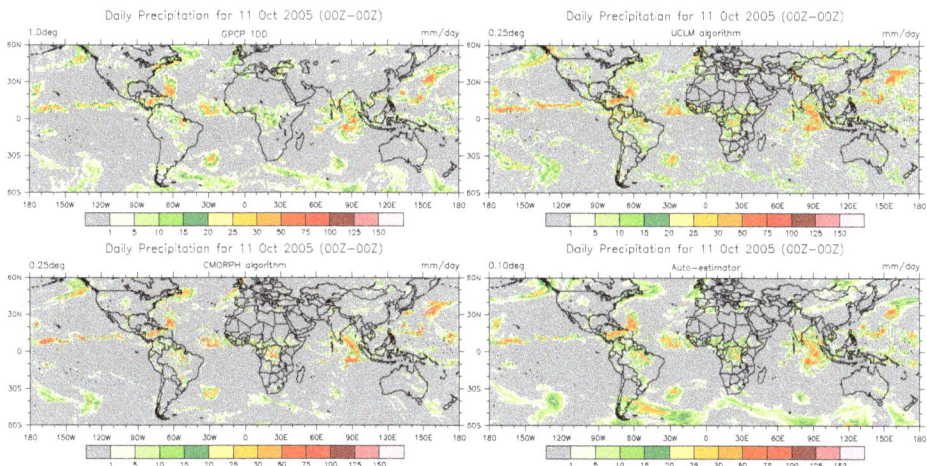

Figure 3. Comparison of daily precipitation estimates for 11 October 2005: GPCP 1DD gauge/multi satellite product and UCLM, CMOPH, and AUTO-ESTIMATOR algorithms.

Indeed, we are not claiming that the Auto-Estimator is a suitable instantaneous precipitation algorithm for the Iberian Peninsula. Our interest is not to describe the actual precipitation field, but to have a realistic precipitation field. The dependence of our results on the actual performance of the Auto-Estimator when compared with gauge data is of a second-order error. The reason for this is that we use the highest spatial and temporal resolution to build our simulated field.

Thus, the analysis presented here is independent of the precipitation algorithm used, providing that the algorithm can generate estimates at suitable spatial and temporal resolution, that is, at ~5 km/15 min. The assumption made here is that those estimates are the real precipitation and that the GMS system is capable of measuring precipitation with no error. Aggregated/subsampled estimates are compared in relative not absolute terms so the effects of varying resolutions can be investigated.

Simulated estimates were generated at 0.05°, with a 15-min resolution for October 2005, resulting in 2976 samples for analyses. These are taken both as real precipitation and as the GMS simulated estimates; that is, the estimates that a perfect GMS instrument would retrieve if on orbit and using a perfect precipitation retrieval algorithm.

Simplifying assumptions for polar instruments are in the form of perfect geometry, a wide swath to cover the Iberian Peninsula, no parallax error or instrumental biases, perfect retrieval and negligible beam-filling effects. All these simplifications can only benefit alternatives to the GMS estimate, as any polar system would suffer from these problems. Therefore, our results are to be considered as baseline estimates, meaning that the loss of performances we observe are the lowest limit over a perfect retrieval. The higher limit is the sum of all possible sources of error as mentioned above (imperfect geometry, narrow swath, etc.) up to the null hypothesis of the satellite retrieval perfectly matching actual precipitation.

To build our simulated estimates at several spatial and temporal resolutions we sampled the reference 0.05° rainrate field at 15-min intervals. The idea was to generate time-degraded estimates. These corresponded with a sensor having the same spatial resolution of the hypothetical GMS, but operating at different temporal resolution, therefore missing a variable number of continuous GMS estimates. Similarly, for the spatial resolution, we upscaled the 0.05° GMS field in 0.15° intervals up to 1.55° to simulate different satellite spatial resolutions. The resulting coarser resolution fields were then processed as before, assuming different temporal resolutions from 15 min to 6 h. This generated a host of different estimates corresponding to many possible satellite sampling/resolution characteristics. A sample of the results is plotted in Figure 4.

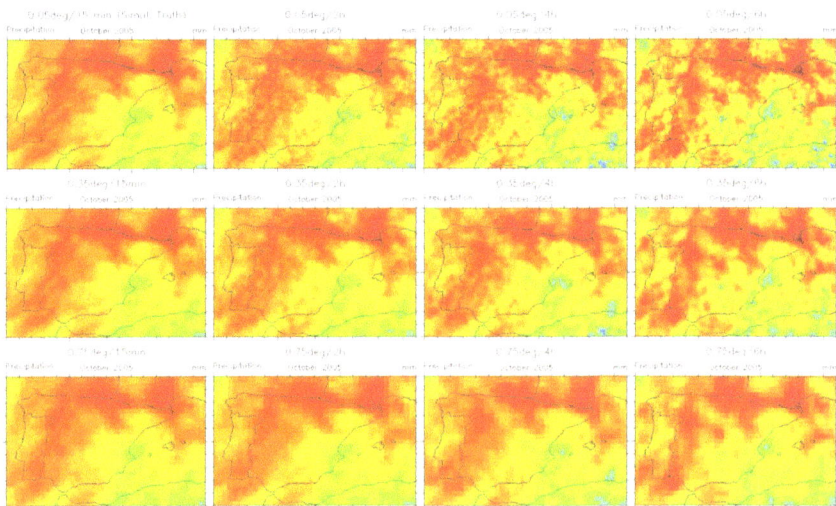

Figure 4. *Cont.*

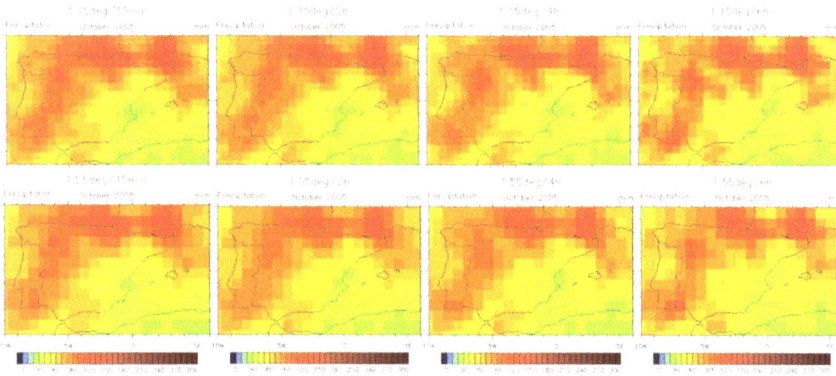

Figure 4. An example of monthly-aggregated precipitation simulations at several spatial resolution and time sampling for the October 2005 case study. The 0.05°/15-min estimate is the reference 'truth'.

By hypothesis, a perfect GMS instrument measuring with no retrieval error would capture all the 2976 samples of the month at the original resolution yielding the totals depicted in the top/left plot of Figure 4. It is worth noting that the rainrate field was used instead of the field of radiances because it is assumed that the algorithm to derive the rainrates from radiances is perfect at each scale and resolution. That allows to isolate the effects of the changes in spatial and temporal resolution in the retrievals and therefore provides the best-case scenario in the event of degrading both variables.

As temporal resolution degrades, more and more samples are missed so the monthly precipitation estimate would deviate from the nominal truth as many relevant but short-living events will not contribute to the final sum. The spatial coarsening also deteriorates the perfect estimate, as small high precipitation events are blurred into the grid mean. It is clear from the figure that degrading the temporal sampling produces a patchy field, while the effect of spatial coarsening is smoothing the field.

Figure 5 explores all the combination between the estimates in terms of correlation, RMSE, bias and entropy. The correlation plot shows the importance of the spatial sampling to maintain the GMS performances. The correlation degrades faster in the spatial resolution direction, going below 0.90 r^2 at half a degree (for a 15-min temporal sampling), which is a working polar MW spatial resolution. In contrast, a revisiting period of 3 h decreased the correlation to 0.87 r^2. In environmental applications where precipitation estimates are integrated into models, multiplicative errors would noticeably worsen the final output in the latest case. In term of correlation, worsening the temporal resolution does not affect that much the estimates: Correlation isolines appear almost parallel to the *x*-axis.

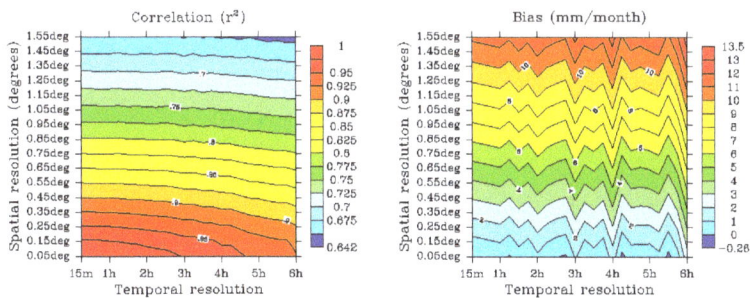

Figure 5. *Cont.*

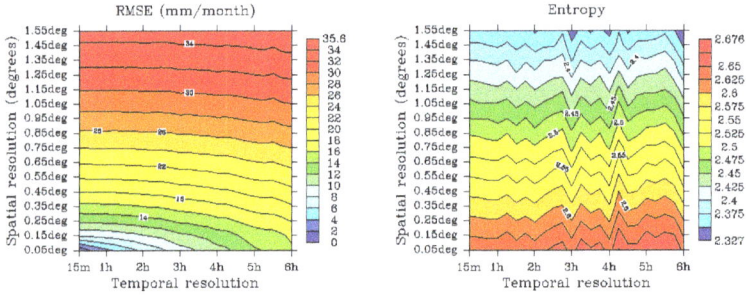

Figure 5. Correlation (r^2), RMSE (mm/month), bias (mm/month) and entropy (arbitrary units) departures from the perfect geostationary microwave simulation at different spatial and temporal resolution.

The RMSE evolution parallels that of the correlation. The bias, however, shows the effects of the spatial aggregation using the pixel average. Negative biases are negligible, and correspond with temporal sampling at original spatial resolution, and the same applies to bias and RMSE. This smoothing in spatial resolution cannot compensate for the missing samples: the differences between the 0.05°/6.0 h estimate and the reference estimate are minute, whereas the 1.55°/6.0 h estimate cannot capture the precipitation features in the reference 0.05°/15 min estimate.

Perhaps more importantly for environmental applications, the entropy plot depicts the expected loose in informational content. The Boltzmann-Gibbs-Shannon entropy is a good estimate of how peaked the histogram of the precipitation estimates is [47]. Flat histograms correspond with a maximum entropy state, giving us little information on the structure of the precipitation, whereas a peaked histogram indicates null uncertainty (zero entropy) on that variable. Therefore, increased entropy would mean a smoother distribution, and less entropy that the reference state would indicate a change to a peaked histogram. Here, Figure 5 shows that temporal resolution is less important for conserving the informational content. Thus, a 0.05°/6 h estimate conserves most of the entropy of the original estimate. If the estimates are to be used to characterize the statistical properties of the precipitation over a region, isentropes mark how to preserve the informational content of the estimates when both spatial and temporal resolution are changed.

To better appreciate the variation in the spatial structure of the precipitation, it is necessary to use a spatial statistic measure such as the variogram. Figure 6 depicts the variograms for the 0.05° and 1.55° spatial resolution. This measurement provides complementary information on the statistical properties of the precipitation, showing the extent at which temporal sampling increases the variance of the field, as some short-living events are missed by the sensor. In both panels, temporal resolution decreases, as does the semivariance.

A practical difference between a GMS and other alternatives can be seen in Figure 7, which gathers the errors committed in the estimation of the maximum precipitation rates location for several spatial/temporal samplings combinations. Contrary to Figures 5 and 6, no clear pattern emerges. While it is true that high spatial resolution avoids large errors, it is also true that the 5 h 30 min sampling seems to work up to a 1.45° spatial resolution. This has implications for natural hazards monitoring, early warning services and ecological models using remotely-sensed data.

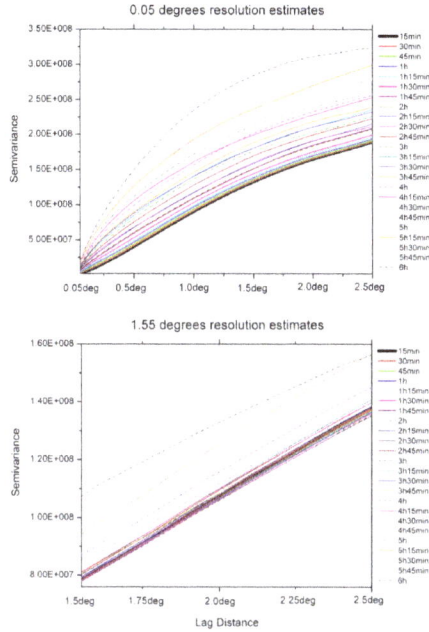

Figure 6. Semivariograms for 0.05° and 1.55° spatial resolution simulations, covering all the temporal samplings.

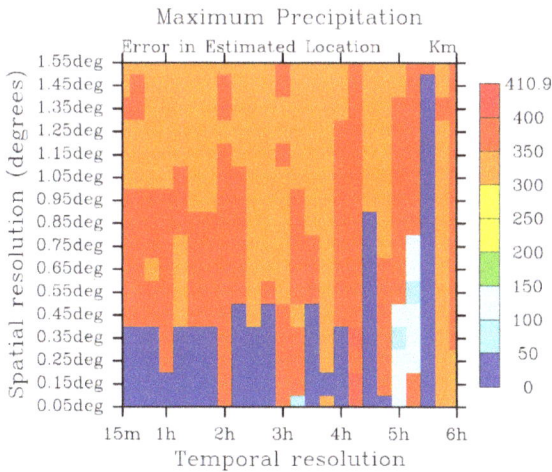

Figure 7. Errors in the estimated location of the maximum precipitation at the spatial and temporal resolutions explored in the paper.

Figure 8 complements Figure 7 in a more visual way. The figure depicts the errors in the location of the maximum monthly precipitation as temporal resolution degrades. The nominal truth locates the maxima over mainland Spain, in Catalonia. A 3-h sampling places the maximum over the Gulf of Leon, and a 6-h sampling over mainland France; the errors are noticeable.

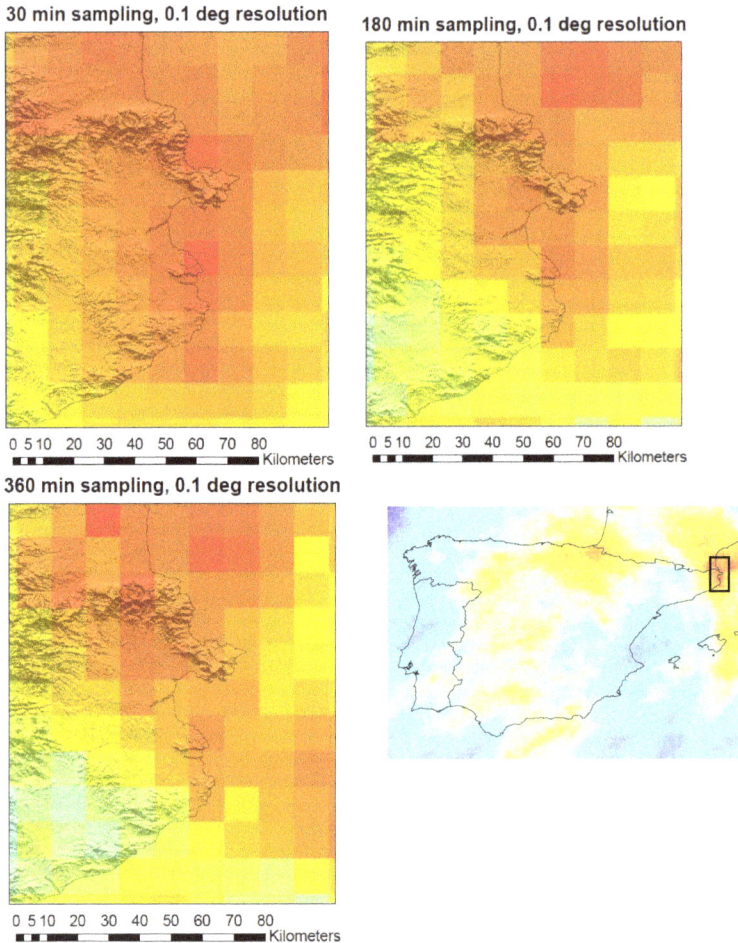

Figure 8. Errors in the location of the maximum monthly precipitation as temporal resolution degrades. The nominal truth (30-min, 0.1° resolution, upper/left) locates the maxima over mainland Spain, in Catalonia. A 3-h sampling places the maximum over the Gulf of Leon (upper/right), and a 6-h sampling over mainland France (bottom/left).

5. Conclusions

We have presented a simulation of an ideal GMS precipitation estimate and have compared it with degraded estimates made up by coarsening the spatial resolution at 0.15° intervals from 0.05–1.55° and by diminishing the temporal resolution from 15 min up to 6 h in 15-min intervals.

The aim of this experiment was to compare the relative merits of GMS estimates with alternatives such as a constellation of MW sensors on polar orbit, since no GMS sensor still exists. The results show that in the best-case scenario (absence of geometry errors and perfect instrumentation and retrieval algorithms) r^2 is expected to worsen up to 0.86, with a 4 mm/month bias and a 17 mm/month RMSE for a 3 h/0.45 deg sensor. The entropy of the alternatives would be slightly smaller, where major differences can be found in the location of monthly maxima: In our case study, a difference of 350 km was observed, albeit this value was deemed as highly dependent on the atmospheric situation.

The results show that spatial resolution is more important than temporal sampling in order to capture the climatology, whereas temporal resolution is critical to identify extreme events. While both the geostationary sounder and the polar constellation strategies can help a better understanding of precipitation and the water cycle, it depends on applications as to whether or not such values and uncertainties are acceptable.

Author Contributions: F.J.T. led the research, outlined the draft of the manuscript and made the amendments suggested by the referees. A.N., C.M., A.J.-A., R.M.G. and J.S. contributed to analysis, plotting and manuscript writing.

Acknowledgments: Funding from projects CGL2013-48367-P, CGL2016-80609-R (Ministerio de Economía y Competitividad, Ciencia e Innovación) is gratefully acknowledged. ANM acknowledges support from grant FPU 13/02798 for carrying out his PhD. We want to thank the anonymous referees for their valuable comments. Their comments have greatly improved the paper under review.

Conflicts of Interest: The authors declare no conflicts of interest.

References

1. Pu, Z.; Tao, W.-K.; Braun, S.; Simpson, J.; Jia, Y.; Halverson, J.; Olson, W.; Hou, A.; Pu, Z.; Tao, W.-K.; et al. The Impact of TRMM Data on Mesoscale Numerical Simulation of Supertyphoon Paka. *Mon. Weather Rev.* **2002**, *130*, 2448–2458. [CrossRef]
2. Zhang, X.; Xiao, Q.; Fitzpatrick, P.J. The Impact of Multisatellite Data on the Initialization and Simulation of Hurricane Lili's (2002) Rapid Weakening Phase. *Mon. Weather Rev.* **2007**, *135*, 526–548. [CrossRef]
3. Xiao, Q.; Zou, X.; Kuo, Y.-H. Incorporating the SSM/I-Derived Precipitable Water and Rainfall Rate into a Numerical Model: A Case Study for the ERICA IOP-4 Cyclone. *Mon. Weather Rev.* **2000**, *128*, 87–108. [CrossRef]
4. Tapiador, F.J.; Behrangi, A.; Haddad, Z.S.; Katsanos, D.; De Castro, M. Disruptions in precipitation cycles: Attribution to anthropogenic forcing. *J. Geophys. Res. Atmos.* **2016**, *121*, 2161–2177. [CrossRef]
5. Tao, W.K.; Chen, J.P.; Li, Z.; Wang, C.; Zhang, C. Impact of aerosols on convective clouds and precipitation. *Rev. Geophys.* **2012**, *50*. [CrossRef]
6. Givati, A.; Rosenfeld, D.; Givati, A.; Rosenfeld, D. Quantifying Precipitation Suppression Due to Air Pollution. *J. Appl. Meteorol.* **2004**, *43*, 1038–1056. [CrossRef]
7. Kucera, P.A.; Ebert, E.E.; Turk, F.J.; Levizzani, V.; Kirschbaum, D.; Tapiador, F.J.; Loew, A.; Borsche, M. Precipitation from space: Advancing earth system science. *Bull. Am. Meteorol. Soc.* **2013**, *94*, 365–375. [CrossRef]
8. De Coning, E. Optimizing satellite-based precipitation estimation for nowcasting of rainfall and flash flood events over the South African domain. *Remote Sens.* **2013**, *5*, 5702–5724. [CrossRef]
9. Li, Y.; Grimaldi, S.; Walker, J.P.; Pauwels, V.R.N. Application of remote sensing data to constrain operational rainfall-driven flood forecasting: A review. *Remote Sens.* **2016**, *8*, 456. [CrossRef]
10. Katsanos, D.; Retalis, A.; Tymvios, F.; Michaelides, S. Analysis of precipitation extremes based on satellite (CHIRPS) and in situ dataset over Cyprus. *Nat. Hazards* **2016**, *83*, 53–63. [CrossRef]
11. Marra, F.; Destro, E.; Nikolopoulos, E.I.; Zoccatelli, D.; Dominique Creutin, J.; Guzzetti, F.; Borga, M. Impact of rainfall spatial aggregation on the identification of debris flow occurrence thresholds. *Hydrol. Earth Syst. Sci.* **2017**, *21*, 4525–4532. [CrossRef]
12. Tapiador, F.J.; Navarro, A.; Jiménez, A.; Moreno, R.; García-Ortega, E. Discrepancies with Satellite Observations in the Spatial Structure of Global Precipitation as Derived from Global Climate Models. *Q. J. R. Meteorol. Soc.* **2018**. [CrossRef]
13. Tapiador, F.J.; Navarro, A.; Levizzani, V.; García-Ortega, E.; Huffman, G.J.; Kidd, C.; Kucera, P.A.; Kummerow, C.D.; Masunaga, H.; Petersen, W.A.; et al. Global precipitation measurements for validating climate models. *Atmos. Res.* **2017**. [CrossRef]
14. Navarro, A.; Moreno, R.; Tapiador, F.J. Improving the representation of anthropogenic CO_2 emissions in climate models: A new parameterization for the Community Earth System Model (CESM). *Earth Syst. Dyn. Discuss.* **2018**, 1–26. [CrossRef]
15. Michaelides, S.; Levizzani, V.; Anagnostou, E.; Bauer, P.; Kasparis, T.; Lane, J.E. Precipitation: Measurement, remote sensing, climatology and modeling. *Atmos. Res.* **2009**, *94*, 512–533. [CrossRef]
16. Kidd, C.; Levizzani, V. Status of satellite precipitation retrievals. *Hydrol. Earth Syst. Sci.* **2011**, *15*, 1109–1116. [CrossRef]

17. Tapiador, F.J.; Turk, F.J.; Petersen, W.; Hou, A.Y.; García-Ortega, E.; Machado, L.A.T.; Angelis, C.F.; Salio, P.; Kidd, C.; Huffman, G.J.; et al. Global precipitation measurement: Methods, datasets and applications. *Atmos. Res.* **2012**, *104–105*, 70–97. [CrossRef]

18. Levizzani, V.; Laviola, S.; Cattani, E. Detection and Measurement of Snowfall from Space. *Remote Sens.* **2011**, *3*, 145–166. [CrossRef]

19. Stephens, G.L.; Kummerow, C.D. The Remote Sensing of Clouds and Precipitation from Space: A Review. *J. Atmos. Sci.* **2007**, *64*, 3742–3765. [CrossRef]

20. Barrett, E.C.; Beaumont, M.J. Satellite rainfall monitoring: An overview. *Remote Sens. Rev.* **1994**, *11*, 23–48. [CrossRef]

21. Wilheit, T.; Adler, R.; Avery, S.; Barrett, E.; Bauer, P.; Berg, W.; Chang, A.; Ferriday, J.; Grody, N.; Goodman, S.; et al. Algorithms for the retrieval of rainfall from passive microwave measurements. *Remote Sens. Rev.* **1994**, *11*, 163–194. [CrossRef]

22. Hou, A.Y.; Kakar, R.K.; Neeck, S.; Azarbarzin, A.A.; Kummerow, C.D.; Kojima, M.; Oki, R.; Nakamura, K.; Iguchi, T. The global precipitation measurement mission. *Bull. Am. Meteorol. Soc.* **2014**, *95*, 701–722. [CrossRef]

23. Kummerow, C.; Hong, Y.; Olson, W.S.; Yang, S.; Adler, R.F.; McCollum, J.; Ferraro, R.; Petty, G.; Shin, D.-B.; Wilheit, T.T. The Evolution of the Goddard Profiling Algorithm (GPROF) for Rainfall Estimation from Passive Microwave Sensors. *J. Appl. Meteorol.* **2001**, *40*, 1801–1820. [CrossRef]

24. Kummerow, C.; Masunaga, H.; Bauer, P. A next-generation microwave rainfall retrieval algorithm for use by TRMM and GPM. In *Measuring Precipitation from Space*; Springer: Dordrecht, The Netherlands, 2007; pp. 235–252. ISBN 13 978-1-4020-5834-9.

25. Kummerow, C.D.; Randel, D.L.; Kulie, M.; Wang, N.Y.; Ferraro, R.; Joseph Munchak, S.; Petkovic, V. The evolution of the goddard profiling algorithm to a fully parametric scheme. *J. Atmos. Ocean. Technol.* **2015**, *32*, 2265–2280. [CrossRef]

26. Sorooshian, S.; Hsu, K.; Coppola, E.; Tomassetti, B.; Verdecchia, M.; Visconti, G. *Hydrological Modelling and the Water Cycle. Coupling the Atmospheric and Hydrological Models*; Springer: Dordrecht, The Netherlands, 2009; ISBN 9783540778424.

27. Vicente, G.A.; Scofield, R.A.; Menzel, W.P. The Operational GOES Infrared Rainfall Estimation Technique. *Bull. Am. Meteorol. Soc.* **1998**, *79*, 1883–1893. [CrossRef]

28. Tapiador, F.J.; Kidd, C.; Levizzani, V.; Marzano, F.S. A Neural Networks–Based Fusion Technique to Estimate Half-Hourly Rainfall Estimates at 0.1° Resolution from Satellite Passive Microwave and Infrared Data. *J. Appl. Meteorol.* **2004**, *43*, 576–594. [CrossRef]

29. Ebert, E.E.; Manton, M.J.; Arkin, P.A.; Allam, R.J.; Holpin, G.E.; Gruber, A. Results from the GPCP algorithm intercomparison programme. *Bull. Am. Meteorol. Soc.* **1996**, *77*, 2875–2887. [CrossRef]

30. Skofronick-Jackson, G.; Petersen, W.A.; Berg, W.; Kidd, C.; Stocker, E.F.; Kirschbaum, D.B.; Kakar, R.; Braun, S.A.; Huffman, G.J.; Iguchi, T.; et al. The Global Precipitation Measurement (GPM) Mission for Science and Society. *Bull. Am. Meteorol. Soc.* **2017**, *98*, 1679–1695. [CrossRef]

31. Bizarro, J.P.S. On the behavior of the continuous-time spectrogram for arbitrarily narrow windows. *IEEE Trans. Signal Process.* **2007**, *55*, 1793–1802. [CrossRef]

32. Staelin, D.H.; Gasiewski, A.J.; Kerekes, J.P.; Shields, M.W.; Solman, F.J., III. *Concept Proposal for a Geostationary Microwave (GEM) Observatory*; prepared for the NASA/NOAA Advanced Geostationary Sensor (AGS) Program; MIT Lincoln Laboratory: Lexington, MA, USA, 1998.

33. Bizzarri, B.; Amato, U.; Bates, J.; Benesch, W.; Bühler, S.; Capaldo, M.; Cervino, M.; Cuomo, V.; De Leonibus, L.; Desbois, M.; et al. Requirements and perspectives for MW/sub-mm sounding from geostationary satellite. In Proceedings of the EUMETSAT Meteorological Satellite Conference, Dublin, Ireland, 2–6 September 2002; pp. 97–105.

34. Gasiewski, A.J. Numerical sensitivity analysis of passive ehf and SMMW channels to tropospheric water vapor, clouds, and precipitation. *IEEE Trans. Geosci. Remote Sens.* **1992**, *30*, 859–870. [CrossRef]

35. Tapiador, F.J.; Gaertner, M.A.; Romera, R.; Castro, M. A multisource analysis of hurricane vince. *Bull. Am. Meteorol. Soc.* **2007**, *88*, 1027–1032. [CrossRef]

36. Ver Hoef, J.M.; Cressie, N. Multivariable spatial prediction. *Math. Geol.* **1993**, *25*, 219–240. [CrossRef]

37. Jaynes, E.T. Information theory and statistical mechanics. *Phys. Rev.* **1957**, *106*, 620–630. [CrossRef]

38. Tapiador, F.J.; Kidd, C.; Hsu, K.-L.; Marzano, F. Neural networks in satellite rainfall estimation. *Meteorol. Appl.* **2004**, *11*, 83–91. [CrossRef]

39. Sorooshian, S.; Hsu, K.L.; Gao, X.; Gupta, H.V.; Imam, B.; Braithwaite, D. Evaluation of PERSIANN system satellite-based estimates of tropical rainfall. *Bull. Am. Meteorol. Soc.* **2000**, *81*, 2035–2046. [CrossRef]
40. Turk, F.J.; Hawkins, J.; Smith, E.A.; Marzano, F.S.; Mugnai, A.; Levizzani, V. Combining SSM/I, TRMM and Infrared Geostationary Satellite Data in a Near-realtime Fashion for Rapid Precipitation Updates: Advantages and Limitations. In Proceedings of the 2000 EUMETSAT Meteorological Satellite Data Users' Conference, Bologna, Italy, 29 May–2 June 2000; pp. 452–459.
41. Marzano, F.S.; Palmacci, M.; Cimini, D.; Giuliani, G.; Tapiador, F.; Turk, J.F. Multivariate probability matching of satellite infrared and microwave radiometric measurements for rainfall retrieval at the geostationary scale. In Proceedings of the 2003 IEEE International Geoscience and Remote Sensing Symposium. Proceedings (IEEE Cat. No.03CH37477), Toulouse, France, 21–25 July 2003; Volume 2, pp. 1151–1153.
42. Joyce, R.J.; Janowiak, J.E.; Arkin, P.A.; Xie, P. CMORPH: A Method that Produces Global Precipitation Estimates from Passive Microwave and Infrared Data at High Spatial and Temporal Resolution. *J. Hydrometeorol.* **2004**, *5*, 487–503. [CrossRef]
43. Turk, F.J.; Bauer, P.; Ebert, E.; Arkin, P.A. Satellite-derived precipitation verification activities within the International Precipitation Working Group (IPWG). In *14th Conference on Satellite Meteorology and Oceanography*; American Meteor Society: Atlanta, GA, USA, 2006.
44. Tapiador, F.J. A physically based satellite rainfall estimation method using fluid dynamics modelling. *Int. J. Remote Sens.* **2008**, *29*, 5851–5862. [CrossRef]
45. Huffman, G.J.; Adler, R.F.; Arkin, P.; Chang, A.; Ferraro, R.; Gruber, A.; Janowiak, J.; McNab, A.; Rudolf, B.; Schneider, U. The Global Precipitation Climatology Project (GPCP) Combined Precipitation Dataset. *Bull. Am. Meteorol. Soc.* **1997**, *78*, 5–20. [CrossRef]
46. Adler, R.F.; Sapiano, M.R.P.; Huffman, G.J.; Wang, J.J.; Gu, G.; Bolvin, D.; Chiu, L.; Schneider, U.; Becker, A.; Nelkin, E.; et al. Bin the Global Precipitation Climatology Project (GPCP) monthly analysis (New Version 2.3) and a review of 2017 global precipitation. *Atmosphere* **2018**, *9*, 138. [CrossRef]
47. Jaynes, E.T. Probability Theory as Logic. In *Maximum Entropy and Bayesian Methods*; Springer: Dordrecht, The Netherlands, 1990; pp. 1–16. ISBN 9789401067928.

remote sensing

MDPI

Article

Variability of Microwave Scattering in a Stochastic Ensemble of Measured Rain Drops

Francisco J. Tapiador [1,*], Raúl Moreno [1], Andrés Navarro [1], Alfonso Jiménez [1], Enrique Arias [2] and Diego Cazorla [2]

1   Earth and Space Sciences Group (ESS), Institute of Environmental Sciences (ICAM), University of Castilla-La Mancha, 45071 Toledo, Spain; raulmorenogaldon@gmail.com (R.M.); Andres.Navarro@uclm.es (A.N.); Alfonso.JAlcazar@uclm.es (A.J.)
2   Albacete Research Institute of Informatics (I3A), University of Castilla-La Mancha, 02071 Albacete, Spain; E.Arias@uclm.es or Enrique.Arias@uclm.es (E.A.); Diego.Cazorla@uclm.es (D.C.)
*   Correspondence: Francisco.Tapiador@uclm.es; Tel.: +34-925-268-800 (ext. 5762)

Received: 16 May 2018; Accepted: 14 June 2018; Published: 15 June 2018

Abstract: While it has been proved that multiple scattering in the microwave frequencies has to be accounted for in precipitation retrieval algorithms, the effects of the random arrangements of drops in space has seldom been investigated. The fact is, a single rain drop size distribution (RDSD) corresponds with many actual 3D distributions of those rain drops and each of those may a priori absorb and scatter radiation in a different way. Each spatial configuration is equivalent to any other in terms of the RDSD function, but not in terms of radiometric characteristics, both near and far from field, because of changes in the relative phases among the particles. Here, using the T-matrix formalism, we investigate the radiometric variability of two ensembles of 50 different 3D, stochastically-derived configurations from two consecutive measured RDSDs with 30 and 31 drops, respectively. The results show that the random distribution of drops in space has a measurable but apparently small effect in the scattering calculations with the exception of the asymmetry factor.

Keywords: precipitation; radar; radiometer; T-Matrix; microwave scattering

---

1. Introduction

The retrieval of precipitation with radars and radiometers is important for a variety of environmental applications and human activities [1]. They provide accurate precipitation estimates that are crucial for monitoring extreme climate events, such as droughts [2–4], floods [5–7], and hailstorms [8,9]. Due to its global coverage and direct measurement, radars have become an essential tool to estimate precipitation, especially in complex terrains [10–12] and in sparsely populated areas affected by poor rain gauge coverage [13–15].

However, regardless of how powerful radars and radiometers are, they are not immune to retrieval biases. One such bias is the effect of multiple scattering in high-frequency radars [16,17]. It has been proved that multiple scattering in the microwave frequencies has to be accounted for in precipitation retrieval algorithms [18–28].

Another important, but sometimes overlooked, problem is the effect of the random arrangements of particles in space. Indeed, a single rain drop size distribution (RDSD) corresponds with many actual 3D distributions of those raindrops. Each spatial configuration is equivalent to any other in terms of the RDSD function; however, they scatter radiation differently both near and far from field due to changes in the relative phases among the particles. This paper investigates the variability of such scattering at microwave frequencies due to the random spatial distribution of drops.

The T-matrix formalism [29] provides a sound approach to calculate radiometric quantities from first principles [30–36]. It permits analysis of the effects of scattering not only of individual particles but also of systems with many interacting scatters. The approach of this paper is to use measured RDSDs from disdrometers and build ensembles of randomly-located drops in a volume to check whether or not there is a difference in the scattering as calculated by the T-matrix method. The aim of the research is to gauge the extent of the variability in a theoretical, first-order setting: the actual scattering depends on the sensor (radar or radiometer), beamwidth, and frequency of operation.

2. Data

Laser disdrometers (Figure 1) provide estimates of the RDSD of precipitation. The RDSD can be built over a 2D instantaneous estimate or over a 2D time sliced estimate, with the extra dimension corresponding to the accumulation time (10 min). The time-integrated slice is deemed as a suitable representation of precipitation within a large 3D box around the instrument under the assumption that the spatial correlation of precipitation decreases only moderately with distance.

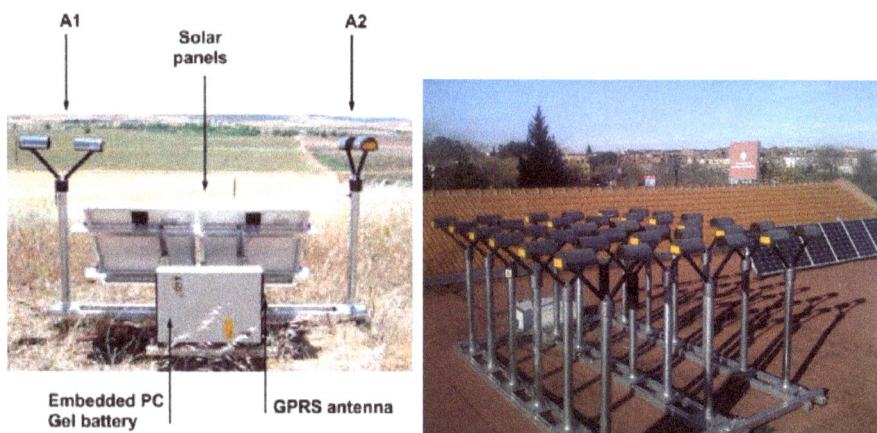

Figure 1. Disdrometers used to measure the Rain Drop Size Distribution (RDSD). (**Left**) Dual, orthogonal setup or autonomous disdrometers used to calculate the spatial variability of rain at hectometre resolution [37]. (**Right**) Calibration array of 18 disdrometers used to measure the small-scale variability of the RDSD [38].

The University of Castilla-La Mancha (UCLM) has been maintaining a network of disdrometers since 2009. The original setup of the instruments consists of a dual arrangement with a solar panel and two batteries so the setup can work autonomously (Figure 1 Left). Data from the sensors were automatically sent to a server every minute and then filtered and processed to standard accumulation periods (1, 5, and 10 min). They have been compared and cross-calibrated (Figure 1 Right). Latest refinement and filtering of the database resulted in the empirical basis for this paper. For reference, the product of the whole process is tagged as the "UCLM disdrometer database v4.0." This 2018 product supersedes previous versions of UCLM's disdrometer data.

Two 10-min accumulated RDSDs were randomly selected from the database to calculate the radiometric quantities associated with several spatial configurations (Figure 2). The small number of drops was chosen to alleviate the computational burden of the simulations since the burden of direct simulation of scattering escalates quickly with the number of scatters.

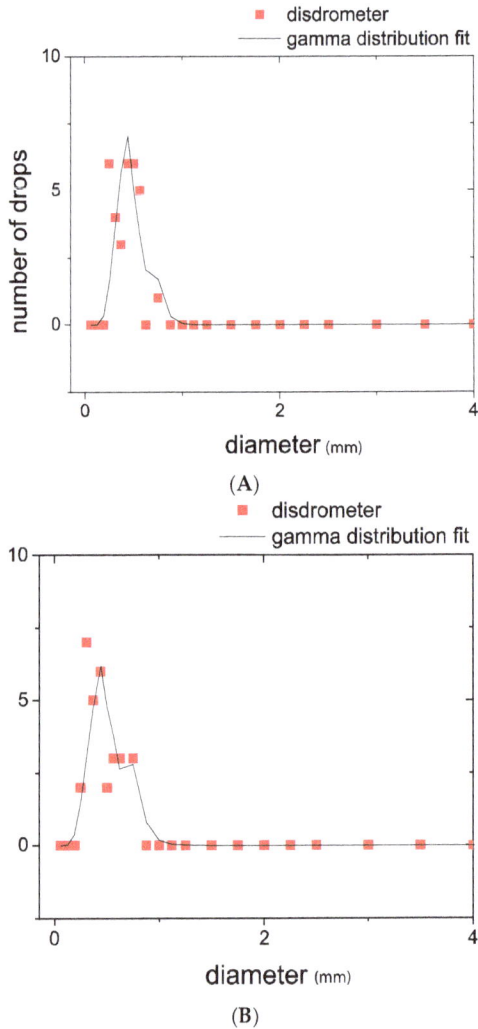

Figure 2. Measured RDSDs (**A,B**) used to build the two cases explored in this paper. Diameters are in mm. The fit is for a three-parameter gamma.

3. Methods

The 10-min data were used to derive an ensemble of 3D configurations. The measured RDSDs were used to derive the concentrations and size distributions in a 1 m³ cube. Such measurement was used as a starting point for which the positions were randomly perturbed. The algorithm used to locate the drops in space was a Monte Carlo generator, as follows. A Pseudo-Random Number Generator (PRNG) was used to generate the locations within the volume. At every iteration, raindrops were sequentially generated in decreasing size order. Collisions are identified and avoided using a relaxation technique.

Drops were assumed to be spherical for simplicity; more complex and realistic geometries will only reinforce the point made—that is, that there is measurable variability because of the actual spatial configuration of the RDSD.

Two tests for randomness were carried out on the PRNG: (i) uniformity test and (ii) independence test. The latter was evaluated through a χ^2 test and a Kolmogorov-Smirnov test. It is worth noting that a simpler algorithm would not satisfy the randomness requirement since the raindrops need to be randomly distributed in a cube and not in a sphere (with the extra condition of not intersecting one to another, Figure 3).

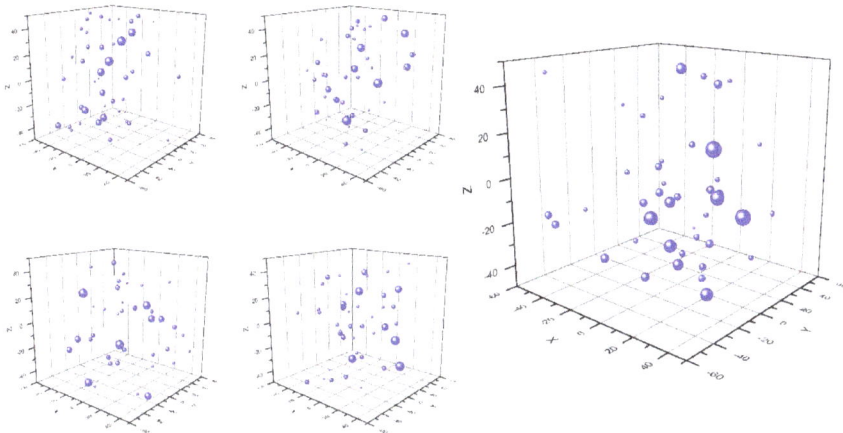

Figure 3. A sample of the different randomly-generated spatial distributions of raindrops consistent with one of the measured RDSDs of Figure 2. The diameters of the raindrops are exaggerated. Dimensions are in centimeters.

The parameters of the Stokes scattering matrix S were calculated using the T-Matrix formalism. The relationship between the incident (*i*) and scattered (*s*) field amplitudes in the far field zone for spheres is given in Mie's theory by Van der Hulst and Bohren [39,40]:

$$\begin{pmatrix} E_{\|s} \\ E_{\perp s} \end{pmatrix} = \frac{e^{ik(r-z)}}{-ikr} \begin{pmatrix} S_2 & 0 \\ 0 & S_1 \end{pmatrix} \begin{pmatrix} E_{\|I} \\ E_{\perp I} \end{pmatrix} \tag{1}$$

where $E_{\|}$ and $E_{\perp}$ are the parallel and perpendicular components of the electric field, $S_{1,2}$ are phase functions, k is the free-space propagation constant, and r the distance from the origin to the observation point (with exp(ikz) the incident plane wave). The phase functions are:

$$S_1 = \sum_{j=1}^{\infty} \frac{2j+1}{j(j+1)} \left(a_j \pi_j + b_j \tau_j \right) \tag{2}$$

$$S_2 = \sum_{j=1}^{\infty} \frac{2j+1}{j(j+1)} \left(a_j \tau_j + b_j \pi_j \right) \tag{3}$$

where $\pi_j = \frac{P_j^1(\theta)}{\sin\theta}$ and $\tau_j = \frac{dP_j^1(\theta)}{d\theta}$, with $P_j^1(\theta)$ the associated Legendre functions.

Since the drops are randomly located, averaging over all orientations results in a scattering matrix with only six independent significant elements (i.e., there is not preferential scattering direction in *xy* when the incident wave comes from -z). The Stokes parameters I, Q, U and V for the scattered wave (S) are derived from the scattering matrix $S_{i,j}$ and the parameters of the incident wave (I) as follows:

$$
\begin{bmatrix} I_s \\ Q_s \\ U_s \\ V_s \end{bmatrix} \propto \begin{bmatrix} S_{11} & S_{21} & \sim 0 & \sim 0 \\ S_{21} & S_{22} & \sim 0 & \sim 0 \\ \sim 0 & \sim 0 & S_{33} & S_{34} \\ \sim 0 & \sim 0 & -S_{34} & S_{44} \end{bmatrix} \begin{bmatrix} I_I \\ Q_I \\ U_I \\ V_I \end{bmatrix} \tag{4}
$$

with:

$$
\begin{aligned}
S_{11} &= \tfrac{1}{2}\left(|S_2|^2 + |S_1|^2\right) \\
S_{12} &= \tfrac{1}{2}\left(|S_2|^2 - |S_1|^2\right) \\
S_{33} &= \tfrac{1}{2}\left(S_2^* S_1 + S_2 S_1^*\right) \\
S_{34} &= \tfrac{1}{2}i\left(S_2^* S_1 + S_2 S_1^*\right)
\end{aligned} \tag{5}
$$

The equations used in the calculations of multiple spheres are those in Mackowski et al. [32]. Spherical drops were assumed in order to generate a first-order analysis. Further work will be devoted to the analyses of the effects of other shapes, including solid precipitation.

The T-matrix v.3.0 code was used for all the members of the two ensembles and for X-band wavelength (3.197 cm = 9.375 GHz; length scale factor of 1.9653). Temperature was set to 5 °C. Refractive index was set according to frequency and temperature. The beamwidth of the incident wave was restricted to the domain. Processing time for each configuration varied (typically, about 72 h).

Once the whole ensemble was processed, the spread of the radiometric outputs was calculated, and the estimates for the two cases were cross-compared. The quantities of interest were total extinction, absorption, scattering efficiency, and asymmetry factor.

4. Results and Discussion

The randomness tests for the location of the drops were satisfied: the histogram of the uniformity test was flat, and the statistics for the independence test also proved that the drops were randomly located in space.

Figures 4–7 gathers the results of calculations of basic radiometric quantities of interest calculated for the two cases in Figure 2. It is clear that the results are consistent and that the actual 3D distribution has an impact, as demonstrated by the existence of a distribution. In spite of the limited number of members in the ensembles, there is a clear shape around a central, more likely value. If there were no effect of different 3D arrangements in the scattering, no spread would appear. A purely random effect ("noise") would generate a random distribution.

Most of the resulting distributions are normal and the skew is low. The tails are relatively large, however, revealing that some configurations produce quite different values. This seems to not be a problem for the retrieval of precipitation from radiometer or radar algorithms, however, as the central tendency is strong. The extent of the variability is low except for the asymmetry factor.

Such variations are consistent with changes in the relative phases among the particles, even under the single scattering hypothesis. In order to ascertain the effect of multiple scattering, different frequencies would need to explored (multiple scattering should increase with frequency). The results, however, show that there is a measurable effect due to the random arrangement of the scatters.

Figure 8 shows the effect of the scattering angle in the retrievals. Here, the S_{11} and S_{12} phase functions are depicted for one of the ensembles. Departures from the mean value appear, albeit the overall behavior is strong and consistent with theory.

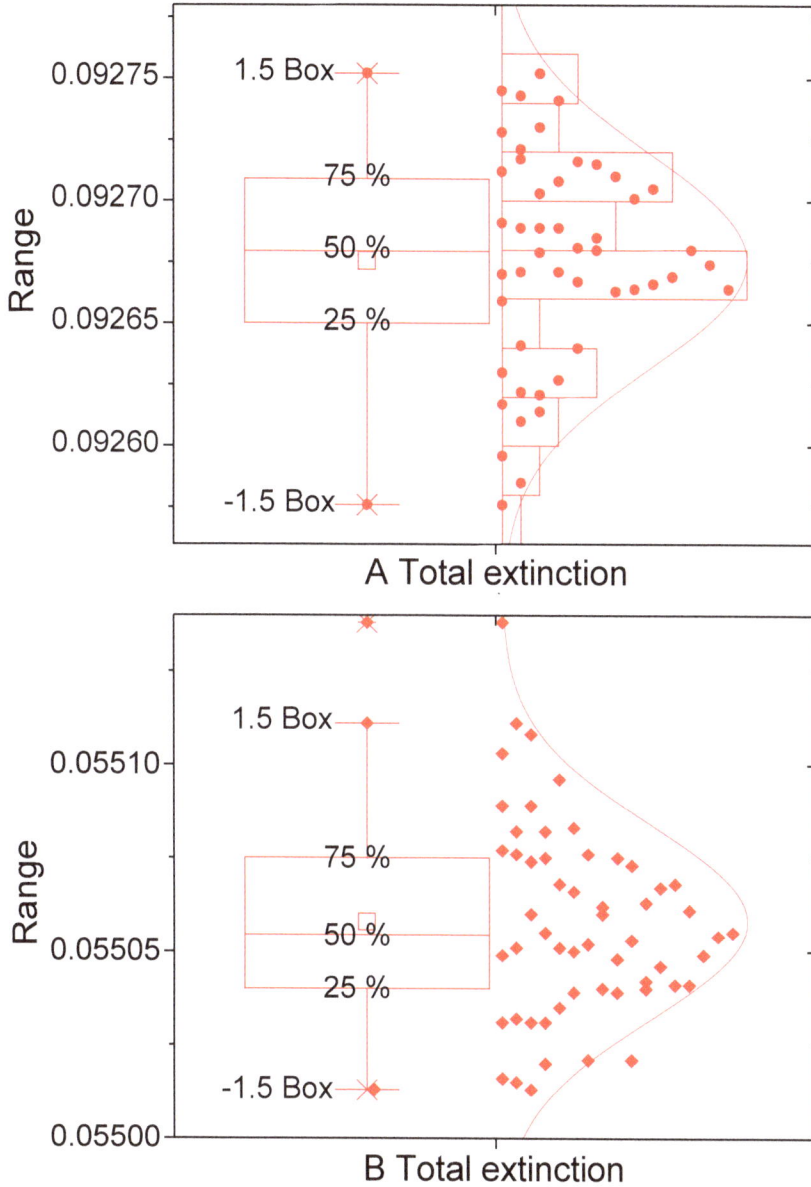

Figure 4. Variability of the total extinction for an ensemble of 50 3D configurations of the (**A**,**B**) RDSDs in Figure 2.

Comparison between the two RDSD cases shows that there is a larger variability over time than within each case. The fact that the RDSDs are consecutive (10 min apart) allows examination of how different the scattered radiation is between the two distributions. Figures 4–7 also show that there is a large difference in spite of the similarities between the two distributions (cf. Figure 2).

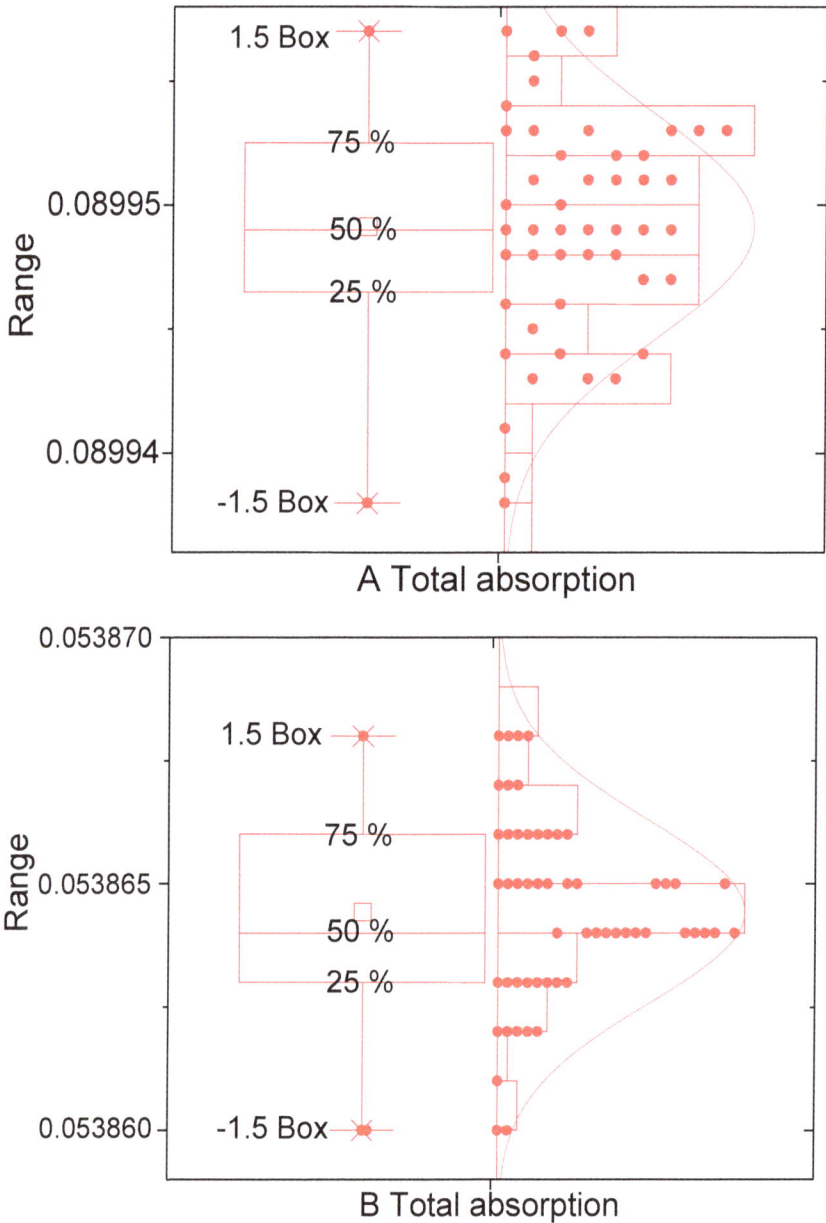

Figure 5. Variability of the total absorption for an ensemble of 50 3D configurations of the (**A**,**B**) RDSDs in Figure 2.

Two ensembles, 10 min apart, for 50 different stochastically-derived 3D configurations from two measured RDSDs, with 30 and 31 drops, respectively, were used (Figure 2). The RDSD are consecutive, and hence, aimed to derive conclusions on microphysical processes [41]. The number of drops is small on purpose to alleviate the computational burden of the quite intensive calculations required.

While the scattering matrix needs to be calculated just once for all of the beam orientations (as the matrix is defined by the properties and distribution of the particles), the many possible distributions require re-calculating a matrix for each configuration.

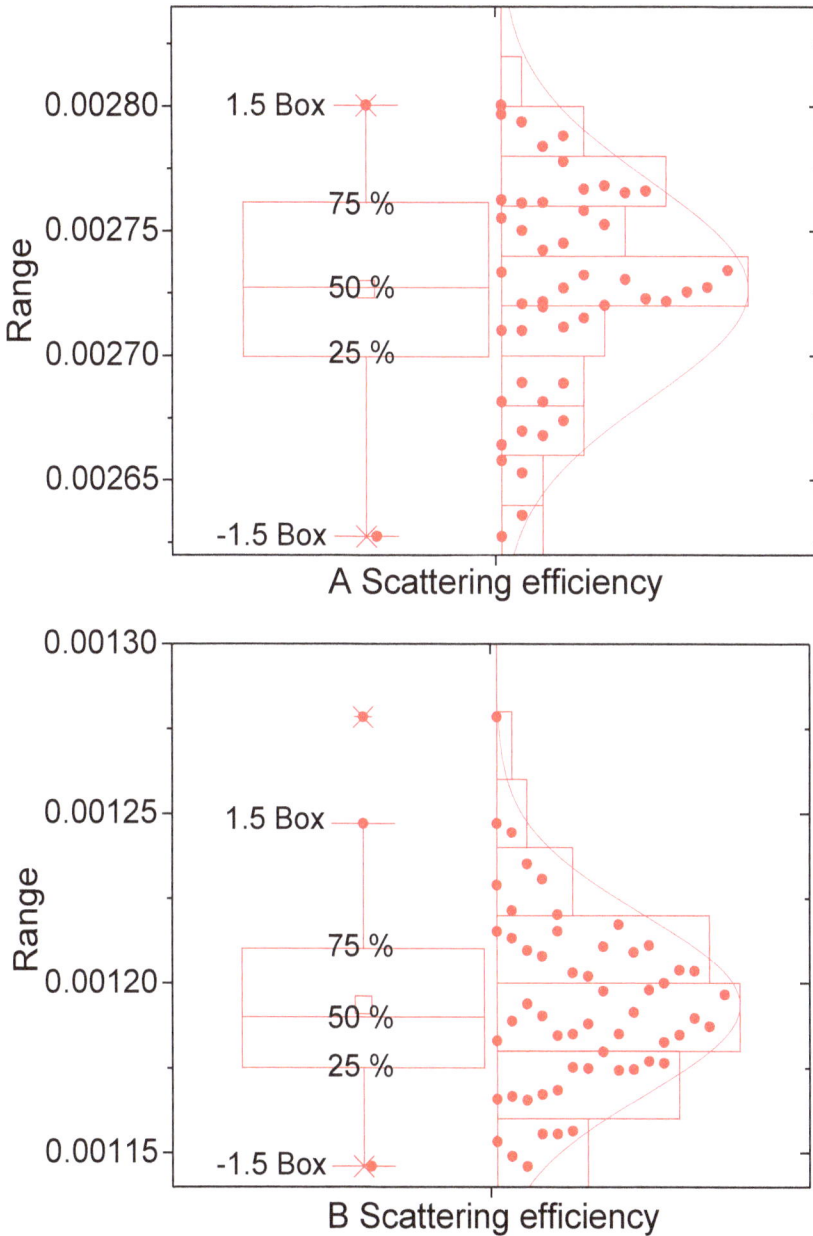

Figure 6. Variability of the scattering efficiency for an ensemble of 50 3D configurations of the (**A**,**B**) RDSDs in Figure 2.

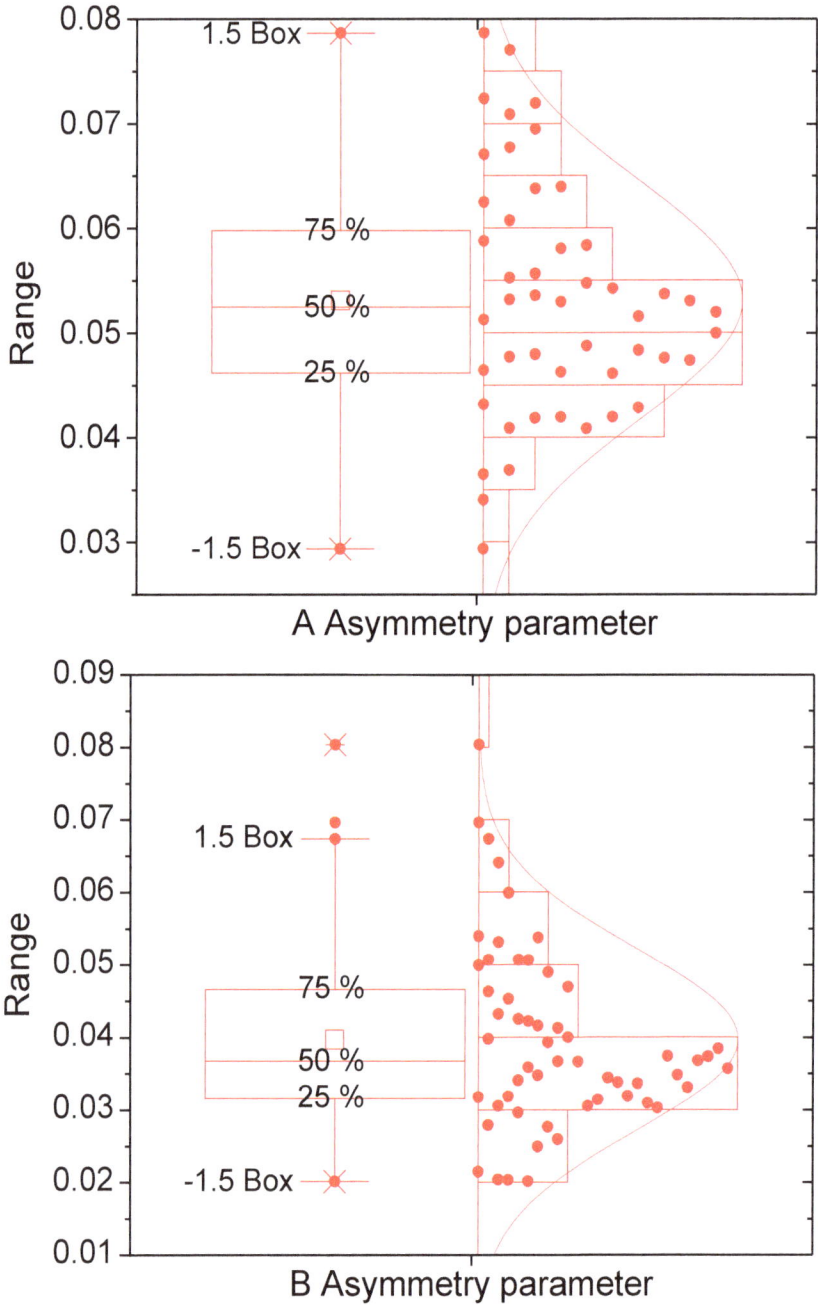

Figure 7. Variability of the asymmetry parameter for an ensemble of 50 3D configurations of the (**A**,**B**) RDSDs in Figure 2.

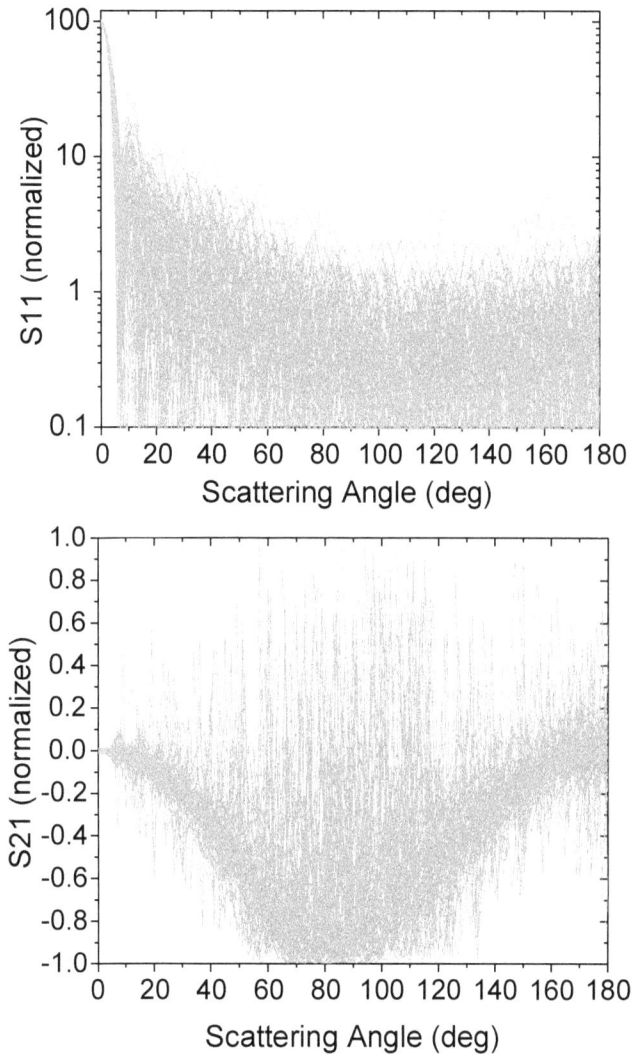

Figure 8. Spread of the dependence of the phase functions, S_{11} and S_{21}, on the scattering angle for all the member of the ensemble.

5. Conclusions and Further Work

Multiple scattering in the microwave frequencies has to be accounted for in precipitation retrieval algorithms [42], but the effects of random arrangements of particles in space has seldom been investigated. Here, such effects have been described using modelled microwave scattering properties using the T-matrix formalism on the simplified case of spherical, non-intersecting raindrops.

The results of measured RDSDs reveal that the random distribution of particles in space have a measurable but small effect on the scattering because of changes in the relative phases among the particles. The effect is only apparent in the asymmetry factor.

Scattering calculations at the microwave frequencies using spherical raindrops are a simplification over the real situations where both ice and mixed-phase exist. The assumption in this research is that a small number of non-interacting spherical water drops are the most favorable possible case in the real atmosphere and thus a good baseline value for the extent of variability induced by random 3D configurations and other shapes and phases of water. By simplifying the water phase, geometry and number of drops, information about the randomness of the scattering quantities has been found, and a first estimate of the expected variability provided. The addition of further realism to the simulations in the form of oblate spheroids [43], melting ice, mixed phase, or other more realistic setups can only increase the differences.

Further research will be devoted to expand the number of cases in order to derive systematic conclusions of the extent of the variability and its effects on the retrievals of precipitation. Indeed, in order to investigate the effects of multiple scattering in precipitation retrievals from radars, the Field of Vision (FOV) would need to be commensurate with the domain of observation of the sensor, and the effects of different frequencies and beamwidths would also need to be investigated. The present paper is only concerned with the first-order variability due to the many possible configurations of a given RDSD.

Author Contributions: F.J.T. lead the research and drafted the manuscript. R.M., E.A., D.C., A.J. and A.N. contributed to analysis, plotting, and literature research.

Acknowledgments: Funding from projects CGL2013-48367-P and CGL2016-80609-R (Ministerio de Economía y Competitividad, Ciencia e Innovación) is gratefully acknowledged. We are indebted to M. Mishchenko of NASA Goddard Institute for Space Studies in New York, NY, USA, for making his T-matrix code publicly available. Thanks are due to the referees for crucial input into the interpretation of the results, especially in that referring to multiple scattering, and to Nabhonil Kar (Princeton University) for editing the manuscript.

Conflicts of Interest: The authors declare no conflict of interest.

Abbreviations

2D	Two-dimensional
3D	Three-dimensional
FOV	Field of Vision
PRNG	Pseudo-random Number Generator
RDSD	Rain Drop Size Distribution
UCLM	University of Castilla-La Mancha

References

1. Hou, A.Y.; Kakar, R.K.; Neeck, S.; Azarbarzin, A.A.; Kummerow, C.D.; Kojima, M.; Oki, R.; Nakamura, K.; Iguchi, T. The global precipitation measurement mission. *Bull. Am. Meteorol. Soc.* **2014**, *95*, 701–722. [CrossRef]
2. Bayissa, Y.; Tadesse, T.; Demisse, G.; Shiferaw, A. Evaluation of satellite-based rainfall estimates and application to monitor meteorological drought for the Upper Blue Nile Basin, Ethiopia. *Remote Sens.* **2017**, *9*. [CrossRef]
3. Zhang, Y.; Gong, J.; Sun, K.; Yin, J.; Chen, X. Estimation of soil moisture index using multi-temporal Sentinel-1 images over Poyang Lake ungauged zone. *Remote Sens.* **2018**, *10*, 12. [CrossRef]
4. Sahoo, A.K.; Sheffield, J.; Pan, M.; Wood, E.F. Evaluation of the Tropical Rainfall Measuring Mission Multi-Satellite Precipitation Analysis (TMPA) for assessment of large-scale meteorological drought. *Remote Sens. Environ.* **2015**, *159*, 181–193. [CrossRef]
5. Ban, H.-J.; Kwon, Y.-J.; Shin, H.; Ryu, H.-S.; Hong, S. Flood Monitoring Using Satellite-Based RGB Composite Imagery and Refractive Index Retrieval in Visible and Near-Infrared Bands. *Remote Sens.* **2017**, *9*, 313. [CrossRef]
6. Alahacoon, N.; Matheswaran, K.; Pani, P.; Amarnath, G. A decadal historical satellite data and rainfall trend analysis (2001-2016) for flood hazard mapping in Sri Lanka. *Remote Sens.* **2018**, *10*. [CrossRef]
7. Yoshimot, S.; Amarnath, G. Applications of satellite-based rainfall estimates in flood inundation modeling-A case study in Mundeni Aru River Basin, Sri Lanka. *Remote Sens.* **2017**, *9*. [CrossRef]

8. Marra, A.C.; Porcù, F.; Baldini, L.; Petracca, M.; Casella, D.; Dietrich, S.; Mugnai, A.; Sanò, P.; Vulpiani, G.; Panegrossi, G. Observational analysis of an exceptionally intense hailstorm over the Mediterranean area: Role of the GPM Core Observatory. *Atmos. Res.* **2017**, *192*, 72–90. [CrossRef]

9. Cecil, D.J. Passive microwave brightness temperatures as proxies for hailstorms. *J. Appl. Meteorol. Climatol.* **2009**, *48*, 1281–1286. [CrossRef]

10. Gabella, M.; Speirs, P.; Hamann, U.; Germann, U.; Berne, A. Measurement of precipitation in the alps using dual-polarization C-Band ground-based radars, the GPM Spaceborne Ku-Band Radar, and rain gauges. *Remote Sens.* **2017**, *9*. [CrossRef]

11. Yuan, F.; Zhang, L.; Wah Win, K.W.; Ren, L.; Zhao, C.; Zhu, Y.; Jiang, S.; Liu, Y. Assessment of GPM and TRMM multi-satellite precipitation products in streamflow simulations in a data sparse mountainous watershed in Myanmar. *Remote Sens.* **2017**, *9*. [CrossRef]

12. Zhao, H.; Yang, B.; Yang, S.; Huang, Y.; Dong, G.; Bai, J.; Wang, Z. Systematical estimation of GPM-based global satellite mapping of precipitation products over China. *Atmos. Res.* **2018**, *201*, 206–217. [CrossRef]

13. Arvor, D.; Funatsu, B.M.; Michot, V.; Dubreui, V. Monitoring rainfall patterns in the southern amazon with PERSIANN-CDR data: Long-term characteristics and trends. *Remote Sens.* **2017**, *9*. [CrossRef]

14. Kimani, M.W.; Hoedjes, J.C.B.; Su, Z. An assessment of satellite-derived rainfall products relative to ground observations over East Africa. *Remote Sens.* **2017**, *9*. [CrossRef]

15. Nogueira, S.M.C.; Moreira, M.A.; Volpato, M.M.L. Evaluating precipitation estimates from Eta, TRMM and CHRIPS data in the south-southeast region of Minas Gerais state-Brazil. *Remote Sens.* **2018**, *10*. [CrossRef]

16. Battaglia, A.; Ajewole, M.O.; Simmer, C. Multiple scattering effects due to hydrometeors on precipitation radar systems. *Geophys. Res. Lett.* **2005**, *32*, 1–5. [CrossRef]

17. Battaglia, A.; Mroz, K.; Tanelli, S.; Tridon, F.; Kirstetter, P.-E. Multiple-scattering-induced "ghost echoes" in GPM DPR observations of a tornadic supercell. *J. Appl. Meteorol. Climatol.* **2016**, *55*, 1653–1666. [CrossRef]

18. Battaglia, A.; Ajewole, M.O.; Simmer, C. Evaluation of radar multiple-scattering effects from a GPM perspective, Part II: Model results. *J. Appl. Meteorol. Climatol.* **2006**, *45*, 1648–1664. [CrossRef]

19. Battaglia, A.; Ajewole, M.O.; Simmer, C. Evaluation of radar multiple scattering effects in Cloudsat configuration. *Atmos. Chem. Phys.* **2007**, *7*, 1719–1730. [CrossRef]

20. Battaglia, A.; Tanelli, S.; Mroz, K.; Tridon, F. Multiple scattering in observations of the GPM dual-frequency precipitation radar: Evidence and impact on retrievals. *J. Geophys. Res.* **2015**, *120*, 4090–4101. [CrossRef] [PubMed]

21. Ishimaru, A.; Cheung, R.L.-T. Multiple scattering effects on wave propagation due to rain. *Ann. Des Télécomm.* **1980**, *35*, 373–379. [CrossRef]

22. Battaglia, A.; Ajewole, M.O.; Simmer, C. Evaluation of radar multiple-scattering effects from a GPM perspective. Part I: Model description and validation. *J. Appl. Meteorol. Climatol.* **2006**, *45*, 1634–1647. [CrossRef]

23. Battaglia, A.; Tanelli, S.; Heymsfield, G.M.; Tian, L. The dual wavelength ratio Knee: A signature of multiple scattering in airborne Ku-Ka observations. *J. Appl. Meteorol. Climatol.* **2014**, *53*, 1790–1808. [CrossRef]

24. Amrullah, F.F.; Setijadi, E.; Hendrantoro, G. Computation of rain attenuation in tropical region with multiple scattering and multiple absorption effects using exponential drop size distribution. In Proceedings of the 2011 XXXth URSI General Assembly and Scientific Symposium, Istanbul, Turkey, 13–20 August 2011.

25. Ito, S.; Oguchi, T.; Iguchi, T.; Meneghini, R. Depolarization of Radar Signals due to Multiple Scattering in Rain. *IEEE Trans. Geosci. Remote Sens.* **1995**, *33*, 1057–1062. [CrossRef]

26. Korotkov, V.A.; Sukhonin, E.V. The radiometric millimeter wave sensing of atmospheric precipitation for measuring attenuation and precipitation parameters with account of the multiple scattering. *Radiotekhnika i Elektron.* **1998**, *43*, 266–270.

27. Bebbington, D.H.O.; Chandra, M.; Watson, R.J. Multiple scattering effects in C-band polarimetric radar observations of intense precipitation. In Proceedings of the 29th International Conference on Radar Meteorology, Montreal, QC, Canada, 12–16 July 1999; pp. 908–909.

28. Oguchi, T.; Ishida, N. Effect of Multiple Scattering on the Estimation of Rainfall Rates Using Dual-Wavelength Radar Techniques. *IEEE Trans. Geosci. Remote Sens.* **1994**, *32*, 943–946. [CrossRef]

29. Waterman, P.C. Symmetry, unitarity, and geometry in electromagnetic scattering. *Phys. Rev. D* **1971**, *3*, 825–839. [CrossRef]

30. Mishchenko, M.I.; Liu, L.; Mackowski, D.W. T-matrix modeling of linear depolarization by morphologically complex soot and soot-containing aerosols. *J. Quant. Spectrosc. Radiat. Transf.* **2013**, *123*, 135–144. [CrossRef]
31. Mishchenko, M.I.; Travis, L.D. Capabilities and limitations of a current FORTRAN implementation of the T-matrix method for randomly oriented, rotationally symmetric scatterers. *J. Quant. Spectrosc. Radiat. Transf.* **1998**, *60*, 309–324. [CrossRef]
32. Mackowski, D.W.; Mishchenko, M.I. A multiple sphere T-matrix Fortran code for use on parallel computer clusters. *J. Quant. Spectrosc. Radiat. Transf.* **2011**, *112*, 2182–2192. [CrossRef]
33. Mishchenko, M.I.; Travis, L.D.; Mackowski, D.W. T-matrix computations of light scattering by nonspherical particles: A review. *J. Quant. Spectrosc. Radiat. Transf.* **1996**, *55*, 535–575. [CrossRef]
34. Mishchenko, M.I.; Travis, L.D.; Mackowski, D.W. T-matrix method and its applications to electromagnetic scattering by particles: A current perspective. *J. Quant. Spectrosc. Radiat. Transf.* **2010**, *111*, 1700–1703. [CrossRef]
35. Mishchenko, M.I.; Martin, P.A. Peter Waterman and T-matrix methods. *J. Quant. Spectrosc. Radiat. Transf.* **2013**, *123*, 2–7. [CrossRef]
36. Bi, L.; Yang, P.; Kattawar, G.W.; Mishchenko, M.I. Efficient implementation of the invariant imbedding T-matrix method and the separation of variables method applied to large nonspherical inhomogeneous particles. *J. Quant. Spectrosc. Radiat. Transf.* **2013**, *116*, 169–183. [CrossRef]
37. Tapiador, F.J.; Checa, R.; De Castro, M. An experiment to measure the spatial variability of rain drop size distribution using sixteen laser disdrometers. *Geophys. Res. Lett.* **2010**, *37*. [CrossRef]
38. Tapiador, F.J.; Navarro, A.; Moreno, R.; Jiménez-Alcázar, A.; Marcos, C.; Tokay, A.; Durán, L.; Bodoque, J.M.; Martín, R.; Petersen, W.; et al. On the optimal measuring area for pointwise rainfall estimation: A dedicated experiment with 14 laser disdrometers. *J. Hydrometeorol.* **2017**, *18*, 753–760. [CrossRef]
39. Van de Hulst, H.C. *Light Scattering by Small Particles*; Dover Publications: Mineola, NY, USA, 1981.
40. Bohren, C.F. *Absorption and Scattering of Light by Small Particles*; John Wiley & Sons: Hoboken, NJ, USA, 1983.
41. Tapiador, F.J.; Haddad, Z.S.; Turk, J. A probabilistic view on raindrop size distribution modeling: A physical interpretation of rain microphysics. *J. Hydrometeorol.* **2014**, *15*, 427–443. [CrossRef]
42. Marzano, F.S.; Ferrauto, G.; Roberti, L.; Di Michele, S.; Mugnai, A.; Tassa, A. Numerical Simulation of Multiple Scattering Effects due to Convective Clouds on Satellite Radar Reflectivity at 14 and 35 GHz. In Proceedings of the 2003 IEEE International Geoscience and Remote Sensing Symposium, Toulouse, France, 21–25 July 2003; Volume 2, pp. 881–883.
43. Seow, Y.-L.; Li, L.-W.; Leong, M.-S.; Kooi, P.-S.; Yeo, T.-S. An efficient TCS formula for rainfall microwave attenuation: T-matrix approach and 3-D fitting for oblate spheroidal raindrops. *IEEE Trans. Antennas Propag.* **1998**, *46*, 1176–1181. [CrossRef]

remote sensing

MDPI

Article

Estimates of the Change in the Oceanic Precipitation Off the Coast of Europe due to Increasing Greenhouse Gas Emissions

Francisco J. Tapiador [1,*], Andrés Navarro [1], Cecilia Marcos [2] and Raúl Moreno [1]

[1] Earth and Space Sciences Group (ESS), Institute of Environmental Sciences (ICAM),
    University of Castilla-La Mancha (UCLM), 45071 Toledo, Spain; Andres.Navarro@uclm.es (A.N.);
    raul.moreno@uclm.es (R.M.)
[2] National Meteorology Agency (AEMET), 28071 Madrid, Spain; cmarcosm@aemet.es
*   Correspondence: Francisco.Tapiador@uclm.es; Tel.: +34-925-268-800 (ext. 5762)

Received: 28 June 2018; Accepted: 25 July 2018; Published: 31 July 2018

Abstract: This paper presents a consensus estimate of the changes in oceanic precipitation off the coast of Europe under increasing greenhouse gas emissions. An ensemble of regional climate models (RCMs) and three gauge and satellite-derived observational precipitation datasets are compared. While the fit between the RCMs' simulation of current climate and the observations shows the consistency of the future-climate projections, uncertainties in both the models and the measurements need to be considered to generate a consensus estimate of the potential changes. Since oceanic precipitation is one of the factors affecting the thermohaline circulation, the feedback mechanisms of the changes in the net influx of freshwater from precipitation are relevant not only for improving oceanic-atmospheric coupled models but also to ascertain the climate signal in a global warming scenario.

Keywords: precipitation; satellites; climate models; regional climate models

1. Introduction

The importance of a precise estimation of oceanic precipitation derives from the freshening effect exerted on the oceanic surface. In certain high European latitudes of the Atlantic Ocean, precipitation exceed evaporation, so precipitation acts as a net source of freshwater that decreases the salinity and thus the density of surface waters. It has been known for some time that the extent in which dense, salty water sinks in the North Atlantic largely affect this circulation [1]. This freshening is one of the processes that has been proposed [2,3] to explain the observed slowing down of the Atlantic meridional overturning circulation in middle latitudes [4].

The precise measurement of such process has been addressed by several authors [5–7]. For instance, Meier has provided insight into future changes in the salinity of the Baltic Sea [6,8] while pointing out that salinity projections have suffered from the large uncertainties of precipitation estimates coming from models [6]. There is certainly agreement in that to address model shortcomings and provide a more precise quantification of the oceanic precipitation, both satellite estimates of precipitation and global circulation/climate models (GCMs) are required.

GCMs are instrumental for improving our understanding of the climate, but their coarse horizontal grid spacing (typically 50–300 km) smooths the highly variable field of precipitation [9]. On the other hand, regional climate models (RCMs) have been proved useful to provide fine-scale, physically-based downscaling [10–12] of global models' outputs. The idea behind an RCM is to improve the simulations by nesting the model on a GCM [13–15]. RCMs have been used to analyze processes (such as orographic enhancement) that GCMs are unable to adequately resolve [9,16,17]. However, criticisms of RCMs remain.

In order to ascertain how reliable an RCM's projection is, present-climate simulations are compared with observed climatologies. If the RCM compares favorably, then it is assumed that the model is suitable to make projections of the future climate. Indeed, this is a necessary, but by no means sufficient, condition.

Several studies have compared RCM simulations with observed climatologies [13,18–20] focusing on changes in mean seasonal values and interannual variability [14,21], including such data for precipitation [22,23]. In these particular cases, comparisons for present or future climate [24] are often done against CRU (climatic research unit) data [25,26]. The CRU dataset, however, is limited to land areas. Moreover, given the differences between oceanic and land precipitation [27] and the dissimilar surface feedbacks between the two cases [28], a separate validation should be required to analyze the uncertainties in future land and oceanic precipitation. Therefore, satellite observations of oceanic precipitation are needed, but this limits comparisons to the 39 years from 1979 to the present. The most reliable estimates of oceanic precipitation require microwave instrumentation, further reducing the comparison window from 1987 at best. On the other hand, PRUDENCE RCM simulations cover a specific interval (1960–1990; cf. Section 2.2 below for the reason of such specific dates), so the comparison has to be limited to the satellite-models overlapping period. This is not optimum, since the observational data includes infrared (IR) estimates which are known to be indirect, but it is a required compromise.

2. Materials and Methods

2.1. Observations

GPCP, CMAP and CPC PREC data (CPC hereafter) were used as the observational databases to determinate the current climate precipitation climatology baseline. Table 1 shows a comparison of the features of the empirical datasets. Subsections below briefly describe relevant information for each product.

Table 1. Summary of the empirical base.

Dataset	Temporal Resolution	Spatial Aggregation	Geographical Coverage	Original Sources	Period Covered
		Observations			
CPC	Monthly	$2.5° \times 2.5°$	Global	Raingauge + EOFs	1948–present
CMAP	Monthly	$2.5° \times 2.5°$	Global	Satellite + Raingauge	1979–present
GPCP [1]	Daily	$2.5° \times 2.5°$	Global	Satellite + Raingauge	1979–present
		Simulations			
PRUDENCE	Daily	$0.5° \times 0.5°$	Europe	RCMs	1960–1990 2070–2100 (A2)

[1] $1.0° \times 1.0°$ data available for 1996–2015.

2.1.1. CPC PREC Data

The Climate Prediction Center (CPC) disseminates a product named as NOAA's Precipitation Reconstruction Dataset (PREC) at a $0.5° \times 0.5°$ spatial resolution ([29], hereafter CPC data). Original data come from rain gauge observations from over 17,000 stations from the Global Historical Climatology Network (GHCN) version 2 and the Climate Anomaly Monitoring System (CAMS) datasets. Interpolation is performed using Gandin's optimal interpolation (OI) technique [30]. The ocean part of the dataset is calculated using an empirical orthogonal function (EOF) reconstruction of historical observations over ocean. The EOF modes are derived from EOF analysis of the satellite estimates for later years (1979–1998) with complete spatial coverage. The current version, 1.0, is based on the first 8 EOF global modes for each of the four seasons and the first four EOF modes of the residual components of the eight EOF global modes over Atlantic Ocean areas [29].

2.1.2. CMAP Data

The US Climate Prediction Center's Merged Analysis of Precipitation (CMAP) has made available monthly land and ocean estimates of global precipitation in which observations from rain gauges are merged with satellite precipitation estimates. The spatial resolution of the dataset is $2.5° \times 2.5°$ for the 1979–2018 period. The merging technique is fully described in [31]. As stated in the CMAP dataset documentation, the methodology first reduces random errors by linearly combining satellite estimates using a maximum likelihood method that gives an inversely proportional weight to the linear combination coefficients in relation to the square of the random error of the individual sources. Over global land areas, the random error is defined for each time period and grid location by comparing the data source with the rain gauge analysis over the surrounding area. Over oceans, the random error is defined by comparing the data sources with the rain gauge observations over the Pacific atolls. Bias is reduced by blending the data sources in the second step using the variational blending technique of Reynolds [32]. The actual CMAP precipitation data version used in this study was provided by the NOAA/OAR/ESRL PSD.

2.1.3. GPCP Data

The International Global Precipitation Climatology Project (GPCP) has produced a comprehensive, mixed rain gauge and satellite global precipitation product for both land and ocean. Several papers [33–37] describe the GPCP product at $2.5° \times 2.5°$ monthly resolution. This resolution is improved to $1.0° \times 1.0°$ daily estimates for the period 1996–2015 (and expected to be continued till present for the near future). Data sources includes the FAO, CRU, GHCN and the additional GPCC original sources seen above. Both land and ocean areas are covered thanks to the use of geo-stationary and low earth orbit satellites, including visible (VIS), infrared (IR) and passive microwave (PMW) sensors.

Satellite information is used to cover the gaps in the rain gauge data, with a physically-based algorithm being used for the retrieval of surface rain from PMW sensors and a statistically-based relationship based on cloud top temperature for VIS and IR data. Screening procedures and separate treatment of outliers ensures the homogeneity of the time series. Only the homogenized and nearly gap-free time series of 9343 stations are taken into account, but the long-term means of over 28,000 stations are used in order to estimate average precipitation fields [38]. All interpolations are performed using ordinary kriging with local and seasonal decorrelation lengths estimated from the observations [38].

2.2. RCMs Simulations

Eight RCMs were used to compare present-climate and increased greenhouse gases scenario simulations. These RCMs were part of the Prediction of Regional Scenarios and Uncertainties for Defining European Climate Change Risks and Effects (PRUDENCE) project, which aimed to provide high-resolution future scenarios of climate change for Europe [39]. The models are the DMI's HIRHAM [40], ETH's CHRM [41], KNMI's RACMO [42], GKSS's CLM [43], Hadley Center's HadRM3H [44], MPI's REMO [45], UCLM's PROMES [46] and the SMHI's RCAO [47–49].

Full 30-year simulations were computed for nominal present (1960–1990) and future (2070–2100) SRES-A2 scenario climates. The present-climate period was selected following WMO's technical regulations, who recognizing the need for a stable base for long-term climate change and variability assessment, fixed a reference, 30-year period from 1 January 1961 to 31 December 1990. This period is used to compare climate change and variability across all countries relative to this standard reference period.

The eight RCM models were forced with the HadAM3H-GCM model [50]. Initial fields of sea surface temperature, sulfate aerosols and other forcings were taken from the GCM to initialize the

RCMs. Greenhouse gases concentrations forcings were obtained from the IPCC [51] for the A2 scenario. The GCM drove the RCMs every 6 h on the corresponding domain contours (typically 8 to 10 points) with perfect boundary conditions.

Current-climate simulation used observed sea surface temperatures (SSTs) from the HadISST database [52], while future period values were obtained from the HadCM3 coupled atmosphere ocean global climate model [53]. More precisely, future monthly SST anomalies were obtained adding present-climate SST to present-climate SST anomalies, as described in [54].

The different domains of the individual RCMs cover most of Europe, the European shore of the Atlantic, the Mediterranean, and Northern Africa. There is a common region covering the center and south of Europe. To make our results fully comparable, this overlapping region was selected as our study area.

Daily precipitation values from the observational datasets were interpolated to a common $0.5° \times 0.5°$ grid by bilinear interpolation, and monthly means were obtained for each cell. The results of the simulations were also interpolated to a common grid of $0.5°$ resolution (roughly 50 km for these latitudes). The first year of each simulation (either 1960 or 2070) was used as spin-up period to allow the RCMs to adjust their large-scale conditions, especially those related with soil moisture conditions.

In order to have a common period for both observations and RCMs simulations, mean climatological values of the 10-years overlapping period of the datasets were used. Although a full climatological characterization would require a 30-year period, shorter periods have been found useful in similar studies [51]. For future climate, the values from January 2089 to January 2099 were used, thus allowing both datasets to be at the same distance from their corresponding spinning periods.

The reason for using the PRUDENCE data instead of more recent datasets is twofold. First, the dataset has withstood time and proved its validity over the years, with a large number of publications analyzing and validating the results. The dataset has been independently cross-validated over a long period of time and no major issues have arisen. More recent datasets, such as CORDEX, still have to pass the test of time. Secondly, the RCMs were nested on a GCM and not in reanalysis which, in a sense, makes the comparison between present and future climates more consistent (since by definition there is no reanalysis of the future).

Also, while the use of SRES scenarios instead of representative concentration pathways (RCP) may seem outdated in 2018, PRUDENCE results are still used to inform policies for climate change adaptation, and therefore it is still worth to examine how they compare with observations.

2.3. Averaging Method

The model data can be thought as a four-dimensional hypercube of longitude, latitude, time and model. Thus, we note

$$PR_{ijkm} \tag{1}$$

for the estimate of precipitation of the *m-eth* RCM for the *k-eth* month, at *j-eth* latitude and *i-eth* longitude. Averaging over the *k* index for all the *n* months gives the average of the model:

$$Average\ of\ the\ model\ \mathrm{m} = \frac{1}{n} \sum_{k=time_1}^{k=time_n} PR_{ijkm} \equiv \overline{PR_{ijm}}^k \tag{2}$$

where the overbar denotes the average over *k*. Further averaging over all the RCMs (*m* = 10) yields the average of the ensemble:

$$Average\ of\ the\ RCM\ ensemble = \frac{1}{10 \cdot n} \sum_{m=1}^{m=10} \sum_{k=time_1}^{k=time_n} PR_{ijkm} \equiv \overline{\overline{PR_{ij}}^k}^m \equiv \langle RCM \rangle. \tag{3}$$

The ensemble mean ⟨RCM⟩ is thus a two-dimension field ($i \times j$). The same averaging procedure is performed with the observational datasets. Thus, for instance for the GCP:

$$GPCP\ Average = \frac{1}{n} \sum_{k=time_1}^{k=time_n} PR_{ijk(GPCP)} \equiv \overline{PR_{ij(GPCP)}}^k \equiv GPCP. \tag{4}$$

The precipitation average of this observational dataset is noted just as "GPCP" in order to keep the notation as clean as possible. The same for CPC and CMAP.

3. Results

Figure 1 shows the mean precipitation estimates for the model ensemble mean ⟨RCMs⟩ and the three observational databases. The agreement of present climate conditions is noticeable albeit differences exist, for instance, in the North Sea and the Adriatic Sea. Table 2 gives a quantitative overview of the cross-correlations between the RCMs, the ensemble mean, and the observations. The ⟨RCMs⟩ compares well with the data being the best correlation with the GPCP (0.842; compared to 0.822 for the CPC, and to 0.792 for the CMAP datasets).

The correlations for present climate are high enough to confirm the relatively good performances of the RCMs and to have confidence in their projections for future climates.

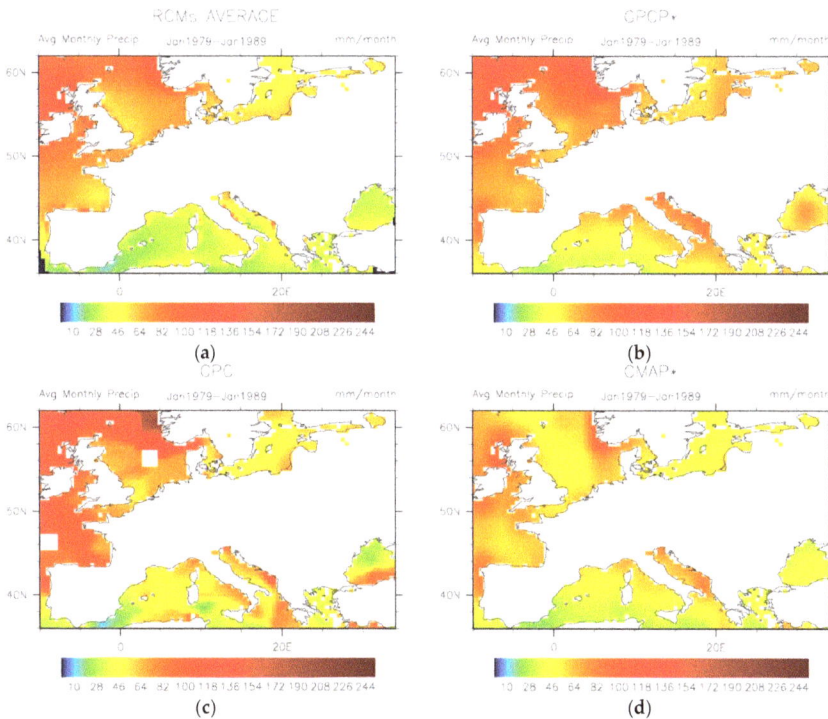

Figure 1. Observational estimates of the mean precipitation for the January 1979–January 1989 period. Units are in mm/month. (**a**) RCM average; (**b**) GPCP estimate; (**c**) CPC estimate; (**d**) CMAP estimate.

An important variable of interest to evaluate the changes in oceanic precipitation is the precipitation climate signal, which is the difference between the precipitation in the future and at present. Such reference precipitation at present can be either the RCM outputs or the observational datasets. Figure 2 shows the climate signal for the ensemble mean in future climate $\langle RCMs \rangle_{FC}$ minus the ensemble mean of the RCMs at present $\langle RCMs \rangle_{PC}$. The figure also shows the climate signal taken separately with reference from the three observational datasets, GPCP, CMAP and CPC.

The differences between $\langle RCMs \rangle_{FC}$ and $\langle RCMs \rangle_{PC}$ are smaller than those between $\langle RCMs \rangle_{FC}$ and the observations. Amongst them all, the CMAP-based climate signal exhibits the largest projected increases in oceanic precipitation with more than a 50% difference above 56° N.

Table 2. Cross-correlations (R) of the monthly mean present-climate precipitation (January 1979–January 1989) for the three observational datasets (CPC, GPCP and CMAP), the eight individual RCMs, and the RCMs ensemble average ($\langle RCMs \rangle$).

	CPC[1]	GPCP[1]	CMAP[1]	HIR.	CHRM	RCAO	CLM	Had.	REMO	PRO.	RAC.	⟨RCMs⟩
					Ocean	*Only*						
CPC[1]	1	0.866	0.808	0.775	0.8	0.645	0.836	0.625	0.743	0.804	0.667	0.822
GPCP[1]		1	0.885	0.788	0.768	0.731	0.832	0.658	0.748	0.741	0.745	0.842
CMAP[1]			1	0.786	0.715	0.634	0.824	0.583	0.701	0.741	0.683	0.792
HIRHAM				1	0.807	0.701	0.848	0.755	0.726	0.798	0.77	0.895
CHRM					1	0.749	0.871	0.728	0.859	0.792	0.77	0.919
RCAO						1	0.725	0.742	0.822	0.655	0.918	0.889
CLM							1	0.685	0.783	0.808	0.772	0.907
HadRM3H								1	0.717	0.704	0.782	0.854
REMO									1	0.742	0.775	0.902
PROMES										1	0.719	0.863
RACMO											1	0.914
⟨RCMs⟩												1

[1] Distance-weighted resampled at 0.5° × 0.5° resolution from the original 2.5° × 2.5° data.

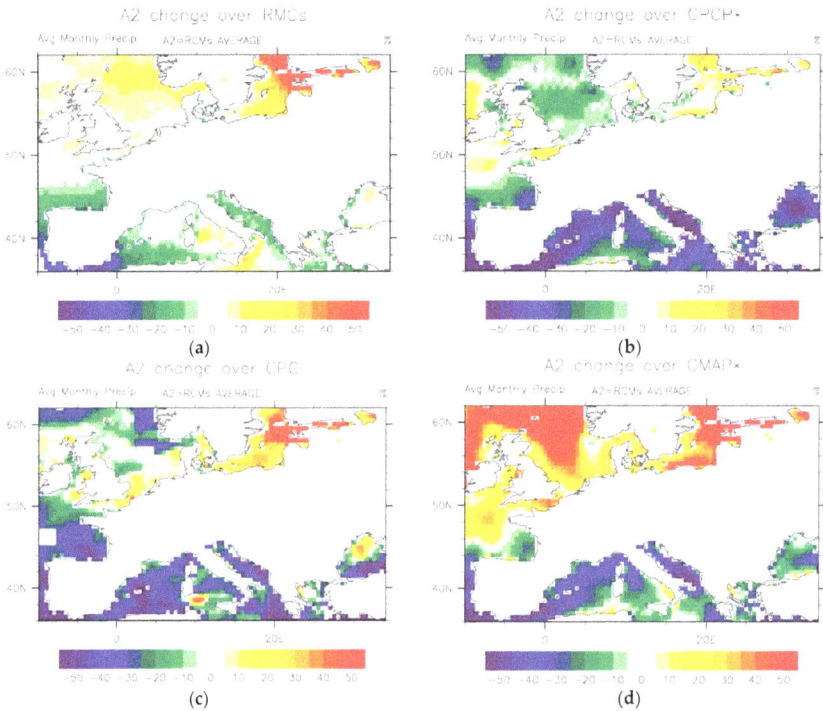

(a)

(b)

(c)

(d)

Figure 2. Differences in oceanic precipitation for the A2 scenario (**a**) taking the RCMs as reference for present climate; (**b**) taking GPCP as reference; (**c**) taking CPC as reference; and (**d**) taking CMAP as reference. Units are percentage over present-climate precipitation in the RCMs.

Since the GPCP is the dataset that presents the best fit with present-climate simulations (0.842 R^2; Table 2), differences with respect to GPCP in the future can be interpreted as the best-case scenario. That is, uncertainties in the climate signal when GPCP is used as the reference are the minimum uncertainties to be found.

Figure 3 shows the distribution of the differences between future- and present-climate when either the average of the RCMs (x-axis) or the GPCP (y-axis) are used as the present-climate climatology. Data in the upper-left and the lower-right quadrants are those in which the precipitation climate signal differs in both reference datasets. In the other two quadrants (1st: upper-right, and 3rd: lower-left), there is a consensus in the sign of the climate signal: positive in the case of the 1st quadrant and negative for the 3rd.

The instances in which the climate signal is negative (less precipitation expected in the future) if the reference data is the RCMs average but positive if the reference data is the GPCP (i.e., upper-left, 2nd quadrant) are limited. However, there is large number of cases for which the climate signal is positive in the first case and negative in the second (lower-right: 4th quadrant); that is, there are many places where the models predict more precipitation in the future when their simulations are compared with their own present-climate simulations, but less when compared with the GPCP reference for present-climate. Indeed, Figure visually represents the underlying reason, model bias, while at the same time providing a succinct account of the consensus in the climate signal between models and observations (GPCP here).

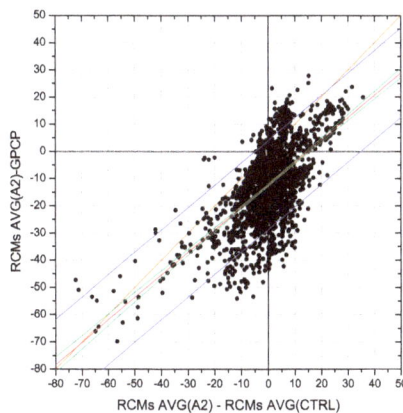

Figure 3. Comparison of the differences in the estimated changes in the mean precipitation of Europe in the A2 scenario between either using the GPCP as reference or the RCMs average for present–climate conditions (control run; CTRL). Lines are the perfect match (orange line), the best linear fit (red line), the 85% confidence interval of the best linear fit (green lines) and the 85% prediction limits (blue lines). Each dot in the plots represent the monthly average for a model grid point. Dots in the 2nd quadrant (upper-left) and in the 4th quadrant (bottom-right) indicate locations in which the reference dataset shows a discrepancy in the precipitation climate signal. Notice that since the plot compares differences, the RCMs AVG (A2) can feature in both axes. Each dot represents the value of the 30-year averages for that grid point (i.e., $\overline{\overline{PR_{ij}}}^{k^m}$).

Therefore, a metric for the consensus in the precipitation climate signal between the RCMs and the references (GPCP, CMAP and CPC) can be built by the segmentation of the grid points into four quadrants. The resulting three categories (Figure 4): consensus (1st and 3rd quadrants), positive-negative dissent (2nd quadrant), and negative-positive dissent (4th quadrant) are helpful to evaluate the changes depicted in Figure 2. Where consensus exists, the changes can be considered more robust while if there is dissent then the model bias may be obscuring the signal.

Figure 4 shows that the RCMs and observations agree in most of the area of interest. Nevertheless, discrepancies appear in the Black Sea, the seas south of Italy, the North Sea above 56° North and, of opposite sign, off North Spain and in the English Channel. If we deem GPCP as the most reliable dataset for reference (Figure 4a), discrepancies are mainly restricted to positive bias in the North Sea, the seas south of Italy and the Black Sea. Conversely, CPC and CMAP (Figure 4b,c) show larger consensus areas than those with the GPCP reference, especially in the North Sea, but differences in the amount of precipitation change are higher than those of RCMs and GPCP (cf. Figure 2).

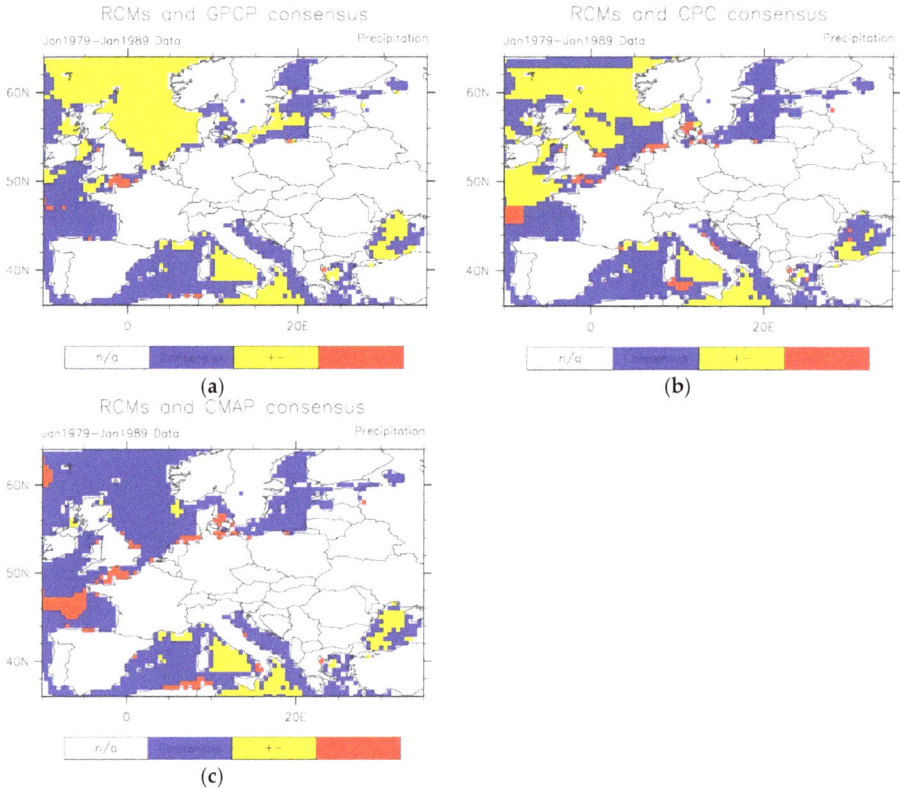

Figure 4. Consensus/dissent for the January 1979–January 1989 mean precipitation between RCMs and GPCP (**a**); RCMs and CPC (**b**); and RCMs and CMAP (**c**). Blues represent consensus between RCMs and observational datasets (1st and 3rd quadrants in Figure 3); yellows dissent (places where the models predict more precipitation in the future if they are compared with the present-climate, model derived reference but less if compared with present-climate GPCP reference, 4th quadrant); and reds those cases for which the climate signal is negative (less precipitation in the future) if the reference data is the RCMs average for present climate, but positive if the reference data is the GPCP (upper-left, 2nd quadrant).

If we take a closer look at Figures 2 and 4, precipitation is expected to decrease with high confidence (i.e., large consensus) in the Mediterranean Basin—the Western Mediterranean Sea, the Adriatic Sea and the Aegean Sea, in the Bay of Biscay, and in the Kattegat Sea area. In those areas, the freshening effect of precipitation is expected to be reduced in the A2 emissions scenario, yielding an increase in salinity. However, a net increase is expected in the Celtic Sea and in the Baltic

Remote Sens. **2018**, *10*, 1198

Sea. The analysis of the North Sea is more complex because of the disagreement between GPCP and RCMs (Figure 4a). Nevertheless, when compared with CMAP (Figure 4c) a net increase in precipitation is observed.

4. Conclusions

In this paper, the oceanic precipitation projections of eight RCMs for the present-climate and the A2-SRES future-climate scenario have been compared with satellite and offshore gauge estimates. Notwithstanding observational errors [55,56] and several other uncertainties, the results show that the RCMs consistently reproduce observed oceanic precipitation of Europe, thus increasing the confidence in such models being capable of estimating changes in the future oceanic precipitation patterns.

By integrating the uncertainties in both the observational and the modeled oceanic precipitation, a consensus estimate is made, resulting in increased confidence of an overall net increase of the freshening above 46° North and a relative decrease in the Mediterranean Basin.

Since oceanic precipitation is one of the factors affecting the thermohaline circulation, the feedback mechanisms of the changes in the net influx of freshwater from precipitation are relevant not only for improving oceanic-atmospheric coupled models but also to ascertain the climate signal in a global warming scenario. Within this context, the use of satellite-derived datasets is crucial as they are the only available means to build the reference datasets required to estimate the magnitude of the changes and their uncertainties. Also, new modeling initiatives such as the Coordinated Regional Climate Downscaling Experiment (CORDEX) are proving to be instrumental to ascertain the effects of precipitation in the global hydrological cycle [57–63], and especially about its impacts in Europe (EURO-CORDEX [64]).

Author Contributions: F.J.T. led the research and outlined the draft of the manuscript. A.N., C.M., and R.M. contributed to analysis, plotting, and manuscript writing.

Funding: This research was funded by the Ministerio de Economía y Competitividad, Ciencia e Innovación under grant numbers CGL2013-48367-P, CGL2016-80609-R, and by the Ministerio de Educación under FPU grant 13/02798.

Acknowledgments: Thanks are due to Nabhonil Kar (Princeton University) for his editing of the manuscript.

Conflicts of Interest: The authors declare no conflicts of interest.

References

1. Hartmann, D.L. *Global Physical Climatology*, 2nd ed.; Elsevier: Cambridge, MA, USA, 2016.
2. Dickson, B.; Yashayaev, I.; Meincke, J.; Turrell, B.; Dye, S.; Holfort, J. Rapid freshening of the deep North Atlantic Ocean over the past four decades. *Nature* **2002**, *416*, 832–837. [CrossRef] [PubMed]
3. Dickson, R.; Rudels, B.; Dye, S.; Karcher, M.; Meincke, J.; Yashayaev, I. Current estimates of freshwater flux through Arctic and subarctic seas. *Prog. Oceanogr.* **2007**, *73*, 210–230. [CrossRef]
4. Cunningham, S.A.; Kanzow, T.; Rayner, D.; Baringer, M.O.; Johns, W.E.; Marotzke, J.; Longworth, H.R.; Grant, E.M.; Hirschi, J.J.-M.; Beal, L.M.; et al. Temporal variability of the Atlantic meridional overturning circulation at 26.5°N. *Science* **2007**, *317*, 935–938. [CrossRef] [PubMed]
5. Brossier, C.L.; Béranger, K.; Drobinski, P. Ocean response to strong precipitation events in the Gulf of Lions (northwestern Mediterranean Sea): A sensitivity study. *Ocean Dyn.* **2012**, *62*, 213–226. [CrossRef]
6. Meier, H.E.M.; Kjellström, E.; Graham, L.P. Estimating uncertainties of projected Baltic Sea salinity in the late 21st century. *Geophys. Res. Lett.* **2006**, *33*. [CrossRef]
7. Zeng, L.; Chassignet, E.P.; Schmitt, R.W.; Xu, X.; Wang, D. Salinification in the South China Sea Since Late 2012: A Reversal of the Freshening Since the 1990s. *Geophys. Res. Lett.* **2018**, *45*, 2744–2751. [CrossRef]
8. Meier, H.E.M. Baltic Sea climate in the late twenty-first century: A dynamical downscaling approach using two global models and two emission scenarios. *Clim. Dyn.* **2006**, *27*, 39–68. [CrossRef]
9. Pan, X.; Li, X.; Cheng, G.; Hong, Y. Effects of 4D-Var data assimilation using remote sensing precipitation products in a WRF model over the complex terrain of an arid region river basin. *Remote Sens.* **2017**, *9*, 963. [CrossRef]

10. Feser, F.; Rockel, B.; von Storch, H.; Winterfeldt, J.; Zahn, M. Regional Climate Models Add Value to Global Model Data: A Review and Selected Examples. *Bull. Am. Meteorol. Soc.* **2011**, *92*, 1181–1192. [CrossRef]

11. Giorgi, F.; Marinucci, M.R.; Visconti, G. A 2XCO2 climate change scenario over Europe generated using a limited area model nested in a general circulation model 2. Climate change scenario. *J. Geophys. Res. Atmos.* **1992**, *97*, 10011–10028. [CrossRef]

12. Kendon, E.J.; Jones, R.G.; Kjellström, E.; Murphy, J.M. Using and designing GCM-RCM ensemble regional climate projections. *J. Clim.* **2010**, *23*, 6485–6503. [CrossRef]

13. Giorgi, F.; Bi, X.; Pal, J.S. Mean, interannual variability and trends in a regional climate change experiment over Europe. I. Present-day climate (1961–1990). *Clim. Dyn.* **2004**, *22*, 733–756. [CrossRef]

14. Giorgi, F.; Bi, X.; Pal, J. Mean, interannual variability and trends in a regional climate change experiment over Europe. II: Climate change scenarios (2071–2100). *Clim. Dyn.* **2004**, *23*, 839–858. [CrossRef]

15. Robertson, A.W.; Qian, J.-H.; Tippett, M.K.; Moron, V.; Lucero, A. Downscaling of Seasonal Rainfall over the Philippines: Dynamical versus Statistical Approaches. *Mon. Weather Rev.* **2012**, *140*, 1204–1218. [CrossRef]

16. Choi, H., II. Application of a land surface model using remote sensing data for high resolution simulations of terrestrial processes. *Remote Sens.* **2013**, *5*, 6838–6856. [CrossRef]

17. Zhang, M.; Luo, G.; De Maeyer, P.; Cai, P.; Kurban, A. Improved atmospheric modelling of the oasis-desert system in central Asia using WRF with actual satellite products. *Remote Sens.* **2017**, *9*, 1273. [CrossRef]

18. Déqué, M.; Jones, R.G.; Wild, M.; Giorgi, F.; Christensen, J.H.; Hassell, D.C.; Vidale, P.L.; Rockel, B.; Jacob, D.; Kjellström, E.; et al. Global high resolution versus Limited Area Model climate change projections over Europe: Quantifying confidence level from PRUDENCE results. *Clim. Dyn.* **2005**, *25*, 653–670. [CrossRef]

19. Räisänen, J.; Hansson, U.; Ullerstig, A.; Döscher, R.; Graham, L.P.; Jones, C.; Meier, H.E.M.; Samuelsson, P.; Willén, U. European climate in the late twenty-first century: Regional simulations with two driving global models and two forcing scenarios. *Clim. Dyn.* **2004**, *22*, 13–31. [CrossRef]

20. Schoetter, R.; Hoffmann, P.; Rechid, D.; Schlünzen, K.H. Evaluation and bias correction of regional climate model results using model evaluation measures. *J. Appl. Meteorol. Climatol.* **2012**, *51*, 1670–1684. [CrossRef]

21. Déqué, M.; Rowell, D.P.; Lüthi, D.; Giorgi, F.; Christensen, J.H.; Rockel, B.; Jacob, D.; Kjellström, E.; De Castro, M.; Van Den Hurk, B. An intercomparison of regional climate simulations for Europe: Assessing uncertainties in model projections. *Clim. Chang.* **2007**, *81*, 53–70. [CrossRef]

22. Beniston, M.; Stephenson, D.B.; Christensen, O.B.; Ferro, C.A.T.; Frei, C.; Goyette, S.; Halsnaes, K.; Holt, T.; Jylhä, K.; Koffi, B.; et al. Future extreme events in European climate: An exploration of regional climate model projections. *Clim. Chang.* **2007**, *81*, 71–95. [CrossRef]

23. Tapiador, F.J.; Sánchez, E.; Gaertner, M.A. Regional changes in precipitation in Europe under an increased greenhouse emissions scenario. *Geophys. Res. Lett.* **2007**, *34*, L06701. [CrossRef]

24. Pal, J.S.; Giorgi, F.; Bi, X. Consistency of recent European summer precipitation trends and extremes with future regional climate projections. *Geophys. Res. Lett.* **2004**, *31*. [CrossRef]

25. New, M.; Lister, D.; Hulme, M.; Makin, I. A high-resolution data set of surface climate over global land areas. *Clim. Res.* **2002**, *21*, 1–25. [CrossRef]

26. Harris, I.; Jones, P.D.D.; Osborn, T.J.J.; Lister, D.H.H. Updated high-resolution grids of monthly climatic observations-the CRU TS3.10 Dataset. *Int. J. Climatol.* **2014**, *34*, 623–642. [CrossRef]

27. Doherty, R.M.; Hulme, M.; Jones, C.G. A gridded reconstruction of land and ocean precipitation for the extended tropics from 1974 to 1994. *Int. J. Climatol.* **1999**, *19*, 119–142. [CrossRef]

28. Koster, R.D.; Suarez, M.J. Relative contributions of land and ocean processes to precipitation variability. *J. Geophys. Res.* **1995**, *100*, 13775. [CrossRef]

29. Chen, M.; Xie, P.; Janowiak, J.E.; Arkin, P.A.; Chen, M.; Xie, P.; Janowiak, J.E.; Arkin, P.A. Global Land Precipitation: A 50-yr Monthly Analysis Based on Gauge Observations. *J. Hydrometeorol.* **2002**, *3*, 249–266. [CrossRef]

30. Reynolds, R.W.; Smith, T.M. Improved global sea surface temperature analyses using optimum interpolation. *J. Clim.* **1994**, *7*, 929–948. [CrossRef]

31. Xie, P.; Arkin, P.A. Global precipitation: A 17-year monthly analysis based on gauge observations, satellite estimates and numerical model outputs. *Bull. Am. Meteorol. Soc.* **1997**, *78*, 2539–2558. [CrossRef]

32. Reynolds, R.W. A Real-Time Global Sea Surface Temperature Analysis. *J. Clim.* **1988**, *1*, 75–87. [CrossRef]

33. Huffman, G.J.; Adler, R.F.; Rudolf, B.; Schneider, U.; Keehn, P.R. Global precipitation estimates based on a technique for combining satellite-based estimates, rain gauge analysis, and NWP model precipitation information. *J. Clim.* **1995**, *8*, 1284–1295. [CrossRef]

34. Huffman, G.J.; Adler, R.F.; Arkin, P.; Chang, A.; Ferraro, R.; Gruber, A.; Janowiak, J.; McNab, A.; Rudolf, B.; Schneider, U. The Global Precipitation Climatology Project (GPCP) Combined Precipitation Dataset. *Bull. Am. Meteorol. Soc.* **1997**, *78*, 5–20. [CrossRef]

35. Huffman, G.J.; Adler, R.F.; Morrissey, M.M.; Bolvin, D.T.; Curtis, S.; Joyce, R.; McGavock, B.; Susskind, J. Global Precipitation at One-Degree Daily Resolution from Multisatellite Observations. *J. Hydrometeorol.* **2001**, *2*, 36–50. [CrossRef]

36. Huffman, G.J.; Adler, R.F.; Bolvin, D.T.; Gu, G. Improving the global precipitation record: GPCP Version 2.1. *Geophys. Res. Lett.* **2009**, *36*, L17808. [CrossRef]

37. Adler, R.F.; Sapiano, M.R.P.; Huffman, G.J.; Wang, J.J.; Gu, G.; Bolvin, D.; Chiu, L.; Schneider, U.; Becker, A.; Nelkin, E.; et al. The Global Precipitation Climatology Project (GPCP) monthly analysis (New Version 2.3) and a review of 2017 global precipitation. *Atmosphere* **2018**, *9*, 138. [CrossRef] [PubMed]

38. Smith, T.M.; Yin, X.; Gruber, A. Variations in annual global precipitation (1979-2004), based on the Global Precipitation Climatology Project 2.5° analysis. *Geophys. Res. Lett.* **2006**, *33*. [CrossRef]

39. Christensen, J.H.; Christensen, O.B. A summary of the PRUDENCE model projections of changes in European climate by the end of this century. *Clim. Chang.* **2007**, *81*, 7–30. [CrossRef]

40. Christensen, J.H.; Christensen, O.B.; Lopez, P.; van Meijgaard, E.; Botzet, M. The HIRHAM 4 regional atmospheric climate model 1996.

41. Vidale, P.L. Predictability and uncertainty in a regional climate model. *J. Geophys. Res.* **2003**, *108*, 4586. [CrossRef]

42. Lenderink, G.; Hurk, B.; Meijgaard, E.; Ulden, A.; Cuijpers, H. Simulation of Present-Day Climate in RACMO2: First Results and Model Developments. 2003. Available online: http://bibliotheek.knmi.nl/knmipubTR/TR252.pdf (accessed on 28 June 2018).

43. Steppeler, J.; Doms, G.; Schättler, U.; Bitzer, H.W.; Gassmann, A.; Damrath, U.; Gregoric, G. Meso-gamma scale forecasts using the nonhydrostatic model LM. *Meteorol. Atmos. Phys.* **2003**, *82*, 75–96. [CrossRef]

44. Buonomo, E.; Jones, R.; Huntingford, C.; Hannaford, J. On the robustness of changes in extreme precipitation over Europe from two high resolution climate change simulations. *Q. J. R. Meteorol. Soc.* **2007**, *133*, 65–81. [CrossRef]

45. Jacob, D. The role of water vapour in the atmosphere. A short overview from a climate modeller's point of view. *Phys. Chem. Earth Part A Solid Earth Geod.* **2001**, *26*, 523–527. [CrossRef]

46. Castro, M.; Fernandez, C.; Gaertner, M.A. Description of a mesoscale atmospheric numerical model. In *Mathematics, Climate and Environment*; Diaz, J.I., Lions, J., Eds.; Masson: Paris, France, 1993; pp. 230–253.

47. Döscher, R.; Willén, U.; Jones, C.; Rutgersson, A.; Meier, H.E.M.; Hansson, U.; Graham, L.P. The development of the regional coupled ocean-atmosphere model RCAO. *Boreal Environ. Res.* **2002**, *7*, 183–192.

48. Jones, C.; Carvalho, L.M.V.; Higgins, R.W.; Waliser, D.E.; Schemm, J.K.E. A statistical forecast model of tropical intraseasonal convective anomalies. *J. Clim.* **2004**, *17*, 2078–2095. [CrossRef]

49. Rummukainen, M.; Räisänen, J.; Bringfelt, B.; Ullerstig, A.; Omstedt, A.; Willén, U.; Hansson, U.; Jones, C. A regional climate model for northern Europe: Model description and results from the downscaling of two GCM control simulations. *Clim. Dyn.* **2001**, *17*, 339–359. [CrossRef]

50. Pope, V.D.; Gallani, M.L.; Rowntree, P.R.; Stratton, R.A. The impact of new physical parametrizations in the Hadley Centre climate model: HadAM3. *Clim. Dyn.* **2000**, *16*, 123–146. [CrossRef]

51. Nakićenović, N. *Intergovernmental Panel on Climate Change. Working Group III*; Cambridge University Press: Cambridge, UK, 2000.

52. Rayner, N.A. Global analyses of sea surface temperature, sea ice, and night marine air temperature since the late nineteenth century. *J. Geophys. Res.* **2003**, *108*, 4407. [CrossRef]

53. Rowell, D.P. A scenario of European climate change for the late twenty-first century: Seasonal means and interannual variability. *Clim. Dyn.* **2005**, *25*, 837–849. [CrossRef]

54. Christensen, J.H.; Carter, T.R.; Rummukainen, M.; Amanatidis, G. Evaluating the performance and utility of regional climate models: The PRUDENCE project. *Clim. Chang.* **2007**, *81*, 1–6. [CrossRef]

55. Gu, G.; Adler, R.F.; Huffman, G.J. Long-term changes/trends in surface temperature and precipitation during the satellite era (1979–2012). *Clim. Dyn.* **2016**, *46*, 1091–1105. [CrossRef]

56. Tapiador, F.J.; Navarro, A.; Jiménez, A.; Moreno, R.; García-Ortega, E. Discrepancies with Satellite Observations in the Spatial Structure of Global Precipitation as Derived from Global Climate Models. *Q. J. R. Meteorol. Soc.* **2018**. [CrossRef]

57. Knist, S.; Goergen, K.; Buonomo, E.; Christensen, O.B.; Colette, A.; Cardoso, R.M.; Fealy, R.; Fernández, J.; García-Díez, M.; Jacob, D.; et al. Land-atmosphere coupling in EURO-CORDEX evaluation experiments. *J. Geophys. Res. Atmos.* **2017**, *122*, 79–103. [CrossRef]

58. Mascaro, G.; Viola, F.; Deidda, R. Evaluation of Precipitation From EURO-CORDEX Regional Climate Simulations in a Small-Scale Mediterranean Site. *J. Geophys. Res. Atmos.* **2018**, *123*, 1604–1625. [CrossRef]

59. Hosseinzadehtalaei, P.; Tabari, H.; Willems, P. Precipitation intensity-duration-frequency curves for central Belgium with an ensemble of EURO-CORDEX simulations, and associated uncertainties. *Atmos. Res.* **2018**, *200*, 1–12. [CrossRef]

60. Cardoso, R.M.; Soares, P.M.M.; Lima, D.C.A.; Semedo, A. The impact of climate change on the Iberian low-level wind jet: EURO-CORDEX regional climate simulation. *Tellus A Dyn. Meteorol. Oceanogr.* **2018**, *68*. [CrossRef]

61. Smiatek, G.; Kunstmann, H.; Senatore, A. EURO-CORDEX regional climate model analysis for the Greater Alpine Region: Performance and expected future change. *J. Geophys. Res. Atmos.* **2016**, *121*, 7710–7728. [CrossRef]

62. Dosio, A. Projections of climate change indices of temperature and precipitation from an ensemble of bias-adjusted high-resolution EURO-CORDEX regional climate models. *J. Geophys. Res. Atmos.* **2016**, *121*, 5488–5511. [CrossRef]

63. Dalelane, C.; Früh, B.; Steger, C.; Walter, A. A pragmatic approach to build a reduced regional climate projection ensemble for Germany using the EURO-CORDEX 8.5 ensemble. *J. Appl. Meteorol. Climatol.* **2018**, *57*, 477–491. [CrossRef]

64. Jacob, D.; Petersen, J.; Eggert, B.; Alias, A.; Christensen, O.B.; Bouwer, L.M.; Braun, A.; Colette, A.; Déqué, M.; Georgievski, G.; et al. EURO-CORDEX: New high-resolution climate change projections for European impact research. *Reg. Environ. Chang.* **2014**, *14*, 563–578. [CrossRef]

remote sensing

MDPI

Article

The Implementation of a Mineral Dust Wet Deposition Scheme in the GOCART-AFWA Module of the WRF Model

Konstantinos Tsarpalis [1], Anastasios Papadopoulos [2], Nikolaos Mihalopoulos [3], Christos Spyrou [1], Silas Michaelides [4] and Petros Katsafados [1,*]

[1] Department of Geography, Harokopio University of Athens, El. Venizelou 70 Str., 17671 Kallithea, Greece; kostastsp@hotmail.com (K.T.); scspir@gmail.com (C.S.)
[2] Institute of Marine Biological Resources and Inland Waters, Hellenic Center of Marine Research, 46.7 km Athens-Sounion Ave., 19013 Anavissos Attikis, Greece; tpapa@hcmr.gr
[3] Institute for Environmental Research and Sustainable Development, National Observatory of Athens (NOA), Metaxa & Vas. Pavlou, 15236 Palea Penteli, Greece; nmihalo@noa.gr
[4] The Cyprus Institute, 20, Konstantinou Kavafi Str., 2121 Aglantzia, Nicosia, Cyprus; s.michaelides@cyi.ac.cy
*   Correspondence: pkatsaf@hua.gr; Tel.: +30-2109549384

Received: 30 July 2018; Accepted: 2 October 2018; Published: 6 October 2018

Abstract: The principal objective of this study is to present and evaluate an advanced dust wet deposition scheme in the Weather and Research Forecasting model coupled with Chemistry (WRF-Chem). As far as the chemistry component is concerned, the Georgia Tech Goddard Global Ozone Chemistry Aerosol Radiation and Transport of the Air Force Weather Agency (GOCART-AFWA) module is applied, as it supports a binary scheme for dust emissions and transport. However, the GOCART-AFWA aerosol module does not incorporate a wet scavenging scheme, nor does it interact with cloud processes. The integration of a dust wet deposition scheme following Seinfeld and Pandis into the WRF-Chem model is assessed through a case study of large-scale Saharan dust transport over the Eastern Mediterranean that is characterized by severe wet deposition over Greece. An acceptable agreement was found between the calculated and measured near surface PM_{10} concentrations, as well as when model estimated atmospheric optical depth (AOD) was validated against the AERONET measurements, indicating the validity of our dust wet deposition scheme.

Keywords: mineral dust; wet deposition; cloud scavenging; dust washout process; Saharan dust transportation; precipitation rate

1. Introduction

Dust particles in the atmosphere are deposited on the Earth's surface via wet (precipitation) or dry deposition (gravitational forcing). Hence, dust wet deposition is one of the most important sink processes of dust aerosol particles. The climatic and meteorological importance of mineral dust particles is significant, as it contributes to the absorption and scattering of solar radiation (shortwave and longwave), both of which modify the albedo of Earth-Atmosphere system [1]. Furthermore, mineral dust contributes to the modification of the optical properties of clouds and ice/snow surfaces [2,3], and even alters the water content of the atmospheric column [4]. Marconi et al. [5] identified two main source areas for intense dust episodes influencing the Mediterranean Basin: one in Algeria-Tunisia, and one in Libya. These extreme outbreaks mainly occur in autumn and spring [6–12]. In order to evaluate the effects of these episodes on global and regional scales, spatial distributions and temporal variations of deposition fluxes must be ascertained [13]. In addition to in-situ measurements, observations using ground-based and space borne lidars provide vertical and spatial distributions of mineral dust particles in the atmosphere [14–18]. However, data related to temporal variations and spatial distributions of mineral dust flux at the ground are still very limited [19,20].

A number of dust modeling studies have been carried out that focus either on dust emissions and advection [3,21,22] or dust budget, including dry/wet deposition [23]. The efficiency of the wet deposition depends on many factors, such as particle size distribution [24], raindrop size distribution [25], the chemical characteristics of the particles [26], scavenging efficiencies [27], and electrical forces [28]. Wet deposition processes are generally represented using a computationally-efficient bulk method with separate treatment of the in cloud (rainout) and below cloud (washout) scavenging [29,30].

Below cloud scavenging by precipitation is the process of aerosol removal from the atmosphere between a precipitating cloud base and the ground. The scavenging of dust particles by falling hydrometeors takes place by Brownian and turbulent shear diffusion, inertial impaction, diffiusiophoresis, thermophoresis, and electrical effects [31]. Guelle et al. [32] estimate an aerosol scavenging coefficient based on a parameterization according to the particle diameter [33] which is integrated over the aerosol size distribution [34]. Some corrections are also applied concerning the mean diameters of dust size distributions, thus providing more accurate scavenging efficiencies of the dust particles. Sensitivity tests have shown that during large-scale precipitation events, below cloud scavenging has been found to be negligible for submicronic aerosols, while the results according to observations show an acceptable reproduction of the annual wet deposition fluxes. Former studies reach the same conclusion [35–37], which shows that there is a minimum in the collection efficiency of aerosol particles with radii of 0.5 to 1 μm called the "Greenfield Gap" [31].

As far as the in-cloud scavenging is concerned, the precipitation formation is related to the cloud droplet number concentration and the liquid water content [38]. In a fundamental study, Giorgy and Chameides [39] estimate the in-cloud scavenging coefficient in terms of the local water vapor condensation (and precipitation) rate. Furthermore, Tsyro [40] and Guelle et al. [32], estimate the in-cloud scavenging as a function of the liquid water content and the particle hygroscopicity. In contrast to below-cloud scavenging, the in-cloud scavenging is important for submicron particles [41].

The available wet deposition parameterizations used for dust modeling can be classified into four types, based on their formulations [41]. The first type calculates the scavenging coefficient as a function of the raindrop-particle collection efficiency and raindrop size distribution [42–44]. In the second type, the wet scavenging coefficient is calculated as a function of a single value, for example relative humidity [45] or precipitation [23,46]. In the third type, the scavenging coefficient is estimated as an empirical relationship based on aerosol size spectrum and precipitation parameters [28]. As far as the fourth type is concerned, the scavenging coefficient is defined as the ratio of the dust concentration in the precipitation divided by the dust concentration in the air [47]. In their study, Jung and Shao [41], through an intercomparison of below cloud dust wet deposition schemes that rely on these four types, show that, apart from the third scheme which is based on field measurements, the other schemes showed similar wet deposition patterns, although the scavenging efficiencies were quite different. The scavenging efficiencies of the first scheme were negligible for submicron particles, which is acceptable, as submicron particle removal is an in-cloud scavenging matter, as mentioned above. The simulated dust concentrations with the third and the fourth schemes are underestimated in comparison to the observations. On the other hand, the dust concentrations concerning the first and the second types are in good agreement with the measured data [41].

In this study, a dust wet deposition scheme is implemented in the Georgia Tech/Goddard Global Ozone Chemistry Aerosol Radiation and Transport (GOCART) module of the fully coupled Weather Research and Forecasting coupled to Chemistry model (WRF/Chem) [48,49]. The original GOCART scheme [50] does not support any dust wet deposition scheme. Thus, a dust wet deposition scheme following Seinfeld and Pandis [44] is fully embedded in the revised GOCART-AFWA (Air Force Weather Agency) module. This scheme belongs to the first type of the abovementioned dust wet deposition schemes in a form adopted by the Comprehensive Air Quality Model with Extensions (CAMx) [22,51,52] for aerosols inside and below clouds. Although the scheme is computationally expensive [53], it does not contribute to systematic underestimation of the simulated dust concentration, as it has negligible scavenging effects on submicron particles for the area below the clouds. It also incorporates an in-cloud scavenging process for submicron particle removal. Moreover, it is a scheme that takes into account parameters that characterize the behavior of particles suspended in a fluid flow, such as the Reynolds, Schmidt, and Stokes numbers, as it represents wet removal using microphysical processes including detailed interactions between hydrometeors and aerosols.

The new scheme has been evaluated in a case study of a severe wet deposition event which affected Central and Eastern Mediterranean. This episode was characterized by large-scale dust transport from the Saharan desert area towards the Eastern Mediterranean, followed by torrential rain over Greece during the early days of June 2014. Due to the fact that ground measurements of the dust wet deposition are scarce, an evaluation using observed data of Aerosol Optical Depth and PM_{10} concentrations has been performed. Furthermore, the impact of the new GOCART wet deposition scheme on the dust vertical profile is also assessed through two sensitivity simulations which are evaluated against ground-based measurements provided by the AERONET network.

2. Materials and Methods

2.1. Model Configuration and Parameterization Schemes

The simulations in this study have been performed using the WRF/Chem (version 3.8) model, which is a fully coupled meteorology-chemistry model that contains a variety of schemes for the simulation of atmospheric chemistry [48]. The meteorological component is fully consistent with the chemical component. Hence, both meteorological and air-quality components use the same physics schemes for the sub-grid scale transport, the same grid on the horizontal and vertical coordinates, and the same transport scheme, which preserves air and scalar mass. WRF/Chem is set up in a domain that covers the area of North Africa, the Mediterranean, the Europe, the Middle East, and the Arabian Peninsula. This domain encompasses the entire subtropical belt of deserts, which act as mineral dust sources emitting particulate matter towards the Mediterranean (see Figure 1). It consists of 400×212 grid points, on a horizontal resolution of 25×25 km, and with 38 vertical levels from the Earth's surface to 50 hPa. The ECMWF (European Centre for Medium-Range Weather Forecasts) analyses are used as initial and boundary conditions at a resolution of $0.5° \times 0.5°$ and with a 6h time increment.

The basic parameterization schemes for the simulations in this study are summarized in Table 1 (see [54–58]). The Rapid Radiative Transfer Model (RRTMG) for both short-wave and long-wave radiation calculations is used [58]. RRTMG has the ability to use prognostic dust fields to calculate the aerosol direct radiative effect; however, this is not used in the present study, and dust is considered as a passive tracer. The Morrison 2-moment microphysics scheme (6-class microphysics scheme with graupel) is used as described in Refs. [55,56]. As a new and advanced 2-moment microphysics scheme, Morrison in its WRF release predicts the mass mixing ratio for five hydrometeor categories: cloud, rain, snow, ice, and graupel. It also predicts the total number concentrations for cloud water, rain, snow, ice, and graupel [59].

Table 1. Primary model configuration settings.

Configuration	
Model	WRF/Chem-3.8
Time step	150 s
Horizontal resolution	25 km × 25 km
Vertical resolution	38 sigma-pressure levels up to 50 hPa
Grid points	400 × 212
Initial and boundary conditions	ECMWF ($0.5° \times 0.5°$)
Emissions scheme	GOCART/AFWA [54]
Microphysics scheme	Morrison [55,56]
Cumulus scheme	Kain-Fritsch [57]
Longwave/Shortwave radiation	RRTMG [58]

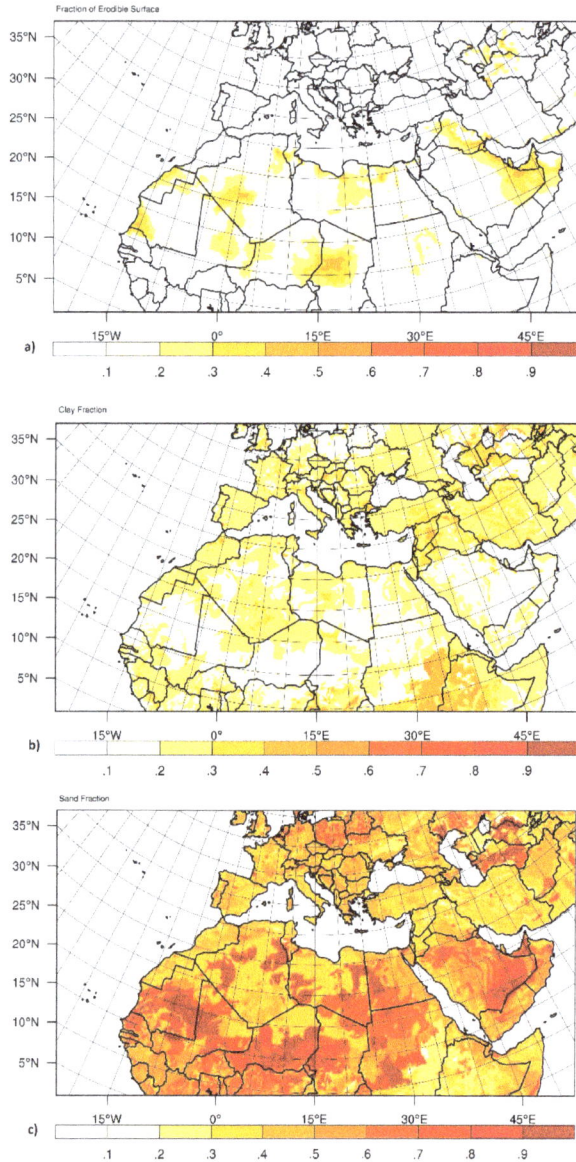

Figure 1. WRF/Chem integration domain with horizontal distribution of: (**a**) Fraction of erodible surface; (**b**) Clay fraction; (**c**) Sand fraction over the domain of integration.

2.2. The Air Quality (Chemistry) Component

The WRF/Chem model includes three alternative packages for mineral dust emission: two (namely, "DUST-GOCART", "DUST-GOCART/AFWA") from the GOCART module, and the third (namely, "DUST-UOC") from the University of Cologne [60]. In "DUST-GOCART", the dust emissions are scaled with the soil erodibility fields, as described by Ref. [61], with the dust emissions activating when the 10-m wind speed exceeds a threshold value proposed by Ref. [62]. In the "DUST-GOCART/AFWA scheme, dust emissions are also scaled with the erodibility fields based

Remote Sens. **2018**, *10*, 1595

on Ref. [61], with the erodibility function being dependent to the particle bin mass fraction. This scheme also parameterizes the initialization of dust production by saltation bombardment. For dry soils, there is the same threshold as that used in "DUST-GOCART", but AFWA uses a different soil moisture correction. On the other hand, the "DUST-UOC" uses the erodible fields of Ref. [61] for the definition of the areas concerning the areas of potential dust emission, but in contrast to the other two previous schemes, the calculated dust emissions are not scaled with the erodibility function [63]. However, "DUST-GOCART/AFWA" produces an important over-prediction of dust concentration [64]. This may be explained in part by the fact that the AFWA scheme only considers vertical dust flux which is related to clay content [60], and it does not support a dust wet deposition scheme, in contrast to the "DUST-UOC", which considers a more realistic soil texture type. Furthermore, the "DUST-UOC" supports a dust wet deposition scheme based on the study of Jung [65], only for below cloud scavenging of aerosols. Thus, it is expected that the implementation of a dust wet deposition scheme in the "DUST-GOCART/AFWA" module that contains both in-cloud and below cloud scavenging processes will contribute to the reduction of the systematic overestimation of dust concentration.

2.3. The GOCART-AFWA Module

The GOCART-AFWA module contains a sectional, non-experimental dust scheme with advanced dust emissions parameterizations [54,66]. In the GOCART-AFWA module, the dust emissions scheme is based on the parameterizations of Ref. [62], incorporating five dust bins of 0.73, 1.4, 2.4, 4.5, and 8.0 μm effective radii. In this scheme, the saltation of large particles is triggered by wind shear, leading to fine particle emissions by disaggregation and bombardment. The dust emissions due to saltation bombardment are parameterized, with the vertical dust emission flux being proportional to the horizontal saltation dust flux, calculated when the friction velocity exceeds a certain threshold. Saltation processes for a given size bin initiate or cease as the friction velocity exceeds or falls below the values of the threshold friction velocity [67]. The proportionality of dust emission and dust flux is empirically related to clay, sand, and erodibility fields that are shown in Figure 1. As stated above, the empirical relationship of the vertical dust flux to the clay content is a reason for the systematic overprediction of the simulated dust concentration [60]. However, as far as soil moisture is concerned, the correction of Ref. [68] is applied. The calculation of the horizontal saltation flux is based on a modification of the expression of Ref. [69]. The distribution of the dust particles follows the brittle fragmentation theory [70]. The dust particles are emitted into the lowest atmospheric model level according to their respective size bins. GOCART-AFWA is able to resolve direct effects concerning aerosol interactions with radiation. However, in this study, no interactions with radiation are taken into account. Moreover, no wet scavenging mechanism is currently supported.

2.4. The Embedded Dust Wet Deposition Scheme

The dust wet removal from the atmosphere can be expressed in two stages: one inside the cloud (in-cloud wet scavenging), and the other below the cloud base (below cloud scavenging). The Seinfeld and Pandis [44] dust wet deposition scheme (hereafter denoted as SP) is embedded in the GOCART-AFWA module in order to estimate the rate of dust scavenging by precipitation for both in and below cloud areas, concerning grid-scale and convective precipitation [55–57]. The wet removal is considered only by rain, and no evaporation mechanism is taken into account. Scavenging is direct and irreversible, and the mass which is collected by rain is effectively removed from the atmosphere. Particles that are captured by the droplets are deposited to the ground. The advantage of this scheme is that it takes into account parameters that characterize the behavior of particles suspended in a fluid flow, such as the Reynolds, Schmidt, and Stokes numbers. Thus, the rate of dust concentration (*C*) removed by the precipitation is represented by

$$\frac{\partial C}{\partial t} = -\Lambda C \tag{1}$$

where, Λ is the scavenging coefficient. For aerosols inside the clouds, the scavenging coefficient is given by:

$$\Lambda = 4.2 \times 10^{-7} \frac{E\,P}{d_c} \tag{2}$$

where E is the collection efficiency ($E = 0.9$), P is the total precipitation (mm h^{-1}), which is the sum of the grid-scale and convective precipitation in each model layer, and d_c is the cloud mean droplet diameter, with values from 2 µm to 50 µm [55]. In order to identify the cloud existence in a grid cell and in a specific model layer, the Cloud Water mixing ratio is utilized. Values greater than zero indicate the presence of clouds, and aerosols are assumed to act as in cloud water. As far as the wet removal below the cloud base is concerned, the same scavenging coefficient is used as before, but d_d instead of d_c is used in Equation (2). Furthermore, the collection efficiency is now dependent on the particle size d_p (effective diameters of 1.46 µm, 2.8 µm, 4.8 µm, 9.0 µm, 16 µm), as shown below [37,44]:

$$E(d_p) = \frac{4}{R_e S_c}\left(1 + 0.4 R_e^{1/2} S_c^{1/3} + 0.16 R_e^{1/2} S_c^{1/2}\right) + 4\varphi\left[\frac{\mu}{\mu_w} + \varphi\left(1 + 2R_e^{1/2}\right)\right] + \left(\frac{S_t - S^*}{S_t - S^* + 2/3}\right) \tag{3}$$

where, R_e is the Reynolds number of the raindrop, S_c is the Schmidt number for the collected particle, S_t is the Stokes number for the collected particle, μ and μ_w are the kinematic viscosity of the air and water respectively, and $\varphi = d_p/d_d$ is the ratio from particle size to raindrop size respectively, with dp being the particle diameter and d_d the raindrop mean diameter with values of 20 µm to 500 µm [55], for below cloud scavenging instances. Parameter S^* is given by Equation (4), and the R_e, S_t and S_c are given by Equations (5)–(7), respectively:

$$S^* = \frac{1.2 + \ln(1 + R_e)/12}{1 + \ln(1 + R_e)} \tag{4}$$

$$R_e = \frac{d_d V_t(d_d)\rho_\alpha}{2\mu} \tag{5}$$

$$S_c = \frac{\mu}{\rho_\alpha D} \tag{6}$$

$$S_t = \frac{2\tau(V_t(d_d) - v_t(d_p))}{d_d} \tag{7}$$

In these relationships, the various variables are defined as follows:

$$D = \frac{k_b T C_c}{3\pi\mu d_p} \tag{8}$$

$$C_c = \frac{1 + 0.167}{d_p} \tag{9}$$

$$\tau = \frac{v_t}{g} \tag{10}$$

$$v_t = \frac{C_c g}{3\pi\mu d_p} \tag{11}$$

$$V_t = ad^b \tag{12}$$

where ρ_α is the air density (kg m^{-3}), V_t and v_t are the terminal velocities of the rain drops and the particles respectively (in m s^{-1}) as they are diagnostically estimated by the model, g is the gravitational acceleration (in m s^{-2}), d is equal to d_c or d_d, C_c is the Cunningham correction factor, $k_b = 1.38066 \times 10^{-23}$ J K^{-1}, and T is the model predicted temperature (in K). For cloud particles, $\alpha = 3 \times 10^7$ m^{1-b}s^{-1} and b = 2, and for rain particles a = 841.997 m^{1-b}s^{-1} and b = 0.8 [55,71,72]. The expressions of D (particle diffusivity) and τ (relaxation time of the collected particle) are dependent on d_p and are given by Refs. [44,73]. The first term in Equation (3) is the contribution to the Brownian diffusion, the second term concerns the collection by interception, and the third term concerns the inertial impaction and it is efficient for large particles ($d_p > 2$ μm) [74] with a restriction when $S_t > S^*$ [75].

3. Results

For its consistency and performance, the WRF/Chem model using the GOCART-AFWA module with the SP scheme has been tested in a case study. To accomplish this, two numerical experiments have been conducted for a case of a large-scale desert dust transportation event which involved severe wet deposition that affected central and eastern Mediterranean from 3 to 5 June, 2014. The two numerical experiments have been designed as follows:

- The control simulation (CTRL): It adopts the default GOCART-AFWA configuration described in Sections 2.1 and 2.2.
- The wet deposition simulation (Wet_Dep): As in CTRL but enabling the dust wet deposition scheme described in Section 2.3.

Both simulations initialized on 2 June in "dust hot started" mode. To this end, a 72hr "dust cold started" simulation initialized on the 30 May, 2014 in order to build the adequate desert dust background for driving CTRL and Wet_Dep. Since the date of interest is the 3rd of June, the spin-up time was 12 h for both simulations.

The inter-comparison between the two simulations demonstrates the impact of the SP scheme on the dust load and the dust concentration vertical profile caused by the vertical dust concentration losses triggered by the rainout and washout processes. Predicted dust is considered as a passive tracer for both simulations as far as interactions with radiation and clouds are concerned. Due to the lack of dust wet deposition measured data, the two simulations are evaluated against observed PM_{10} concentrations and AOD data provided by the Finokalia station of Crete and the AERONET network respectively.

3.1. Description of Synoptic Conditions

On 1 June 2014, an upper-air trough, transferring cold air masses towards central Europe and the Mediterranean, was associated with the development of a barometric low over Northern Algeria (Figure 2a,b). As the trough propagated eastward on 2 June, the surface low triggered updraft motions due to the formation of a well-organized warm front over Tunis and Libya (not shown). The prevailing upper air southwesterly synoptic flow favored the advection of the suspended particles towards Italy and Greece. On 3 June, the torrential rains that occurred over Italy and Greece originated mainly from the warm front passage over Sicily and Greece, transferring warm and moist air masses over the Mediterranean Sea (Figure 2c,d). On 2 June, the enhanced instability led to cloud formation and triggered dust uptake mechanisms over Northern Africa (Figure 3a). On 3 June, the barometric low is located over the Gulf of Sirte, while the northward advected dust is washed out over western and central Greece (Figure 3b). On 4 June, the upper air conditions supported the eastward propagation of the barometric low over western Turkey, followed by a steady rise in surface pressure and moderate precipitation.

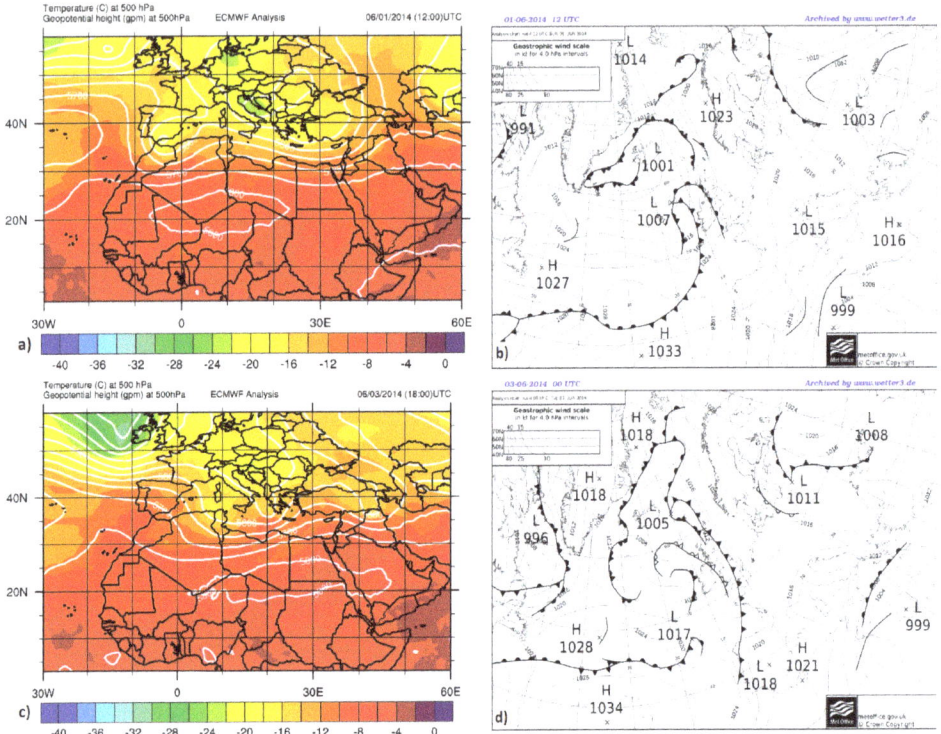

Figure 2. (a) ECMWF analysis for geopotential height (contours in gpm) and temperature at 500 hPa (color shaded in °C) for 12:00 UTC 1 June 2014; **(b)** UK MetOffice surface pressure analysis map (hPa) for 12:00 UTC 1 June 2014; **(c)** Same as for **(a)** but for 18:00 UTC 3 June 2014; **(d)** Same as for **(b)** but for 00:00 UTC 3 June 2014.

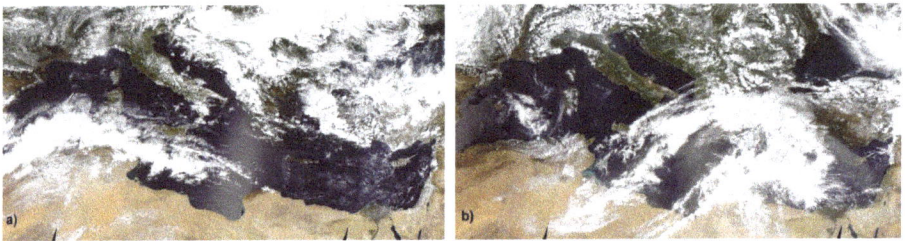

Figure 3. Dust mobilization over Libya and transport to Greek Peninsula for **(a)** 2 June 2014 and **(b)** 3 June 2014 (NASA satellite snapshots, EOSDIS-WORLDVIEW, 5 km resolution per pixel).

3.2. Dust Uptake

Dust uptake processes and vertical profiles have been investigated in an area between points A and B, which denote the major dust sources of this case study (see Figure 4 top left). In Figure 4a, the dust dispersion is shown as a combination of mechanisms that transfer the dust plume downstream from its source region (area between A and B points). The sharp downward slope of the isotheta contours in the area 10°–20° E confirms the presence of a warm front, which favours dust uptake. The synergy of the synoptic and mesoscale motions suspends dust particles with effective radii in the range of 0.5–4.5 µm (bins 1–4) up to the height of 6 km, while coarser particles reach up to the height of 5 km (bin 5) (Figure 4c–g). The enhanced buoyancy can also be confirmed by the positive values of the vertical wind velocity component which dominates the dust plume area and exceeds 0.14 m s^{-1} at 3 km, as shown in the cross-section of Figure 4b. The suspended particles in the lower troposphere are then transported horizontally downwind following the strong prevailed westerly flow.

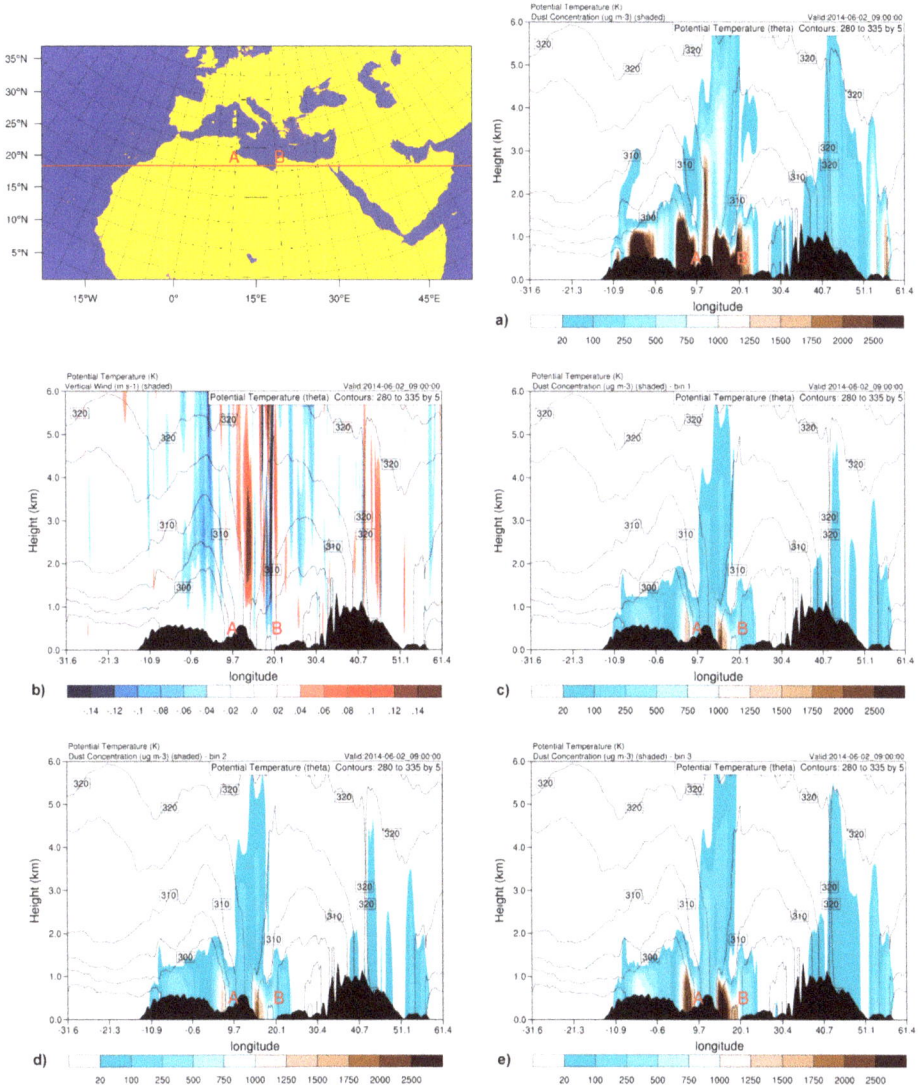

Figure 4. *Cont.*

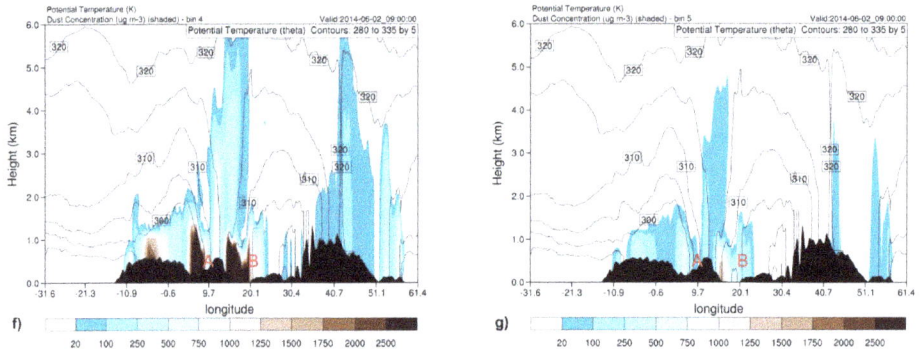

Figure 4. Cross-section of major dust emission areas between points of A and B (10°E–20°E) **[top left]** at 09:00 UTC 2 June 2014: (**a**) Simulated dust concentration (μg m^{-3}) (shaded), potential temperature (K) (contours); (**b**) Potential temperature (K) (contours), and vertical winds (m s^{-1}) (shaded). Cross-section of the simulated dust concentration (μg m^{-3}) (shaded), potential temperature (K) (contours); (**c**) for bin 1; (**d**) for bin 2; (**e**) for bin 3; (**f**) for bin 4; (**g**) for bin 5.

The spatial distribution of the dust load (mg m^{-2}) and wind speed at 3 km (m s^{-1}) over the model's area for 18:00 UTC 3 June 2014 are shown in Figure 5; from this figure, it can be seen that the simulated dust load over the Greek Peninsula reached 6000 mg m^{-2} at 18:00 UTC on 3 June 2014.

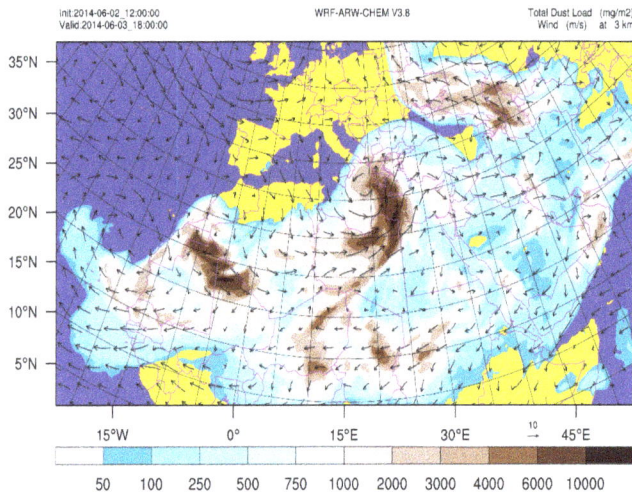

Figure 5. Spatial distribution of the total dust load (mg m^{-2}) and wind speed at 3 km (m s^{-1}) over the model's area for 18:00 UTC 3 June 2014.

3.3. Modeled and Remotely Sensed Precipitation Distribution

In order to assess the model's precipitation performance, a qualitative comparison is made between the model's simulation results and remotely sensed precipitation rate from the EUMETSAT historical imagery archive (with 15 minutes frequency). Figure 6a,b show the precipitation rate on 3 June 2014 at 12:00 and 15:00 UTC, respectively, over the Greek Peninsula, with values reaching approximately 4–5 mm/hr. Figure 6c shows the simulated 6h precipitation on 3 June 2014 for 12:00–18:00 UTC. The 6h accumulated precipitation over Greece exceeded 36 mm in the northern areas of the country. The comparison shows satisfactory spatial agreement of the rainfall that occurred over the Greek Peninsula and the southeast Aegean Sea. Central and Eastern Mediterranean are dominated by a large-scale desert dust transport due to the prevailing synoptic conditions during the period from 3–5 June 2014.

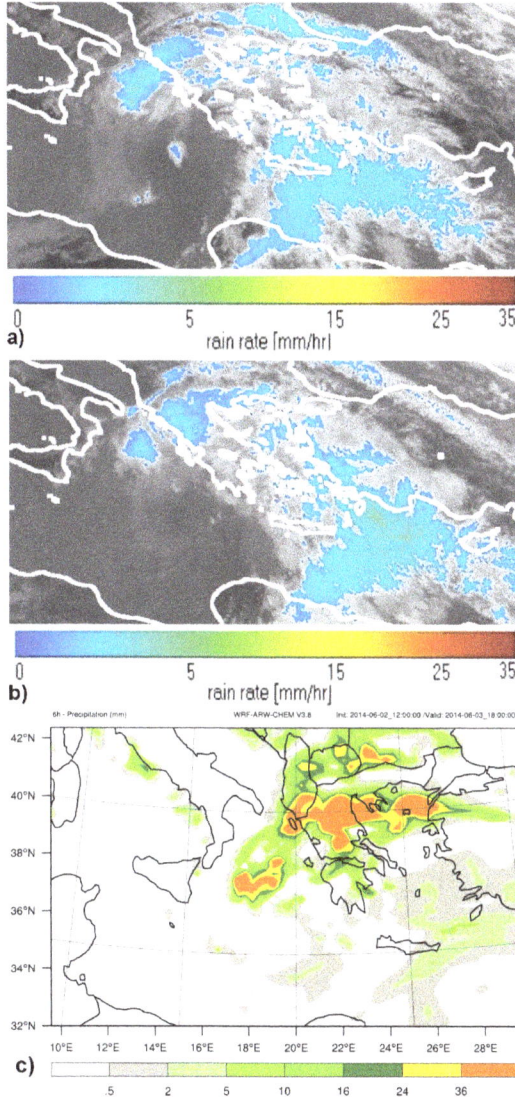

Figure 6. Remotely sensed METEOSAT precipitation rate (mm/hr) for (**a**) 12:00 UTC 3 June and for (**b**) 1500 UTC 3 June; (**c**) WRF/Chem accumulated precipitation (mm) for 12:00–18:00 UTC 3 June 2014.

3.4. Dust Transport and Wet Deposition

The simulated dust wet deposition over central Greece was 400 mg m^{-2} in the period 06:00–12:00 UTC on 3 June (see Figure 7a). The values of dust wet deposition further increased in the period 12:00–18:00 UTC 3 June and 18:00 UTC 3 June–00:00 UTC June 2014, exceeding 800 mg m^{-2}

and 1200 mg m^{-2}, respectively (see Figure 7b,c). During those periods, dust is mainly deposited over the areas with precipitation maxima. In the period from 00:00-06:00 UTC 4 June, the simulated dust wet deposition is restricted over the northern Aegean Sea, reaching values of 400 mg m^{-2} due to the eastward propagation of the barometric low, which is followed by moderate precipitation over Aegean Sea (Figure 7d). In the periods 18:00 UTC 4 June–00:00 UTC 5 June 2014 and 00:00–06:00 UTC (Figure 7e,f, respectively), the precipitation is confined over the island of Crete, with the 6hr accumulated dust deposition reaching 100 mg m^{-2}.

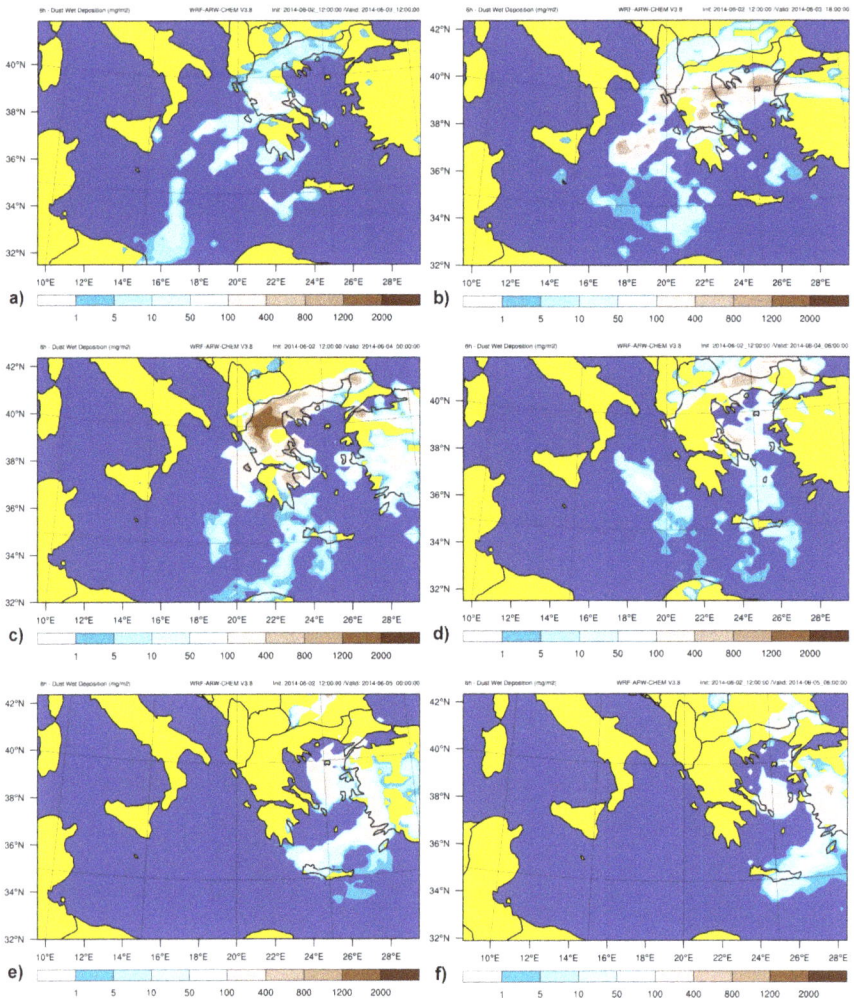

Figure 7. 6 h dust wet deposition (mg m^{-2}) simulated by the WRF-Chem model (**a**) for 06:00–12:00 UTC 3 June 2014 (**b**) for 12:00–18:00 UTC 3 June 2014, (**c**) for 18:00 UTC 3 June–00:00 UTC 4 June 2014, (**d**) for 00:00–06:00 UTC 4 June 2014, (**e**) for 18:00 UTC 4 June–00:00 UTC 5 June 2014, (**f**) for 00:00–06:00 UTC 5 June 2014.

The impact of the ingested SP scheme on the loss of the dust mass in the atmospheric column is also assessed through vertical cross-sections of the dust concentration with latitudinal orientation

C–D (20°E–30°E) [see Figure 8 top left] and meridional orientation E–F (35°N–41°N) (see Figure 9 top left]. Cross sections are valid at the dust concentration peak (15:00 UTC 3 June 2014), as is shown in Figures 8a and 9a, respectively covering the area of Central and Northern Greece, where significant dust wet deposition occurred. In Figure 8b, CTRL simulation estimates a maximum of concentration that exceeds the 2000 µg m^{-3} around the height of 2 km over the area of central and northern Greece (C–D points). The downward slope of the isotheta contours confirms the existence of the warm front, as has been shown in Figure 4a, which is the major synoptic feature responsible for the advection of warm air and the enhanced ascent motions in the area. During this time, the torrential rain over the Central and Northern Greek Peninsula coincided with the main core of the dust plume extending (C–D area) up to 6 km height. The simulation with the SP scheme enabled (Wet_Dep) represents significant losses of the suspended dust mass (Figure 8c,d). Local maxima of the dust loss reach approximately 1400 µg m^{-3} at heights greater than 3.5 km. More specifically, the maximum reduction in dust concentration occurs in the layer between 3.0 and 3.5 km, coinciding with the local maxima of relative humidity. This indicates that an important percentage of the dust wet removal occurred in-cloud, where the finer particles play a leading role in the wet scavenging processes. In Figure 9, the meridional structure of the dust plume extends in the area of 27°–40°N. CTRL simulation in Figure 9a exhibits a maximum concentration of 2000 µg m^{-3}, with the suspended particles extending up to 6 km (Figure 9a). In Figure 9b,c, which concern the Wet_Dep simulation, an almost 100% dust loss of approximately 1250 µg m^{-3} over the mountainous areas of Northern Greece (40°N above point F) at the height of about 3.5 km was observed. At this level, the high values of relative humidity (i.e., 90%) indicate that the greatest part of the wet scavenging process occurred in-cloud (Figure 9d). In Figures 8e and 9e, the cross sections of cloud water and the relative humidity confirm the existence of clouds in most areas of high relative humidity, which are also characterized by significant wet scavenging.

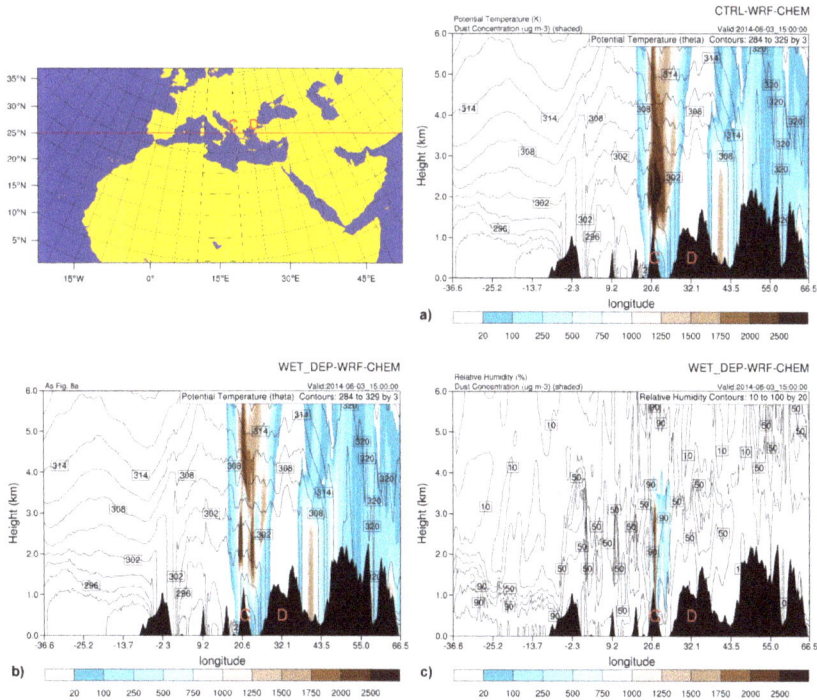

Figure 8. *Cont.*

d)

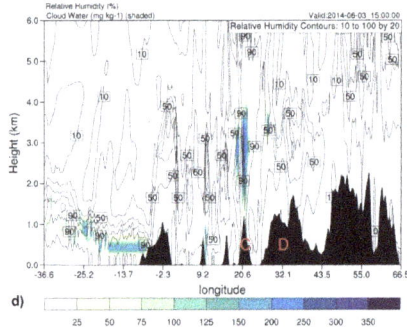

Figure 8. Latitudinal cross-section between points of C and D (20°E–30°E) [**top left**]. (**a**) Cross-section of dust concentration (μg m^{-3}) (shaded), potential temperature (K) (contours), in CTRL simulation; (**b**) as a in WET_DEP simulation; (**c**) Cross-section of dust concentration loss (μg m^{-3}) (shaded) and relative humidity (%) (contours), in WET_DEP simulation for 15:00 UTC 3 June 2014; (**d**) Cross-section of cloud water (mg kg^{-1}) (shaded) and relative humidity (%) (contours), in WET_DEP simulation for 15:00 UTC 3 June 2014.

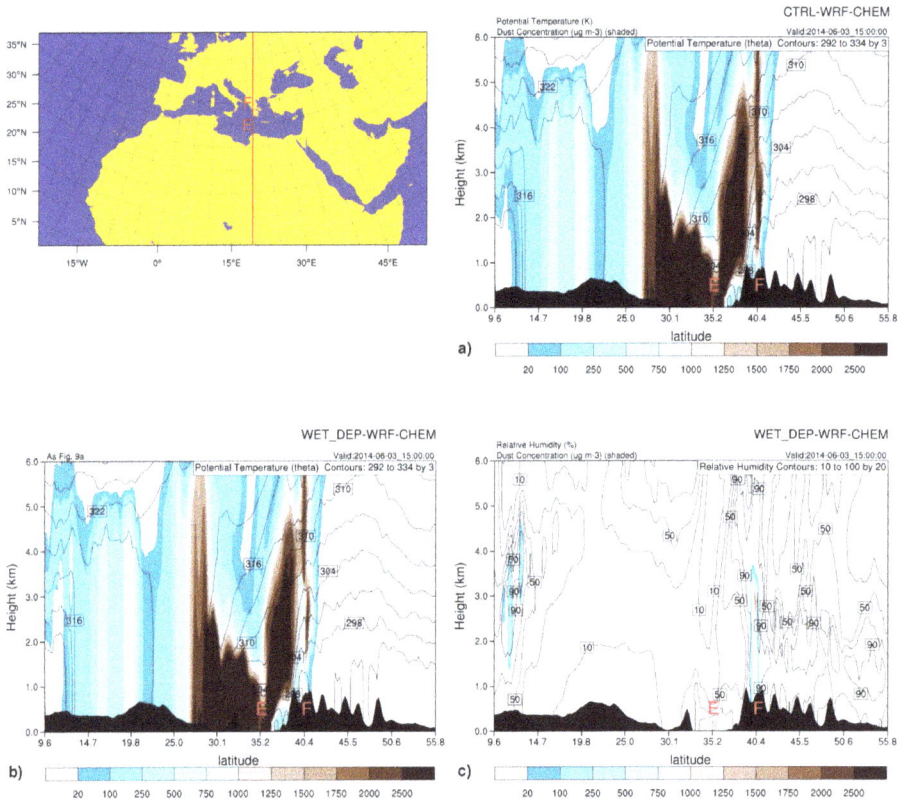

a)

b)

c)

Figure 9. *Cont.*

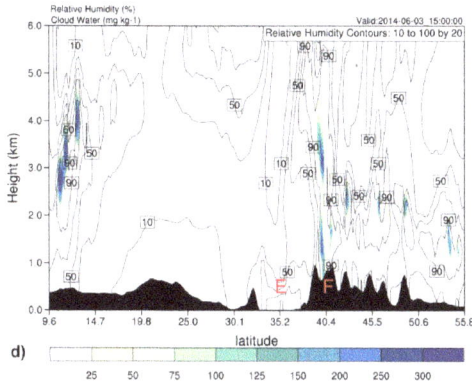

Figure 9. Meridional cross-section between points of E and F (35°E–40°E) [**top left**]. (**a**) Cross-section of dust concentration ($\mu g\ m^{-3}$) (shaded), potential temperature (K) (contours); (**b**) as a in WET_DEP simulation; (**c**) Cross-section of dust concentration loss ($\mu g\ m^{-3}$) (shaded) and relative humidity (%) (contours), in WET_DEP simulation for 15:00 UTC 3 June 2014; (**d**) Cross-section of cloud water ($mg\ kg^{-1}$) (shaded) and relative humidity (%) (contours), in WET_DEP simulation for 15:00 UTC 3 June 2014.

3.5. Dust Load Timeseries

The consistency of the SP scheme is evaluated through the dust load timeseries of CTRL and WET_DEP simulations, which are modeled over the cities Thessaloniki (22.960°E–40.630°N) and Athens (23.735°E–37.975°N), and shown in Figure 10a,d, respectively. In particular, the estimated dust load at Thessaloniki is identical for both simulations up to 09:00 UTC 3 June. At this time, the initiation of the precipitation coincides with the beginning of the dust particles wet removal (Figure 10a,c). The removal rate remained almost constant with 250 mg m^{-2} per 3 h in average until 18:00 UTC 3 June. At 00:00 UTC 4 June, as the precipitation strengthens up to its peak with approximately 25 mm, the dust loss also increases with a maximum value of 1000 mg m^{-2}, showing also the contribution of the wet deposition to this specific dust mass loss. The dust removal mechanism shows an almost direct response to the enhanced precipitation during this period.

In contrast to Thessaloniki, Athens is characterized by negligible 6hr dust wet deposition values of 5–10 mg m^{-2} for 12:00-18:00 UTC 3 June, 50–100 mg m^{-2} for 18:00 UTC 3 June–00:00 UTC 4 June, and 10–50 mg m^{-2} for 00:00–06:00 UTC 4 June, which is mainly attributed to the moderate precipitation, as shown in Figure 6. However, a dust load loss of approximately 200 mg m^{-2} is observed at 18:00 UTC 3 June 2014, coinciding with the initialization of precipitation (Figure 10f).

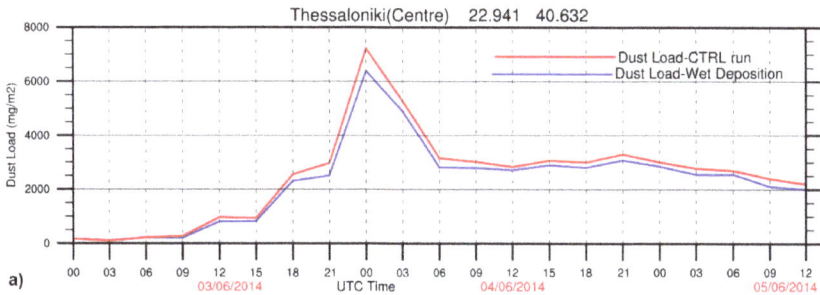

Figure 10. *Cont.*

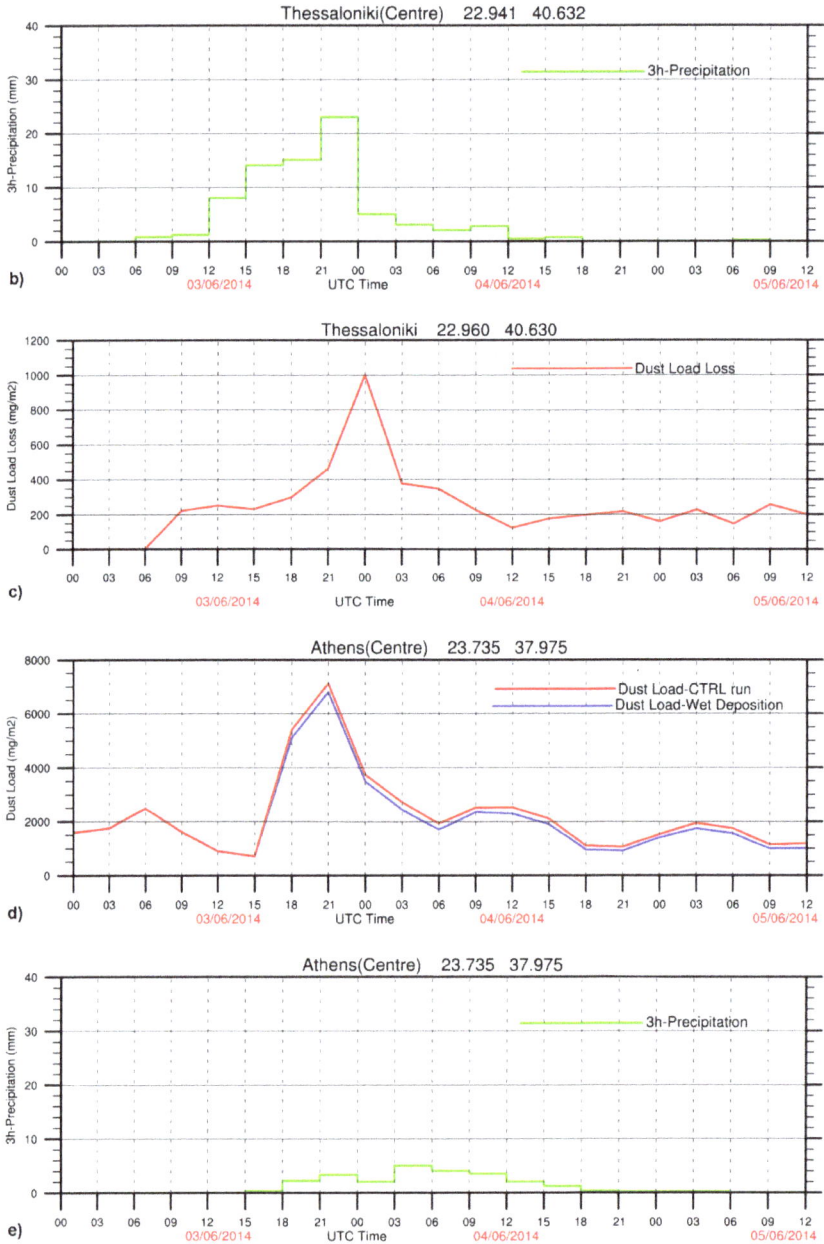

Figure 10. *Cont.*

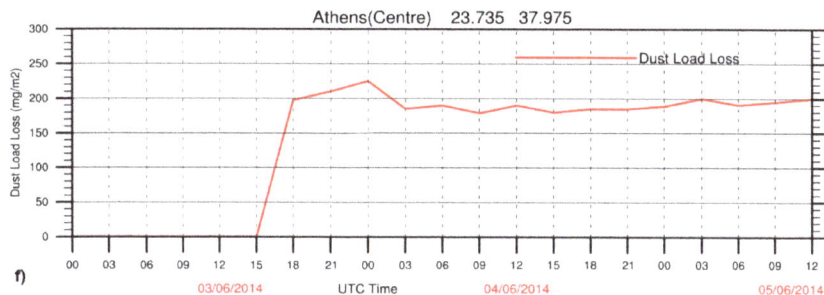

Figure 10. (**a–c**) timeseries of the simulated dust load (mg m^{-2}), 3h accumulated precipitation (mm) and dust load loss (mg m^{-2}) over Thessaloniki (22.960°E–40.630°N) and (**d–f**) over Athens (23.735°E–37.975°N). Red and blue lines in dust load diagrams (a, d) correspond to CTRL and Wet_Dep simulations.

Additionally, the difference in dust loads remains almost constant after the event; this is true for both time series, confirming that dust wet removal processes modify the spatiotemporal distribution of the total dust mass.

3.6. Validation With in situ Measurements

In order to evaluate the consistency of the SP scheme and its impact on the performance of WRF-Chem, a comparison is made between hourly values of PM$_{10}$ concentrations measured at Finokalia surface station and the dust concentration values (µg m^{-3}) simulated by the CTRL and Wet_Dep simulations. The Finokalia environmental research station is located in Southern Crete; additional information can be found in Ref. [3]. As far as PM$_{10}$ concentration is concerned, both simulations reproduce a pattern similar to the measurements for the entire period under consideration (Figure 11a). Obviously, Wet_Dep estimates lower concentrations than the CTRL ones, which are more prominent around the peaks at 03:00–06:00 UTC 4 June. However, Wet_Dep reproduces better the general dust concentration pattern throughout the period of evaluation, which is reflected in the statistical scores (Table 2).

Table 2. BIAS and RMSE between dust concentration (CTRL and Wet_DEP) produced by WRF/Chem model and observed PM$_{10}$ concentration by Finokalia station and between AOD (CTRL and Wet_DEP) and observed AOD concentration by Thessaloniki station.

		CTRL-OBSERVED	WET_DEP-OBSERVED
Dust-PM$_{10}$	BIAS	1.90	−1.50
	RMSE	30.20	28.50
Dust-AOD	BIAS	0.15	−0.05
	RMSE	0.23	0.17

Indeed, Wet Dep underestimates the dust concentration by −1.50 µg m^{-3}, having an overall lower RMSE of 28.50 µg m^{-3}. In contrast, the bias score of the CTRL dust concentration reveals a systematical overestimation of PM$_{10}$ measurements, despite the fact that the latter one also includes non-dust origin species such as sea salt, anthropogenic particles, etc.

A comparison is also made between the estimated values of aerosol optical depth (AOD) from AERONET's database (500 nm, Level 1.5) for Thessaloniki station (Figure 11b) and the simulated ones. As the dust loss begins after 09:00 UTC 3 June 2014, a significant reduction in AOD is also shown in Figure 11b, with an approximately difference of 0.1–0.2 between the two time series. Despite the lack of AOD measurements for 4 June 2014, it seems that Wet Dep is in better accordance with the observed one for 5 June 2014 compared to the CTRL simulation, which clearly overestimates AOD by a value of 0.1–0.2 (Figure 11b). The statistical scores in Table 2 also confirm this statement. Indeed, the systematic overestimation of CTRL by 0.15 turns to a slight underestimation of Wet Dep by −0.05. Additionally, Wet Dep substantially improves the AOD forecasting skill with a RMSE of 0.17 instead of 0.23 of the CTRL run. It has to be mentioned that these results cannot be considered as significant, due to the limited sample of AOD measurements. However, in a case study that concerns a long-range dust transport event over the Italian Peninsula during 18–26 May 2014, Rizza et al. [60] reached similar statistical scores.

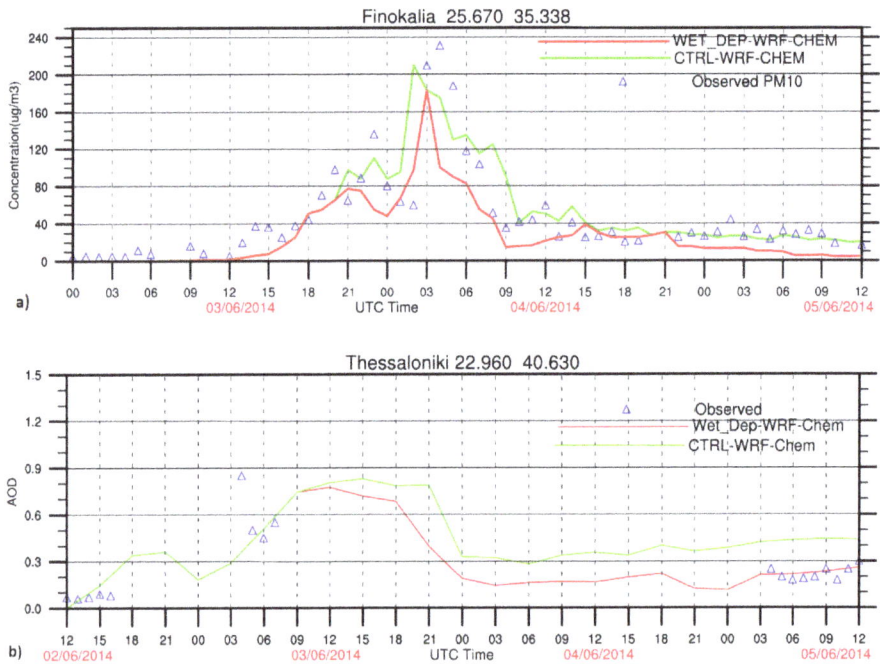

Figure 11. Time series (**a**) of the dust concentration ($\mu g\ m^{-3}$) over Finokalia–Crete station (25.670°E–35.338°N) and (**b**) of AOD over Thessaloniki station (22.960°E–40.630°N) for CTRL (green) and Wet_Dep (red) simulations which are compared against the observed PM_{10} concentrations by Finokalia and the AOD values by Thessaloniki stations, respectively.

4. Discussion and Conclusions

In this study, the Seinfeld and Pandis [44] dust wet deposition scheme is embedded in the fully coupled atmospheric-chemistry model WRF/Chem with the GOCART/AFWA module, which uses an advanced dust emissions parameterization scheme in comparison to the original GOCART. As the GOCART/AFWA scheme does not support a dust wet deposition scheme, the implementation of the Seinfeld and Pandis (SP) scheme is assumed to be an important upgrade, with an advanced wet scavenging scheme.

The performance of the GOCART/AFWA including the SP scheme has been assessed in a long-range dust transport case study which occurred on 3–5 June 2014. To this end, two sensitivity simulations with the default GOCART/AFWA configuration (CTRL) and the GOCART/AFWA including the SP scheme (Wet_Dep) have been performed. The sensitivity simulations revealed that the dust wet deposition mainly occurs at heights greater than 3.5 km, with the in-cloud area playing an important role in the scavenging of significant amounts of the finer suspended particles. Furthermore,

Remote Sens. **2018**, *10*, 1595

the dust loss due to the wet scavenging mechanisms reduces its total load in the atmosphere up to 6 km, and ultimately affects the entire spatiotemporal distribution of the suspended dust.

The incorporation of the advanced SP scheme also improves the model's performance by limiting the overestimated values of the simulated dust concentration and dust load. Indeed, statistical evaluation reveals that the Wet_Dep simulation turns the CTRL systematic overestimation of dust to underestimation, while at the same time, reduces the forecast error. Wet_Dep, in comparison to observed PM_{10} values, underestimates surface concentration on a scale that could be generally acceptable, as the measured concentration of PM_{10} does not include mineral dust particles exclusively, but it also detects additional species such as sea salt, anthropogenic particles, etc. Similar conclusions can also be drawn through the comparison of the observed AOD values with those estimated by both simulations. Thus, Wet_Dep improves the AOD bias score by slightly underestimating it; additionally, it reduces RMSE. Such improvements substantially correct the spatiotemporal distribution of the suspended dust throughout the simulation period and are contributing factors to the reduction of its well-known dust concentration overestimation problem [60,64,76–78].

Author Contributions: K.T. contributed to the methodology and formal analysis, investigation, visualization and writing; A.P.; N.M. and S.M. contributed to the investigation writing, review and editing; C.S. contributed to the formal analysis, investigation, writing, review and editing; P.K. contributed to the conceptualization, methodology, supervision, investigation, writing, review and editing.

Funding: This research received no external funding.

Acknowledgments: We acknowledge the use of imagery from the NASA Worldview application (https://worldview.earthdata.nasa.gov/) operated by the NASA/Goddard Space Flight Center Earth Science Data and Information System (EOSDIS) project. We gratefully thank the EUMETSAT Historical Browse Imagery Archive, Darmstadt, Germany. British MetOffice is also acknowledged for the provision of the surface analysis charts. We thank the PI investigators and their staff for establishing and maintaining the AERONET sites used in this study.

Conflicts of Interest: The authors declare no conflict of interest.

References

1. She, L.; Xue, Y.; Yang, X.; Guang, J.; Li, Y.; Che, Y.; Fan, C.; Xie, Y. Dust detection and intensity estimation using Himawari-8/AHI observation. *Remote Sens.* **2018**, *10*, 490. [CrossRef]

2. Tegen, I. Modeling the mineral dust aerosol cycle in the climate system. *Quat. Sci. Rev.* **2003**, *22*, 1821–1834. [CrossRef]

3. Solomos, S.; Kalivitis, N.; Mihalopoulos, N.; Amiridis, V.; Kouvarakis, G.; Gkikas, A.; Binietoglou, I.; Tsekeri, A.; Kazadzis, S.; Kottas, M.; et al. From Tropospheric Folding to Khamsin and Foehn Winds: How Atmospheric Dynamics Advanced a Record-Breaking Dust Episode in Crete. *Atmosphere* **2018**, *9*, 240. [CrossRef]

4. Spyrou, C. Direct radiative impacts of desert dust on atmospheric water content. *Aerosol Sci. Technol.* **2018**, *52*, 693–701. [CrossRef]

5. Marconi, M.; Sferlazzo, D.M.; Becagli, S.; Bommarito, C.; Calzolai, G.; Chiari, M.; Di Sara, A.; Ghendini, J.; Gomez-Amo, L.; Lucarelli, F.; et al. Saharan dust aerosol over the central Mediterranean Sea: PM_{10} chemical composition and concentration versus optical columnar measurements. *Atmos. Chem. Phys.* **2014**, *14*, 2039–2054. [CrossRef]

6. Ginoux, P.; Clarisse, L.; Clerbaux, C.; Coheur, P.-F.; Dubovik, O.; Hsu, N.C.; Van Damme, M. Mixing of dust and NH3 observed globally over anthropogenic dust sources. *Atmos. Chem. Phys.* **2012**, *12*, 7351–7363. [CrossRef]

7. Israelevich, P.; Ganor, E.; Alpert, P.; Kishcha, P.; Stupp, A. Predominant transport paths of Saharan dust over the Mediterranean Sea to Europe. *J. Geophys. Res.* **2012**, *117*, D02205. [CrossRef]

8. Gkikas, A.; Hatzianastassiou, N.; Mihalopoulos, N.; Katsoulis, V.; Kazadzis, S.; Pey, J.; Querol, X.; Torres, O. The regime of intense desert dust episodes in the Mediterranean based on contemporary satellite observations and ground measurements. *Atmos. Chem. Phys.* **2013**, *13*, 12135–12154. [CrossRef]

9. Gkikas, A.; Basart, S.; Hatzianastassiou, N.; Marinou, E.; Amiridis, V.; Kazadzis, S.; Pey, J.; Querol, X.; Jorba, O.; Gassó, S.; et al. Mediterranean intense desert dust outbreaks and their vertical structure based on remote sensing data. *Atmos. Chem. Phys.* **2016**, *16*, 8609–8642. [CrossRef]

10. Georgoulias, A.K.; Alexandri, G.; Kourtidis, K.A.; Lelieveld, J.; Zanis, P.; Amiridis, V. Differences between the MODIS Collection 6 and 5.1 aerosol datasets over the greater Mediterranean region. *Atmos. Environ.* **2016**, *147*, 310–319. [CrossRef]

11. Marinou, E.; Amiridis, V.; Binietoglou, I.; Tsikerdekis, A.; Solomos, S.; Proestakis, E.; Konsta, D.; Papagiannopoulos, N.; Tsekeri, A.; Vlastou, G.; et al. Three-dimensional evolution of Saharan dust transport towards Europe based on a 9-year EARLINET-optimized CALIPSO dataset. *Atmos. Chem. Phys.* **2017**, *17*, 5893–5919. [CrossRef]

12. Tsikerdekis, A.; Zanis, P.; Steiner, A.L.; Solmon, F.; Amiridis, V.; Marinou, E.; Katragkou, E.; Karacostas, T.; Foret, G. Impact of dust size parameterizations on aerosol burden and radiative forcing in RegCM4. *Atmos. Chem. Phys.* **2017**, *17*, 769–791. [CrossRef]

13. Osada, K.; Ura, S.; Kagawa, M.; Mikami, M.; Tanaka, T.Y.; Matoba, S.; Aoki, K.; Shinoda, M.; Kurosaki, U.; Hayashi, M.; et al. Wet and dry deposition of mineral dust particles in Japan: factors related to temporal variation and spatial distribution. *Atmos. Chem. Phys.* **2014**, *14*, 1107–1121. [CrossRef]

14. Shimizu, A.; Sugimoto, N.; Matsui, I.; Arao, K.; Uno, I.; Murayama, T.; Kagawa, N.; Aoki, K.; Uchiyama, A.; Yamazaki, A. Continuous observations of Asian dust and other aerosols by polarization lidars in China and Japan during ACE-Asia. *J. Geophys. Res. Atmos.* **2004**, *109(D19)*. [CrossRef]

15. Su, L.; Toon, O.B. Saharan and Asian dust: similarities and differences determined by CALIPSO, AERONET, and a coupled climate-aerosol microphysical model. *Atmos. Chem. Phys.* **2011**, *11*, 3263. [CrossRef]

16. Mona, L.; Liu, Z.; Müller, D.; Omar, A.; Papayannis, A.; Pappalardo, G.; Sugimoto, N.; Vaughan, M. Lidar measurements for desert dust characterization: an overview. *Adv. Meteorol.* **2012**, *2012*, 1–36. [CrossRef]

17. Xie, Y.; Zhang, W.; Qu, J.J. Detection of Asian Dust Storm Using MODIS Measurements. *Remote Sens.* **2017**, *9*, 869. [CrossRef]

18. Di, A.; Xue, Y.; Yang, X.; Leys, J.; Guang, J.; Mei, L.; Wang, J.; She, L.; Hu, Y.; He, X.; et al. Dust aerosol optical depth retrieval and dust storm detection for Xinjiang region using Indian National Satellite Observations. *Remote Sens.* **2016**, *8*, 702. [CrossRef]

19. Washington, R.; Wiggs, G.S.F. Desert dust. In *Arid Zone Geomorphology: Process, Form and Change in Drylands*, 3rd ed.; Thomas, D.S.G., Ed.; John Wiley & Sons: Hoboken, NJ, USA, 2011; Chapter 20; pp. 517–537. ISBN 9780470519080.

20. Schulz, M.; Prospero, J.M.; Baker, A.R.; Dentener, F.; Ickes, L.; Liss, P.S.; Mahowald, N.M.; Nickovic, S.C.; Garcia-Pando, P.; Rodriguez, S.; et al. Atmospheric transport and deposition of mineral dust to the ocean: implications for research needs. *Environ. Sci. Technol.* **2012**, *46*, 10390–10404. [CrossRef] [PubMed]

21. In, H.-J.; Park, S.-U. A simulation of long-range transport of Yellow Sand observed in April 1998 in Korea. *Atmos. Environ.* **2002**, *36*, 4173–4187. [CrossRef]

22. Spyrou, C.; Mitsakou, C.; Kallos, G.; Louka, P.; Vlastou, G. An improved limited area model for describing the dust cycle in the atmosphere. *J. Geophys. Res. Atmos.* **2010**, *115(D17)*. [CrossRef]

23. Nickovic, S.; Kallos, G.; Papadopoulos, A.; Kakaliagou, O. A model for prediction of desert dust cycle in the atmosphere. *J. Geophys. Res. Atmos.* **2001**, *106*, 18113–18129. [CrossRef]

24. Jaffrezo, J.L.; Colin, J.L. Rain-aerosol coupling in urban area: scavenging ratio measurement and identification of some transfer processes. *Atmos. Environ. (1967)* **1988**, *22*, 929–935. [CrossRef]

25. Zender, C.S.; Bian, H.; Newman, D. Mineral Dust Entrainment and Deposition (DEAD) model: Description and 1990s dust climatology. *J. Geophys. Res. Atmos.* **2003**, *108(D14)*. [CrossRef]

26. Harrison, R.M.; Pio, C.A. Size-differentiated composition of inorganic atmospheric aerosols of both marine and polluted continental origin. *Atmos. Environ. (1967)* **1983**, *17*, 1733–1738. [CrossRef]

27. Loosmore, G.A.; Cederwall, R.T. Precipitation scavenging of atmospheric aerosols for emergency response applications: testing an updated model with new real-time data. *Atmos. Environ.* **2004**, *38*, 993–1003. [CrossRef]

28. Laakso, L.; Rannik, Ü.; Grönholm, T.; Kosmale, M.; Fiedler, V.; Vehkamäki, H.; Kulmala, M. Ultrafine particle scavenging coefficients calculated from 6 years field measurements. *Atmos. Environ.* **2003**, *37*, 3605–3613. [CrossRef]

29. Draxler, R.R.; Hess, G.D. An overview of the HYSPLIT_4 modelling system for trajectories. *Aust. Meteorol. Mag.* **1998**, *47*, 295–308.

30. Webster, H.N.; Thomson, D.J. The NAME Wet Deposition Scheme. Met Office, Met Office Forecasting Research Technical Report, No: 584; 43p, 2014. Available online: https://www.metoffice.gov.uk/binaries/content/assets/mohippo/pdf/c/a/frtr584.pdf (accessed on 20 June 2018).

31. Gong, S.L.; Barrie, L.A.; Blanchet, J.P.; Von Salzen, K.; Lohmann, U.; Lesins, G.; Spacel, L.; Zhang, L.M.; Girard, E.; Lin, H.; et al. Canadian Aerosol Module: A size-segregated simulation of atmospheric aerosol processes for climate and air quality models 1. Module development. *J. Geophys. Res. Atmos.* **2003**, *108(D1)*. [CrossRef]

32. Guelle, W.; Balkanski, Y.J.; Schulz, M.; Dulac, F.; Monfray, P. Wet deposition in a global size-dependent aerosol transport model: 1. Comparison of a 1 year 210Pb simulation with ground measurements. *J. Geophys. Res. Atmos.* **1998**, *103(D10)*, 11429–11445. [CrossRef]

33. Dana, M.T.; Hales, J.M. Statistical aspects of the washout of polydisperse aerosols. *Atmos. Environ.* **1976**, *10*, 45–50. [CrossRef]

34. Butcher, S.S.; Charlson, R.J. *An Introduction to Air Chemistry*; Academic: San Diego, CA, USA, 1972; 241p, ISBN 978-0-12-148250-3.

35. Greenfield, S.M. Rain scavenging of radioactive particulate matter from the atmosphere. *J. Meteorol.* **1957**, *14*, 115–125. [CrossRef]

36. Beheng, K.D.; Herbert, F. Mathematical studies on the aerosol concentration in drops changing due to particle scavenging and redistribution by coagulation. *Meteor. Atmos. Phys.* **1986**, *35*, 212–219. [CrossRef]

37. Slinn, W.G.N. Precipitation scavenging. In *Atmospheric Sciences and Power Production—1979*; Division of Biomedical Environmental Research, US Department of Energy: Washington, DC, USA, 1983; Chapter 11; pp. 57–90.

38. Beheng, K.D. A parameterization of warm cloud microphysical conversion processes. *Atmos. Res.* **1994**, *33*, 193–206. [CrossRef]

39. Giorgi, F.; Chameides, W.L. Rainout lifetimes of highly soluble aerosols and gases as inferred from simulations with a general circulation model. *J. Geophys. Res. Atmos.* **1986**, *91(D13)*, 14367–14376. [CrossRef]

40. Tsyro, S. *First Estimates of the Effect of Aerosol Dynamics in the Calculation of PM_{10} and $PM_{2.5}$*; EMEP/MSC-W: Oslo, Norway, 2002; Note 4, 40p, ISSN 0332-9879.

41. Jung, E.; Shao, Y. An intercomparison of four wet deposition schemes used in dust transport modeling. *Global Planet. Chang.* **2006**, *52*, 248–260. [CrossRef]

42. Wang, P.K.; Pruppacher, H.R. An experimental determination of the efficiency with which aerosol particles are collected by water drops in subsaturated air. *J. Atmos. Sci.* **1977**, *34*, 1664–1669. [CrossRef]

43. Grover, S.N.; Pruppacher, H.R. The effect of vertical turbulent fluctuations in the atmosphere on the collection of aerosol particles by cloud drops. *J. Atmos. Sci.* **1985**, *42*, 2305–2318. [CrossRef]

44. Seinfeld, J.H.; Pandis, S.N. *Atmospheric Chemistry and Physics: From Air Pollution to Climate Change*; John Willey and Sons, Inc.: Hoboken, NJ, USA, 1998.

45. Pudykiewicz, J. Simulation of the Chernobyl dispersion with a 3-D hemispheric tracer model. *Tellus B* **1989**, *41*, 391–412. [CrossRef]

46. Brandt, J.; Christensen, J.H.; Frohn, L.M. Modelling transport and deposition of caesium and iodine from the Chernobyl accident using the DREAM model. *Atmos. Chem. Phys.* **2002**, *2*, 397–417. [CrossRef]

47. Tegen, I.; Fung, I. Modeling of mineral dust in the atmosphere: Sources, transport, and optical thickness. *J. Geophys. Res. Atmos.* **1994**, *99(D11)*, 22897–22914. [CrossRef]

48. Grell, G.A.; Peckham, S.E.; Schmitz, R.; McKeen, S.A.; Frost, G.; Skamarock, W.C.; Eder, B. Fully coupled "online" chemistry in the WRF model. *Atmos. Environ.* **2005**, *39*, 6957–6976. [CrossRef]

49. Skamarock, W.C.; Klemp, J.B.; Dudhia, J.; Gill, D.O.; Barker, D.M.; Duda, M.G.; Huang, X.-Y.; Wang, W.; Powers, J.G. *A Description of the Advanced Research WRF Version 3*; NCAR Tech. Note NCAR/TN-475+STR; 2008; 113p. [CrossRef]

50. Chin, M.; Rood, R.B.; Lin, S.J.; Müller, J.F.; Thompson, A.M. Atmospheric sulfur cycle simulated in the global model GOCART: Model description and global properties. *J. Geophys. Res. Atmos.* **2000**, *105(D20)*, 24671–24687. [CrossRef]

51. Solomos, S.; Kallos, G.; Kushta, J.; Astitha, M.; Tremback, C.; Nenes, A.; Levin, Z. An integrated modeling study on the effects of mineral dust and sea salt particles on clouds and precipitation. *Atmos. Chem. Phys.* **2011**, *11*, 873–892. [CrossRef]

52. Environ. *User's Guide to the Comprehensive Air Quality Model with Extensions (CAMx), Version 4*; ENVIRON International Corporation: Novato, CA, USA, 2006.

53. Dare, R.A.; Potts, R.J.; Wain, A.G. Modelling wet deposition in simulations of volcanic ash dispersion from hypothetical eruptions of Merapi, Indonesia. *Atmos. Environ.* **2016**, *143*, 190–201. [CrossRef]

54. Jones, S.L.; Creighton, G.A.; Kuchera, E.L.; Rentschler, S.A. Adapting WRF-CHEM GOCART for Fine-Scale Dust Forecasting. In AGU Fall 2011 Meeting Abstracts, Abstract id: U14A-06. 2011. Available online: http://adsabs.harvard.edu/abs/2011AGUFM.U14A..06J (accessed on 20 June 2018).

55. Morrison, H.; Gettelman, A. A new two-moment bulk stratiform cloud microphysics scheme in the Community Atmosphere Model, version 3 (CAM3). Part I: Description and numerical tests. *J. Clim.* **2008**, *21*, 3642–3659. [CrossRef]

56. Morrison, H.; Thompson, G.; Tatarskii, V. Impact of cloud microphysics on the development of trailing stratiform precipitation in a simulated squall line: Comparison of one-and two-moment schemes. *Mon. Weather Rev.* **2009**, *137*, 991–1007. [CrossRef]

57. Kain, J.S. The Kain–Fritsch convective parameterization: an update. *J. Appl. Meteorol.* **2004**, *43*, 170–181. [CrossRef]

58. Mlawer, E.J.; Taubman, S.J.; Brown, P.D.; Iacono, M.J.; Clough, S.A. Radiative transfer for inhomogeneous atmospheres: RRTM, a validated correlated-k model for the longwave. *J. Geophys. Res. Atmos.* **1997**, *102(D14)*, 16663–16682. [CrossRef]

59. Morrison, H.; Milbrandt, J. Comparison of two-moment bulk microphysics schemes in idealized supercell thunderstorm simulations. *Mon. Weather Rev.* **2010**, *4*, 1103–1130. [CrossRef]

60. Rizza, U.; Barnaba, F.; Miglietta, M.M.; Mangia, C.; Di Liberto, L.; Dionisi, D.; Costabile, F.; Grasso, F. Gobbi, G.P. WRF-Chem model simulations of a dust outbreak over the central Mediterranean and comparison with multi-sensor desert dust observations. *Atmos. Chem. Phys.* **2017**, *17*, 93. [CrossRef]

61. Ginoux, P.; Chin, M.; Tegen, I.; Prospero, J.M.; Holben, B.; Dubovik, O.; Lin, S.-J. Sources and distributions of dust aerosols simulated with the GOCART model. *J. Geophys. Res. Atmos.* **2001**, *106(D17)*, 20255–20273. [CrossRef]

62. Marticorena, B.; Bergametti, G. Modeling the atmospheric dust cycle: 1. Design of a soil-derived dust emission scheme. *J. Geophys. Res. Atmos.* **1995**, *100(D8)*, 16415–16430. [CrossRef]

63. Flaounas, E.; Kotroni, V.; Lagouvardos, K.; Klose, M.; Flamant, C.; Giannaros, T.M. Sensitivity of the WRF-Chem (V3.6.1) model to different dust emission parametrisation: assessment in the broader Mediterranean region. *Geosci. Model Dev.* **2017**, *10*, 2925–2945. [CrossRef]

64. Fountoukis, C.; Ackermann, L.; Ayoub, M.A.; Gladich, I.; Hoehn, R.D.; Skillern, A. Impact of atmospheric dust emission schemes on dust production and concentration over the Arabian Peninsula. *Model. Earth Syst. Environ.* **2016**, *2*, 1–6. [CrossRef]

65. Jung, E. Numerical Simulation of Asian Dust Events: The Impacts of Convective Transport and Wet Deposition. Ph.D. Thesis, The University of New South Wales, Sydney, Australia, 2005.

66. Jones, S.L.; Adams-Selin, R.; Hunt, E.D.; Creighton, G.A.; Cetola, J.D. Update on Modifications to WRF-CHEM GOCART for Fine-Scale Dust Forecasting at AFWA. In AGU Fall 2012 Meeting Abstracts, Abstract id: A33D-0188. 2012. Available online: http://adsabs.harvard.edu/abs/2012AGUFM.A33D0188J (accessed on 20 June 2018).

67. LeGrand, S.L.; Polashenski, C.; Letcher, T.W.; Creighton, G.A.; Peckham, S.E.; Cetola, J.D. The AFWA emissions Scheme for the GOCART Aerosol Model in WRF-Chem. *Geosci. Model Dev. Discuss.* **2018**, 1–57. [CrossRef]

68. Fécan, F.; Marticorena, B.; Bergametti, G. Parametrization of the increase of the aeolian erosion threshold wind friction velocity due to soil moisture for arid and semi-arid areas. *Ann. Geophys.* **1998**, *17*, 149–157. [CrossRef]

69. White, B.R. Soil transport by winds on Mars. *J. Geophys. Res. Sol. Ea.* **1979**, *84(B9)*, 4643–4651. [CrossRef]

70. Kok, J.F. Does the size distribution of mineral dust aerosols depend on the wind speed emission? *Atmos. Chem. Phys.* **2011**, *11*, 10149–10156. [CrossRef]

Remote Sens. **2018**, *10*, 1595

71. Liu, J.Y.; Orville, H.D. Numerical modeling of precipitation and cloud shadow effects on mountain-induced cumuli. *J. Atmos. Sci.* **1969**, *26*, 1283–1298. [CrossRef]

72. Ikawa, M.; Saito, K. *Description of the Non Hydrostatic Model Developed at the Forecast Research Department of the MRI*; Technical Report 28; Meteorological Research Institute (MRI), Japan Meteorological Agency: Tsukuba, Japan, 1990; 238p.

73. Sportisse, B. A review of parameterizations for modelling dry deposition and scavenging of radionuclides. *Atmos. Environ.* **2007**, *41*, 2683–2698. [CrossRef]

74. Berthet, S.; Leriche, M.; Pinty, J.P.; Cuesta, J.; Pigeon, G. Scavenging of aerosol particles by rain in a cloud resolving model. *Atmos. Res.* **2010**, *96*, 325–336. [CrossRef]

75. Feng, J. A size-resolved model for below-cloud scavenging of aerosols by snowfall. *J. Geophys. Res. Atmos.* **2009**, *114(D8)*. [CrossRef]

76. Mona, L.; Papagiannopoulos, N.; Basart, S.; Baldasano, J.; Binietoglou, I.; Cornacchia, C.; Pappalardo, G. EARLINET dust observations vs. BSCDREAM8b modeled profiles: 12-year-long systematic comparison at Potenza, Italy. *Atmos. Chem. Phys.* **2014**, *14*, 8781–8793. [CrossRef]

77. Binietoglou, I.; Basart, S.; Alados-Arboledas, L.; Amiridis, V.; Argyrouli, A.; Baars, H.; Baldasano, J.M.; Balis, D.; Belegante, L.; Bravo-Aranda, J.; et al. A methodology for investigating dust model performance using synergistic EARLINET/AERONET dust concentration retrievals. *Atmos. Meas. Tech.* **2015**, *8*, 3577–3600. [CrossRef]

78. Georgoulias, A.K.; Tsikerdekis, A.; Amiridis, V.; Marinou, E.; Benedetti, A.; Zanis, P.; Alexandri, G.; Mona, L.; Kourtidis, K.A.; Lelieveld, J. A 3-D evaluation of the MACC reanalysis dust product over Europe, northern Africa and Middle East using CALIOP/CALIPSO dust satellite observations. *Atmos. Chem. Phys.* **2018**, *18*, 8601–8620. [CrossRef]

remote sensing

MDPI

Article

A Multi-Platform Hydrometeorological Analysis of the Flash Flood Event of 15 November 2017 in Attica, Greece

George Varlas [1,2], Marios N. Anagnostou [3,4,5], Christos Spyrou [1], Anastasios Papadopoulos [2,*], John Kalogiros [3], Angeliki Mentzafou [2], Silas Michaelides [6], Evangelos Baltas [5], Efthimios Karymbalis [1] and Petros Katsafados [1]

[1] Department of Geography, Harokopio University of Athens, HUA, 17671 Athens, Greece; gvarlas@hcmr.gr (G.V.); spyrou@hua.gr (C.S.); karymba@hua.gr (E.K.); pkatsaf@hua.gr (P.K.)
[2] Institute of Marine Biological Resources and Inland Waters, Hellenic Centre for Marine Research, HCMR, 19013 Anavyssos, Greece; angment@hcmr.gr
[3] National Observatory of Athens, IERSD, 15236 Athens, Greece; managn@noa.gr (M.N.A.); jkalog@noa.gr (J.K.)
[4] Department of Environmental Sciences, Ionian University, 29100 Zakynthos, Greece
[5] Department of Water Resources, School of Civil Engineering, NTUA, 10682 Athens, Greece; baltas@central.ntua.gr
[6] The Cyprus Institute, 20, Konstantinou Kavafi Str., Aglantzia, CY2121 Nicosia, Cyprus; s.michaelides@cyi.ac.cy
* Correspondence: tpapa@hcmr.gr; Tel.: +30-22910-76399

Received: 13 November 2013; Accepted: 21 December 2018; Published: 28 December 2018

Abstract: Urban areas often experience high precipitation rates and heights associated with flash flood events. Atmospheric and hydrological models in combination with remote-sensing and surface observations are used to analyze these phenomena. This study aims to conduct a hydrometeorological analysis of a flash flood event that took place in the sub-urban area of Mandra, western Attica, Greece, using remote-sensing observations and the Chemical Hydrological Atmospheric Ocean Wave System (CHAOS) modeling system that includes the Advanced Weather Research Forecasting (WRF-ARW) model and the hydrological model (WRF-Hydro). The flash flood was caused by a severe storm during the morning of 15 November 2017 around Mandra area resulting in extensive damages and 24 fatalities. The X-band dual-polarization (XPOL) weather radar of the National Observatory of Athens (NOA) observed precipitation rates reaching 140 mm/h in the core of the storm. CHAOS simulation unveils the persistent orographic convergence of humid southeasterly airflow over Pateras mountain as the dominant parameter for the evolution of the storm. WRF-Hydro simulated the flood using three different precipitation estimations as forcing data, obtained from the CHAOS simulation (CHAOS-hydro), the XPOL weather radar (XPOL-hydro) and the Global Precipitation Measurement (GMP)/Integrated Multi-satellitE Retrievals for GPM (IMERG) satellite dataset (GPM/IMERG-hydro). The findings indicate that GPM/IMERG-hydro underestimated the flood magnitude. On the other hand, XPOL-hydro simulation resulted to discharge about 115 m³/s and water level exceeding 3 m in Soures and Agia Aikaterini streams, which finally inundated. CHAOS-hydro estimated approximately the half water level and even lower discharge compared to XPOL-hydro simulation. Comparing site-detailed post-surveys of flood extent, XPOL-hydro is characterized by overestimation while CHAOS-hydro and GPM/IMERG-hydro present underestimation. However, CHAOS-hydro shows enough skill to simulate the flooded areas despite the forecast inaccuracies of numerical weather prediction. Overall, the simulation results demonstrate the potential benefit of using high-resolution observations from a X-band dual-polarization radar as an additional forcing component in model precipitation simulations.

Remote Sens. **2019**, *11*, 45

Keywords: XPOL radar; GPM/IMERG; WRF-Hydro; CHAOS; hydrometeorology; flash flood; Mandra

1. Introduction

Floods are considered one of the most frequent natural disasters, causing many fatalities and damages every year. In 2011, six out of 10 of the biggest natural disasters, on a global scale, were flash floods. The frequency of flooding events has increased in recent decades and a warmer climate is expected to aggravate their destruction potential and the effects on human life [1]. All climate model projections show that more frequent precipitation extremes are expected in warmer climates [2], particularly in the populated mid and high latitudes [3]. This is expected in turn to increase the flash flooding risk over urban areas with negative consequences.

Accurate estimation of precipitation has always been one of the most challenging physical based tasks due to its large spatial and temporal variability in regional and global scale [4]. Recent technological advances in ground- and space-borne remote-sensing precipitation measurements allows us to produce near-real-time rainfall estimates at high spatial and temporal resolutions, from hundreds of kilometers up to quasi-global coverage, making this data potentially useful for hydrological and other applications. Ground-based remote-sensing observations are usually performed by either one or a network of meteorological radars, which provide real-time high spatiotemporal-resolution precipitation monitoring. Nevertheless, the weather radars are also suffering from limitations and uncertainties, including limited coverage, variable accuracy, and limited utility in complex terrains [5,6]. Past studies [7–12] have shown that locally deployed X-band dual-polarization (XPOL) radar systems can contribute to higher-resolution rain rate estimations and improved rainfall quantification accuracies than the lower frequency (C-band and S-band) long-range operational radar systems. These short-range radar systems could potentially be used to fill in coverage gaps of operational weather radar networks, which is essential for early warning of precipitation driven hydrological hazards (flash floods, landslides, debris flows, etc.) in urban and small mountainous basins [13–15].

Technological advances over the past two decades on satellite precipitation products have been developed and extensively used for large-scale hydrological and precipitation studies [16–18]. In the past 20 years, two international precipitation missions, the Tropical Rainfall Measuring Mission (TRMM), which launched in 1997 and lasted through 2015, and its successor, the Global Precipitation Measurement (GPM) mission, Core Observatory (CO) satellite, launched in 2014, contributed to provide high-resolution precipitation radar (PR) measurements. The GPM mission initiated by the National Aeronautics and Space Administration (NASA) and the Japan Aerospace Exploration Agency (JAXA) was launched on 27 February 2014 to globally observe rain and snow with improved accuracy [19]. Subsequent to the GPM Core Observatory launch, an advanced high-resolution multi-satellite-based precipitation product that combines the advantages of PERSIANN-CCS (Precipitation Estimation from Remotely Sensed Information using Artificial Neural Networks-Cloud Classification System), CMORPH (Climate Prediction Center MORPHing technique), and TMPA (TRMM Multisatellite Precipitation Analysis) was incrementally released since late 2014, i.e., the Integrated Multi-satellitE Retrievals for GPM (IMERG). Although the accuracy of satellite precipitation products has improved over the past few decades, they always suffer from significant errors associated with indirect measurements of ground precipitation [6]. All these methods cannot provide ground precipitation directly but rely on monitoring or modeling the precipitation-related variables to estimate precipitation indirectly.

Flash floods are frequent over several parts of the Mediterranean region due to the local climate, which is prone to intense storms [20]. The Mediterranean basin is a transitional zone between the cold and wet climates of northern Europe and the hot and dry climates of low latitudes (North Africa).

It is also characterized by high diversity in local climatic conditions [21]. Therefore, the types of precipitation that can generate flooding events vary along its coasts [20]. In general, convective thunderstorms and showers, which usually occur during late spring, summer and early autumn, are the main reason of flash flooding episodes [22]. In addition, the morphology of the drainage basins, the geographical characteristics of stream networks, and various anthropogenic interventions modify the response and the physical properties of the catchments [23]. Thus, these factors affect water cycle even exacerbating flash floods and their negative effects.

Greece experiences a variety of catastrophic weather events that are frequently followed by severe consequences on social and economic activity. Flash floods have caused tremendous loss of life and property over the past decades [24]. In Greece, they are primarily connected with strong convective storms developed during the warm season, especially over continental areas. These are mesoscale weather phenomena characterized by small spatial coverage (100 km^2) and small-time scales, usually 1–2 h [25]. The potential of these storms to become hazardous for human life and infrastructure depends on whether the convection organizes into mesoscale convective systems which are characterized by spatial coverage of about 9000 km^2 and time scales of about 5–8 h over Europe and the Mediterranean [26]. Sometimes, they are upgraded into long-lived mesoscale convective cells (supercells) with spatial coverage even exceeding 10,000 km^2 and time scale reaching 2–3 days [27]. In any case, they distribute large amounts of water in a limited area very fast. In some cases, the response of watersheds to this type of rainfall events and their runoff rates are not fast enough, leading to flooding events [20]. The negative effects of these phenomena are aggravated in built riverside areas as prevention is a parameter usually ignored during urban planning [28].

A significant part of the drainage network of Greece consists of mountain torrents and small- to medium-size drainage basins with a limited amount of discharge capabilities for most of the year [29]. Even worse, several parts of the network are diminished, turned into streets or built upon, thus cutting off critical river cross sections [30]. The combination of highly intense convective storms and bad urban planning has led to a series of severe flash flood occurrences in Greece over the past years. In November 1993, several areas of southern Attica (mainly Glyfada and Voula) experienced severe flooding phenomena due to a sudden thunderstorm [31]. The flood caused serious property damage along with the destruction of public infrastructure (roads, bridges etc). Most of the damage was attributed to urbanization and road construction that had not considered the specific conditions around the area. On October 1994 and July 1995, two extreme rainfall events caused extensive flooding, damage to streets, houses and commercial areas, and overflow of water courses in a big part of Athens [32]. On 11 and 12 January 1997 a severe flood event caused loss of life and damage to houses, cultivation and constructions in the broader area of the city of Corinth due to an extreme rainfall and the human interference in the geomorphological characteristics of the local drainage network [33]. Mazi and Koussis [34] studied the overflowing of the Kifissos River in Athens due to a storm on the 8 July 2002, which was attributed to the hydraulic works underway in the lower part of the river at that time. More recently on 22 October 2015, Athens's suburban areas were affected by a severe storm system, due to the passage of a barometric low associated with a cold front. The rainstorm caused extended flash flood incidents, mainly in the northern part of Athens, and led to the death of four people [35]. Summarizing, Diakakis et al. [29] performed a statistical analysis of available data and showed 686 fatalities due to a total of 545 flood events during the period 1880–2010 in Greece. In general, one of the most important factors for flash floods in Greece is the outbreak of intense rainfalls during the autumn and summer periods [24]. Geomorphology, soil characteristics (such as soil moisture), land use and human interference are also important factors.

Many studies present the efforts to simulate flash flooding events using numerical models, applying a multitude of systems like the hydrological model WRF-Hydro [36], DRiFt (Discharge River Forecast) [37], and CREST (Coupled Routing and Excess Storage) [38], for the hydrological part, and AROME (Applications of Research to Operations at Mesoscale model) [39], WRF (Weather Research and Forecasting model) [40,41], and others for the atmospheric component. The main aim of these models is the successful prediction of high-intensity rainfall often associated with convection and the simulation of channel- and surface-water runoff [22,42–44]. There are studies which rely on the combination of rain gauges and meteorological radar data [22,45,46]. Other studies use WRF model results, data assimilation techniques and satellite data to make forcing data for hydrological models like WRF-Hydro [43]. In general, a successful numerical simulation of a flash flood event requires the combination of an accurate atmospheric model, a suitable hydrological model and rainfall data of adequate spatial and temporal resolution [46].

The main aim of this study is to present a hydrometeorological analysis of an extreme flash flood event took place in the suburban area of Mandra, western Attica, Greece, using an integrated remote-sensing and observation-modeling system. This destructive flash flooding event occurred on 15 November 2017 causing 24 fatalities and extensive damage to property and infrastructure. The atmospheric conditions and the hydrological response of the drainage basin during the flood were simulated by the state-of-the-art modeling system CHAOS (Chemical Hydrological Atmospheric Ocean wave System [47,48], including the WRF-Hydro hydrological model [36]). To this end, three sensitivity tests using WRF-Hydro were performed in order to assess its hydrological response to different precipitation forcings. The first precipitation dataset was based on the results of the atmospheric component of CHAOS, the second on X-band dual-polarization experimental ground radar estimations and the third on GMP/IMERG satellite estimations. Post-survey and remote sensing maps were used to compare the extension of the flood [49–51]. The rest of the manuscript is organized as follows:

In Section 2, the geographical characteristics of study area and the methodology of remote sensing estimations are presented. Moreover, the set-up of CHAOS modeling system and the design of three different WRF-Hydro simulations are described. In Section 3, the meteorological conditions before and during the flood are analyzed through surface and upper air analysis charts. Section 4 presents the results of the three simulations focusing on the role of spatiotemporal characteristics of precipitation data on the hydrological characteristics of the flood. An evaluation of results using post-survey and satellite remote sensing maps is also presented. In Section 5, the main conclusions of this study are presented.

2. Materials and Methods

2.1. Study Area

Attica is an administrative region in Greece, situated at the southeasternmost point of central Greece. The whole region covers 3808 km^2 and it is currently the most populated region of Greece, reaching 3.8 million inhabitants in 2011, with most of the population living in Athens. Attica experiences a typical Mediterranean climate with the mean annual precipitation ranging between 350 and 390 mm in the southwestern low-lying coastal region and 500 mm in the northeastern mountainous region [52]. Precipitation is distributed relatively unevenly with about 75% of it occurring between the months of October and March.

Western Attica hosts one of the largest industrial units in Greece including refineries, metallurgical industries, factories, shipyards as well as the waste landfill of all Attica prefecture. This industrial zone is included in Thriasio Plateau area covering a total range of 812.95 km^2. The area is bounded by Pateras mountain (1016 m) from the west, Parnitha mountain (1413 m) from the north and Aigaleo mountain (468 m) from the east (Figure 1). The intense presence of heavy industry, urbanization and other land changes have worsened environmental pressures increasing the risk for flooding episodes. The residential and industrial development was lacking any appropriate plan regarding the infrastructure network for flood protection and drainage rainwater collection [53].

Figure 1. Map of study area depicting Thriasio Plateau among Pateras, Parnitha and Aigaleo mountains and the towns of Mandra, Nea Peramos and Elefsina. Source: Google Maps.

More specifically, the town of Mandra (Figure 1) is among the most damaged areas by the devastating November 2017 flood event. Mandra is a small industrial town, with a population of about 13,000 people, located 40 km west of Athens, which has significantly grown during recent decades. In term of lithology, the catchments consist exclusively of highly permeable geological formations (limestone and dolomites of Middle-Upper Triassic to Upper Cretaceous age which belong to the Subpelagonic geotectonic zone) [54] and streams of the drainage networks that are developed from alluvial deposits (mainly sands, silt and clays, gravels and pebbles). The town of Mandra is located at the apex of an alluvial fan formed by the two streams at the western part of the Thriassio plain. As depicted in Figure 2, the two main small ephemeral streams of the town are Soures and Agia Aikaterini which drain an area of 23.0 and 22.0 km^2 respectively. These two streams meet at the southeastern edge of the town to form a river that discharges into the Gulf of Elefsina. Their drainage basins are elongated along an almost northwestern-southeastern trending axis and reach the elevation of 659 m and 800 m, respectively, at their western ends. Intensive construction activities in the greater area during the last decades resulted in significant morphological changes of the streams' channels with most important the significant decrease of their available cross-sectional areas while in certain areas the streams practically vanished.

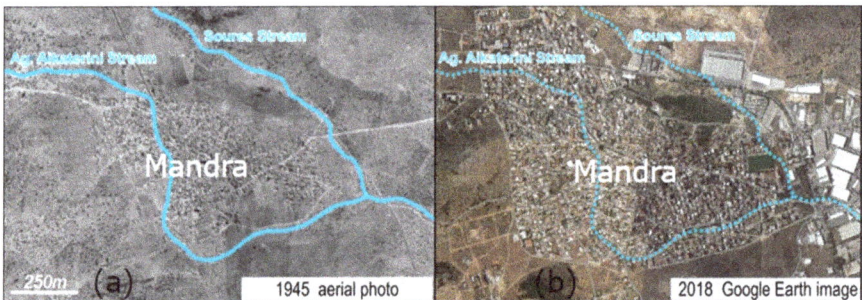

Figure 2. (**a**) Aerial photo taken in 1945 (Hellenic Military Geographical Service); (**b**) 2018 Google Earth image of the broader area of the city of Mandra showing the natural channels of streams Soures and Agia Aikaterini.

The comparative observation of an aerial photo taken in 1945 (obtained from the Hellenic Military Geographical Service, Figure 2a) with the recent Google Earth image (Figure 2b) of the broader area of Mandra is indicative of the effects of human activities on the streams' channels. In 1945 the Soures stream has a totally natural channel without human interference, which runs east and out of the city, while Agia Aikaterini stream is already passing through a part of the city and has a natural bed downstream. The recent Google Earth image shows that the Soures stream channel has been affected by major individual buildings, road works, and the eastward extension of the city which sometimes interrupt its flow, while the stream channel of Agia Aikaterini has totally disappeared under the main streets, which have a meandering pattern like the covered, previously natural, channel.

2.2. Ground Precipitation Measurements

During this event, precipitation data were collected from the XPOL experimental ground radar of the National Observatory of Athens located on Penteli mountain (approximately 35 km east of Mandra), as shown in Figure 3.

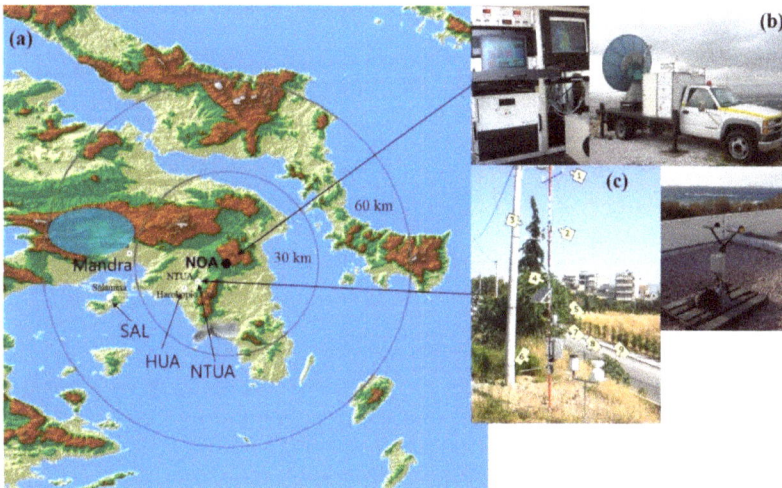

Figure 3. (**a**) Topographic map of the area covered from the X-band dual polarization (XPOL) radar encompassing (the shading circle) the experimental basin, the in situ instruments, and the town of Mandra. Pictures from the deployed sensors are also shown; (**b**) at the upper right is the XPOL radar; (**c**) meteorological station with the Parsivel disdrometer at the National Technical University of Athens (NTUA) site.

The XPOL radar operates during rain events in plan position indicator (PPI) mode taking measurements in a sector scan of 180°, at 0.5°, 1.5° and 2.5° elevation sweeps with a range resolution of 120 m for the total range of 65 km. Antenna rotation rate was 6 deg s^{-1} and the time-period for a full volume scan was less than 3 min. For the current study, two land surface in situ meteorological stations including typical tipping buckets for rainfall (with 0.2 mm resolution) at Salamina Island (SAL) and Harokopio University of Athens (HUA) and one meteorological station including tipping bucket with 0.1 mm resolution (Young and Campbell model) at the National Technical University of Athens (NTUA) are available. The last station is associated with one laser type, Autonomous Parsivel Unit (APU), disdrometer. Figure 4 shows the rainfall rates measured by each rain gauge at the three different sites. These measurements are used as reference for the assessment of the radar rainfall estimation. The period from 14 November to 26 December 2017 was selected for the assessment due to the availability of the XPOL radar observations. Unfortunately, the radar was not available on more days before the flooding event in order to exploit longer time series. It is important to note here that the precipitation rates illustrated in this figure are too low in comparison with the ones observed by XPOL over Pateras mountain. This fact verifies the large variation of precipitation in space and time and amplifies the significance of radar precipitation observations either on ground or from space.

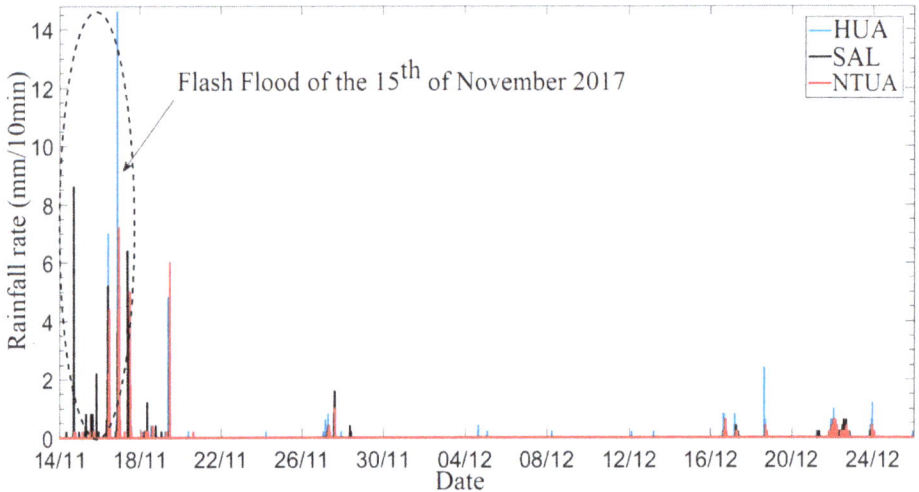

Figure 4. Time series of precipitation rate (mm/10min) from the three different in situ sites used for the evaluation of the XPOL precipitation algorithm. The dashed line indicates the flash flood event took place on 15 November 2017.

2.3. X-Band Dual Polarization (XPOL) Radar Rainfall Estimation Algorithm

In comparison with in situ measurements, ground radar (GR) observations provide horizontal distribution of the rainfall field including information even over ungauged areas. Due to their smaller size, X-band radars can be deployed in mountainous areas and have better coverage of specific areas than permanently installed long-range operational radars (C- and S-bands). To retrieve precipitation and microphysical estimates, we applied the Self-Consistent Optimal Parameterization-Microphysics Estimation (SCOPE-ME) X-band dual-polarization algorithm, consisting of new polarimetric techniques for bright band identification and vertical profile of reflectivity correction [15,55,56], attenuation correction [7,57] and precipitation microphysics retrievals [58].

The algorithm is based on theoretical analysis of T-matrix electromagnetic simulations in rain and minimizes the parameterization error compared to other parameterizations. It is applied in liquid precipitation and once the radar volume has been quality controlled and then corrected for attenuation (only in the selected liquid precipitation regimes) and bright band effects [55], the precipitation rate is estimated using the parameterization given by [12,59] ... The radar precipitation which is estimated by the SCOPE-ME polarimetric algorithm shows good accuracy at the full range of precipitation values [60]. The SCOPE-ME algorithm has been validated in various experimental studies [7,12,56,59] using different radar instruments [61,62]. Radar-precipitation estimates were verified using drop-size distribution and rain rate observations collected by in situ disdrometers and rain gauges.

In this study, before the application of radar algorithms, disdrometer data at a close range from the radar (~10 km from the NTUA disdrometer site) were used for the validation of the calibration of radar reflectivity (horizontal polarization). Differential reflectivity was calibrated using an average theoretical relationship between reflectivity and differential reflectivity as described in [57,59]. After the bias was removed, this algorithm was used to process the raw XPOL data and produce precipitation estimates.

The basic statistical metrics [6,56,59] for the evaluation of the radar rainfall estimates from 14 November to 26 December 2017 are the following:

The correlation coefficient (CR):

$$CR = \frac{\sum_i^N \left(R_{ref}(i) - \overline{R_{ref}} \right) \left(R_{est}(i) - \overline{R_{est}} \right)}{\sqrt{\sum_i^N \left(R_{ref}(i) - \overline{R_{ref}} \right)^2 \sum_i^N \left(R_{est}(i) - \overline{R_{est}} \right)^2}} \tag{1}$$

between the hourly radar rainfall estimates (R_{est}) and reference (raingauges) rainfall (R_{ref}), the bias ratio (BR):

$$BR = \frac{\sum_i^N (R_{est}(i))}{\sum_i^N \left(R_{ref}(i) \right)} \tag{2}$$

which is defined as the ratio of total radar-precipitation estimates during a storm event to the corresponding total reference values and the normalized error (NE):

$$NE = \frac{\sum_i^N \left(R_{est}(i) - R_{ref}(i) \right)}{\sum_i^N \left(R_{ref}(i) \right)} \tag{3}$$

defined as the mean difference of the estimate minus the reference divided by the mean reference values.

The precipitation observations from the rain gauges that are shown in Figure 5 match with the XPOL precipitation rate estimates. All the statistical metrics are performed only for liquid precipitation, for less than 5% radar beam occlusion from ground clutter and for hourly precipitation greater than or equal to 0.01 mm. Thus, a significant amount of temporal smoothing has been achieved resulting in less random error and higher correlation with rain gauge data. The CR of XPOL precipitation estimates compared to the rain gauge observations are over 0.9 and it is noted that estimates are almost unbiased (BR = 0.91) and with a very small normalized error (NE = 0.05).

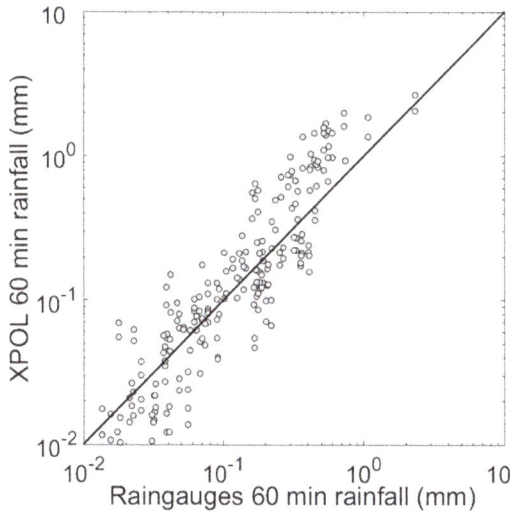

Figure 5. The scatterplot of XPOL hourly precipitation estimates versus the three rain gauge observations, where the "1-1" line is the bold black line.

2.4. Satellite-Based Precipitation Estimations

The GPM Core Observatory (CO) is currently the primary rain-measuring satellite equipped with a dual-frequency PR (DPR) (consisting of a Ku-band radar at 13.6 GHz and a Ka-band radar at 35.5 GHz) and the GPM Microwave Imager (GMI), which is a high resolution, conically scanning multichannel (frequencies range 10–183 GHz) MW radiometer [10,63], providing a reference standard to precipitation measurements. The retrieval technique that have been developed, which uses empirical and/or physically-based schemes to estimate precipitation from these satellite observations, are the Integrated Multisatellite Retrievals for Global Precipitation Measurements (IMERG; [64]). The onboard radar on the CO is similar to that on TRMM. The radar scans a swath that is approximately 245 km wide across the satellite track, measuring some 49 footprints of approximately 5 km in diameter and with 250 m vertical resolution at the nadir. The coarsening of the vertical resolution along the distance from the nadir increases from the 250 m up to about 2 km (at 17° from the nadir). The GPM data used in this study is the latest version 5B (V05B) IMERG precipitation product (2.5 months Research/final run) with $0.1° \times 0.1°$ spatial and 30 min temporal resolution. The IMERG is a level 3 gauge-calibrated GPM surface precipitation accumulation estimation product and benefits from the prior precipitation retrieval algorithms including PERSIANN-CCS [65], CMORPH [66], and the TMPA [67].

2.5. Model Set Up

In order to investigate the hydrometeorological characteristics of the flood, the integrated modeling system CHAOS [47,48] is used. The modeling system is configured to perform the simulation from 14 November at 12:00 UTC to 15 November at 12:00 UTC to represent the meteorological conditions during the life-cycle of the severe storm which occurred early in the morning on 15 November. CHAOS consists of two components: the atmospheric model WRF-ARW version 4.0 [40,41] and the ocean wave model WAM version 4.5.4 [68,69]. CHAOS was selected since its atmospheric component offers advanced capabilities in simulating severe weather phenomena [48,70–73]. The atmospheric component is two-way coupled with the ocean wave component through the OASIS3-MCT version 3.0 coupler [74] to better represent sea surface roughness which plays an important role in the atmospheric surface layer processes offering improvements in forecasting skill [48,75–77]. The advantage of CHAOS is the capability to simulate hydrological processes using the WRF-Hydro version 3.0 [36] at defined drainage basins. WRF-Hydro is currently one of the most growing hydrological models. It is noteworthy that WRF-Hydro is used as a framework for connecting atmospheric and hydrologic modeling at the National Water Center of United States [78].

In this study, three experiments using the hydrological component of CHAOS (WRF-Hydro) are performed in order to assess its hydrological response to different precipitation forcing data. The first precipitation dataset is based on the results of the atmospheric component of CHAOS (WRF-ARW), the second on X-band dual-polarization experimental ground radar (XPOL) estimations and the third on GMP/IMERG satellite estimations. CHAOS is configured to produce meteorological forcing fields for the hydrological component with a time step of 1 h. The forcing fields are presented in Table 1. In order to simulate the atmospheric conditions during the flood, the atmospheric component is set up in fine horizontal resolution using multiple nests. This is configured on 4 domains with horizontal resolutions of 9 km × 9 km, 3 km × 3 km, 1 km × 1 km and 0.25 km × 0.25 km (Figure 6). Time steps of 45, 15, 5 and 1 s are employed for the 4 domains, respectively. The initial conditions for 14 November at 12:00 UTC are based on the Global Forecasting System (GFS) operational analyses of the National Centers for Environmental Prediction (NCEP) with a horizontal resolution of $0.25° \times 0.25°$. The boundary conditions are also based on the GFS operational analyses with a time step of 6 h. The initial sea surface temperature (SST) field is based on the real time global (RTG) SST analyses with a horizontal resolution of $0.083° \times 0.083°$ produced by the NCEP. The Global Multi-resolution Terrain Elevation Data (GMTED 2010 30-arc-sec USGS) [79], the vegetation data MODIS FPAR [80] and the land-use data 21-class IGBP MODIS [81] are used as static input data in the pre-processing stage of WRF model.

Table 1. The seven meteorological forcing fields [34].

Meteorological Forcing Fields	Units
Incoming shortwave radiation (SR)	(W/m^2)
Incoming longwave radiation (LR)	(W/m^2)
Air specific humidity at 2 m (Q_2)	(kg/kg)
Air temperature at 2 m (T_2)	(K)
Surface pressure (PSFC)	(Pa)
Near surface wind at 10 m in the u- and v-components (U_{10}, V_{10})	(m/s)
Liquid water precipitation rate (PREC)	(mm/s)

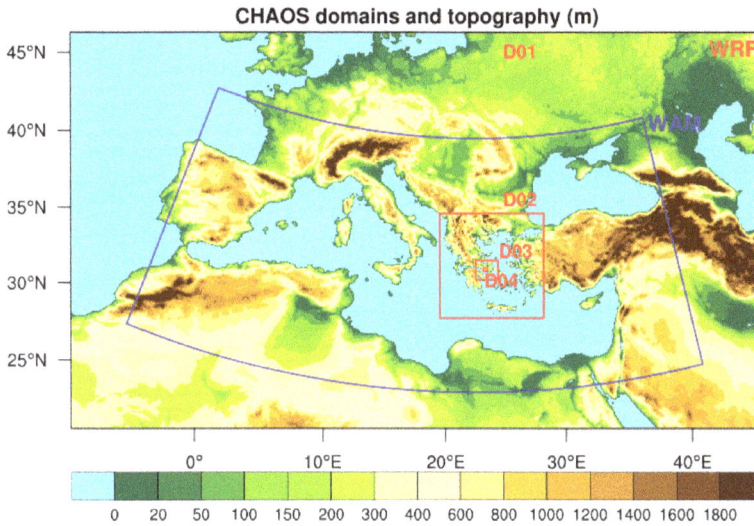

Figure 6. Chemical Hydrological Atmospheric Ocean Wave System (CHAOS) domains and topography (m). The nested domains of atmospheric component are depicted with red polygons and the domain of the wave component with blue polygon.

The revised Monin–Obukhov scheme [82] is employed to simulate the processes in the atmospheric surface layer, involving a number of modifications to encapsulate wave-dependent sea surface roughness information [47,48]. For the simulation of planetary boundary layer (PBL) processes the Yonsei University scheme (YSU) [65] is used. The ground processes are simulated using the unified Noah land surface model [83]. In order to resolve the long-wave and short-wave radiation processes the rapid radiative transfer model (RRTM) [84] and the Dudhia's [85] scheme are used, respectively. In order to simulate the microphysics and the convective processes, the Lin [86] and the Grell-Freitas ensemble [87] are used, respectively.

The hydrological model was configured on the extent of the 4th domain of the atmospheric model covering the drainage basin of Mandra on a 5-times finer horizontal resolution (50 m × 50 m). The Shuttle Radar Topographic Mission (SRTM) digital elevation model (DEM) [88] data of NASA in the native horizontal resolution of 90 m × 90 m was used. More specifically, the void-filled version [89] of this dataset distributed by the Hydrological Data and Maps Based on Shuttle Elevation Derivatives at Multiple Scales (HydroSHEDS; https://hydrosheds.cr.usgs.gov/index.php) was selected to be used. The dataset was resampled using nearest-neighbor interpolation on the 50 m × 50 m horizontal grid to introduce information about topography (Figure 7a), flow direction (Figure 7b,c), channel grid and stream order (Figure 7d) in the hydrological model (for more technical details see [36]. Initial conditions for the land surface parameters such as land use, soil type, vegetation, soil moisture and soil temperature required by the land surface model (LSM) NOAH of WRF-Hydro (more information in [36]), were prepared in the pre-processing stage of WRF-ARW model. The land surface data at 30-arc-sec native horizontal resolution was regridded to the grid of the forth domain of WRF-ARW model on horizontal resolution of 250 m × 250 m and, afterwards, to the grid of WRF-Hydro domain at horizontal resolution of 50 m × 50 m. For the two regridding procedures, the bilinear and the nearest-neighbor interpolation methods were used, respectively. As far as land-use is concerned, the USGS 24-category land use categorization was employed.

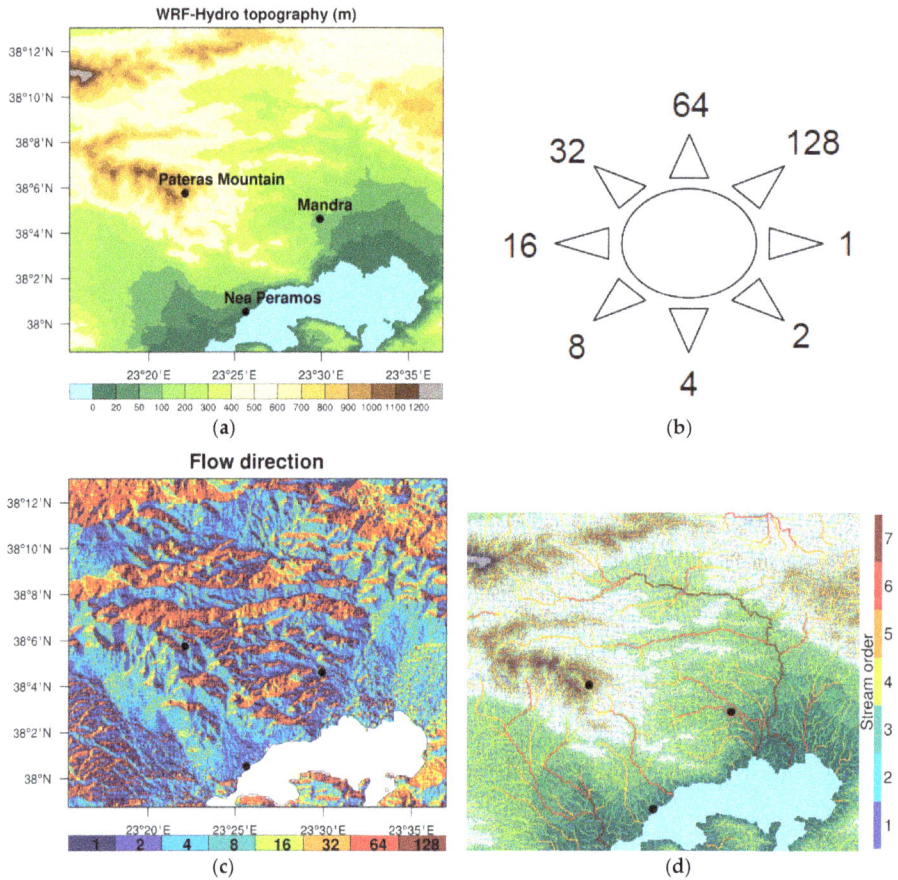

Figure 7. (**a**) hydrological model WRF-Hydro 50 m × 50 m topography and three areas represented by black dots: Pateras mountain, Mandra and Nea Peramos; (**b**) flow direction discretization. The flow direction grid has integer values of 1, 2, 4, 8, 16, 32, 64 and 128 which are oriented in the depicted directions; (**c**) flow direction; (**d**) hydrographic and stream order map of the water basin used for the hydrological runs.

The hydrographic network (Figure 7d) of the water basin drains surface runoff from the steep mountains to the Thriasio plateau (in the city of Elefsina). In general, the hydrographic network follows a north-south direction, except the Soures and Agia Aikaterini streams (see Figure 2) that initially have a west–east direction and then turns to the north–south direction. A number of smaller streams (5th and 6th order) flowing through the sub-urban area of Elefsina (Figure 7d) drainage from

Pateras mountain (Figure 7a) on Sarantapotamos, which functions as the major drainage channel reaching 7th stream order. Sarantapotamos water basin has a length of 31.5 km covering an area of 245.03 km², surrounded by the mountains Pateras, Pastra (1016 m), Cithaeron (1408 m) and Parnitha (Figure 7d).

For the configuration of the hydrological model, Manning roughness coefficient, channel bottom width, initial water depth and slide slope for channels are set for each stream order as shown in Table 2. These values are chosen after a series of calibration tests which are generally used for a good approximation of river modeling [36]. The initial calibration tests were based on discharge and flood extent assessment.

Table 2. Manning roughness coefficient (Manning), channel bottom width (CBW) in m, initial water depth (IWD) in m and slide slope (CSS) of channels for each stream order.

Stream Order	Manning	CBW (m)	IWD (m)	CSS
1	0.3	1	0.05	1.0
2	0.3	2	0.05	0.8
3	0.25	3	0.1	0.6
4	0.2	4	0.1	0.4
5	0.15	6	0.1	0.2
6	0.1	8	0.2	0.1
7	0.05	10	0.2	0.05

The hydrographic network of Soures and Agia Aikaterini streams is not monitored at all due to the lack of installed instrumentation. Thus, in order to evaluate the simulated flood extent, post-survey maps created just after the end of the flood event on 15 November and satellite remote sensing images (WorldView-4 in the very high-resolution of 0.31 m) during the period 21–23 November were used; the observed flood extent was mapped by the group of the FloodHub service of BEYOND Center of Excellence for EO-based monitoring of Natural Disasters. The simulated flood extent is delineated using the peak water level along the streams [49]. Regarding the XPOL and Global Precipitation Measurement/Integrated Multi-satellitE Retrievals for GPM (GPM/IMERG) forced hydrological simulations, the water level results on 15 November at 06:00 UTC are used. However, regarding CHAOS forced simulation, the water level results at 03:00 UTC are used because the model predicted the peak of water level 3 h before the 2 other simulations. The topography used for the evaluation procedure is based on the 50 m × 50 m horizontal grid (Figure 7a) produced using nearest-neighbor interpolation for resampling the SRTM DEM 90 m × 90 m data distributed by HydroSHEDS. The interpolation method used for the extraction of the water level surface and consequently the inundated areas is the inverse distance weighted (IDW). Figure 8 illustrates a flowchart summarizing data used, model setup and experimental procedure.

2.6. Evaluation Methodology

Dichotomous forecasts (occurrence or non-occurrence) of the event of interest, in this case a flood, can be considered as "completely confident" forecasts. These forecasts can be treated as categorical forecasts and are usually verified using contingency tables (Table 3) and various scores defined by them [90,91]. A combination of verification measures and scores can be used in comparative evaluation of forecasts, by applying the statistical concept of "sufficiency" [92]. An example of pairs of sufficient statistics for the dichotomous completely confident probability forecasts include probability of detection (PoD) and false alarm ratio (FAR; [93]). In this study, additionally critical success index (CSI) and frequency bias (FB) verification measures was chosen to be included to the evaluation procedure.

Table 3. Contingency table for evaluation.

Simulated Event	Observed Event	
	Yes	No
Yes	Hit (a)	False alarm (b)
No	Miss (c)	Correct non-event (d)

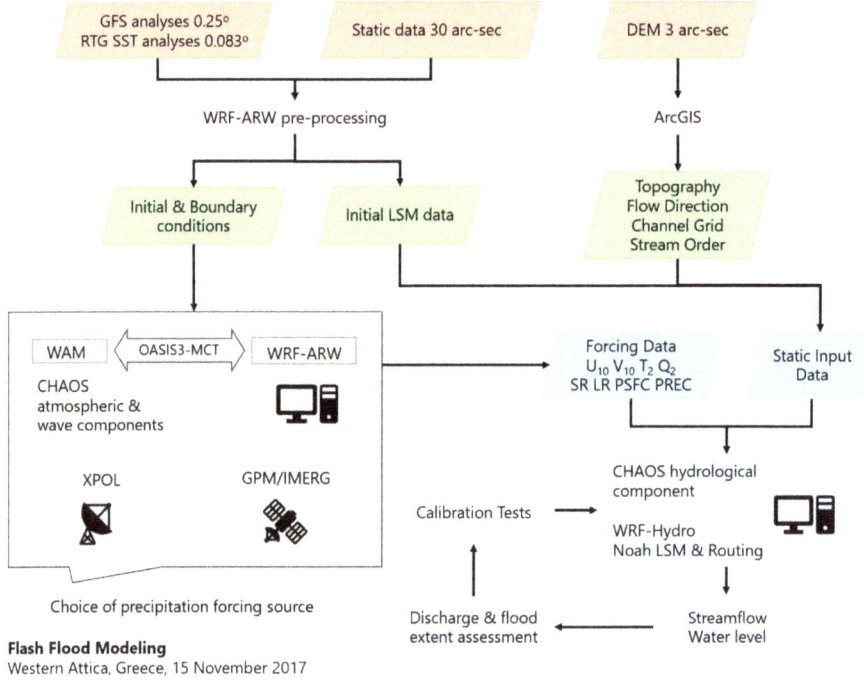

Figure 8. Flowchart of data used, model configuration and experimental procedure. Data, parameters and models are thoroughly described in the text.

The Probability of Detection (PoD or hit rate) and the Critical Success Index (CSI or threat score) range between 0 and 1, with 1 representing a perfect forecast, and can be computed based on Equations (4) and (5), respectively.

$$PoD = \frac{a}{a+c} \tag{4}$$

$$CSI = \frac{a}{a+b+c} \tag{5}$$

The FAR ranges between 0 and 1, with 0 representing a perfect forecast. This score is not sensitive to missed events. The FAR can be improved by systematically under forecasting rare events. It also is an incomplete score and should be used in connection with the PoD. FAR can be computed with the following relationship:

$$FAR = \frac{b}{a+b} \tag{6}$$

The FB is not a true verification measure, as it only compares the forecast and observed frequencies of occurrence of the event in the sample. Frequency bias of 1 represents the best score, values higher than 1 indicate over forecasting and values less than 1 indicate under forecasting. It can be computed using the following relationship [94]:

$$FB = \frac{a+b}{a+c} \tag{7}$$

3. Results

3.1. Meteorological Analysis of the Weather Event

A notable extreme weather event, leading to one of the most devastating flash floods in Attica, occurred on 15 November 2017. The synoptic meteorological conditions over the Mediterranean Sea that led to this destructive flash flood are presented here. On 13 November 2017, a cold air mass was associated with a well-defined trough extending from northwest Europe to the central Mediterranean Sea, as can be seen from the 850 and 700 hPa maps (Figure 9a,b respectively), leading to the generation of a quite deep and extended barometric low centered over northern Italy and a secondary barometric low located to the southeast of Sicily, (Figure 9c). On 14 November, both systems merged into one barometric low over the Tyrrhenian Sea accompanied by widespread warm and cold fronts extending from the Balkan Peninsula to the Gulf of Sirte (Figure 9d).

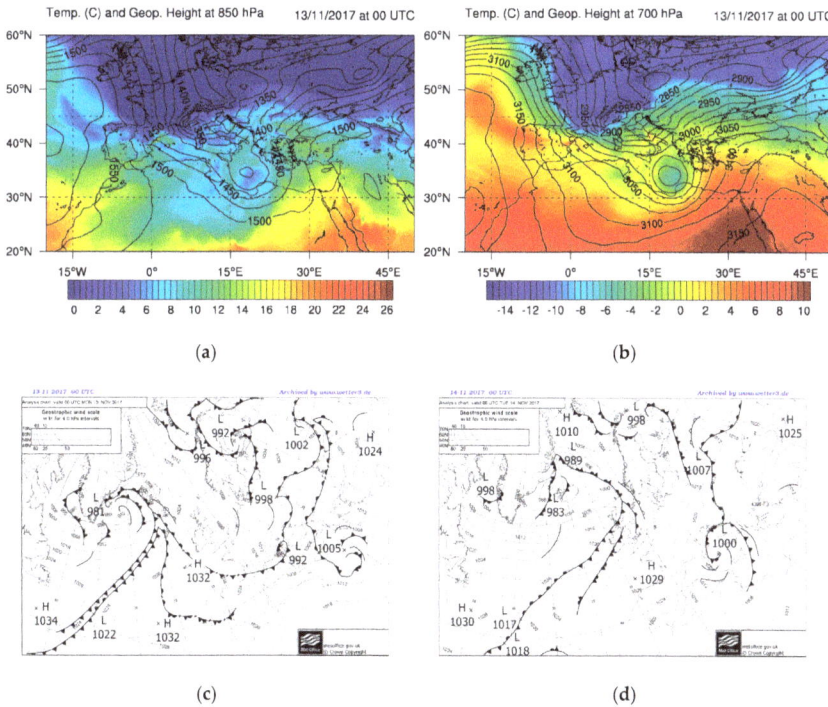

Figure 9. Temperature and geopotential height based on Global Forecasting System (GFS) analyses for 13 November 2017 (00:00 UTC) at: (**a**) 850 hPa and (**b**) 700 hPa. UK Met Office surface analysis maps on: (**c**) 13 November 2017 (00:00 UTC) and (**d**) 14 November 2017 (00:00 UTC).

Subsequently, the barometric system continued to be supported by the upper-level trough. It is important to note that the trough cut off and reinforced the core of the low by introducing very cold air masses over relatively warmer sea (Figure 10a–c). It was stayed almost stagnant during the period from 14 to 19 November under the impact of an upper-air blocking, caused by a well-defined succession of warm air-cold air-warm air (ridge-trough-ridge). This blocking was additionally intensified by the cut-off of the trough forcing the low to stay in place and cause extreme weather phenomena, associated with the floods examined in this study. In the morning of 15 November when the floods at Mandra occurred, the low was well-organized, and it was associated with fronts (Figure 10d).

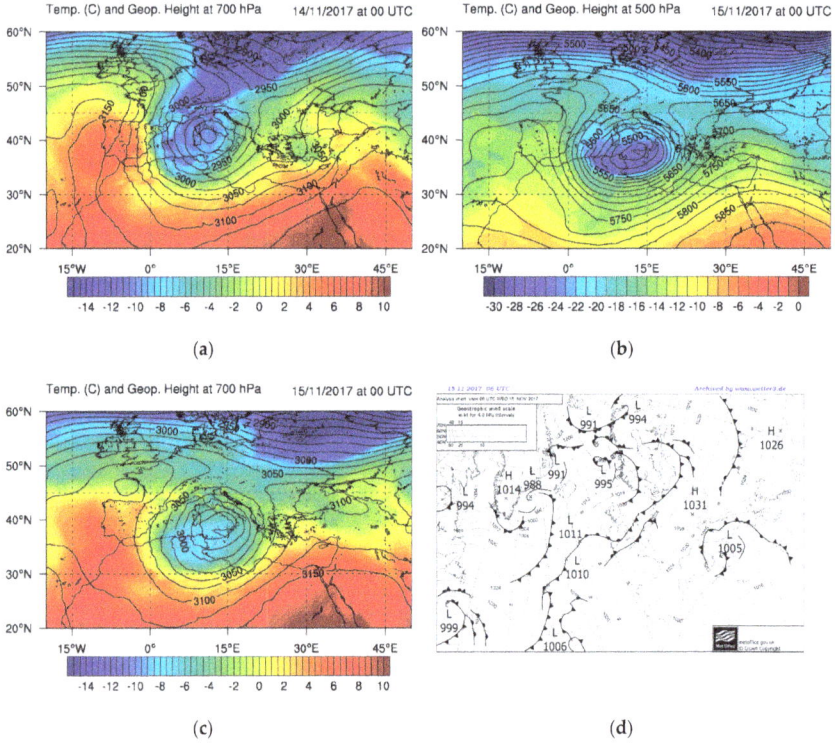

(a)

(b)

(c)

(d)

Figure 10. (a) Temperature and geopotential height based on GFS analyses for 14 November 2017 (00:00 UTC) at 700 hPa; temperature and geopotential height on 15 November 2017 (00:00 UTC) at (b) 500 hPa and (c) 700 hPa; (d) UK Met Office surface analysis map for 15 November 2017 (06:00 UTC).

However, the low and the associated fronts were located far away the flooded area to directly cause storms (Figure 10d). So, it is important to analyze in depth the factors caused such an intense flooding storm. The dominant mechanism is the orographic convergence of humid southeasterly airflow over the southeastern slopes of Pateras mountain (Figure 11a,b). The orographic lifting and the humidity advection are depicted on a vertical cross section up to 6 km for airflow and hydrometeors specific humidity along Pateras mountain and Soures stream head (Figure 11c). Figure 11d–f shows updrafts of humid air over the eastern slopes of Pateras mountain and around Mandra area. The local convergence is collocated with the dry downdrafts over the western slopes of Pateras mountain which are attributed to the middle- and upper-level air circulation. This effect intensified the persistence of rainfall at the east of Pateras mountain increasing surface water runoff and stream discharge as will be shown in the following section. The rainfall was characterized by a long duration of approximately 5 h over this area because it was also indirectly supported by the stagnant spinning barometric low which preserved both the humid southeasterly airflow near the surface and the dry westerly airflow at the middle troposphere. An analysis of the storm focusing on its precipitation characteristics and the hydrological response of the drainage basin of Mandra are presented in the following section.

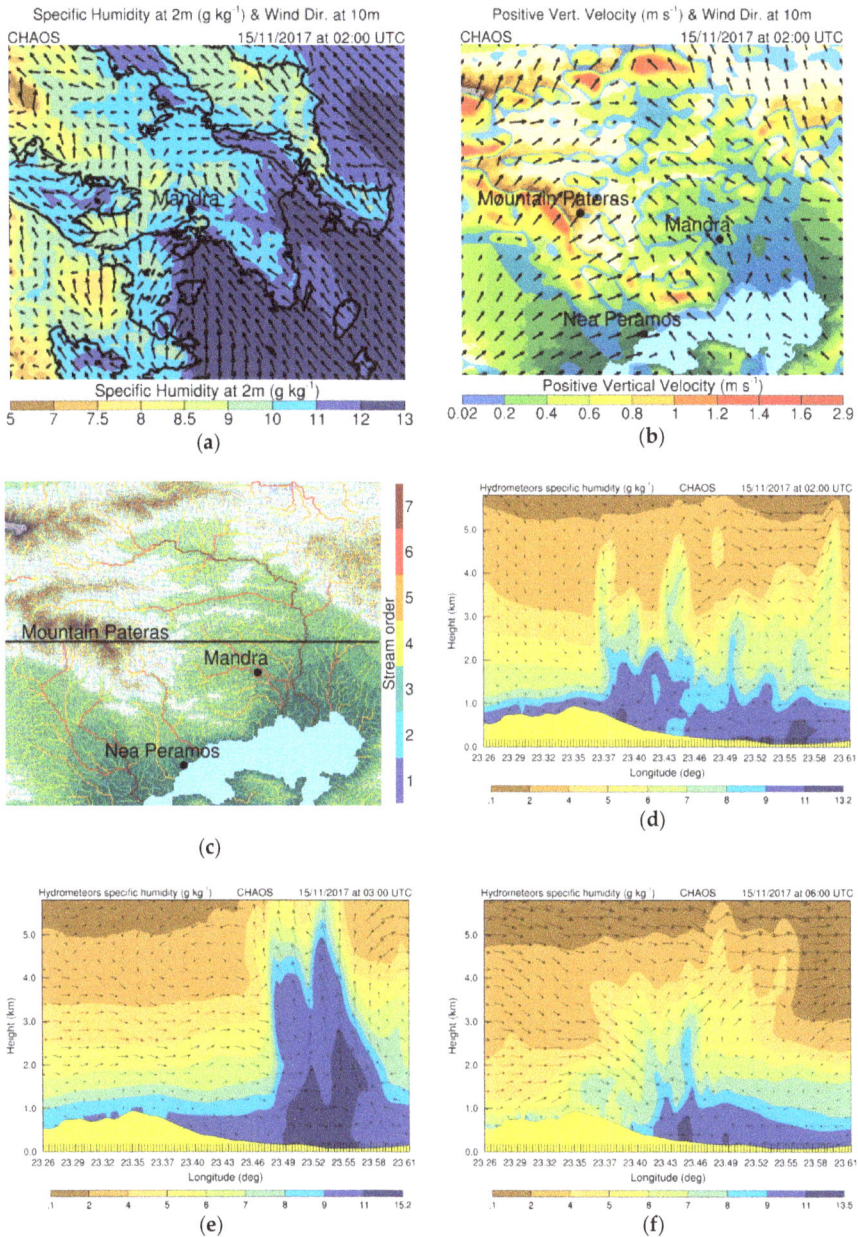

Figure 11. (**a**) Specific humidity at 2 m and horizontal wind vectors at 10 m; (**b**) positive vertical velocity and horizontal wind vectors at 10 m (on the right) on 15 November 2017 (02:00 UTC); The two maps (**a,b**) are based on CHAOS simulation results obtained from the third and fourth nested domains, respectively. (**c**) Hydrographic and stream order map of the water basin illustrating the vertical cross section area (black line). Vertical cross section up to 6 km for hydrometeors specific humidity (g kg^{-1}) and wind on 15 November 2017 at (**d**) 02:00 UTC, (**e**) 03:00 UTC and (**f**) 06:00 UTC based on CHAOS simulation results obtained from the fourth nested domain.

3.2. Quantitative Comparison of Precipitation, Discharge and Water Level

This section presents the quantitative comparison of XPOL, GPM/IMERG and CHAOS precipitation estimates and the respective response of the hydrological model to reproduce the flood event. The high precipitation rates of the 15 November that triggered the high peak runoff at the Agia Aikaterini and Soures streams are analyzed. In addition, this section shows a qualitative comparison between the three different forcing precipitation estimates using hourly simulated discharge maps. Statistics and flood extent maps based on combined simulated data, post-surveys and satellite images as a reference are used.

During the period from 14 November at 12:00 UTC to 15 November at 12:00 UTC, XPOL radar observed up to 300 mm accumulated precipitation in the core of the storm (Figure 12a). During the peak of the storm, from 02:00 to 06:00 UTC, the radar observed basin-average precipitation rates up to 57 mm/h over Soures stream (Figure 12b) and up to 41 mm/h over Agia Aikaterini stream (Figure 12d). Impressive was the fact that the highest precipitation rates in the core of the storm during 02:00–06:00 UTC reached 140 mm/h (not shown here). Moreover, the basin-average 24 h accumulated precipitation reached 194 mm over Soures stream (Figure 12c) and 153 mm over Agia Aikaterini stream (Figure 12e). The local intense rainfall drove the sub-basin network to overflow, flooding the area of Mandra and causing 24 fatalities and extensive million-euro damages to property and infrastructure.

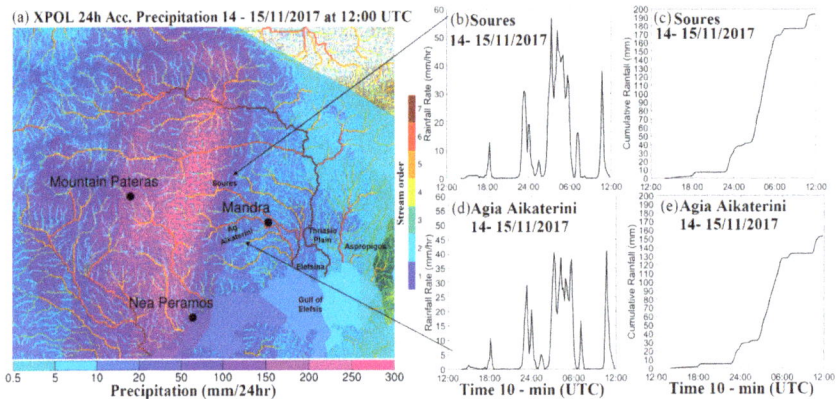

Figure 12. (**a**) Hydrographic and stream order map of the water basin with superimposed the XPOL 24 h precipitation accumulation; Time series of basin-average (**b**) precipitation rate and (**c**) accumulated precipitation for Soures stream; Time series of basin-average (**d**) precipitation rate and (**e**) accumulated precipitation for Agia Aikaterini stream. Figures refer to the period from 14 November at 12:00 UTC to 15 November at 12:00 UTC.

The results of hydrometeorological simulations of the flash flood at western Attica are presented here. Simulated discharge from 02:00 to 04:00 UTC of 15 November is shown in Figure 13a–c, superimposed by the corresponding 1 h accumulated precipitation (mm) estimated by the National Observatory of Athens' (NOA) XPOL (left), GPM/IMERG (middle) and CHAOS (right). Figure 13 presents different characteristics in the spatial pattern and intensities of the three precipitation estimates, and, subsequently, in the simulated discharge. In the case of XPOL precipitation estimate, a persistence over Pateras mountain is observed. This is attributed to the orographic lifting of warm and humid air masses originating from the Aegean Sea (following the southeasterly atmospheric flow which was induced by the stagnant barometric low). It is noteworthy that the spatiotemporal distribution of the precipitation affects the available surface and soil moisture determining surface- and channel-water runoff and, therefore, the streams' routing [42–44]. The relatively increased discharge at Soures and Agia Aikaterini streams at 04:00 UTC presented on the XPOL-hydro simulation is attributed to the increased persistence of intense rainfall over the steep slopes of Pateras mountain (Figure 13c). On the other hand, it seems that the satellite product missed most of the peaks and stage of the storm as it only indicates precipitation at 03:00 UTC (Figure 13b). Yucel et al. [43] also observed an underestimation of WRF-Hydro response on satellite Multi-sensor Precipitation Estimates (MPEs) of EUMETSAT. This may be attributed to poor representation of high precipitation rates by satellites. Possible reasons are the coarse horizontal resolution of satellite data (~11 km for GPM/IMERG and ~4 km for MPEs) which make difficult to capture local extreme weather events as well as various inaccuracies in satellite algorithms. On the other hand, the CHAOS modeling system is closer to the radar precipitation estimates especially in the total duration and the intensity of the rain. An overall comparison of the discharge simulations indicates that the simulation forced by the radar precipitation estimates produced the highest and most accurate in time discharge, which finally led to the flooding.

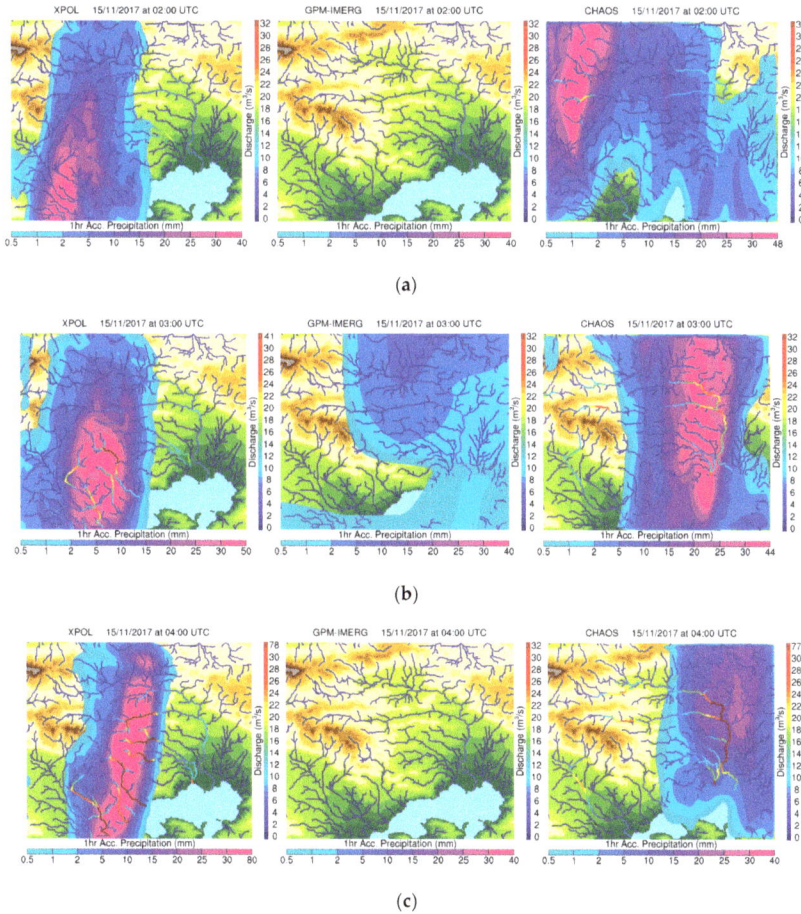

Figure 13. 1 h accumulated precipitation estimations from the National Observatory of Athens' (NOA) XPOL radar (**left**), the Global Precipitation Measurement/Integrated Multi-satellitE Retrievals for GPM (GPM/IMERG) (**middle**) and CHAOS (**right**). On the same maps the simulated discharges forced from each of the three precipitation estimations are superimposed. Only streams with order higher than 4 are presented. The maps refer to 15 November at (**a**) 02:00, (**b**) 03:00 and (**c**) 04:00 UTC.

Concerning the estimated water level, similar outcomes are found from the three differently forced hydrological simulations (Figure 14). Water reaches the level of 130 cm at the Agia Aikaterini and Soures streams when simulated by XPOL-hydro and CHAOS-hydro simulations. However, GPM/IMERG-hydro simulation does not increase water level. Combining Figures 13 and 14, it is obvious that this flash flood event is characterized by a small delay of about 1–2 h after the high precipitation peaks. This is attributed to the limited spatial scale of hydrographic network of the drainage basin of western Attica. Another factor is the high steepness of the eastern slopes of Pateras mountain which accelerates the runoff towards the Agia Aikaterini and Soures streams (see maps in Figure 7).

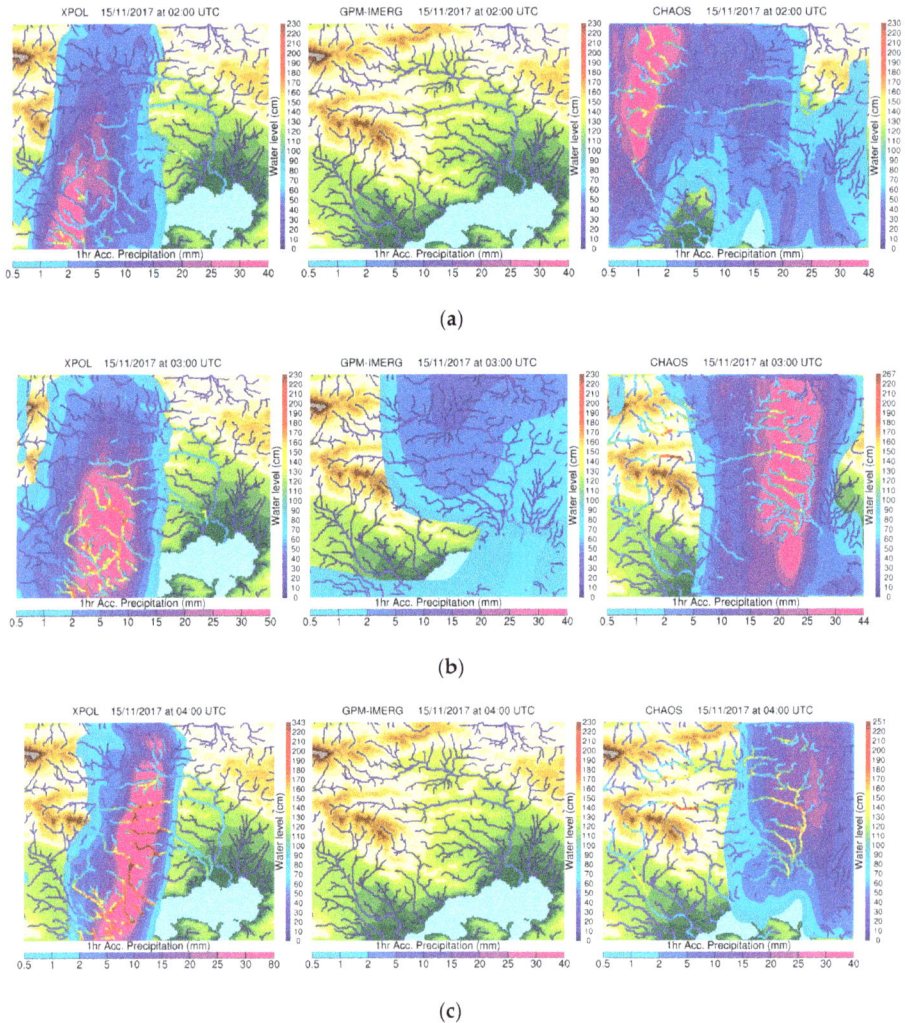

Figure 14. 1 h accumulated precipitation estimations from the NOA's XPOL radar (**left**), the GPM/IMERG (**middle**) and CHAOS (**right**). On the same maps the simulated water levels forced from each of the three precipitation estimations are superimposed. Only streams with order higher than 4 are presented. The maps refer to 15 November at (**a**) 02:00, (**b**) 03:00 and (**c**) 04:00 UTC.

Simulated discharge and 1 h accumulated precipitation for the period from 05:00 to 07:00 UTC of 15 November are shown in Figure 15. During this period, GPM/IMERG-hydro and CHAOS-hydro simulations underestimate both precipitation and discharge especially at 05:00 UTC when no precipitation was observed or produced. However, CHAOS-hydro has a better performance at Nea Peramos flooding which is not captured by the GPM/IMERG-hydro simulation. XPOL-hydro simulation preserves the storm for hours extending the flooding at Mandra and Nea Peramos areas. XPOL-hydro simulation results in discharge values about 115 m^3/s in Soures and Agia Aikaterini streams reaching a peak of 195 m^3/s in Sarantapotamos river.

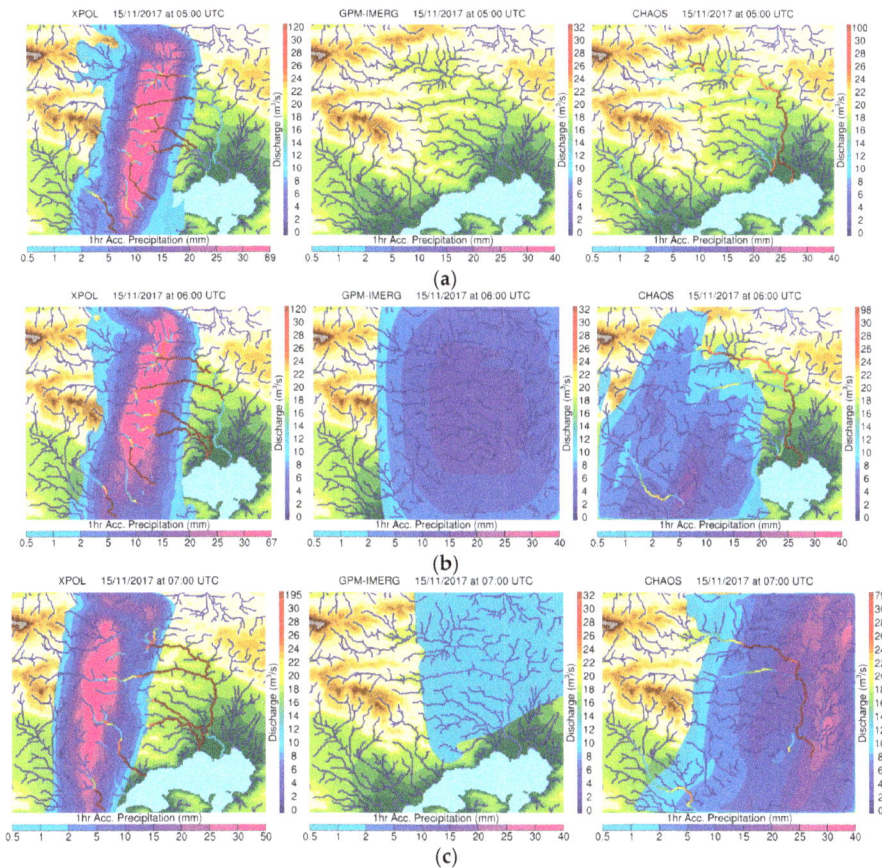

Figure 15. 1 h accumulated precipitation estimations from the NOA's XPOL radar (**left**), the GPM/IMERG (**middle**) and CHAOS (**right**). On the same maps the simulated discharges forced from each of the three precipitation estimations are superimposed. Only streams with order higher than 4 are presented. The maps refer to 15 November at (**a**) 05:00, (**b**) 06:00 and (**c**) 07:00 UTC.

Diakakis et al. [51] estimated a peak discharge for Soures stream about 170 m^3/s and for Agia Aikaterini stream about 140 m^3/s, using the same XPOL radar rainfall data to force the kinematic local excess model (KLEM; more information in [51]). Diakakis et al. [51] also proposed a semi-empirical method to estimate peak discharge values based on site detailed post-surveys of flood extent and impact, which resulted to wide ranges of peak discharge values (30–40% error based on their analysis) for Soures and Agia Aikaterini streams around the values produced by the KLEM model. The XPOL-based WRF-Hydro simulation is characterized by the minimum difference about 20–30% in the estimated discharge peak with respect to [51]. This difference may be attributed to the configuration of the hydrological models which is usually based on calibration procedures using measured streamflow data [36,43,95–98] which are not available in this study. It is important to note here that an existing, enclosed rectangular conduit (L = 2.27 km and A = 3.4 m^2) in the town has a maximum discharge about 10 m^3/s. So, it was inadequate to cope with the extreme discharge during the flash flood event since the minimum required cross sectional area of the streams varies between 20 and 40 m^2 [99].

Moreover, Figure 16 shows the simulated water level and 1 h accumulated precipitation for the period from 05:00 to 07:00 UTC of 15 November. During this period, XPOL-hydro simulates water level exceeding 300 cm in Agia Aikaterini and Soures streams as well as in Sarantapotamos river. CHAOS-hydro estimates approximately half the water levels having a good agreement with XPOL-hydro. GPM/IMERG-hydro simulates a weak increase of water level.

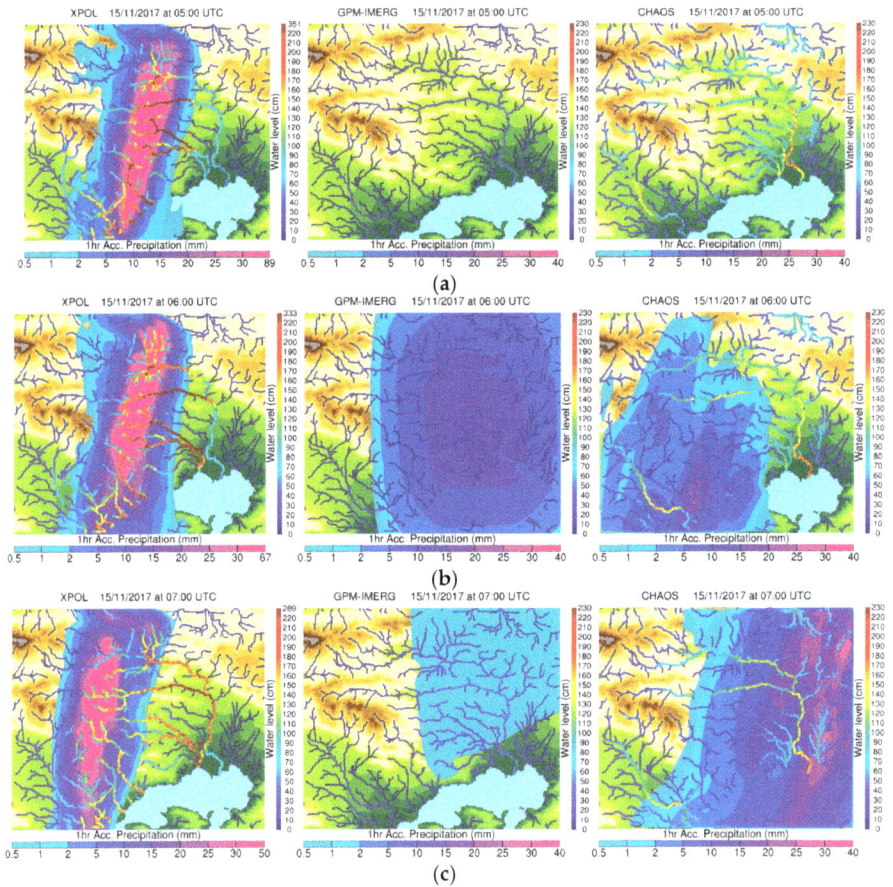

Figure 16. 1 h accumulated precipitation estimations from the NOA's XPOL radar (**left**), the GPM/IMERG (**middle**) and CHAOS (**right**). On the same maps the simulated water levels forced from each of the three precipitation estimations are superimposed. Only streams with order higher than 4 are presented. The maps refer to 15 November at (**a**) 05:00, (**b**) 06:00 and (**c**) 07:00 UTC.

Figure 17a depicts the hydrographic and stream order map of the water basin where black dots represent the points used to make hydrograph of each simulation for the period from 14 November at 12:00 UTC to 15 November at 12:00 UTC (Figure 17b). P1 is used to export 1 h accumulated precipitation from the XPOL dataset, while P2 is used for GPM/IMERG and CHAOS datasets. Different points are used here to represent the precipitation maxima of the three datasets avoiding in that manner mismatches of the pattern and the phase speed of the barometric low. P3 is used to export discharge from the datasets that resulted from all three hydrological simulations. As shown in Figure 17b, GPM/IMERG data systematically estimate much less precipitation compared to XPOL while CHAOS hourly precipitation is closer to the XPOL one. The increased underestimation of the GPM/IMERG, comparing to the XPOL and CHAOS, is mainly attributed to its coarser resolution which did not support enough representation of such a local extreme precipitation event. In addition, due to the lack of passive microwave or radar data in this specific area at the time of the event the IMERG algorithm uses interpolation of available data following the feature motion which is estimated from infrared satellite data. As far as the discharge is concerned, CHAOS matches better with the XPOL radar estimations in comparison to the satellite which does not cause flooding. The differences between CHAOS and XPOL precipitation data are attributed to forecast inaccuracies of numerical weather prediction.

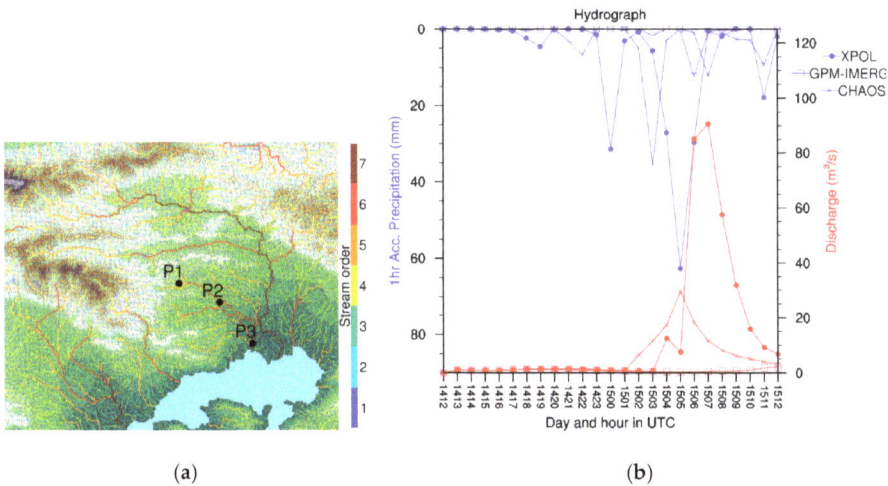

Figure 17. (**a**) Hydrographic and stream order map of the water basin as Figure 7d. Black dots represent the points used for the time series. P1 is used to export 1 h accumulated precipitation from XPOL dataset, while P2 is respectively used for GPM/IMERG and CHAOS datasets. P3 is used to export discharge from the datasets that resulted from all the three hydrological simulations. (**b**) Hydrograph: time series of 1 h accumulated precipitation and discharge estimations using XPOL radar, the GPM/IMERG and CHAOS forcing on hydrological model.

As far as streams' overflow is concerned, Soures and Agia Aikaterini were not enough to carry the extremely high discharges and they overflowed [49]. Various public infrastructure and private constructions were erected over the two streams while the necessary water drainage pipes were either too small or they were not built at all [100]. One of the reasons for the extensive damage during the flood event was the human interference on Mandra's streams (see Figure 2). Moreover, the artificially confined channels of the streams and the drainage pipes were inefficient at facilitating the flow of the run-off water [99]. As a result, the flood additionally caused significant soil erosion by carrying away large quantities of sediment particles due to high flow and stream-bottom shear stresses. Fine sand, silt and clay of considerable thickness were deposited along the roads of Mandra as well as at the mouth of the streams along the shoreline of the Gulf of Elefsina, indicating the high sediment load of the drainage networks during the event [50]. Additionally, debris and heavy objects, such as cars and trees were carried away by the torrents. The large amounts of solid materials transported by the streams may have resulted in an additional rise of the water level in the streams.

3.3. Qualitative Evaluation of Estimated Flood Extent

In this section a comparison of simulated and observed flood extent is presented. Due to the high inaccuracies of the flooded areas estimation, the assessment of flood extent is qualitative. The evaluation is restricted to the area upstream of the "Attiki Odos" highway (Figure 18), the construction of which has altered the hydro-morphological characteristics of the area but simultaneously provides adequate flood protection at most of the adjacent areas [101]. Only the 6th and 7th order streams are used during this analysis, while the upstream areas are not included in the analysis.

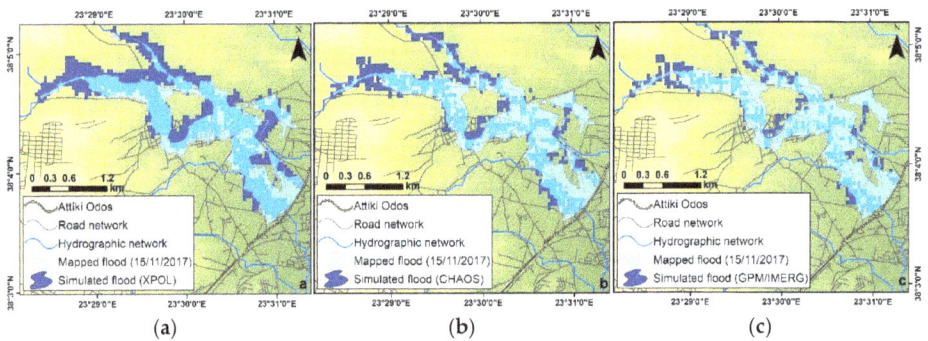

Figure 18. Flood extent produced by the three hydrological simulations (dark blue) using precipitation forcing by: (**a**) XPOL; (**b**) CHAOS; (**c**) GPM/IMERG datasets. Flood extent mapped by the FloodHub group of BEYOND (Building a Centre of Excellence for Earth Observation-based monitoring of Natural Disasters) project during the 15 November flood event (light blue).

Based on the evaluation of the different statistical scores that are summarized in Table 4, the best PoD score was accomplished by the XPOL rainfall estimation forcing to the hydrological model (0.66), meaning that 66% of the observed flooded area was correctly predicted. The worst PoD score was accomplished by the GPM/IMERG-hydro (0.30). The higher CSI score was again accomplished by XPOL (0.43), indicating that less than half of the flooded area (observed and/or predicted) was correctly forecasted. The worst CSI score was again accomplished by the GPM/IMERG-hydro (0.24). FAR score was practically identical among the three experiments confirming in that manner their similar capabilities to simulate flooded areas that failed to materialize. FAR scores (0.43–0.45) indicate that in less than the half-flooded area simulated, flood was eventually not observed. The FB score was greater than 1 in the case of XPOL-hydro, indicating over-forecast, and lower than 1 in the cases of CHAOS-hydro and GPM/IMERG-hydro, indicating under-forecast, especially in the case of GPM/IMERG-hydro. Overall, XPOL-hydro managed to predict the flood event with satisfactory accuracy, although it over estimated the inundated area (Figure 18a). CHAOS-hydro showed a marginally acceptable skill to simulate the flooded areas despite the fact of the significant uncertainties and forecast errors usually posed by the numerical weather prediction models (Figure 18b). GPM/IMERG-hydro, on the other hand, did not predict the flood event accurately and under estimated the flooded area (Figure 18c).

Table 4. Results of the scores used for the validation of the three hydrological simulations.

	XPOL-Hydro	CHAOS-Hydro	GPM/IMERG-Hydro
PoD	0.66	0.43	0.30
CSI	0.43	0.33	0.24
FAR	0.45	0.43	0.43
FB	1.21	0.75	0.53

Possible inaccuracies affected the entire set of simulations can be attributed to the coarse resolution of the digital elevation map (DEM), which does not allow the extraction of detailed cross sections along the streams. Especially upstream where the riverbeds are narrow, the flooded areas could not be predicted with accuracy due to the lack of detailed information concerning the geometry of the streams. It should be also noted that the flood took place in an extremely dense urban fabric that was not properly designed. This information was not possible to be included in the simulation in detail.

4. Discussion

The comparison of the results of the three hydrological simulations reveals XPOL and CHAOS precipitation forcing data estimated water level and discharge in a more efficient way than GMP/IMERG data. The rapid increase of these hydrological parameters indicated a flash flood event, which is attributed to the persistence of high precipitation rates over the steep slopes of Pateras mountain for approximately 5 h (as observed by the XPOL radar). Basin-average precipitation rates reached 57 mm/h over Soures stream, 41 mm/h over Agia Aikaterini stream and 140 mm/h in the core of the storm. Moreover, the XPOL-radar observed 24 h accumulated precipitation up to 300 mm in the core of the storm. Time series of basin-average accumulated precipitation for the two sub-basins of Soures and Agia Aikaterini streams indicated high cumulative precipitation reaching 194 mm and 153 mm, respectively. CHAOS-hydro estimated approximately the half water levels and even lower discharges compared to XPOL-hydro simulation, while, having good agreement in the representation of spatial and temporal characteristics of the storm, and, subsequently of the flooding event. GPM/IMERG-hydro simulated a weak increase of water level and discharge due to the low precipitation estimation. It is noteworthy that XPOL-hydro simulation resulted in discharge values about 115 m^3/s and water level values exceeding 3 m in Soures and Agia Aikaterini streams which caused the flash flooding around the Mandra area.

It is difficult to make a comprehensive representation of reference discharge during this flood event, due to lack of instrumentation. Diakakis et al. [51] proposed a semi-empirical method to estimate peak discharge values based on site detailed post-surveys of flood extent and impact, which yielded relatively broad ranges of peak discharge values (30–40% error) for Soures and Agia Aikaterini streams around the values produced by the KLEM model forced by the same XPOL radar rainfall estimates. More specifically, Diakakis et al. [51] estimated a peak discharge for Soures stream about 170 m^3/s and for Agia Aikaterini stream about 140 m^3/s. All the three experiments based on WRF-Hydro underestimated discharge peak compared with KLEM results. The XPOL-hydro simulation is characterized by the minimum difference in the estimation of discharge peak per 20–30% in respect to the KLEM estimation. The differences may be attributed to inaccuracies in the precipitation estimates and the configuration of the hydrological model, which demands high calibration effort [36,43,96–98] using measurements which are not available at this area. Moreover, an explanation for the dominance of the XPOL-hydro simulation is the higher spatial and temporal resolution of precipitation estimates (120 m and 3 min) than the respective resolutions of CHAOS (250 m and 1 h) and GPM/IMERG (~11 km and 30 min) precipitation datasets.

The combination of the increased volume of run-off water with the human pressures on the streams interrupting their flow caused wide flood extent. The flash flood affected almost the entire town of Mandra [49]. Comparing with the observed flooded areas, the XPOL-hydro simulation presented overestimation. On the other hand, GPM/IMERG-hydro did not predict the flood event accurately and underestimated the flooded area while CHAOS-hydro showed sufficient skill to simulate the flooded areas despite the forecast inaccuracies of numerical weather prediction.

Another reason for the extensive damage during this flood event was the fact that the artificially confined channels of the streams and the drainage pipes were inefficient at facilitating the run-off water. An existing, enclosed rectangular conduit with capability of discharge about 10 m^3/s was inadequate to cope with the extreme discharge during the flash flood event [99]. Moreover, the streets of Mandra sometimes played the role of the streams of Soures and Agia Aikaterini concentrating large quantities of water and sediments.

5. Conclusions

The aim of this work is to perform a hydrometeorological analysis of the extreme flash flood event of the 15 November 2017 at the sub-urban area of Mandra, western Attica, Greece, applying an integrated remote sensing and data-modeling system. The flood event occurred at Soures and Agia Aikaterini streams and affected the suburban town of Mandra with landslides, extensive damage and a total of 24 fatalities.

This case study has been simulated by the integrated modeling system CHAOS including an advanced hydrological model (WRF-Hydro) able to offer improved simulation of land surface hydrology in a fine spatial resolution. Analyzing CHAOS meteorological results, an intense storm over Pateras mountain during the morning of 15 November resulted in a tremendous flooding event. The dominant meteorological mechanism for this storm was the orographic convergence of warm and humid southeasterly airflow over the steep slopes of Pateras mountain which caused persistent rainfall and increased the volume of surface run-off water. This effect was characterized by a long duration of approximately 5 h over this area as it was indirectly supported by an almost stagnant spinning Mediterranean cyclone over the Tyrrhenian and Ionian Seas.

The WRF-Hydro model simulated this case study using three different precipitation forcing datasets. The first was based on CHAOS hindcasts (CHAOS-hydro), the second one was based on XPOL radar precipitation estimates (XPOL-hydro) and the third one, on GMP/IMERG estimated precipitation rates (GMP/IMERG-hydro). The XPOL precipitation forcings were evaluated against surface rain gauge and disdrometer observations close to the area of interest with encouraging results. The comparison of the precipitation estimation with in situ rain gauge observations revealed high correlations (~0.90) and bias ratio around 10% with a normalized error of 5%.

Overall, the findings of this study from the hydrologic simulations demonstrate the potential benefit of using high-resolution observations from a locally deployed X-band dual-polarization radar as an additional forcing component in model precipitation simulations. The use of X-band weather radars is rapidly increasing with the advance of technology and data correction and processing algorithms, as well as their reduced costs for installation and maintenance with respect to the operational long-range C- and S-band systems. The high resolution and the increased sensitivity of X-band radars make them quite suitable for nowcasting purposes [102–104]. The range limitation of X-band radars due to the larger attenuation in rain in this microwave band can be surpassed using radar networks, whose setup improves also the accuracy of radar products [105–107].

In future, radar data assimilation in CHAOS modeling system and radar tracking algorithms, which are available for operational radars but are equally applied to X-band radars [103], can be implemented for nowcasting purposes, supporting better estimations of both atmospheric and hydrological characteristics during flash flooding situations.

Author Contributions: G.V., M.N.A. and C.S. contributed to the methodology, formal analysis, investigation, visualization and writing-original draft preparation; A.P., J.K., A.M. and S.M. contributed to the methodology, investigation, writing, review and editing; E.B. and E.K. contributed to the methodology and investigation; P.K. contributed to the conceptualization, methodology, supervision, investigation, review and editing.

Funding: This research received no external funding.

Acknowledgments: The Greek Research and Technology Network (GRNET) is gratefully acknowledged for the access to the high-performance computer (HPC) ARIS (https://hpc.grnet.gr/), where the simulations of this study have been performed. The National Centers for Environmental Prediction (NCEP) is gratefully acknowledged for the provision of the Global Forecasting System (GFS) operational analyses and the real time global (RTG) sea surface temperature (SST) analyses. The National Aeronautics and Space Administration (NASA) and the Hydrological Data and Maps Based on Shuttle Elevation Derivatives at Multiple Scales (HydroSHEDS) are fruitfully acknowledged for the kind provision of the Shuttle Radar Topographic Mission (SRTM) digital elevation model (DEM) data. UK Met Office and www1.wetter3.de are gratefully acknowledged for the kind provision of surface analysis maps.

Conflicts of Interest: The authors declare no conflict of interest.

References

1. Hirabayashi, Y.; Mahendran, R.; Koirala, S.; Konoshima, L.; Yamazaki, D.; Watanabe, S.; Kim, H.; Kanae, S. Global flood risk under climate change. *Nat. Clim. Chang.* **2013**, *3*, 816–821. [CrossRef]

2. Kundzewicz, Z.W.; Mata, L.J.; Arnell, N.W.; Döll, P.; Kabat, P.; Jiménez, B.; Miller, K.A.; Oki, T.; Sen, Z.; Shiklomanov, I.A. *Freshwater Resources and Their Management. Climate Change 2007: Impacts, Adaptation and Vulnerability. Contribution of Working Group II to the Fourth Assessment Report of the Intergovernmental Panel on Climate Change*; Parry, M.L., Canziani, O.F., Palutikof, J.P., van der Linden, P.J., Hanson, C.E., Eds.; Cambridge University Press: Cambridge, UK, 2007; pp. 173–210.

3. Meehl, G.A.; Arblaster, J.M.; Tebaldi, C. Understanding future patterns of precipitation extremes in climate model simulations. *Geophys. Res. Lett.* **2005**, *32*, L18719. [CrossRef]

4. Conti, F.L.; Hsu, K.L.; Noto, L.V.; Sorooshian, S. Evaluation and comparison of satellite precipitation estimates with reference to a local area in the Mediterranean sea. *Atmos. Res.* **2014**, *138*, 189–204. [CrossRef]

5. Dinku, T.; Anagnostou, E.N.; Borga, M. Improving radar-based estimation of rainfall over complex terrain. *J. Appl. Meteorol.* **2002**, *41*, 1163–1178. [CrossRef]

6. Derin, Y.; Anagnostou, E.N.; Anagnostou, M.N.; Kalogiros, J.; Casella, D.; Marra, A.C.; Panegrossi, G.; Sanò, P. Passive microwave rainfall error analysis using high-resolution X-band dual-polarization radar observations in complex terrain. *IEEE Trans. Geosci. Remote Sens.* **2018**, *56*, 2565–2586. [CrossRef]

7. Anagnostou, M.N.; Kalogiros, J.; Anagnostou, E.N.; Papadopoulos, A. Experimental results on rainfall estimation in complex terrain with a mobile X-band polarimetric radar. *Atmos. Res.* **2009**, *94*, 579–595. [CrossRef]

8. Wang, Y.; Chandrasekar, V. Quantitative precipitation estimation in the CASA X-band dual-polarization radar network. *J. Atmos. Ocean. Technol.* **2010**, *27*, 1665–1676. [CrossRef]

9. Matrosov, S.Y.; Cifelli, R.; Gochis, D. Measurements of heavy convective rainfall in the presence of hail in flood-prone areas using an X-band polarimetric radar. *J. Appl. Meteorol. Climatol.* **2013**, *52*, 395–407. [CrossRef]

10. Koffi, A.K.; Gosset, M.; Zahiri, E.-P.; Ochou, A.D.; Kacou, M.; Cazenave, F.; Assamoi, P. Evaluation of X-band polarimetric radar estimation of rainfall and rain drop size distribution parameters in West Africa. *Atmos. Res.* **2014**, *143*, 438–461. [CrossRef]

11. Vulpiani, G.; Baldini, L.; Roberto, N. Characterization of Mediterranean hail-bearing storms using an operational polarimetric X-band radar. *Atmos. Meas. Tech.* **2015**, *8*, 4681–4698. [CrossRef]

12. Anagnostou, M.N.; Nikolopoulos, E.I.; Kalogiros, J.; Anagnostou, E.N.; Marra, F.; Mair, E.; Bertoldi, G.; Tappeiner, U.; Borga, M. Advancing precipitation estimation and streamflow simulations in complex terrain with X-band dual-polarization radar observations. *Remote Sens.* **2018**, *10*, 1258. [CrossRef]

13. Chandrasekar, V.; Chen, H.; Philips, B. DFW urban radar network observations of floods, tornadoes and hail storms. In Proceedings of the IEEE Radar Conference, Oklahoma City, OK, USA, 23–27 April 2018; pp. 765–770.

14. Shakti, P.C.; Maki, M.; Shimizu, S.; Maesaka, T.; Kim, D.; Lee, D.; Iida, H. Correction of Reflectivity in the Presence of Partial Beam Blockage over a Mountainous Region Using X-Band Dual Polarization Radar. *J. Hydrometeorol.* **2013**, *14*, 744–764.

15. Chen, H.; Chandrasekar, V. The quantitative precipitation estimation system for Dallas-Fort Worth (DFW) urban remote sensing network. *J. Hydrol.* **2015**, *531*, 259–271. [CrossRef]

16. Tobin, K.J.; Bennett, M.E. Adjusting satellite precipitation data to facilitate hydrologic modeling. *J. Hydrometeorol.* **2010**, *11*, 966–978. [CrossRef]

17. Cohen Liechti, T.; Matos, J.P.; Boillat, J.L.; Schleiss, A.J. Comparison and evaluation of satellite derived precipitation products for hydrological modeling of the Zambezi River Basin. *Hydrol. Earth Syst. Sci.* **2012**, *16*, 489–500. [CrossRef]

18. Anagnostou, E.N. Overview of overland satellite rainfall estimation for hydro-meteorological applications. *Surv. Geophys.* **2004**, *25*, 511–537. [CrossRef]

19. Hou, A.Y.; Kakar, R.K.; Neeck, S.; Azarbarzin, A.A.; Kummerow, C.D.; Kojima, M.; Oki, R.; Nakamura, K.; Iguchi, T. The global precipitation measurement mission. *Bull. Am. Meteorol. Soc.* **2014**, *95*, 701–722. [CrossRef]

20. Gaume, E.; Borga, M.; Llassat, M.C.; Maouche, S.; Lang, M.; Diakakis, M. Mediterranean extreme floods and flash floods. In *The Mediterranean Region under Climate Change—A Scientific Update. IRD Editions*; Coll. Synthèses: Marseille, France, 2016; pp. 133–144, ISBN 978-2-7099-2219-7. Available online: https://hal.archives-ouvertes.fr/hal-01465740/document (accessed on 29 November 2018).

21. Camarasa-Belmonte, A.; Segura, F. Flood events in Mediterranean ephemeral streams (ramblas) in Valencia region, Spain. *Catena* **2001**, *45*, 229–249. [CrossRef]

22. Llasat, M.C.; Marcos, R.; Turco, M.; Gilabert, J.; Llasat-Botija, M. Trends in flash flood events versus convective precipitation in the Mediterranean region: The case of Catalonia. *J. Hydrol.* **2016**, *541*, 24–37. [CrossRef]

23. Runge, J.; Nguimalet, C.R. Physiogeographic features of the Oubangui catchment and environmental trends reflected in discharge and floods at Bangui 1911–1999, Central African Republic. *Geomorphology* **2005**, *70*, 311–324. [CrossRef]

24. Papagiannaki, K.; Lagouvardos, K.; Kotroni, V. A database of high-impact weather events in Greece: A descriptive impact analysis for the period 2001–2011. *Nat. Hazards Earth Syst. Sci.* **2013**, *13*, 727–736. [CrossRef]

25. Wallace, J.M.; Hobbs, P.V. *Atmospheric Science: An Introductory Survey*, 2nd ed.; Academic Press: Cambridge, MA, USA, 2006; ISBN 978-0-12-732951-2.

26. Morel, C.; Senesi, S. A climatology of mesoscale convective systems over Europe using satellite infrared imagery. II. Characteristics of European mesoscale convective systems. *Q. J. R. Meteorol. Soc.* **2002**, *128*, 1973–1995. [CrossRef]

27. Kolios, S.; Feidas, H. A warm season climatology of mesoscale convective systems in the Mediterranean basin using satellite data. *Theor. Appl. Climatol.* **2010**, *102*, 29–42. [CrossRef]

28. Despiniadou, V.; Athanasopoulou, E. Flood prevention and sustainable spatial planning. The case of the river Diakoniaris in Patras. In Proceedings of the 46th Congress of the European Regional Science Association (ERSA), Volos, Greece, 30 August–3 September 2006. Available online: https://www.econstor.eu/bitstream/10419/118466/1/ERSA2006_672.pdf (accessed on 9 November 2018).

29. Diakakis, M.; Mavroulis, S.; Deligiannakis, G. Floods in Greece, a statistical and spatial approach. *Nat. Hazards* **2012**, *62*, 485–500. [CrossRef]

30. Baltas, E.A.; Mimikou, M.A. Considerations for the optimum location of a C-band weather radar in the Athens area. In Proceedings of the 3rd European Conference on radar Meteorology and Hydrology, ERAD 2002, Delft, The Netherlands, 18–22 November 2002; pp. 348–351. Available online: https://www.copernicus.org/erad/online/erad-348.pdf (accessed on 9 November 2018).

31. Skilodimou, H.; Livaditis, G.; Bathrellos, G.; Verikiou-Papaspiridakou, E. Investigating the flooding events of the urban regions of Glyfada and Voula, Attica, Greece: A contribution to Urban Geomorphology. *Geogr. Ann. A* **2003**, *85*, 197–204. [CrossRef]

32. Mimikou, M.; Baltas, E.; Varanou, E. *A Study of Extreme Storm Events in the Greater Athens Area, Greece. The Extremes of the Extremes, Extraordinary Floods*; IAHS-AISH Publication: Reykjavik, Iceland, 2002; Volume 271, pp. 161–166.

33. Karymbalis, E.; Katsafados, P.; Chalkias, C.; Gaki-Papanastassiou, K. An integrated study for the evaluation o of natural and anthropogenic causes of flooding in small catchments based on geomorphological and meteorological data and modeling techniques: The case of the Xerias torrent (Corinth, Greece). *Z. Geomorphol.* **2012**, *56*, 45–67. [CrossRef]

34. Mazi, K.; Koussis, A.D. The 8 July 2002 storm over Athens: Analysis of the Kifissos River/Canal overflows. *Adv. Geosci.* **2006**, *7*, 301–306. [CrossRef]

35. Papagiannaki, K.; Kotroni, V.; Lagouvardos, K.; Ruin, I.; Bezes, A. Urban areas response to flash flood-triggering rainfall, featuring human behavioural factors: The case of 22 October 2015, in Attica, Greece. *Weather Clim. Soc.* **2017**, *9*, 621–638. [CrossRef]

36. Gochis, D.J.; Yu, W.; Yates, D.N. *The WRF-Hydro Model Technical Description and User's Guide, version 3.0*; NCAR Technical Document; NCAR: Boulder, CO, USA, 2015. Available online: https://ral.ucar.edu/sites/default/files/public/images/project/WRF_Hydro_User_Guide_v3.0.pdf (accessed on 9 November 2018).

37. Giannoni, F.; Roth, G.; Rudari, R. A semi-distributed rainfall–runoff model based on a geomorphologic approach. *Phys. Chem. Earth B* **2000**, *25*, 665–671. [CrossRef]

38. Shen, X.; Hong, Y.; Zhang, K.; Hao, Z.; Wang, D. CREST v2.1 Refined by a Distributed Linear Reservoir Routing Scheme. In Proceedings of the American Geophysical Union, Fall Meeting 2014, San Francisco, CA, USA, 15–19 December 2014; abstract #H33G-0918.

39. Seity, Y.; Brousseau, P.; Malardel, S.; Hello, G.; Benard, P.; Bouttier, F.; Lac, C.; Masson, V. The AROME–France convective-scale operational model. *Mon. Weather Rev.* **2011**, *139*, 976–991. [CrossRef]

40. Skamarock, W.C.; Klemp, J.B.; Dudhia, J.; Gill, D.O.; Barker, D.M.; Dudha, M.G.; Huang, X.; Wang, W.; Powers, Y. *A Description of the Advanced Research WRF Ver. 3.0*; NCAR Technical Note; NCAR: Boulder, CO, USA, 2008. Available online: http://www2.mmm.ucar.edu/wrf/users/docs/arw_v3.pdf (accessed on 9 November 2018).

41. Powers, J.G.; Klemp, J.B.; Skamarock, W.C.; Davis, C.A.; Dudhia, J.; Gill, D.O.; Coen, J.L.; Gochis, D.J.; Ahmadov, R.; Peckham, S.E.; et al. The weather research and forecasting model: Overview, system efforts, and future directions. *Bull. Am. Meteorol. Soc.* **2017**, *98*, 1717–1737. [CrossRef]

42. Senatore, A.; Mendicino, G.; Gochis, D.J.; Yu, W.; Yates, D.N.; Kunstmann, H. Fully coupled atmosphere-hydrology simulations for the central Mediterranean: Impact of enhanced hydrological parameterization for short and long time scales. *J. Adv. Model. Earth Syst.* **2015**, *7*, 1693–1715. [CrossRef]

43. Yucel, I.; Onen, A.; Yilmaz, K.K.; Gochis, D.J. Calibration and evaluation of a flood forecasting system: Utility of numerical weather prediction model, data assimilation and satellite-based rainfall. *J. Hydrol.* **2015**, *523*, 49–66. [CrossRef]

44. Givati, A.; Gochis, D.; Rummler, T.; Kunstmann, H. Comparing one-way and two-way coupled hydrometeorological forecasting systems for flood forecasting in the Mediterranean region. *Hydrology* **2016**, *3*, 19. [CrossRef]

45. Berne, A.; Delrieu, G.; Creutin, J.; Obled, C. Temporal and spatial resolution of rainfall measurements required for urban hydrology. *J. Hydrol.* **2004**, *299*, 166–179. [CrossRef]

46. Atencia, A.; Mediero, L.; Llasat, M.C.; Garrote, L. Effect of radar rainfall time resolution on the predictive capability of a distributed hydrologic model. *Hydrol. Earth Syst. Sci.* **2011**, *15*, 3809–3827. [CrossRef]

47. Varlas, G. Development of an Integrated Modeling System for Simulating the Air-Ocean Wave Interactions. Ph.D. Dissertation, Harokopio University of Athens (HUA), Athens, Greece, 2017. Available online: https://www.didaktorika.gr/eadd/handle/10442/41238 (accessed on 9 November 2018).

48. Varlas, G.; Katsafados, P.; Papadopoulos, A.; Korres, G. Implementation of a two-way coupled atmosphere-ocean wave modeling system for assessing air-sea interaction over the Mediterranean Sea. *Atmos. Res.* **2018**, *208*, 201–217. [CrossRef]

49. FloodHub. Analysis of the Flood in Western Attica on 15/11/2017 Using Satellite Remote Sensing. Available online: http://www.beyond-eocenter.eu/images/news-events/20180430/Mandra-Report-BEYOND.pdf (accessed on 29 November 2018). (In Greek)

50. Environmental, Disasters and Crises Management (EDCM). Flash Flood in West Attica (Mandra, Nea Peramos) Newsletter #5. 15 November 2017. Available online: http://www.elekkas.gr/index.php/en/epistimoniko-ergo/edcm-newsletter/1603-edcm-newsletter-5-flash-flood-in-west-attica-mandra-nea-peramos-november-15-2017 (accessed on 9 November 2018).

51. Diakakis, M.; Andreadakis, E.; Spyrou, N.I.; Gogou, M.E.; Nikolopoulos, E.I.; Deligiannakis, G.; Katsetsiadou, N.K.; Antoniadis, Z.; Melaki, M.; Georgakopoulos, A.; et al. The flash flood of Mandra 2017, in West Attica, Greece—Description of impacts and flood characteristics. *Int. J. Disaster Risk Reduct.* **2018**. [CrossRef]

52. Katsafados, P.; Kalogirou, S.; Papadopoulos, A.; Korres, G. Mapping long-term atmospheric variables over Greece. *J. Maps* **2012**, *8*, 181–184. [CrossRef]

53. Mavrakis, A.; Theoharatos, G.; Asimakopoulos, D.; Christides, A. Distribution of the trace metals in sediments of Eleusis Gulf. *Mediterr. Mar. Sci.* **2004**, *5*, 151–158. [CrossRef]

54. Institute of Geology and Mineral Exploration (IGME). Geological Map of Greece (scale 1:50,000), Sheet Erithrai. 1971. Available online: http://portal.igme.gr/geoportal/ (accessed on 9 November 2018).

55. Kalogiros, J.; Anagnostou, M.N.; Anagnostou, E.N.; Montopoli, M.; Picciotti, E.; Marzano, F.S. Optimum estimation of rain microphysical parameters using X-band dual-polarization radar observables. *IEEE Trans. Geosci. Remote Sens.* **2013**, *51*, 3063–3076. [CrossRef]

56. Anagnostou, M.N.; Kalogiros, J.; Anagnostou, E.N.; Tarolli, M.; Papadopoulos, A.; Borga, M. Performance evaluation of high-resolution rainfall estimation by X-band dual-polarization radar for flash flood applications in mountainous basin. *J. Hydrol.* **2010**, *394*, 4–16. [CrossRef]

57. Kalogiros, J.; Anagnostou, M.N.; Anagnostou, E.N.; Montopoli, M.; Picciotti, E.; Marzano, F.S. Correction of polarimetric radar reflectivity measurements and rainfall estimates for apparent vertical profile in stratiform rain. *J. Appl. Meteorol. Climatol.* **2013**, *52*, 1170–1186. [CrossRef]

58. Kalogiros, J.; Anagnostou, M.N.; Anagnostou, E.N.; Montopoli, M.; Picciotti, E.; Marzano, F.S. Evaluation of a new polarimetric algorithm for rain-path attenuation correction of X-band radar observations against disdrometer. *IEEE Trans. Geosci. Remote Sens.* **2014**, *52*, 1369–1380. [CrossRef]

59. Anagnostou, M.N.; Kalogiros, J.; Marzano, F.S.; Anagnostou, E.N.; Montopoli, M.; Picciotti, E. Performance evaluation of a new dual-polarization microphysical algorithm based on long-term X-band radar and disdrometer observations. *J. Hydrometeorol.* **2013**, *14*, 560–576. [CrossRef]

60. Habib, E.; Krajewski, W.F.; Kruger, A. Sampling errors of tipping bucket rain gauge measurements. *J. Hydrol. Eng.* **2001**, *6*, 159–166. [CrossRef]

61. Porcacchia, L.; Kirstetter, P.-E.; Gourley, J.J.; Maggioni, V.; Cheong, B.L.; Anagnostou, M.N.; Kalogiros, J. Toward a radar polarimetric classification scheme for warm-rain precipitation: Application to complex terrain. *J. Hydrometeorol.* **2017**, *18*, 3199–3215. [CrossRef]

62. Erlingis, J.M.; Gourley, J.J.; Kirstetter, P.-E.; Anagnostou, E.N.; Kalogiros, J.; Anagnostou, M.N.; Peterseni, W. Evaluation of operational and experimental precipitation algorithms and microphysical insights during IPHEx. *J. Hydrometeorol.* **2018**, *19*, 113–125. [CrossRef]

63. Sun, Q.; Miao, C.; Duan, Q.; Ashouri, H.; Sorooshian, S.; Hsu, K.-L. A review of global precipitation datasets: Data sources, estimation, and intercomparisons. *Rev. Geophys.* **2018**, *56*, 79–107. [CrossRef]

64. Huffman, G.J.; Bolvin, D.T.; Braithwaite, D.; Hsu, K.L.; Joyce, R.; Xie, P. *NASA Global Precipitation Measurement (GPM) Integrated Multi-Satellite Retrievals for GPM (IMERG)*; Algorithm Theor. Basis Doc. (ATBD) Version 5.1; NASA GSFC: Greenbelt, MD, USA, 2014. Available online: https://pmm.nasa.gov/sites/default/files/document_files/IMERG_ATBD_V5.1b.pdf (accessed on 9 November 2018).

65. Hong, Y.; Hsu, K.L.; Sorooshian, S.; Gao, X. Precipitation estimation from remotely sensed imagery using an Artificial Neural Network cloud classification system. *J. Appl. Meteorol.* **2004**, *43*, 1834–1852. [CrossRef]

66. Joyce, R.J.; Janowiak, J.E.; Arkin, P.A.; Xie, P. CMORPH: A method that produces global precipitation estimates from passive microwave and infrared data at high spatial and temporal resolution. *J. Hydrometeorol.* **2004**, *5*, 487–503. [CrossRef]

67. Huffman, G.J.; Bolvin, D.T.; Nelkin, E.J.; Wolff, D.B.; Adler, R.F.; Gu, G.; Hong, Y.; Bowman, K.P.; Stocker, E.F. The TRMM Multisatellite Precipitation Analysis (TMPA): Quasi-global, multiyear, combined-sensor precipitation estimates at fine scales. *J. Hydrometeorol.* **2007**, *8*, 38–55. [CrossRef]

68. Hasselmann, S.; Hasselmann, K.; Bauer, E.; Janssen, P.A.E.M.; Komen, G.J.; Bertotti, L.; Lionello, P.; Guillaume, A.; Cardone, V.C.; Greenwood, J.A.; et al. The WAM model—A third generation ocean wave prediction model. *J. Phys. Oceanogr.* **1988**, *18*, 1775–1810. [CrossRef]

69. Komen, G.J.; Cavaleri, L.; Donelan, M.; Hasselmann, K.; Hasselmann, S.; Janssen, P.A.E.M. *Dynamics and Modeling of Ocean Waves*; Cambridge University Press: Cambridge, UK, 1994.

70. Christakos, K.; Varlas, G.; Reuder, J.; Katsafados, P.; Papadopoulos, A. Analysis of a low-level coastal jet off the western coast of Norway. *Energy Procedia* **2014**, *53*, 162–172. [CrossRef]

71. Christakos, K.; Cheliotis, I.; Varlas, G.; Steeneveld, G.J. Offshore wind energy analysis of cyclone Xaver over North Europe. *Energy Procedia* **2016**, *94*, 37–44. [CrossRef]

72. Cheliotis, I.; Varlas, G.; Christakos, K. The impact of cyclone Xaver on hydropower potential in Norway. In *Perspectives on Atmospheric Sciences*; Springer: Cham, Germany, 2017; pp. 175–181.

73. Varlas, G.; Papadopoulos, A.; Katsafados, P. An analysis of the synoptic and dynamical characteristics of hurricane Sandy (2012). *Meteorol. Atmos. Phys.* **2018**, 1–11. [CrossRef]
74. Valcke, S.; Craig, T.; Coquart, L. *OASIS3-MCT_3.0 Coupler User Guide*; CERFACS/CNRS: Toulouse, France, 2015. Available online: http://www.cerfacs.fr/oa4web/oasis3-mct_3.0/oasis3mct_UserGuide.pdf (accessed on 9 November 2018).
75. Katsafados, P.; Papadopoulos, A.; Korres, G.; Varlas, G. A fully coupled atmosphere-ocean wave modeling system for the Mediterranean Sea: Interactions and sensitivity to the resolved scales and mechanisms. *Geosci. Model Dev.* **2016**, *9*. [CrossRef]
76. Katsafados, P.; Varlas, G.; Papadopoulos, A.; Korres, G. Implementation of a Hybrid Surface Layer Parameterization Scheme for the Coupled Atmosphere-Ocean Wave System WEW. In *Perspectives on Atmospheric Sciences*; Springer: Cham, Germany, 2017; pp. 159–165.
77. Katsafados, P.; Varlas, G.; Papadopoulos, A.; Spyrou, C.; Korres, G. Assessing the implicit rain impact on sea state during hurricane Sandy (2012). *Geophys. Res. Lett.* **2018**, *45*, 12015–12022. [CrossRef]
78. Maidment, D.R. Conceptual framework for the national flood interoperability experiment. *J. Am. Water Resour. Assoc.* **2017**, *53*, 245–257. [CrossRef]
79. Danielson, J.J.; Gesch, D.B. *Global Multi-Resolution Terrain Elevation Data 2010 (GMTED2010) (No. 2011-1073)*; US Geological Survey: Reston, VA, USA, 2011. Available online: https://pubs.er.usgs.gov/publication/ofr20111073 (accessed on 9 November 2018).
80. Myneni, R.B.; Hoffman, S.; Knyazikhin, Y.; Privette, J.L.; Glassy, J.; Tian, Y.; Wang, Y.; Song, X.; Zhang, Y.; Smith, G.R.; et al. Global products of vegetation leaf area and fraction absorbed PAR from year one of MODIS data. *Remote Sens. Environ.* **2002**, *83*, 214–231. [CrossRef]
81. Friedl, M.A.; Sulla-Menashe, D.; Tan, B.; Schneider, A.; Ramankutty, N.; Sibley, A.; Huang, X. MODIS Collection 5 global land cover: Algorithm refinements and characterization of new datasets. *Remote Sens. Environ.* **2010**, *114*, 168–182. [CrossRef]
82. Jiménez, P.A.; Dudhia, J.; González-Rouco, J.F.; Navarro, J.; Montávez, J.P.; García-Bustamante, E. A revised scheme for the WRF surface layer formulation. *Mon. Weather Rev.* **2012**, *140*, 898–918. [CrossRef]
83. Tewari, M.; Chen, F.; Wang, W.; Dudhia, J.; LeMone, M.A.; Mitchell, K.; Gayno, G.; Wegiel, J.; Cuenca, R.H. Implementation and verification of the unified NOAH land surface model in the WRF model. In Proceedings of the 20th Conference on Weather Analysis and Forecasting/16th Conference on Numerical Weather Prediction, Seattle, WA, USA, 12–16 January 2004; pp. 2165–2170. Available online: https://ams.confex.com/ams/84Annual/techprogram/paper_69061.htm (accessed on 9 November 2018).
84. Mlawer, E.J.; Taubman, S.J.; Brown, P.D.; Iacono, M.J.; Clough, S.A. Radiative transfer for inhomogeneous atmospheres: RRTM, a validated correlated-k model for the longwave. *J. Geophys. Res. Atmos.* **1997**, *102*, 16663–16682. [CrossRef]
85. Dudhia, J. Numerical study of convection observed during the winter monsoon experiment using a mesoscale two-dimensional model. *J. Atmos. Sci.* **1989**, *46*, 3077–3107. [CrossRef]
86. Lin, Y.-L.; Farley, R.D.; Orville, H.D. Bulk Parameterization of the Snow Field in a Cloud Model. *J. Clim. Appl. Met.* **1983**, *22*, 1065–1092. [CrossRef]
87. Grell, G.A.; Freitas, S.R. A scale and aerosol aware stochastic convective parameterization for weather and air quality modeling. *Atmos. Chem. Phys.* **2014**, *14*, 5233–5250. [CrossRef]
88. Jarvis, A.; Reuter, H.I.; Nelson, A.; Guevara, E. Hole-Filled SRTM for the Globe Version 4. CGIAR-CSI SRTM 90m Database. Available online: http://srtm.csi.cgiar.org (accessed on 9 November 2018).
89. Lehner, B.; Verdin, K.; Jarvis, A. New global hydrography derived from spaceborne elevation data. *Eos Trans. Am. Geophys. Union* **2008**, *89*, 93–94. [CrossRef]
90. Wilks, D.S. *Statistical Methods in the Atmospheric Sciences*; Academic Press: Cambridge, MA, USA, 2011.
91. Wilson, L.J. Verification of Precipitation Forecasts: A Survey of Methodology. Part I: General Framework and Verification of Continuous Variables. In Proceedings of the WWRP/WMO Workshop on the Verification of Quantitative Precipitation Forecasts, Prague, Czech Republic, 14–16 May 2001.
92. Ehrendorfer, M.; Murphy, A.H. Comparative evaluation of weather forecasting systems: Sufficiency, quality, and accuracy. *Mon. Weather Rev.* **1988**, *116*, 1757–1770. [CrossRef]
93. Brown, B.G. Verification of Precipitation Forecasts: A Survey of Methodology. Part II: Verification of Probability Forecasts at Points. In Proceedings of the WWRP/WMO Workshop on the Verification of Quantitative Precipitation Forecasts, Prague, Czech Republic, 14–16 May 2001.

94. World Meteorological Organization. *Forecast Verification for the African Severe Weather Forecasting Demonstration Projects*; No. 1132; World Meteorological Organization: Geneva, Switzerland, 2014. Available online: https://www.wmo.int/pages/prog/www/Documents/1132_en.pdf (accessed on 9 November 2018).

95. Arnault, J.; Wagner, S.; Rummler, T.; Fersch, B.; Bliefernicht, J.; Andresen, S.; Kunstmann, H. Role of runoff–infiltration partitioning and resolved overland flow on land–atmosphere feedbacks: A case study with the WRF-Hydro coupled modeling system for West Africa. *J. Hydrometeorol.* **2016**, *17*, 1489–1516. [CrossRef]

96. Krajewski, W.F.; Ceynar, D.; Demir, I.; Goska, R.; Kruger, A.; Langel, C.; Mantilla, R.; Niemeier, J.; Quintero, F.; Seo, B.C.; et al. Real-time flood forecasting and information system for the state of Iowa. *Bull. Am. Meteorol. Soc.* **2017**, *98*, 539–554. [CrossRef]

97. Ryu, Y.; Lim, Y.J.; Ji, H.S.; Park, H.H.; Chang, E.C.; Kim, B.J. Applying a coupled hydrometeorological simulation system to flash flood forecasting over the Korean Peninsula. *Asia-Pacific. J. Atmos. Sci.* **2017**, *53*, 421–430. [CrossRef]

98. Silver, M.; Karnieli, A.; Ginat, H.; Meiri, E.; Fredj, E. An innovative method for determining hydrological calibration parameters for the WRF-Hydro model in arid regions. *Environ. Model. Softw.* **2017**, *91*, 47–69. [CrossRef]

99. Stamou, A.I. The disastrous flash flood of Mandra in Attica-Greece and now what? *Civ. Eng. Res. J.* **2018**, *6*, 1–6. [CrossRef]

100. Greek City Times. Local Authorities and Bureaucracy Blamed for Mandra Floods. Available online: https://greekcitytimes.com/2017/12/29/local-authorities-bureaucracy-blamed-mandra-floods (accessed on 9 November 2018).

101. Serbis, D.; Papathanasiou, C.; Mamassis, N. Mitigating flooding in a typical urban area in North Western Attica in Greece. In Proceedings of the Conference on Changing Cities: Spatial Design, Landscape and Socio-economic Dimensions, Porto Heli, Peloponnese, Greece, 22–26 June 2015. Available online: http://www.itia.ntua.gr/en/getfile/1563/1/documents/P588-Changing_Cities2015_Full_paper.pdf (accessed on 9 November 2018).

102. Picciotti, E.; Marzano, F.S.; Anagnostou, E.N.; Kalogiros, J.; Fessas, Y.; Volpi, A.; Cazac, V.; Pace, R.; Cinque, G.; Bernardini, L.; et al. Coupling X-band dual-polarized mini-radar and hydro-meteorological forecast models: The HYDRORAD project. *Nat. Hazards Earth Syst. Sci.* **2013**, *13*, 1229–1241. [CrossRef]

103. Conti, F.L.; Francipane, A.; Pumo, D.; Noto, L.V. Exploring single polarization X-band weather radar potentials for local meteorological and hydrological applications. *J. Hydrol.* **2015**, *531*, 508–522. [CrossRef]

104. Shah, S.; Notarpietro, R.; Branca, M. Storm identification, tracking and forecasting using high-resolution images of short-range X-Band radar. *Atmosphere* **2015**, *6*, 579–606. [CrossRef]

105. McLaughlin, D.; Pepyne, D.; Chandrasekar, V.; Philips, B.; Kurose, J.; Zink, M.; Droegemeier, K.; Cruz-Pol, S.; Junyent, F.; Brotzge, J.; et al. Short-wavelength technology and the potential for distributed networks of small radar systems. *Bull. Am. Meteorol. Soc.* **2009**, *90*, 1797–1817. [CrossRef]

106. Chandrasekar, V.; Wang, Y.; Chen, H. The CASA quantitative precipitation estimation system: A five year validation study. *Nat. Hazards Earth Syst. Sci.* **2012**, *12*, 2811–2820. [CrossRef]

107. Lengfeld, K.; Clemens, M.; Munster, H.; Ament, F. Performance of high-resolution X-band weather radar networks—The PATTERN example. *Atmos. Meas. Tech.* **2014**, *7*, 4151–4166. [CrossRef]

remote sensing

MDPI

Article

Comparison of TMPA-3B42RT Legacy Product and the Equivalent IMERG Products over Mainland China

Lei Wu [1,2]**, Youpeng Xu** [1,]*** and Siyuan Wang** [1]

[1]   School of Geographic and Oceanographic Sciences, Nanjing University, Nanjing 210023, China;
     wulei@smail.nju.edu.cn (L.W.); siyuanwang6365@163.com (S.W.)
[2]   Fenner School of Environment and Society, Australian National University, Canberra, ACT 0200, Australia
*   Correspondence: xuyp305@163.com; Tel.: +86-025-8968-5069

Received: 24 October 2018; Accepted: 3 November 2018; Published: 9 November 2018

Abstract: The near-real-time legacy product of Tropical Rainfall Measuring Mission Multi-satellite Precipitation Analysis (3B42RT) and the equivalent products of Integrated Multi-satellite Retrievals for Global Precipitation Measurement mission (IMERG-E and IMERG-L) were evaluated and compared over Mainland China from 1 January 2015 to 31 December 2016 at the daily timescale, against rain gauge measurements. Results show that: (1) Both 3B42RT and IMERG products overestimate light rain (0.1–9.9 mm/day), while underestimate moderate rain (10.0–24.9 mm/day) to heavy rainstorm ($\geq$250.0 mm/day), with an increase in mean (absolute) error and a decrease in relative mean absolute error (RMAE). The IMERG products perform better in estimating light rain to heavy rain (25.0–49.9 mm/day), and heavy rainstorm, while 3B42RT has smaller error magnitude in estimating light rainstorm (50.0–99.9 mm/day) and moderate rainstorm (100.0–249.9 mm/day). (2) Higher rainfall intensity associates with better detection. Threshold values are <2.0 mm/day, below which 3B42RT is unreliable at detecting rain; and <1.0 mm/day, below which both 3B42RT and IMERG products are more likely to cause false alarms. (3) Generally, both 3B42RT and IMERG products perform better in wet areas with relatively heavy rainfall intensity and/or during wet season than in dry areas with relatively light rainfall intensity and/or during dry season. Compared with 3B42RT, IMERG-E and IMERG-L constantly improve performance in space and time, but it is not obvious in dry areas and/or during dry season. The agreement between IMERG products and rain gauge measurements is low and even negative for different rainfall intensities, and the RMAE is still at a high level (>50%), indicating the IMERG products remain to be improved. This study will shed light on research and application during the transition in multi-satellite rainfall products from TMPA to IMERG and future algorithms improvement.

Keywords: TRMM-era TMPA; GPM-era IMERG; satellite rainfall estimate; Mainland China

---

1. Introduction

Accurate and timely knowledge of when, where and how much rain falls is essential to water resource management, natural hazard monitoring and decision support [1–3]. Despite its importance, rainfall measurement at high resolution across regional and global scale remains a challenge for scientific community [4,5]. Common approaches for measuring rainfall are rain gauges, weather radars and satellite-based sensors [6]. Rain gauges provide direct physical measurements of surface rainfall at specific sites. While areal rainfall estimates may be obtained by interpolating rain gauge measurements, the accuracy of rainfall estimates is largely areas with dense rain gauge network coverage. However, such networks are not feasible in the vast expanses of oceanic, desert and mountainous areas, and sparsely distributed in remote regions [7–9]. Weather radars estimate rainfall from reflectivity signal strength via hydrometeors (i.e., raindrops and ices) that result in surface rainfall, within ~250 km distance of the station in minutes [10]. However, some deficiencies also remain, such as beam blockage in mountains, and limited spatial coverage [11,12].

Satellite-based sensors have become a viable approach to address the problem of comprehensive rainfall coverage. In recent years, some algorithms have been developed to combine the respective advantages of the range available satellite sensors for estimating rainfall, namely geostationary-orbit infrared (geo-IR) sensors, low-Earth-orbit passive microwave (leo-PMW) sensors, and precipitation radars (PR) [13,14]. Many quasi-global satellite rainfall products with various temporal and spatial resolutions have been released to the public, such as the Tropical Rainfall Measuring Mission (TRMM) Multi-satellite Precipitation Analysis (TMPA) [15], the Precipitation Estimation from Remotely Sensed Information using Artificial Neural Networks (PERSIANN) [16], the Climate Prediction Center Morphing (CMORPH) technique [17], and the Integrated Multi-satellite Retrievals for Global Precipitation Measurement (GPM) mission (IMERG) [4].

The TRMM satellite was launched on 27 November 1997 as the first Earth Science mission dedicated to studying tropical and subtropical rainfall. TMPA was intended to provide the best multi-satellite rainfall estimates in TRMM-era, and has undergone three upgrades (i.e., Versions 5, 6 and 7) since its inception as results of new sensors integration and upgrades to retrieval algorithms [18]. After over 17 years of productive data gathering, the Precipitation Radar (PR, 13.8 GHz) and TRMM Microwave Imager (TMI, 10.65, 19.35, 37.0 and 85.5 GHz at dual polarization, and 22.235 GHz at single polarization) aboard TRMM satellite were decommissioned in October 2014 and April 2015, respectively. Nevertheless, the TMPA continues to produce merged multi-satellite rainfall analyses until the equivalent IMERG products are deemed satisfactory [19]. The GPM Core Observatory was launched on 27 February, 2014, designed to unify and advance rainfall measurement from a constellation of research and operational satellites. The GPM Core Observatory carries the first space-borne Ku-band and Ka-band (13 and 35 GHz) Dual-frequency Precipitation Radar (DPR) and a multi-channel (10–183 GHz) GPM Microwave Imager (GMI), which extend the measurement range attained by TRMM to include light-intensity rainfall (i.e., <0.5 mm/h) and snowfall [4].

There have been a multitude of statistical and hydrological studies evaluating and comparing the performance of TMPA and IMERG products on varying surface backgrounds, latitudes, elevations and seasons. Kim et al. [20] demonstrated that IMERG "final" product (hereinafter referred to as IMERG-F) performs ~8% better than TMPA-3B42V7 during the pre-monsoon and monsoon seasons over far-east Asia. Xu et al. [21] showed that IMERG-F performs better than 3B42V7 in high-elevation region while the opposite is true in low-elevation regions over southern Tibetan Plateau. Tang et al. [22] found that IMERG-F performs better than 3B42V7 legacy product at mid- and high-latitudes, as well as dry climate regions over Mainland China. Prakash et al. [1] indicated IMERG-F shows a notable improvement in detecting heavy rainfall event frequency across India at a daily time scale during the southwest monsoon season over 3B42V7. The above research results provide important insights into the performance of the TMPA and IMERG research quality products, as well as their suitability for hydrological applications. However, the near-real-time TMPA legacy product (currently Version 7) and the equivalent IMERG products (currently Version 5) remain largely unstudied. It is necessary to investigate the products' performance, which will shed light on research and application during the transition from TMPA to IMERG, and future algorithm improvement.

The objective of this study was to conduct a thorough quantitative evaluation of the performance near-real-time products from TMPA and IMERG over the common overlap period between TRMM and GPM eras. We specifically evaluated TMPA-3B42RT legacy product and the equivalent IMERG products (see Section 2.2) over Mainland China between 1 January 2015 and 31 December 2016. We conducted our evaluation of daily aggregated quantities against data from rain gauge measurements, to see whether the shifts in input data and algorithms impact on the product's performance. We specifically focused on the dependence of product's performance for different levels of rainfall intensity.

Remote Sens. **2018**, *10*, 1778

2. Materials and Methods

2.1. Study Area

The study area is Mainland China, located within 73°–135°E and 18°–53°N (Figure 1). The landscape of Mainland China includes alluvial plains in the east, hills and low mountains in the south, high mountains in the west, and high plateaus in the north. The climate of Mainland China includes from tropical in the far south, subarctic in the far north, and alpine in the Tibetan Plateau. The diverse topography and climate make Mainland China a good test-bed for evaluating the performance of satellite rainfall products, as these factors affect the accuracy of the retrieval methods.

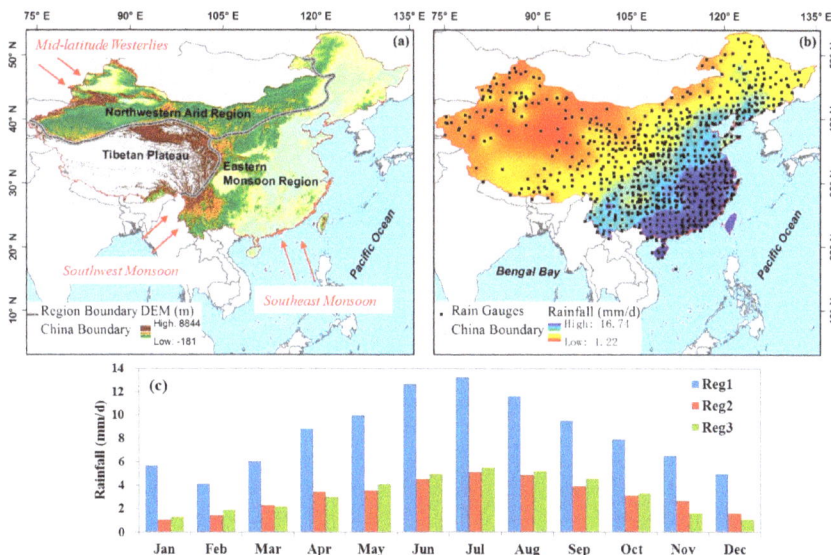

Figure 1. Study area of the evaluation of TMPA-3B42RT legacy product and the equivalent IMERG products: (**a**) terrain elevation and sub-regions; (**b**) rain gauge network (830 gauges) overlaid on a grid of daily average rainfall; and (**c**) monthly average daily rainfall, based on rain gauge measurements from 2015 to 2016.

Mainland China is divided into three sub-regions based on topographic and climatic features: the Eastern Monsoon Region (Reg1), the Northwestern Arid Region (Reg2) and the Tibetan Plateau (Reg3) [23]. Reg1 is primarily controlled by the East Asian monsoon, with warm and wet summers caused by summer monsoon, and cold and dry winters caused by the Siberian High. Reg2 is influenced by the temperate continental climate, with annual rainfall from ~400 mm (semi-arid climate) in the east to <200 mm (arid climate) in the west, mostly concentrated in summer [24]. Reg3 is dominated by the plateau mountain climate, characterized by high elevation, complex topography and variable climate.

2.2. TMPA-3B42RT Legacy Product and the Equivalent IMERG Products

Figure 2 shows the flowcharts of TMPA and IMERG near-real-time rainfall estimation algorithms. The common feature in both workflows is that brightness temperature (Tb) observations from all available PMW sensors are calibrated using the TMI or GMI as reference. These data are subsequently calibrated to combined microwave imager and PR, i.e., TRMM Combined Instrument (TCI) product or GPM Combined Radar-Radiometer (CORRA) product. These PMW estimates are then combined into 3-hourly or half-hourly rainfall fields for each near-real-time product, respectively.

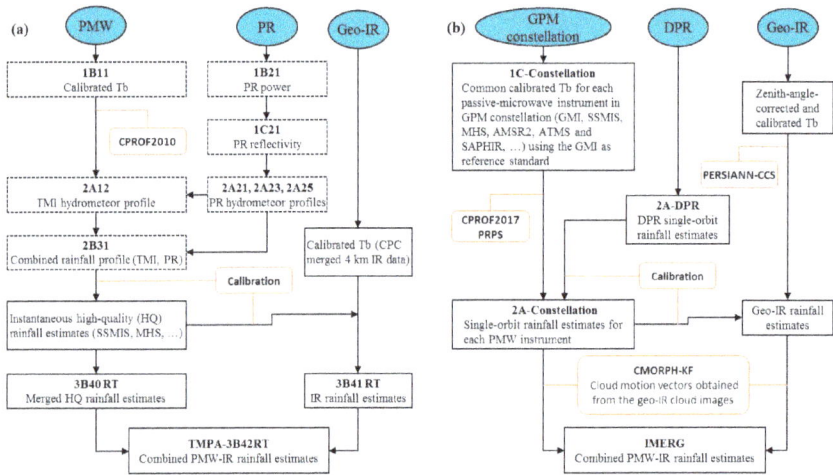

Figure 2. Flowcharts of: (**a**) TMPA-3B42RT algorithm; and (**b**) IMERG algorithm. The dashed boxes indicate the profiles were terminated since October 2014, and the calibrator for HQ shifted from using coincident data to past data.

For the TMPA-3B42RT algorithm, calibrated geo-IR Tb are converted to geo-IR estimates, and then combined with PMW estimates into TMPA-3B42RT product (hereinafter referred to as 3B42RT) with ~8 h latency and 0.25°/3 h resolution over the latitudes 50°N–50°S.

For the IMERG algorithm, the calibrated geo-IR Tb are provided to both the PERSIANN-Cloud Classification System (PERSIANN-CCS) [25] and CMORPH-Kalman Filter (CMORPH-KF) [26] algorithms. PERSIANN-CCS is designed to improve the relationship between geo-IR Tb and rainfall rate. The Tb field is segmented into separable cloud patches, and then classified into different groups based on the similarities of patch features. Every group is specified a cloud rainfall function by a training set of PMW estimates, using histogram-matching and exponential regression to derive the parameters. CMORPH-KF is developed to mitigate the gaps in PMW coverage by propagating the instantaneous PMW estimates though time to the closest analysis time (within a 90-min window) using cloud motion vectors computed from consecutive geo-IR images. This approach is run once forward in time to generate the near-real-time IMERG "Early" run estimates (hereinafter referred to as IMERG-E), and then both forward and backward in time for the IMERG "Late" run estimates (hereinafter referred to as IMERG-L). IMERG-E and –L have latencies of ~4 h and ~12 h, respectively. Products are gridded to 0.1°/30-min resolution over the latitudes 60°N–60°S.

The daily 3B42RT legacy product, and the equivalent IMERG-E and IMERG-L from 1 January 2015 to 31 December 2016 were used in this study, downloaded from the Precipitation Measurement Missions website (https://pmm.nasa.gov/data-access/downloads).

2.3. Rain Gauge Measurement

Rain gauge measured rainfalls are normally employed as reference to validate satellite rainfall products [27]. In this study, we used a network of 830 rain gauges (Figure 1b) which provided rainfall measurements from 1 January 2015 to 31 December 2016. Data were obtained from the China Meteorological Data Service Center (http://data.cma.cn/) which is an authoritative and unified shared service platform for China Meteorological Administration [22]. Rain gauges record 24-h accumulated rainfall twice daily, namely, at 08:00 and 20:00 local time, or 00:00 and 12:00 UTC. These data are released three months after the day of interest. To have a consistent definition of daily rainfall between gauge and satellite precipitation products, we chose accumulations to 00:00 UTC. As shown in Figure 1b, the rain gauges are densely spaced in the eastern and southern parts of Mainland China, while sparsely spaced in the northern and western parts, especially in the Tibetan Plateau.

2.4. Methods

Cell values of satellite rainfall products were extracted based on the nearest rain gauge location. For grids with two or more rain gauges, an average of the gauge data was used as reference in the evaluation.

To investigate the performance of satellite rainfall products across different rainfall regimes, rainfall intensity was categorized according to the national standard of China (GB/T 28592-2012) as: light rain (0.1–9.9 mm/day), moderate rain (10.0–24.9 mm/day), heavy rain (25.0–49.9 mm/day), light rainstorm (50.0–99.9 mm/day), moderate rainstorm (100.0–249.9 mm/day) and heavy rainstorm (≥250.0 mm/day).

Seven performance metrics were used to evaluate satellite rainfall products against rain gauge measurements. Metrics are divided into two categories, i.e., continuous and categorical metrics (Table 1). The continuous metrics include correlation coefficient (CC), mean error (ME), mean absolute error (MAE) and relative mean absolute error (RMAE), which are used to measure the accuracy of satellite rainfall estimates. CC denotes the agreement between satellite rainfall estimates and rain gauge measurements. ME describes the systematic error, and can be either positive (denoting overestimation) or negative (denoting underestimation). MAE and RMAE represent the average and relative error magnitude, respectively. The categorical metrics include probability of detection (POD), false alarm ratio (FAR) and critical success index (CSI), which are used to measure the detection capability of satellite rainfall products. POD denotes the fraction of rainfall occurrences that are correctly detected. FAR measures the fraction of detected rainfall events that did not occur [21]. As function of POD and FAR, CSI gives more balanced score. A value of 0.1 mm/day was set for the rain/no rain threshold in calculation of the metrics.

Table 1. List of the metrics used to quantify the performance of satellite rainfall products.

Metrics	Equation	Perfect Value	Unit
Correlation Coefficient	$CC = \dfrac{\sum_{i=1}^{n}(E_i - \overline{E})(G_i - \overline{G})}{\sqrt{\sum_{i=1}^{n}(E_i - \overline{E})^2}\sqrt{\sum_{i=1}^{n}(G_i - \overline{G})^2}}$	1	NA
Mean Error	$ME = \frac{1}{n}\sum_{i=1}^{n}(E_i - G_i)$	0	mm
Mean Absolute Error	$MAE = \frac{1}{n}\sum_{i=1}^{n}|E_i - G_i|$	0	mm
Relative Mean Absolute Error	$RMAE = \frac{MAE}{\overline{G}} \times 100\%$	0	%
Probability of Detection	$POD = \frac{Hits}{Hits+Misses}$	1	NA
False Alarm Ratio	$FAR = \frac{False}{Hits+False}$	0	NA
Critical Success Index	$CSI = \frac{Hits}{Hits+Misses+False}$	1	NA

where *n* is the number of samples; E_i is the estimate; G_i is the observation; $\overline{G}$ is the mean observation; *Hits* is the number of observed rainfall that are detected; *Misses* is the number of observed rainfall that are not detected; *False* is the number of rainfall that are detected but there is no observed rainfall.

3. Results

3.1. Performance of Satellite Rainfall Products at Different Rainfall Intensities

Continuous metrics for satellite rainfall products for different rainfall intensities over Mainland China are shown in Table 2. For all satellite rainfall products, the CC values were low and even negative (−0.19 to 0.23 for 3B42RT, −0.12 to 0.29 for IMERG-E, and −0.08 to 0.32 for IMERG-L), indicating that overall the satellite rainfall estimates showed a lack of agreement with rain gauge measurements. In other words, there is considerable non-linear error between satellite and rain gauge rainfall [28,29]. The ME showed a clear trend of increasing negative value with increasing rainfall intensity, from a base of relatively small positive values for light rain. The MAE similarly showed a systematic increasing in error with rainfall intensity. The ME and MAE in Table 2 indicated that all satellite rainfall products tended to overestimate light rain, while underestimating moderate rain to heavy rainstorm. In contrast, the RMAE decreased with increasing rainfall intensity as a whole. It is interesting that the IMERG products significantly reduced error magnitude (including MAE and RMAE) in estimating light rainfall compared with 3B42RT, which may be largely because the enhanced sensor characteristics which extend the measurement range over that used in 3B42RT to include light-intensity rainfall and snowfall.

Table 2. Continuous metrics for satellite rainfall products at different rainfall intensities.

Observation (mm/day)	3B42RT				IMERG-E				IMERG-L			
	CC	ME	MAE	RMAE	CC	ME	MAE	RMAE	CC	ME	MAE	RMAE
0.1–9.9	0.23	2.13	4.89	190.23	0.29	1.05	3.51	136.88	0.32	0.90	3.32	129.69
10.0–24.9	0.15	−0.66	13.96	88.80	0.16	−2.85	11.58	73.66	0.17	−2.83	11.30	71.89
25.0–49.9	0.15	−6.72	22.87	66.52	0.14	−10.33	21.38	62.21	0.15	−10.19	21.04	61.23
50.0–99.9	0.17	−19.74	36.37	55.03	0.14	−26.19	37.54	56.81	0.15	−25.72	37.07	56.10
100.0–249.9	0.20	−58.26	71.71	53.13	0.20	−66.07	76.07	56.35	0.19	−65.37	76.23	56.47
≥250.0	-0.19	−201.02	201.02	65.92	−0.12	−180.68	190.00	62.30	−0.08	−184.26	186.51	61.16

Chen and Li [30] suggested that monthly satellite rainfall estimates are considered reliable when the relative root-mean-square error is less than 50%. Pipunic et al. [29] observed large differences between satellite and rain gauge rainfall, with the satellite rainfall estimates often ranging from twice to less than half that of the rain gauge record. Based on the 50% criteria alone, the results in Table 2 suggest that the estimation accuracy of satellite rainfall products for each rainfall intensity over Mainland China remains a challenge.

Categorical metrics for satellite rainfall products at different rainfall intensities over Mainland China are shown in Tables 3 and 4. For the detection of rainfall occurrence, POD and FAR continuously improved along with increasing rainfall intensity. Compared with 3B42RT, the IMERG products significantly improved the detection skill in POD for light rain, while performing slightly poorer in FAR. It is worth noting that POD/FAR for light rain was significantly smaller/larger than that for moderate rain to heavy rainstorm. Hence, the prospect of satellite rainfall products detecting rainfall for intensity ≥10.0 mm/day are very good, while light rainfall (<10.0 mm/day) is an issue.

Table 3. POD for satellite rainfall products at different rainfall intensities.

Observation (mm/day)	Hits + Misses			Hits			POD		
	3B42RT	IMERG-E	IMERG-L	3B42RT	IMERG-E	IMERG-L	3B42RT	IMERG-E	IMERG-L
0.1–9.9	148,089	149,877	149,877	70,524	103,995	106,527	0.48	0.69	0.71
10.0–24.9	29,695	29,920	29,920	23,856	28,254	28,632	0.80	0.94	0.96
25.0–49.9	11,848	11,893	11,893	10,910	11,677	11,722	0.92	0.98	0.99
50.0–99.9	3918	3923	3923	3800	3876	3889	0.97	0.99	0.99
100.0–249.9	673	674	674	661	673	674	0.98	1.00	1.00
≥250.0	22	22	22	22	22	22	1.00	1.00	1.00
All Samples	194,245	196,309	196,309	109,773	148,497	151,466	0.57	0.76	0.77

Table 4. FAR for satellite rainfall products at different rainfall intensities.

Estimate (mm/day)	Hits + False			False			FAR		
	3B42RT	IMERG-E	IMERG-L	3B42RT	IMERG-E	IMERG-L	3B42RT	IMERG-E	IMERG-L
0.1–9.9	109,084	216,613	214,820	46,619	108,990	103,820	0.43	0.50	0.48
10.0–24.9	32,598	28,276	27,682	6308	2134	1439	0.19	0.08	0.05
25.0–49.9	15,353	11,063	10,175	1567	649	150	0.10	0.06	0.01
50.0–99.9	6298	3609	3468	305	28	9	0.05	0.01	0.00
100.0–249.9	1216	671	675	25	1	1	0.02	0.00	0.00
≥250.0	20	25	25	0	0	0	0.00	0.00	0.00
All Samples	164,569	260,257	256,845	54,824	111,802	105,419	0.33	0.43	0.41

The POD and FAR were calculated for ten 1-mm intervals for rainfall intensity <10.0 mm/day, as shown in Figure 3. Generally, the POD increased with increasing rainfall intensity, in contrast to a decreasing trend for FAR. IMERG-L showed the best performance in detecting rainfall occurrence for each sub-sample, followed by IMERG-E and 3B42RT. We assumed a threshold of 0.5 for POD and FAR to determine whether the satellite rainfall estimates could have acceptable detection capability. It is evident from Figure 3 that for rainfall intensity <2.0 mm/day 3B42RT had little to no skill in detecting rainfall occurrence, and for rainfall intensity <1.0 mm/day both 3B42RT and IMERG products were more likely to cause false alarms.

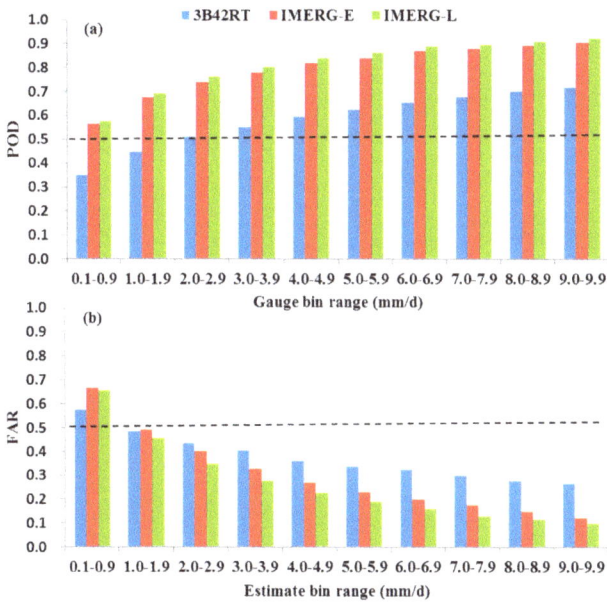

Figure 3. (**a**) POD; and (**b**) FAR for satellite rainfall products at rainfall intensity <10.0 mm/day.

3.2. Spatial Differences of the Performance of Satellite Rainfall Products

Spatial distributions of metrics for satellite rainfall products over Mainland China are shown in Figure 4. Boxplots of the distribution of the statistics over three sub-regions are shown in Figure 5. The performance of satellite rainfall products has significant spatial differences, especially among sub-regions.

The distributions of CC for three satellite rainfall products were identical in general pattern. The values of the first/third quartiles and upper/lower end of outliers were greatest in Reg1, followed by Reg3 and Reg2. The IMERG-E and IMERG-L constantly improved the agreement with rain gauge measurements compared with 3B42RT (Figure 5a). Besides, the weak and even negative correlation (CC < 0.1) mainly occurred along the junction of Reg2 and Reg3, where are rain shadow deserts formed by the Tibetan Plateau blocking summer monsoon reaching the inland areas (Figure 4a–c). The ME for 3B42RT was generally high over most parts of Mainland China, and characterized by an alternate distribution of positive and negative values in Reg1, and predominantly positive values in Reg2 and Reg3. For IMERG products, positive values mainly located in central and northern Reg1, and positive and negative values were, respectively, dominant in Reg2 and Reg3 (Figure 4d–f and 5b). Figure 4g–i shows that the MAE for all satellite rainfall products gradually decreased from the southeast coast to the northwest inland, showing a similar pattern to the spatial distribution of rainfall intensity (Figure 1b). This phenomenon is reasonable because the MAE increased with increasing rainfall intensity (Table 2). The first/third quartiles and upper/lower end of outliers were closer to the median for IMERG products than that for 3B42RT, and the corresponding values were lower, indicating that IMERG products significantly reduce the average error magnitude compared with 3B42RT (Figure 5c). Similarly, IMERG-E and IMERG-L constantly reduced the relative error magnitude over three sub-regions, but the improvements were not obvious in western Reg2, and the junction of Reg2 and Reg3 (Figure 4j–l). As highlighted in Figure 5d, all satellite rainfall products were unreliable over Mainland China, adopting 50% as threshold for RMAE to determine whether the satellite estimates can be trusted. In terms of CSI, all satellite rainfall products showed higher value in the southeastern and southwestern parts of Mainland China, while lower value in the northwestern part (Figure 4m–o). Figure 5e shows that IMERG products could capture rainfall events better than 3B42RT over Mainland China. Nonetheless, both 3B42RT and IMERG products in Reg2 were more likely to cause misjudgments.

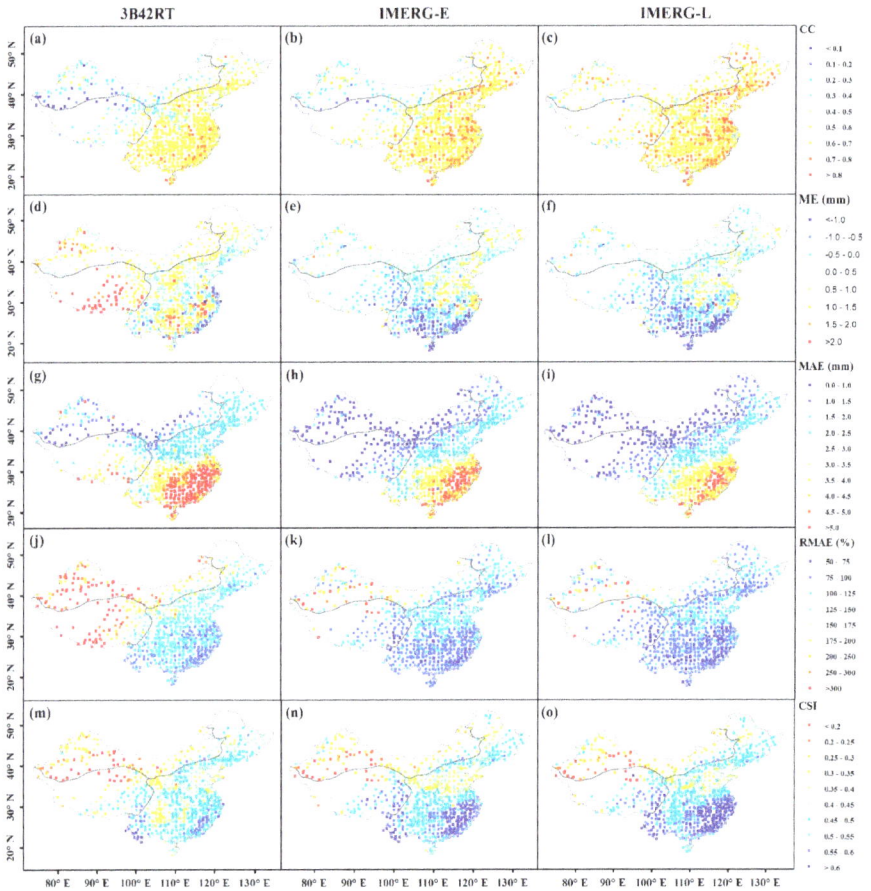

Figure 4. Spatial distributions of metrics for satellite rainfall products over Mainland China: (a–c) CC; (d–f) ME; (g–i) MAE; (j–l) RMAE; and (m–o) CSI.

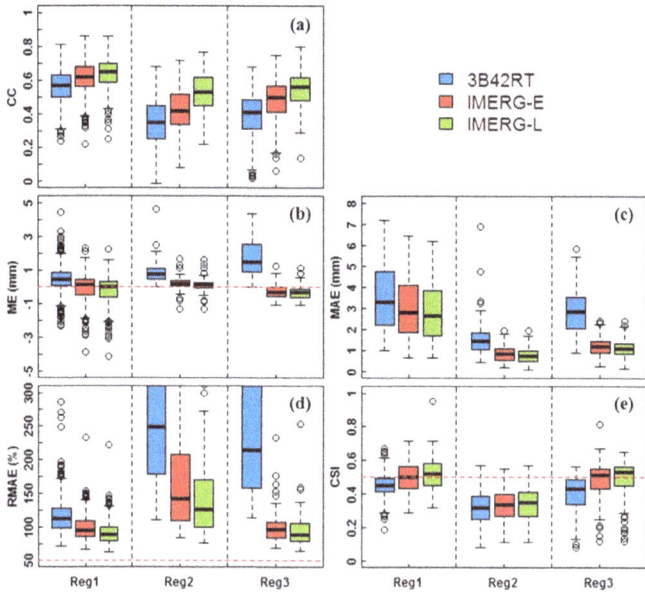

Figure 5. Boxplots of metrics for satellite rainfall products over three sub-regions: (**a**) CC; (**b**) ME; (**c**) MAE; (**d**) RMAE; and (**e**) CSI. The boxes indicate the 25th, 50th, and 75th percentiles of the distribution, and the vertical lines indicate the 5th and 95th percentiles. Note that some metric values are beyond the scope of vertical axis.

3.3. Temporal Characteristics of the Performance of Satellite Rainfall Products

The daily time series of metrics for satellite rainfall products from 1 January 2015 to 31 December 2016 are shown in Figure 6. All metrics showed symmetrical distribution along the timeline of 1 January 2016 (as indicated by a dashed black line), indicating the performance of satellite rainfall estimates have an annual periodic variation. Furthermore, the metrics generally demonstrated seasonal variation as well.

The CC values for all satellite rainfall products were low and unstable in dry season (October to April), and reached relatively high and stable in wet season (May to September). In general, IMERG-L showed the best agreement with rain gauge measurements, followed by IMERG-E and 3B42RT (Figure 6a–c). 3B42RT was characterized by overestimation, and the systematic error was larger in wet season than in dry season, especially in Reg3. In contrast, IMERG products showed a smaller systematic error (Figure 6d–f). The MAE presented an arched shape with vault in summer (Figure 6g–i). Compared with 3B42RT, the IMERG products significantly reduced average error magnitude, but with obvious differences among seasons and sub-regions. More specifically, the improvements mainly occurred during wet season in Reg1, while throughout the year in Reg2 and Reg3. Figure 6j–l shows that IMERG products outperformed 3B42RT in wet season, while both 3B42RT and IMERG products were violently fluctuant in dry reason. Generally, the use of three satellite rainfall products remains great challenge throughout the year over sub-regions, adopting 50% as threshold for RMAE to determine whether the satellite estimates can be trusted. Figure 6m–o shows temporal patterns of CSI were similar to those of MAE, but the differences among satellite rainfall products were small in each sub-region.

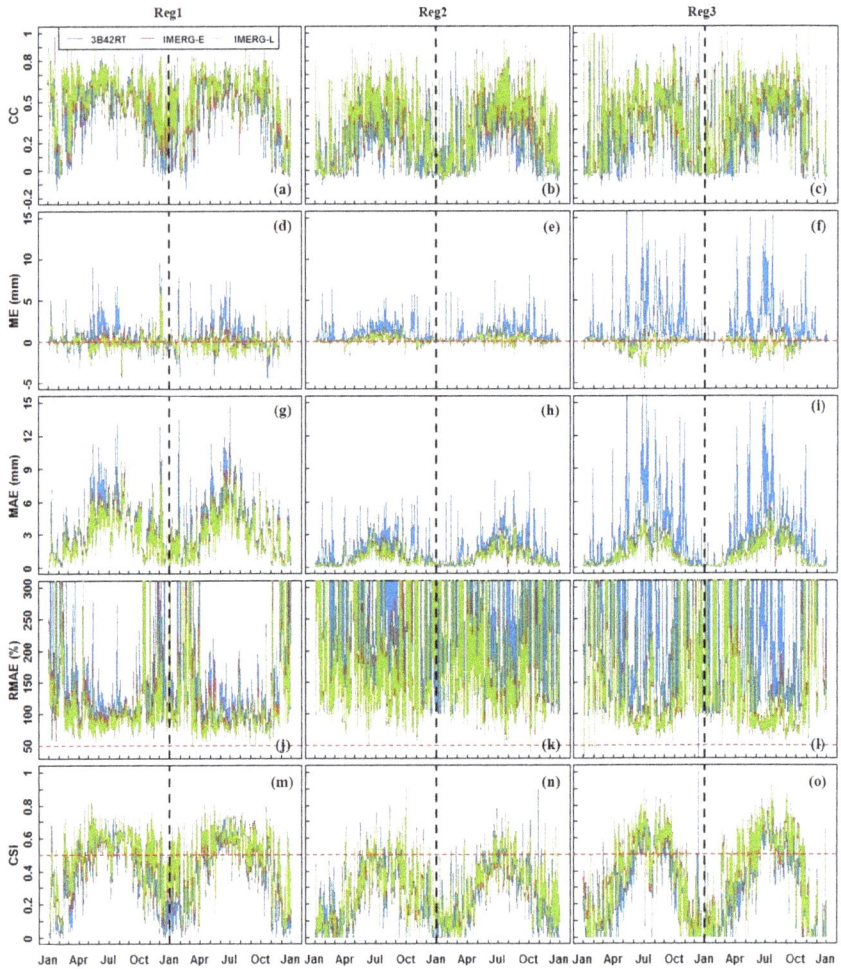

Figure 6. Daily time series of metrics for satellite rainfall products over sub-regions from 1 January 2015 to 31 December 2016: (**a–c**) CC; (**d–f**) ME; (**g–i**) MAE; (**j–l**) RMAE; and (**m–o**) CSI. Note that some metric values are beyond the scope of vertical axis.

4. Discussion

4.1. Dependence of the Performance of Satellite Rainfall Products

Our evaluation and comparison of 3B42RT and the equivalent IMERG products have provided insights into how different errors vary with observed rainfall intensities, climate zones and rainfall seasonality. Results for the different rainfall bin ranges give a general indication of the likely error characteristics for certain rainfall intensities, that is, satellite rainfall products overestimated light rain, while underestimated moderate rain to heavy rainstorm (Table 2), which is consistent with the study of 3B42RT v7 over Mainland Australia [29]. However, the finding is different with the results in Southeast Asia, which suggested that 3B42, 3B42RT v7 and IMERG-F v4 products underestimate <1.0 mm/day rainfall and overestimate 1.0–20.0 mm/day rainfall over Singapore [31], while 3B42 and 3B42RT v7 products underestimate <1.0 mm/day rainfall and overestimate 1.0–50.0 mm/day rainfall over Malaysia [32]. This highlights the varying performance of satellite rainfall estimates over regions. In this study, the possible reason for overestimating light rainfall is that hydrometeors that are detected by infrared and microwave sensors as well as precipitation radars might partially or even totally be evaporated before they are observed by rain gauges [33–35]. Moreover, sensitivity to attenuation in the vertical direction for microwave wavelengths has been reported to result in underestimate heavy-intensity rainfall [36].

Higher rainfall intensity associated with increased mean (absolute) error and decreased relative mean absolute error, which could explain the spatial-temporal distributions of the corresponding metrics to some extent. However, 3B42RT overestimated rainfall over most of Mainland China (Figure 4d–f). Similar results were also seen in the conterminous United States [37]. Amitai et al. [38] indicated that the overestimation is probably due to overcorrection of attenuation by the TRMM PR algorithms.

Furthermore, higher rainfall intensity associated with better detection given the relationships between rainfall intensity and detection capability highlighted in Section 3.1. The results indicate threshold values of <2.0 mm/day, below which 3B42RT is unreliable at detecting rain, and <1.0 mm/day, below which both 3B42RT and IMERG products are more likely to cause false alarms. Recall that, since light rain and snowfall account for significant fractions of rainfall occurrences in middle and high latitudes, a key advancement of GPM over TRMM is the extended capability to measure light-intensity rainfall (i.e., <0.5 mm/h) and solid precipitation, which could lead to the improved probability of detection for IMERG products compared with 3B42RT, as shown in Figure 7a–c. However, it raises false alarms as well (Figure 7d–f), and several factors may contribute to this phenomenon: (1) hydrometeors that are detected by sensors as shallow rainfall might totally be evaporated before they are observed by rain gauges, especially over hot and/or dry regions; and (2) desert and Gobi, with wide distribution in the western Reg2, scatter the upwelling microwave radiation in manner similar to light rain [39], which makes it more likely to detect the signals as rainfall for IMERG products compared with 3B42RT. Based on the analysis of POD and FAR above, it can explain the spatial distributions of CSI.

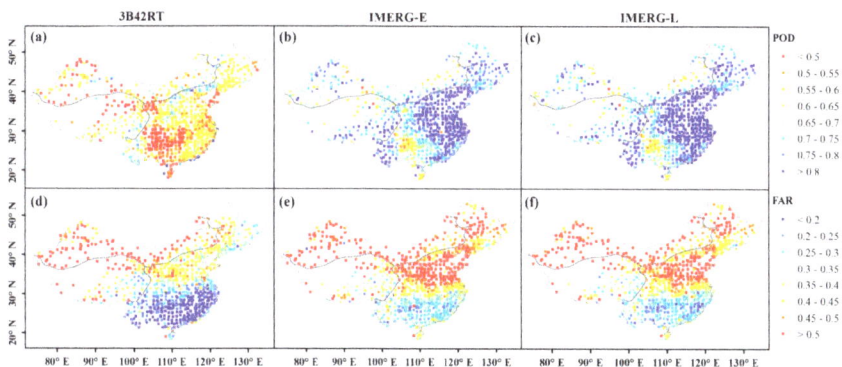

Figure 7. Spatial distributions of metrics for satellite rainfall products over Mainland China: (**a**–**c**) POD; and (**d**–**f**) FAR.

4.2. Cause of the Performance Differences

In this study, the IMERG products generally performed better than 3B42RT, which could be attributed to the satellite overpasses and sensor capability. As near-real-time systems, the individual satellite overpasses are the basis for data used to estimate instantaneous rainfall rate, and result in data coverage gaps. The PMW Tb have strong physical relationship with hydrometeors, but with poor sampling frequency [20,40]. In contrast, geo-satellites give frequent sampling, while the IR

Tb are related to cloud-top features (temperature and albedo) rather than to surface precipitation directly [41,42]. Therefore, PMW estimates are used in preference to IR estimates, meaning that deficiency is minimized in satellite rainfall products.

Recall that 3B42RT draws on data from SSMIS, MHS and geo-IR sensors [19]. Additional GMI, AMSR2, ATMS and SAPHIR sensor data are involved in the IMERG products [43]. That is, the IMERG products narrow PMW coverage gaps compared with 3B42RT. Besides, the CMORPH-KF morphing scheme used in IMERG algorithm interpolates PMW estimates with the cloud motion vectors, which further mitigates the PMW coverage gaps [37].

Furthermore, the PMW estimates are calibrated to TCI or CORRA products, and then provide calibration for geo-IR estimates in the TMPA or IMERG algorithms, respectively. Thus, the time series of the completed data tend to follow that of calibrator. The TCI and CORRA products are derived from post and coincident data, respectively. Experience shows that calibrators computed from coincident data perform better than those from post data (i.e., climatological calibration) [19].

4.3. Uncertainty of the Evaluation Results

As rain gauges measure rainfall directly by physical methods, rain gauge measurements are normally employed as reference to validate satellite rainfall products. However, the rain gauge measurements are subject to errors, due to wind-induced error, wetting loss, evaporation loss and trace amount [44,45]. Besides, rain gauge measurements at point-scale might deviate far from true areal rainfall. Jensen and Pedersen [46] found that variation in accumulated rainfall is up to 100% between neighboring rain gauges within a 500 m × 500 m pixel over a four-day period. Villarini et al. [47] indicated that there is a tendency for the error (averaging the rainfall measurements from rain gauges to represent the true areal rainfall) to decrease for increasing number of rain gauges. There are at most two rain gauges in a 0.25° × 0.25° pixel in this study. Therefore, the rain gauge measurements based on scarce rain gauges could not perfectly represent areal rainfall, and the evaluation results may be impacted by the mismatch between satellite rainfall estimates and rain gauge measurements. It is worth noting that Tang et al. [27] suggested that the actual quality of satellite rainfall estimates could be better than the evaluation results, when using rain gauge measurements at point-scale to evaluate areal satellite rainfall estimates.

5. Conclusions

In this study, we evaluated and compared the performance of the daily 3B42RT legacy product and the equivalent IMERG products, using rainfall data from 830 gauges as reference across Mainland China. We particularly concerned about the accuracy of satellite rainfall estimates at different rainfall intensities, and the spatial-temporal distribution patterns of their performance. The major conclusions are summarized as follows:

(1) Both 3B42RT and IMERG products overestimated light rain, while underestimated moderate rain to heavy rainstorm, with an increase in mean (absolute) error and a decrease in relative mean absolute error. The 3B42RT had smaller error magnitude in estimating light rainstorm and moderate rainstorm, while the equivalent IMERG products performed better in estimating light rain to heavy rain, and heavy rainstorm.

(2) Higher rainfall intensity associated with better detection. Threshold values of <2.0 mm/day, below which 3B42RT is unreliable at detecting rain, and <1.0 mm/day, below which both 3B42RT and IMERG products are more likely to cause false alarms, were found.

(3) Generally, both 3B42RT and IMERG products performed better in wet areas with relatively heavy rainfall intensity and/or during wet season than in dry areas with relatively light rainfall intensity and/or during dry season. Compared with 3B42RT, the IMERG-E and IMERG-L constantly improved the performance in space and time, but it is not obvious in dry areas and/or during dry season.

Our work suggests that IMERG products generally perform better than 3B42RT legacy product during the overlap period, but the agreement between IMERG products and rain gauge measurements is low and even negative for different rainfall intensities, and the RMAE is still at a high level (>50%), indicating that IMERG products remain to be improved. Experiments demonstrated that numerical weather prediction (NWP) models show greatest skill for shallow convection rain and/or in winter [41]. Therefore, an important research focus for future work will be combining the respective advantages of IMERG products (or algorithms) and NWP models [48], to estimate more accurate near-real-time rainfall.

Author Contributions: Conceptualization, L.W.; Formal analysis, L.W.; Funding acquisition, Y.X.; Supervision, Y.X.; Writing—original draft, L.W.; and Writing—review and editing, L.W. and S.W.

Funding: This study was financially supported by the National Key Research and Development Program of China (2016YFC0401502), the National Natural Science Foundation of China (41771032), the Water Conservancy Science and Technology Foundation of Jiangsu Province (2015003) and the State Foundation for Studying Abroad (201706190207).

Acknowledgments: We are very much indebted to Luigi Renzullo for his supervision, review and editing in this research effort. We thank Lingjie Li, Yongqing Bai, Qingfang Hu and Juanle Wang for technical assistance in data processing. We acknowledge the editors and three anonymous reviewers for their insightful and constructive comments which improved the manuscript substantially.

Conflicts of Interest: The authors declare no conflict of interest.

References

1. Prakash, S.; Mitra, A.K.; Pai, D.S.; AghaKouchak, A. From TRMM to GPM: How well can heavy rainfall be detected from space? *Adv. Water Resour.* **2016**, *88*, 1–7. [CrossRef]
2. Tekeli, A.E.; Fouli, H. Evaluation of TRMM satellite-based precipitation indexes for flood forecasting over Riyadh City, Saudi Arabia. *J. Hydrol.* **2016**, *541*, 471–479. [CrossRef]
3. Li, D.; Christakos, G.; Ding, X.X.; Wu, J.P. Adequacy of TRMM satellite rainfall data in driving the SWAT modeling of Tiaoxi catchment (Taihu lake basin, China). *J. Hydrol.* **2018**, *556*, 1139–1152. [CrossRef]
4. Hou, Y.; Kakar, R.K.; Neeck, S.; Azarbarzin, A.A.; Kummerow, C.D.; Kojima, M.; Oki, R.; Nakamura, K.; Iguchi, T. The global precipitation measurement mission. *Bull. Am. Meteorol. Soc.* **2014**, *95*, 701–722. [CrossRef]
5. Guo, H.; Chen, S.; Bao, A.M.; Behrangi, A.; Hong, Y.; Ndayisaba, F.; Hu, J.J.; Stepanian, P.M. Early assessment of Integrated Multi-satellite Retrievals for Global Precipitation Measurement over China. *Atmos. Res.* **2016**, *176–177*, 121–133. [CrossRef]
6. Li, Z.; Yang, D.; Hong, Y. Multi-scale evaluation of high-resolution multi-sensor blended global precipitation products over the Yangtze River. *J. Hydrol.* **2013**, *500*, 157–169. [CrossRef]
7. Collischonn, B.; Collischonn, W.; Tucci, C.E.M. Daily hydrological modeling in the Amazon basin using TRMM rainfall estimates. *J. Hydrol.* **2008**, *360*, 207–216. [CrossRef]
8. Islam, T.; Rico-Ramirez, M.A.; Han, D.W.; Srivastava, P.K.; Ishak, A.M. Performance evaluation of the TRMM precipitation estimation using ground-based radars from the GPM validation network. *J. Atmos. Sol.-Terr. Phys.* **2012**, *77*, 194–208. [CrossRef]
9. Pombo, S.; de Oliveira, R.P. Evaluation of extreme precipitation estimates from TRMM in Angola. *J. Hydrol.* **2015**, *523*, 663–679. [CrossRef]
10. Mahmoud, M.T.; Al-Zahrani, M.A.; Sharif, H.O. Assessment of global precipitation measurement satellite products over Saudi Arabia. *J. Hydrol.* **2018**, *559*, 1–12. [CrossRef]
11. Chappell, A.; Renzullo, L.J.; Raupach, T.H.; Haylock, M. Evaluating geostatistical methods of blending satellite and gauge data to estimate near real-time daily rainfall for Australia. *J. Hydrol.* **2013**, *493*, 105–114. [CrossRef]
12. Schneebeli, M.; Dawes, N.; Lehning, M.; Berne, A. High-resolution vertical profiles of X-band polarimetric radar observables during snowfall in the Swiss Alps. *J. Appl. Meteorol. Clim.* **2013**, *52*, 378–394. [CrossRef]
13. Kidd, C.; Kniveton, D.R.; Todd, M.C.; Bellerby, T.J. Satellite Rainfall Estimation Using Combined Passive Microwave and Infrared Algorithms. *J. Hydrometeorol.* **2003**, *4*, 1088–1104. [CrossRef]
14. Turk, F.J.; Miller, S.D. Toward improved characterization of remotely sensed precipitation regimes with MODIS/AMSR-E blended data techniques. *IEEE Trans. Geosci. Remote Sens.* **2005**, *43*, 1059–1069. [CrossRef]

15. Huffman, G.J.; Adler, R.F.; Bolvin, D.T.; Gu, G.J.; Nelkin, E.J.; Bowman, K.P.; Hong, Y.; Stocker, E.F.; Wolff, D.B. The TRMM multi-satellite precipitation analysis (TMPA): Quasi-global, multiyear, combined-sensor precipitation estimates at fine scale. *J. Hydrometeorol.* **2007**, *8*, 38–55. [CrossRef]

16. Hsu, K.L.; Gupta, H.V.; Gao, X.G.; Sorooshian, S. Estimation of physical variables from multichannel remotely sensed imagery using a neural network: Application to rainfall estimation. *Water Resour. Res.* **1999**, *35*, 1605–1618. [CrossRef]

17. Joyce, R.J.; Janowiak, J.E.; Arkin, P.A.; Xie, P.P. CMORPH: A method that produces global precipitation estimates from passive microwave and infrared data at high spatial and temporal resolution. *J. Hydrometeorol.* **2004**, *5*, 487–503. [CrossRef]

18. Yong, B.; Liu, D.; Gourley, J.J.; Tian, Y.D.; Huffman, G.J.; Ren, L.L.; Hong, Y. Global view of real-time TRMM Multisatellite Precipitation Analysis: Implications for its successor Global Precipitation Measurement Mission. *Bull. Am. Meteorol. Soc.* **2015**, *96*, 1–15. [CrossRef]

19. Huffman, G.J.; Bolvin, D.T. Real-Time TRMM Multi-Satellite Precipitation Analysis Data Set Documentation. NASA Goddard Space Flight Center. Available online: https://pmm.nasa.gov/sites/default/files/document_files/3B4XRT_doc_V7_180426.pdf (accessed on 28 May 2018).

20. Kim, K.Y.; Park, J.M.; Baik, J.J.; Choi, M.H. Evaluation of topographical and seasonal feature using GPM IMERG and TRMM 3B42 over Far-East Asia. *Atmos. Res.* **2017**, *187*, 95–105. [CrossRef]

21. Xu, R.; Tian, F.Q.; Yang, L.; Hu, H.C.; Lu, H.; Hou, A.Z. Ground validation of GPM IMERG and TRMM 3B42V7 rainfall products over southern Tibetan Plateau based on a high-density rain gauge network. *J. Geophys. Res. Atmos.* **2017**, *122*, 1–15. [CrossRef]

22. Tang, G.Q.; Ma, Y.Z.; Long, D.; Zhong, L.Z.; Hong, Y. Evaluation of GPM Day-1 IMERG and TMPA Version-7 legacy products over Mainland China at multiple spatiotemporal scales. *J. Hydrol.* **2016**, *533*, 152–167. [CrossRef]

23. Zhao, S.Q. A new scheme for comprehensive physical regionalization in China. *Acta Geogr. Sin.* **1983**, *38*, 1–10.

24. Bothe, O.; Fraedrich, K.; Zhu, X.H. Precipitation climate of Central Asia and the large-scale atmospheric circulation. *Theor. Appl. Climatol.* **2012**, *108*, 345–354. [CrossRef]

25. Hong, Y.; Hsu, K.L.; Sorooshian, S.; Gao, X. Precipitation estimation from remotely sensed imagery using an Artificial Neural Network Cloud Classification System. *J. Appl. Meteorol.* **2004**, *43*, 1834–1852. [CrossRef]

26. Joyce, R.J.; Xie, P.P. Kalman Filter-based CMORPH. *J. Hydrometeorol.* **2011**, *12*, 1547–1563. [CrossRef]

27. Tang, G.Q.; Behrangi, A.; Long, D.; Li, C.M.; Hong, Y. Accounting for spatiotemporal errors of gauges: A critical step to evaluate gridded precipitation products. *J. Hydrol.* **2018**, *559*, 294–306. [CrossRef]

28. Hong, Y.; Hsu, K.L.; Moradkhani, H.; Sorooshian, S. Uncertainty quantification of satellite precipitation estimation and Monte Carlo assessment of the error propagation into hydrologic response. *Water Resour. Res.* **2006**, *42*, 1–15. [CrossRef]

29. Pipunic, R.C.; Ryu, D.; Costelloe, J.F.; Su, C.H. An evaluation and regional error modeling methodology for near-real-time satellite rainfall data over Australia. *J. Geophys. Res. Atmos.* **2015**, *120*, 10767–10783. [CrossRef]

30. Chen, F.R.; Li, X. Evaluation of IMERG and TRMM 3B43 monthly precipitation products over Mainland China. *Remote Sens.* **2016**, *8*, 472. [CrossRef]

31. Tan, M.L.; Duan, Z. Assessment of GPM and TRMM Precipitation Products over Singapore. *Res. Atmos.* **2017**, *9*, 720. [CrossRef]

32. Tan, M.L.; Ibrahim, A.L.; Duan, Z.; Cracknell, A.P.; Chaplot, V. Evaluation of six high-resolution satellite and ground-based precipitation products over Malaysia. *Remote Sens.* **2015**, *7*, 1504–1528. [CrossRef]

33. Villarini, G.; Krajewski, W.F. Evaluation of the research version TMPA three hourly 0.25° × 0.25° rainfall estimates over Oklahoma. *Geophys. Res. Lett.* **2007**, *34*, 1–5. [CrossRef]

34. Chen, S.; Hong, Y.; Cao, Q.; Gourley, J.J.; Kirstetter, P.E.; Yong, B.; Tian, Y.D.; Zhang, Z.X.; Shen, Y.; Hu, J.J.; et al. Similarity and difference of the two successive V6 and V7 TRMM multisatellite precipitation analysis performance over China. *J. Geophys. Res. Atmos.* **2013**, *118*, 13060–13074. [CrossRef]

35. Zhao, H.G.; Yang, B.G.; Yang, S.T.; Huang, Y.C.; Dong, G.T.; Bai, J.; Wang, Z.W. Systematical estimation of GPM-based global satellite mapping of precipitation products over China. *Atmos. Res.* **2018**, *201*, 206–217. [CrossRef]

36. Zhao, H.G.; Yang, S.T.; Wang, Z.W.; Zhou, X.; Luo, Y.; Wu, L.N. Evaluating the suitability of TRMM satellite rainfall data for hydrological simulation using a distributed hydrological model in the Weihe River catchment in China. *J. Geogr. Sci.* **2015**, *25*, 177–195. [CrossRef]

Remote Sens. **2018**, *10*, 1778

37. Gebregiorgis, A.S.; Kirstetter, P.E.; Hong, Y.E.; Gourley, J.J.; Huffman, G.J.; Petersen, W.A.; Xue, X.W.; Schwaller, M.R. To what extent is the day 1 GPM IMERG satellite precipitation estimate improved as compared to TRMM TMPART? *J. Geophys. Res. Atmos.* **2018**, *123*, 1694–1707.

38. Amitai, E.; Liort, X.; Liao, L.; Meneghini, R. A framework for global verification of space-borne radar estimates of precipitation based on rain type classification. In Proceedings of the 2nd TRMM International Science Conference, Nara, Japan, 6–10 September 2004.

39. Grody, N.C.; Weng, F.Z. Microwave emission and scattering from deserts: Theory compared with satellite measurements. *IEEE Trans. Geosci. Remote Sens.* **2008**, *46*, 361–375. [CrossRef]

40. Kidd, C.; Levizzani, V. Status of satellite precipitation retrievals. *Hydrol. Earth Syst. Sci.* **2011**, *15*, 1109–1116. [CrossRef]

41. Pretty, G.W. The status of satellite-based rainfall estimation over land. *Remote Sens. Environ.* **1995**, *51*, 125–137. [CrossRef]

42. Ebert, E.E.; Janowiak, J.E.; Kidd, C. Comparison of near-real-time precipitation estimates from satellite observations and numerical models. *Bull. Am. Meteorol. Soc.* **2007**, *88*, 1–18. [CrossRef]

43. Huffman, G.J.; Bolvin, D.T.; Nelkin, E.J. Integrated Multi-Satellite Retrievals for GPM (IMERG) Technical Documentation. NASA Goddard Space Flight Center. Available online: https://pmm.nasa.gov/sites/default/files/document_files/IMERG_doc_180207.pdf (accessed on 28 May 2018).

44. Adam, J.C.; Lettenmaier, D.P. Adjustment of global gridded precipitation for systematic bias. *J. Geophys. Res. Atmos.* **2003**, *108*, 1–9. [CrossRef]

45. Ma, Y.Z.; Zhang, Y.S.; Yang, D.Q.; Farhan, S.B. Precipitation bias variability versus various gauges under different climatic conditions over the Third Pole Environment (TPE) region. *Int. J. Climatol.* **2015**, *35*, 1201–1211. [CrossRef]

46. Jensen, N.E.; Pedersen, L. Spatial variability of Rainfall: Variations within a single radar pixel. *Atmos. Res.* **2005**, *77*, 269–277. [CrossRef]

47. Villarini, G.; Mandapaka, P.V.; Krajewski, W.F.; Moore, R.J. Rainfall and sampling errors: A rain gauge perspective. *J. Geophys. Res. Atmos.* **2008**, *113*, 1–12. [CrossRef]

48. Zhang, X.X.; Anagnostou, E.N.; Schwartz, C.S. NWP-based adjustment of IMERG precipitation for flood-inducing complex terrain storms: evaluation over CONUS. *Remote Sens.* **2018**, *10*, 642. [CrossRef]

remote sensing

MDPI

Article

Hydrologic Evaluation of Multi-Source Satellite Precipitation Products for the Upper Huaihe River Basin, China

Zhiyong Wu [1,2,*], Zhengguang Xu [1,2], Fang Wang [3], Hai He [1,2], Jianhong Zhou [1,2], Xiaotao Wu [1,2] and Zhenchen Liu [1,2]

[1]   Institute of Water Problems, Hohai University, 1 Xikang Road, Nanjing 210098, China; xzghhu@hhu.edu.cn (Z.X.); hehai_hhu@hhu.edu.cn (H.H.); zhoujianhong@hhu.edu.cn (J.Z.); zjyelaoma@126.com (X.W.); liuzhenchen90@163.com (Z.L.)
[2]   College of Hydrology and Water Resources, Hohai University, 1 Xikang Road, Nanjing 210098, China
[3]   Editorial Department of Journal, Hohai University, 1 Xikang Road, Nanjing 210098, China; wangfang990@126.com
*   Correspondence: wzyhhu@gmail.com; Tel.: +86-138-1387-9528

Received: 16 April 2018; Accepted: 24 May 2018; Published: 28 May 2018

Abstract: To evaluate the performance and hydrological utility of merged precipitation products at the current technical level of integration, a newly developed merged precipitation product, Multi-Source Weighted-Ensemble Precipitation (MSWEP) Version 2.1 was evaluated in this study based on rain gauge observations and the Variable Infiltration Capacity (VIC) model for the upper Huaihe River Basin, China. For comparison, three satellite-based precipitation products (SPPs), including Climate Hazards Group InfraRed Precipitation with Station data (CHIRPS) Version 2.0, Climate Prediction Center MORPHing technique (CMORPH) bias-corrected product Version 1.0, and Tropical Rainfall Measuring Mission (TRMM) Multi-satellite Precipitation Analysis (TMPA) 3B42 Version 7, were evaluated. The error analysis against rain gauge observations reveals that the merged precipitation MSWEP performs best, followed by TMPA and CMORPH, which in turn outperform CHIRPS. Generally, the contribution of the random error in all four quantitative precipitation estimates (QPEs) is larger than the systematic error. Additionally, QPEs show large uncertainty in the mountainous regions, with larger systematic errors, and tend to underestimate the precipitation. Under two parameterization scenarios, the MSWEP provides the best streamflow simulation results and TMPA forced simulation ranks second. Unfortunately, the CHIRPS and CMORPH forced simulations produce unsatisfactory results. The relative error (RE) of QPEs is the main factor affecting the RE of simulated streamflow, especially for the results of Scenario I (model parameters calibrated by rain gauge observations). However, its influence on the simulated streamflow can be greatly reduced by recalibration of the parameters using the corresponding QPEs (Scenario II). All QPEs forced simulations underestimate the streamflow with exceedance probabilities below 5.0%, while they overestimate the streamflow with exceedance probabilities above 30.0%. The results of the soil moisture simulation indicate that the influence of the precipitation input on the RE of the simulated soil moisture is insignificant. However, the dynamic variation of soil moisture, simulated by precipitation with higher precision, is more consistent with the measured results. The simulation results at a depth of 0–10 cm are more sensitive to the accuracy of precipitation estimates than that for depths of 0–40 cm. In summary, there are notable advantages of MSWEP and TMPA with respect to hydrological applicability compared with CHIRPS and CMORPH. The MSWEP has a greater potential for basin–scale hydrological modeling than TMPA.

Keywords: CHIRPS; CMORPH; TMPA; MSWEP; statistical evaluation; VIC model; hydrological simulation

1. Introduction

Precipitation is an important component of the hydrological cycle and the most primary forcing data of hydrological models [1–3]. Accurate and reliable precipitation records are crucial, not only to investigate the spatial pattern and temporal change of precipitation but also to improve the accuracy of hydrological simulation [4,5].

Conventional rain gauge stations provide the most accurate point–based precipitation data. However, due to the high spatial heterogeneity of precipitation, it is inadequate to capture the spatial-temporal variability of precipitation systems based on unevenly and sparsely distributed rain gauge stations and hardly meets the needs of hydrological models and other related research [6,7]. A number of satellite-based precipitation products (SPPs) became available to the public such as the Climate Hazards Group Infrared Precipitation with Station data (CHIRPS) [8], Climate Prediction Center (CPC) MORPHing technique bias-corrected product (CMORPH CRT) [9], and Tropical Rainfall Measuring Mission (TRMM) Multi-satellite Precipitation Analysis (TMPA) 3B42 [10]. More recently, a lot of finer and more accurate SPPs have been released, such as Global Satellite Mapping of Precipitation (GSMaP) version 6 product [11] and Integrated Multi-satellite Retrievals for Global Precipitation Measurement (GPM) (IMERG) product [12]. A few preliminary assessments of GSMaP and IMERG suggest a great potential for the hydrological applications [13–16]. The SPPs have a wide spatial coverage and high spatiotemporal resolution, effectively making up deficiencies of the conventional rain gauge observations and greatly enriching alternative precipitation data sources, especially in data-scarce or ungauged regions [17,18]. Benefitting from these advantages, the SPPs have been extensively applied in many fields such as hydrological simulations [2,14,17,19], extreme events analysis [20–23], and water resource management [24]. However, the SPPs are inevitably subject to errors resulting from sampling uncertainties and retrieval algorithms; furthermore, the error characteristics change depending on the climate regions, seasons, altitudes, and other factors [3,7,25,26].

To minimize the limitations of individual precipitation products, many researchers focused on merging different precipitation datasets to obtain a higher-quality gridded precipitation product [27–31]. Recently, the global gridded precipitation dataset Multi-Source Weighted-Ensemble Precipitation (MSWEP) that optimally merges gauge, satellite, and reanalysis data has been produced by Beck et al. [1,28]. The latest version MSWEP V2.1 with 3-hourly temporal and 0.1° spatial resolution has been available to the public since 20 November 2017. Tong et al. [32] pointed out that the SPPs are most accurate during summer and at lower latitudes, while atmospheric model reanalysis datasets perform better than SPPs during winter and at higher latitudes. Therefore, the MSWEP V2.1 can take full advantage of the complementary nature of satellite and reanalysis data and offers the prospect of increasing the precision of precipitation estimates. Due to its high spatial resolution, long time span, and great potential for hydrological applications, the MSWEP V2.1 has received the wide attention of hydrologists since its release. The research on the applicability and efficiency of MSWEP V2.1 in hydrological simulations is in progress. To evaluate the error characteristics of merged precipitation products and their performance in hydrological simulations at the current technical level of integration, the MSWEP V2.1 is used as a typical representative of merged precipitation products in this study and its applicability of MSWEP V2.1 is evaluated based on a typical semi-humid and semi-dry climatic transitional region of China.

Generally, the error characteristics of quantitative precipitation estimates (QPEs) can be directly quantified based on sufficient rain gauge observations. The direct evaluation for the error characteristics of QPEs is very important to improve the gauge adjustment scheme or merging procedure and can further be used to determine the impact of the error characteristics of QPEs on hydrological simulations. The quality of QPEs can impact the hydrologic outputs through the rainfall-runoff response [19]. Thus, the applicability analysis of QPEs in hydrological modeling serves as an alternative validation method and is more efficient than using limited rain gauge observations as a reference to better understand the properties and optimal hydrological usage of QPEs in data-scarce regions [2]. The applicability of QPEs to streamflow simulation has been evaluated in numerous studies. With the exception of streamflow, macroscale land surface hydrological models also provide an alternative way to reconstruct and continually update the spatial and temporal distribution of soil moisture over a large area [33]. Thus, indirect evaluation work can also be accomplished by evaluating the potential of QPEs in soil moisture simulation.

Based on these considerations, this study aims to statistically assess the MSWEP V2.1 against a relatively dense network of rain gauge stations and then evaluate its hydrological performance from multiple perspectives under two parameterization scenarios in the upper Huaihe River Basin. Subsequently, the impacts of QPE errors on the simulated streamflow and soil moisture are analyzed. For comparison, CHIRPS V2.0, CMORPH CRT V1.0, and TMPA 3B42 V7 are also evaluated in this study. The results provide a perspective on error characteristics and the hydrological applicability of the MSWEP V2.1 in regions with similar terrains and climates.

The rest of the paper is organized as follows: The study area and datasets used in this study are introduced in Section 2. The methodology, including the Variable Infiltration Capacity (VIC) hydrologic model and some evaluation metrics, is described in Section 3. The results of the statistical evaluation and hydrological simulation are presented in Section 4, followed by the discussion (Section 5) and main conclusions (Section 6).

2. Study Area and Data

2.1. Study Area

The Huaihe River Basin in eastern China has a drainage area of approximately 270,000 km^2; it is also the line of demarcation between Chinese southern and northern climates. The Huaihe River Basin suffers from frequent drought and flood disasters due to the uneven spatiotemporal distribution of precipitation influenced by complex and changeable climate. The Wangjiaba Station in the main stream of the Huaihe River Basin plays a vital role in flood control and management [34]. The study area is the upper region of Wangjiaba Station in the Huaihe River Basin, which has a drainage area of 30,630 km^2. Its terrain gradually lowers from west to east (Figure 1). The flood season ranges from June to September.

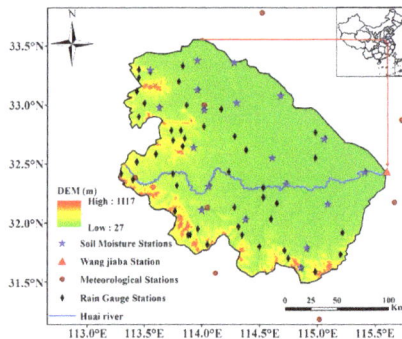

Figure 1. The location of the upper region of the Wangjiaba Station in the Huaihe River Basin and the distribution of stations used in this study.

2.2. Data

2.2.1. Precipitation

This study aims to assess the error characteristics of four QPE products and their hydrological application. A brief introduction to the four QPE products and rain gauge observations is provided in this section.

The CHIRPS is a quasi-global rainfall dataset, which was specially developed for drought monitoring. As of February 2015, version 2.0 of CHIRPS is complete and available to the public. Several data sources were introduced for the construction of CHIRPS product such as the monthly precipitation climatology CHPclim, infrared measurements from geostationary satellites, TMPA 3B42 product, and rain gauge data [8]. The daily CHIRPS V2.0 with a spatial resolution of 0.25° was selected for this study.

NOAA's CPC CMORPH contains global satellite–based precipitation generated by integrated microwave data and infrared data [9]. The CMORPH combines the superior retrieval accuracy of passive microwave estimates and the higher temporal and spatial resolution of infrared data. The latest CMORPH V1.0 product includes CMORPH RAW, CMORPH CRT, and CMORPH BLD, covering 60°S–60°N, 180°W–180°E. The CMORPH CRT with 3 h temporal and 0.25° spatial resolutions were used in this study.

The TMPA products were designed based on a wide variety of satellite datasets and supplied by NASA [10]. The two latest version 7 products of the TMPA (quasi-near-real-time 3B42RT V7 and post-real-time 3B42 V7) are the most prevalent at present. The 3B42V7 product was adjusted by monthly rain gauge precipitation data from the Global Precipitation Climatology Center (GPCC) and is superior to the TMPA 3B42RT V7. The TMPA 3B42V7 with a spatiotemporal resolution of 3 h and 0.25° were used in this study.

The MSWEP version 2.1 is a new global historic precipitation dataset (1979–2016) with 3-hourly temporal and 0.1° spatial resolution, which was specially designed for hydrological modeling. The design philosophy of MSWEP was to optimally merge the highest quality precipitation data sources as a function of timescale and location, the temporal variability of MSWEP was determined by weighted averaging. For each grid cell, the weight varies according to the network density and the accuracy of satellite- and reanalysis-based estimates [28].

The daily precipitation data of 57 rain gauges from 2000 to 2010 provided by the Hydrology Bureau of the Huaihe Water Conservancy Commission were used to evaluate the four QPEs mentioned above. For convenience, the precipitation datasets used in this study are given short names, as described in Table 1.

In this study, the VIC model is run at a spatial resolution of 0.05° × 0.05° using a daily time step. This means that, before starting the evaluation, all four QPEs and rain gauge observations are aggregated into a 0.05° × 0.05° spatial grid using the bilinear interpolation and the inverse distance weighted method, respectively. The precipitation products with 3 h temporal resolution are integrated over time to daily accumulated values.

Table 1. The summary of the precipitation datasets used in this study.

Full Name and Details	Short Name	Data Sources	Spatiotemporal Resolution Used in This Study
Climate Hazards Group Infrared Precipitation with Station data Version2.0 (CHIRPS V2.0)	CHIRPS	CHPclim, thermal infrared data, the TMPA 3B42 V7, rain gauge stations data	Daily, 0.25°
NOAA's Climate Prediction Center (CPC) MORPHing technique (CMORPH) bias-corrected product (CMORPH CRT V1.0)	CMORPH	TMI, SSM/I, AMSR-E, AMSU-B, Meteosat, GOES, MTSAT and CPC	3 hourly, 0.25°
NASA's Tropical Rainfall Measuring Mission (TRMM) Multi-satellite Precipitation Analysis (TMPA) 3B42 Version 7	TMPA	TMI, TRMM Combined Instrument, SSM/I, SSMIS, AMSR-E, MHS, AMSU-B, GEO IR and GPCC	3 hourly, 0.25°
Multi-Source Weighted-Ensemble Precipitation Version 2.1(MSWEP)	MSWEP	CPC, GPCC, CMORPH, GSMaP-MVK, TMPA, ERA-Interim, JRA-55	Daily, 0.25°
daily precipitation interpolated from 57 rain gauges provided by Hydrology Bureau of the Huaihe Water Conservancy Commission	HWCC	Rain gauges	Daily

2.2.2. Streamflow and Soil Moisture

Daily streamflow records of the Wangjiaba Station were derived from hydrologic year books published by the Hydrologic Bureau of the Ministry of Water Resources of China.

The observed soil moisture data were downloaded from the Ministry of Water Resources of China. The soil moisture has been routinely measured at three different depths (10, 20, and 40 cm) three times per month, on the 1st, 11th, and 21st, since 2008 using the oven–drying method. A total of 19 soil moisture observation stations are located in the study area (Figure 1). To maintain consistency with the model simulated soil moisture, gravimetric soil moisture data at a depth of 0–10 cm and 0–40 cm were translated to volumetric soil moisture data using the soil bulk density of each 0.05° grid.

3. Methodology

3.1. VIC Hydrological Model and Calibration Methods

The VIC grid-based macroscale semi-distributed hydrological model was developed by Liang et al. [35]. The VIC model uses a spatial probability distribution function to represent the subgrid variability of the soil moisture storage capacity, which more realistically treats the hydrological processes within a model grid cell [33]. In this study, the VIC model was applied on each grid point for the calculation of the water balance using a daily time step. The observed daily maximum and minimum air temperature data from 2000 to 2010 required for the VIC model were derived from eight meteorological stations within and around the upper Huaihe River Basin provided by the China Meteorological Data Sharing Service System. For each grid, the global 1 km land cover classification dataset [36] and the global 10 km soil types dataset [37] were used to define the vegetation and soil parameters. The geographical parameters were calculated using meteorological variables and the location of the study region.

The hydrological parameters of the VIC model are closely related to the runoff yielding process and are difficult to directly determine, thus, it should be calibrated using observed hydrographs. These parameters include the exponent of variable infiltration capacity curve (B), the maximum velocity of the base flow (Dsmax), the fraction of Dsmax where non-linear base flow begins (Ds), the fraction of maximum soil moisture when non-linear base flow occurs (Ws), and the depth of the second and third soil layer (d2 and d3, the thickness of the first soil moisture layer being fixed at 0.1 m). Based on the method that integrated the Rosenbrock algorithm [38] and manual intervention, the hydrological parameters were optimized and determined. The manual intervention is to determine the initial values and a reasonable range of the parameters according to the physical meaning of the parameters and the characteristics of the basin. Then the hydrological parameters were automatically calibrated using the Rosenbrock algorithm by maximizing the Nash-Sutcliffe coefficient of Efficiency (NSCE) and simultaneously minimizing the relative error (RE) between observed and simulated streamflow.

The hydrological simulation was conducted under two parameterization scenarios to evaluate the hydrological application of the four QPEs. In Scenario I, the VIC model calibration and validation were forced by the HWCC to obtain optimal hydrological parameters. The four QPEs were then used as forcing data to simulate the streamflow and soil moisture. Scenario I was mainly used to directly compare the influence of different precipitation products on the accuracy of the hydrological simulation. In Scenario II, all these hydrological parameters were recalibrated separately using the precipitation from CMORPH, CHIRPS, TMPA, and MSWEP. Subsequently, the hydrological simulation was conducted using both parameters recalibrated under scenario II and the corresponding precipitation estimate. Scenario II can be used to determine if the evaluated precipitation products have the potential to be alternative data sources for hydrological simulations in data–poor or ungauged basins [14,19].

Considering that the study area was dominated by the low-flow years after 2008, 2005–2010 and 2002–2004 were selected as the calibration and validation periods in both scenarios, respectively, to balance the high-flow years and low-flow years and prevent overfitting during high-flow years.

To eliminate the influence of the initial state on the simulation results, the VIC model was prerun for two additional years before the calibration and validation period to initiate the model state in both scenarios.

3.2. Evaluation Metrics

Several verification indices were used to quantitatively assess the error characteristics of the QPE, including the relative error (RE), root mean square error (RMSE), and correlation coefficient (CC). The mean square error (square of RMSE, MSE) is composed of the systematic error (MSE$_{sys}$) and random error (MSE$_{ran}$) [4]; thus, two error components were assessed for each QPE in this study. In addition, the frequency bias (FBI), the probability of detection (POD), the false alarm ratio (FAR), and the equitable threat score (ETS) were calculated to quantitatively evaluate the accuracy of the four QPEs under different precipitation thresholds.

Specifically, the NSCE and exceedance probabilities (P_m) were added to the RE to evaluate the simulated discharge. The simulated soil moisture was assessed based on several metrics including the RE, CC, and unbiased root mean square error (ubRMSE). The ubRMSE removes the bias from the RMSE and can reasonably reflect the random error of the simulated soil moisture [39]. The equations to calculate the above-mentioned indices are listed in Table 2.

Table 2. The list of statistical evaluation indices to evaluate the quantitative precipitation estimates (QPEs) and their hydrological applicability.

Evaluation Indexes	Formulas	Comments	Perfect Value
RE	$RE = \dfrac{\sum_{i=1}^{n}(S_i - O_i)}{\sum_{i=1}^{n} O_i} \times 100\%$		0
RMSE	$RMSE = \sqrt{\dfrac{\sum_{i=1}^{n}(S_i - O_i)^2}{n}}$		0
CC	$CC = \dfrac{\sum_{i=1}^{n}(S_i - \bar{S})(O_i - \bar{O})}{\sqrt{\sum_{i=1}^{n}(S_i - \bar{S})^2 \sum_{i=1}^{n}(O_i - \bar{O})^2}}$	S_i and O_i are the evaluated and observed values; $\bar{S}$ and $\bar{O}$ are the mean values of S_i and O_i; n is the number of samples.	1
NSCE	$NSCE = 1 - \sum_{i=1}^{n}(S_i - Q_i)^2 / \sum_{i=1}^{n}(O_i - \bar{O})^2$		1
ubRMSE	$ubRMSE = \sqrt{\dfrac{1}{n}\sum_{i=1}^{n}\left((S_i - \bar{S}) - (O_i - \bar{O})\right)^2}$		0
MSE$_{sys}$	$MSE_{sys} = \dfrac{\sum_{i=1}^{n}(S_i^{*} - O_i)^2}{n}$	S_i^{*} is defined using a linear regression error model (S_i^{*} = $a \times O_i + b$) with a as the slope and b as the intercept.	0
MSE$_{ran}$	$MSE_{ran} = \dfrac{\sum_{i=1}^{n}(S_i - S_i^{*})^2}{n}$		0
P_m	$P_m = \dfrac{m}{n+1}$	P_m is the frequency of discharge of no less than O_m; m is the number of discharge not less than O_m.	
FBI	$FBI = \dfrac{N_A + N_B}{N_A + N_C}$	N_A, number of observed and detected rainfall events; N_B, number of detected but not observed rainfall events; N_C, number of observed but not detected rainfall events; N_D, number of rainfall events that were not observed and not detected. $N_{Aref} = \dfrac{(N_A + N_B)(N_A + N_C)}{(N_A + N_B + N_C + N_D)}$	1
POD	$POD = \dfrac{N_A}{N_A + N_C}$		1
FAR	$FAR = \dfrac{N_B}{N_A + N_B}$		0
ETS	$ETS = \dfrac{(N_A - N_{Aref})}{(N_A + N_B + N_C - N_{Aref})}$		1

4. Results

4.1. Evaluation and Comparison of Different Precipitation Products

4.1.1. Spatiotemporal Characteristics

Figure 2 presents the spatial distribution of the average annual precipitation of the HWCC and four QPEs during 2000 and 2010. All four QPEs generally show a consistent spatial pattern with the precipitation decreasing from south to north, which agrees with the HWCC.

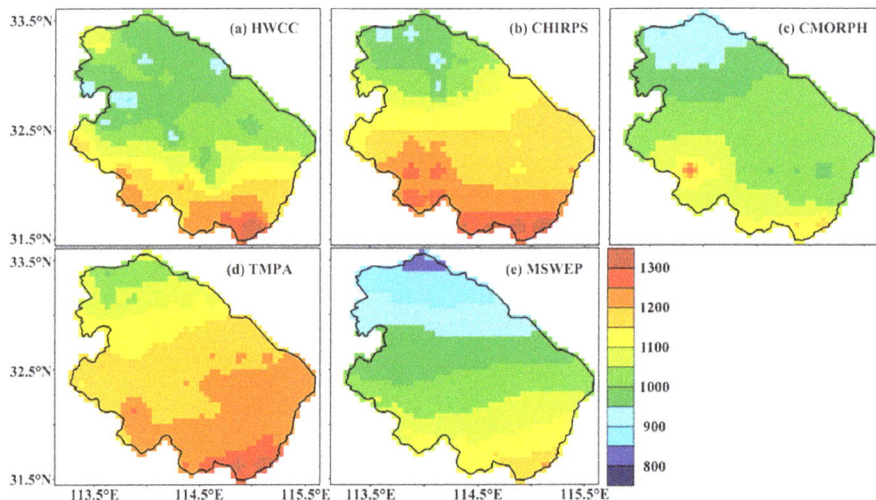

Figure 2. The spatial distribution of the average annual precipitation (mm/year) of daily precipitation interpolated from 57 rain gauges (HWCC), Climate Hazards Group Infrared Precipitation with Station data Version2.0 (CHIRPS), NOAA's Climate Prediction Center (CPC) MORPHing technique bias-corrected product (CMORPH), NASA's Tropical Rainfall Measuring Mission (TRMM) Multi-satellite Precipitation Analysis 3B42 Version 7 (TMPA), and Multi-Source Weighted-Ensemble Precipitation Version 2.1 (MSWEP) from 2000 to 2010.

The spatial distribution and seasonal variability of the RE, RMSE, and CC of the four QPEs are listed in Figure 3 and Table 3. Based on the RE for the whole study period (Figure 3a1–a4), CHIRPS and TMPA generally overestimate the precipitation with a grid–averaged RE of 7.1% and 10.4%, respectively. However, the RE of the CHIRPS and TMPA in the northwestern and southern mountain areas (less than 6% or underestimate) are significantly lower than the RE in the central parts of the basin. The similar error features of CHIRPS and TMPA may be attributed to the use of the TMPA for the generation of CHIRPS [8,40]. Conversely, CMORPH and MSWEP are dominated by an overall negative RE, especially in the northwestern and southern parts of the basin; the grid–averaged RE of CMORPH and MSWEP are −3.3% and −5.4%, respectively. The significant difference in the RE between TMPA and CMORPH is consistent with the results of Sun et al. [19], which might partly be due to the difference in the retrieval algorithms. With respect to seasonal variations (Table 3), negative REs of CMORPH and MSWEP are significant in spring and winter and the TMPA shows a considerable overestimation in winter, which can be attributed to the low amount of precipitation in these two seasons and the weak detection to the snow and light rainfall events of the satellite sensors [13,14]. The CHIRPS shows a consistent overestimation, especially in summer with more intense precipitation than in other seasons, which is similar to the results of Duan et al. [41].

The differences of the RMSE (Figure 3b1–b4) and CC (Figure 3c1–c4) for different QPEs are significant; however, the spatial distributions of CC and RMSE for a specific precipitation product are homogeneous. Throughout the whole period, the MSWEP is superior to the other three QPEs, with the lowest RMSE ($\leq$6 mm/day in nine out of ten of the study areas) and the highest CC (almost all above 0.8), which may be due to the use of the reanalysis datasets in MSWEP, providing a unique advantage over other QPEs [42]. The performances of CMORPH and TMPA are very similar, with the RMSE ranging from 5.0 to 8.0 mm/day and CC varying from 0.7 to 0.8 in most parts of the basin, which may be due to the similar data sources used in these two QPEs [9,10]. The CHIRPS exhibits the worst performance, with an RMSE above 7.0 mm/day and CC mainly between 0.6 and 0.7. The poor performance of CHIRPS may be attributed to cloud–top infrared (IR) observations [41]. As shown in Table 3, the RMSE and CC values generally demonstrate seasonal variations, except for the CC of MSWEP. All four QPEs have the smallest and largest RMSE values in spring and autumn. However, the QPEs perform poorly in spring in terms of CC, except for MSWEP. Additionally, the MSWEP is superior to the other three QPEs based on the CC and RMSE values in every season. In contrast, CHIRPS provides the worst estimates based on the RMSE and CC.

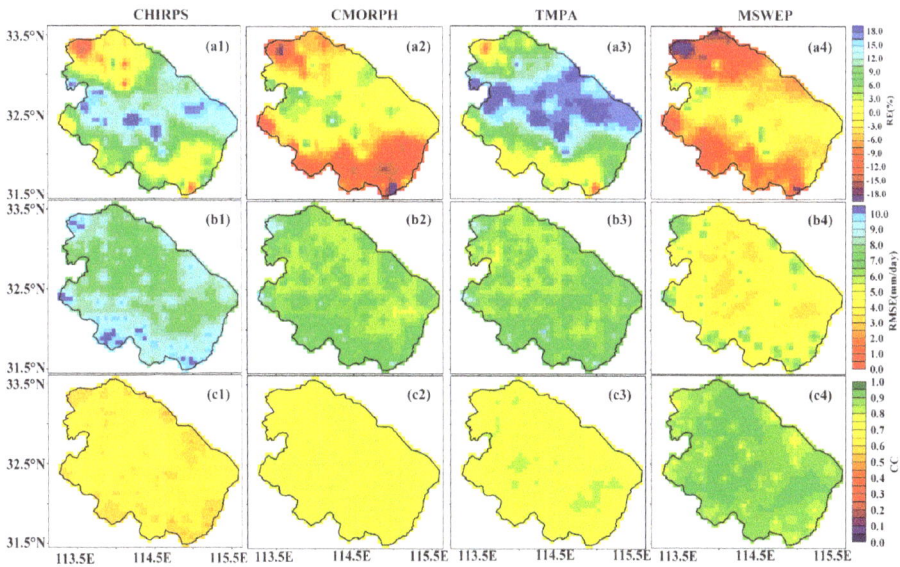

Figure 3. The spatial patterns of the relative error (RE) (%), root mean squared error (RMSE) (mm/day), and correlation coefficient (CC) of the four quantitative precipitation estimates (QPEs) and HWCC (from top to bottom) from 2000 to 2010.

Table 3. The grid-averaged relative error (RE) (%),root mean squared error (RMSE) (mm/day), and correlation coefficient (CC) of the four QPEs and daily precipitation interpolated from 57 rain gauges (HWCC) for the whole period, spring, summer, autumn, and winter.

	Criteria	CHIRPS	CMORPH	TMPA	MSWEP
	RE (%)	7.1	−3.3	10.4	−5.4
Whole period	RMSE (mm/day)	8.3	6.5	6.6	4.5
	CC	0.63	0.76	0.77	0.88
	RE (%)	1.6	−35.7	−0.2	−24.6
Spring (MAM)	RMSE (mm/day)	3.5	3.0	3.4	1.3
	CC	0.45	0.50	0.51	0.93
	RE (%)	15.2	2.8	12.1	−4.8
Summer (JJA)	RMSE (mm/day)	7.4	4.7	4.7	3.0
	CC	0.54	0.80	0.80	0.91
	RE (%)	6.0	1.7	10.7	1.2
Autumn (SON)	RMSE (mm/day)	12.9	10.8	10.7	7.9
	CC	0.66	0.75	0.77	0.87
	RE (%)	5.4	−7.5	15.6	−16.5
Winter (DJF)	RMSE (mm/day)	6.4	4.5	4.9	2.7
	CC	0.58	0.68	0.73	0.89

The contributions of systematic and random error components to the mean square error are presented in Figure 4. The spatial distribution of the systematic (random) error has a similar pattern for all QPEs. The random error component clearly dominates the four QPEs, which is due to the adjustment of all precipitation products by rain gauge observations. Mountainous regions exhibit larger systematic errors than plain areas, which can be attributed to the higher uncertainty of precipitation estimates in mountainous regions [4]. This shows that the existing methods of bias correction are not as effective at higher altitudes as at lower altitude areas and the integration of different precipitation products does not seem to be a solution to this problem. Therefore, the effect of the terrain should be further considered during bias correction or multi–source precipitation fusion.

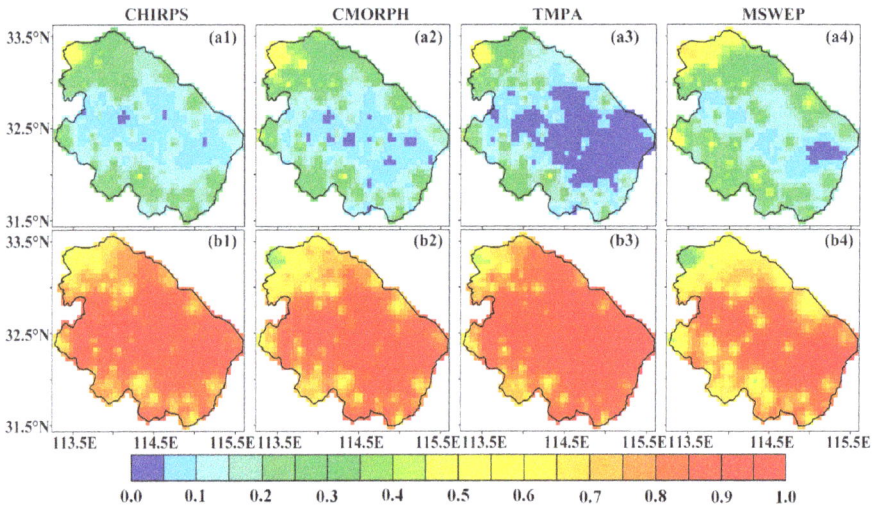

Figure 4. The contribution of systematic error (**a1–a4**) and random error (**b1–b4**) components to the mean square error of the four QPEs from 2000 to 2010.

All four QPEs tend to underestimate the precipitation in the northwestern and southern mountainous areas, which is probably caused by the higher temporal variability of precipitation in the mountainous regions. However, limited by the revisiting period of the satellite, satellite-based precipitation estimates are generally incapable of reflecting temporal variability well in mountainous precipitation. The standard deviation (SD) of grid-based daily precipitation can be used as a measure of temporal variability of daily precipitation, and the spatial distribution of the SD of the daily precipitation reflects the spatial heterogeneity of regional precipitation. The SD of each VIC grid was calculated based on the HWCC. Subsequently, the relationships between the REs of the four QPEs and SD were analyzed. As presented in Figure 5, there are remarkable negative correlations between the REs of the QPEs and SD (CC = −0.55, −0.62, −0.77, and −0.66 for CHIRPS, CMORPH, TMPA, and MSWEP, respectively), which indicate that the QPEs tend to underestimate the precipitation in regions with higher temporal variability of the daily precipitation and the spatial heterogeneity of daily precipitation will affect the spatial distribution of the RE of the QPEs.

4.1.2. Results for Different Rainfall Intensity

The accuracy of the four QPEs under different precipitation thresholds was evaluated based on four statistical scores (Figure 6). The results of the FBI are related to the RE of the QPEs. The analyzed precipitation with positive (negative) RE generally overestimates (underestimates) the occurrence of rain events across most thresholds. All four QPEs underestimate the occurrence of rain events when the daily precipitation is above 85.0 mm, especially the MSWEP. The overall precipitation accuracy of the four QPEs (based on POD, FAR, and ETS) declines with increasing precipitation threshold, indicating that the QPEs are less capable of depicting intense precipitation. As a whole, the MSWEP exhibits a more stable discrimination skill across all precipitation thresholds with the lowest FAR and highest POD and ETS. The TMPA shows slightly better skills than the CMORPH at most thresholds and CHIRPS performs the worst.

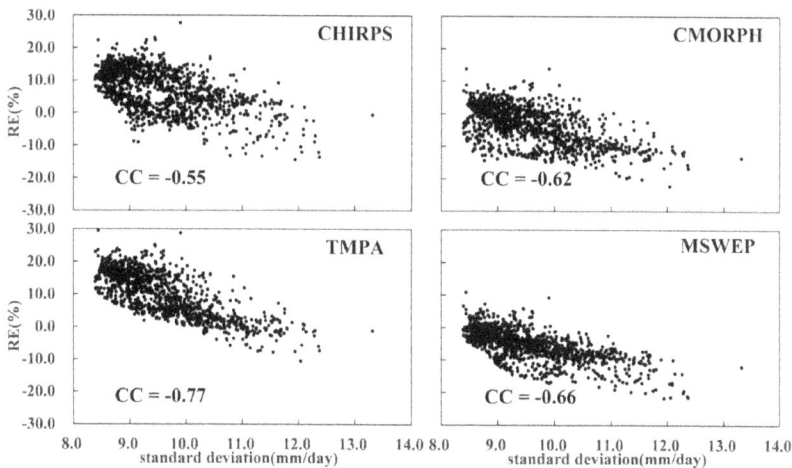

Figure 5. The scatter plots of the standard deviation (SD) versus RE for the four QPEs.

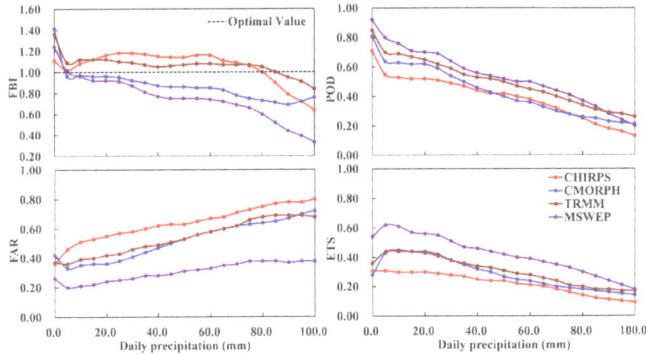

Figure 6. The four statistical scores of different QPEs versus HWCC at different precipitation thresholds from 2000 to 2010.

4.2. Evaluation of the Streamflow Simulation under Two Parameterization Scenarios

In this section, we evaluate and inter-compare the hydrological applicability of the four QPEs by performing a streamflow simulation using the VIC model under two parameterization scenarios.

4.2.1. Scenario I: Model Parameters Calibrated by Rain Gauge Observations

Figure 7 shows the observed and simulated streamflow driven by the HWCC for the calibration and validation period. Generally, the streamflow simulated by HWCC is in good agreement with the observed streamflow with high NSCE of 0.93 and almost no RE for the calibration period, and still maintains a high NSCE of 0.90 and small RE of 4.7% for the validation period. The simulated results indicate that the adaptability of the VIC model and model parameters calibrated by the HWCC are reasonable. Thus, it is suitable to evaluate the applicability of the four QPEs.

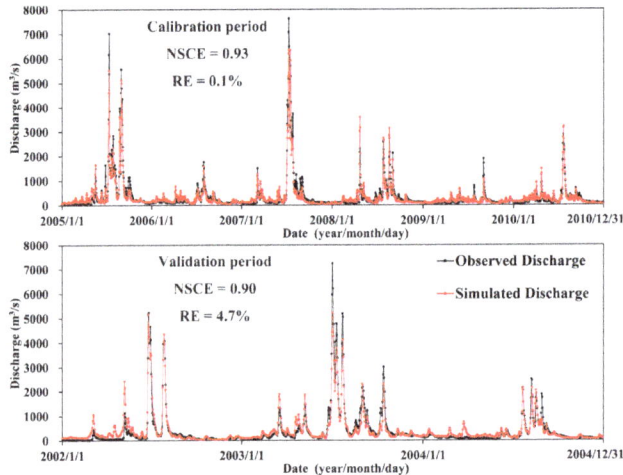

Figure 7. The comparison between the daily discharge simulated using HWCC precipitation and the observed data at the Wangjiaba Station.

The VIC model was forced by four QPEs using the model parameters calibrated by the HWCC to simulate the streamflow from 2002 to 2010. As shown in Table 4, the CMORPH and MSWEP forced simulations underestimate the streamflow, while the CHIRPS and TMPA forced simulations overestimate the streamflow, which is related to the RE of the corresponding forcing data and indicates that the biases of the QPEs can directly propagate to streamflow simulations [7]. The MSWEP performs best in the streamflow simulation with the largest NSCE (0.87 and 0.74, respectively). The TMPA takes the second place in terms of NSCE (0.66 and 0.69, respectively), as well as with highest RE (30.8% and 19.1%, respectively). The CHIRPS generally performs poorly throughout the whole period. The simulated results of CMORPH for the calibration and validation period significantly differ, which indicates that the stability of CMORPH is poor.

Table 4. The RE and Nash-Sutcliffe coefficient of Efficiency (NSCE) of the hydrological simulation results of the four QPEs under two scenarios.

Precipitation Products	Scenario I				Scenario II			
	Calibration Period		Validation Period		Calibration Period		Validation Period	
	RE (%)	NSCE	RE (%)	NSCE	RE (%)	NSCE	RE (%)	NSCE
HWCC	0.1	0.93	4.7	0.90				
CHIRPS	23.2	0.60	5.8	0.56	19.0	0.72	−0.1	0.55
CMORPH	0.4	0.69	−21.3	0.53	9.2	0.77	−10.5	0.52
TMPA	30.8	0.66	19.1	0.69	18.1	0.79	1.1	0.73
MSWEP	−12.0	0.87	−19.4	0.74	−4.1	0.89	−11.6	0.78

4.2.2. Scenario II: Model Parameters Separately Recalibrated Using the QPEs

The simulation results under Scenario II (Table 4 and Figure 8) improved compared with the simulations under Scenario I, which is consistent with former studies [5,19]. For the calibration period, MSWEP presents the best performance with the highest NSCE (0.89) and lowest RE (−4.1%). However, the CHIRPS, CMORPH, and TMPA forced simulations tend to overestimate streamflow, especially for CHIRPS with the highest RE (19.0%) and lowest NSCE (0.72). TMPA performs slightly better than CMORPH in terms of the NSCE, but TMPA also significantly overestimates streamflow (RE = 18.1%). For the validation period, all the QPE forced simulations tend to underestimate the streamflow, except for TMPA (RE = 1.1%), and NSCEs decreases with a different extent compared with the calibration period, especially as the NSCE of CMORPH declined to 0.52 from 0.77. However, the MSWEP forced simulations are still better than the other QPEs, with the highest NSCE (0.78), an obvious but acceptable RE of −11.6%, followed by TMPA (NSCE = 0.73). The good agreement between the observed streamflow and MSWEP forced simulations reveals the strong streamflow simulation capability of the MSWEP product.

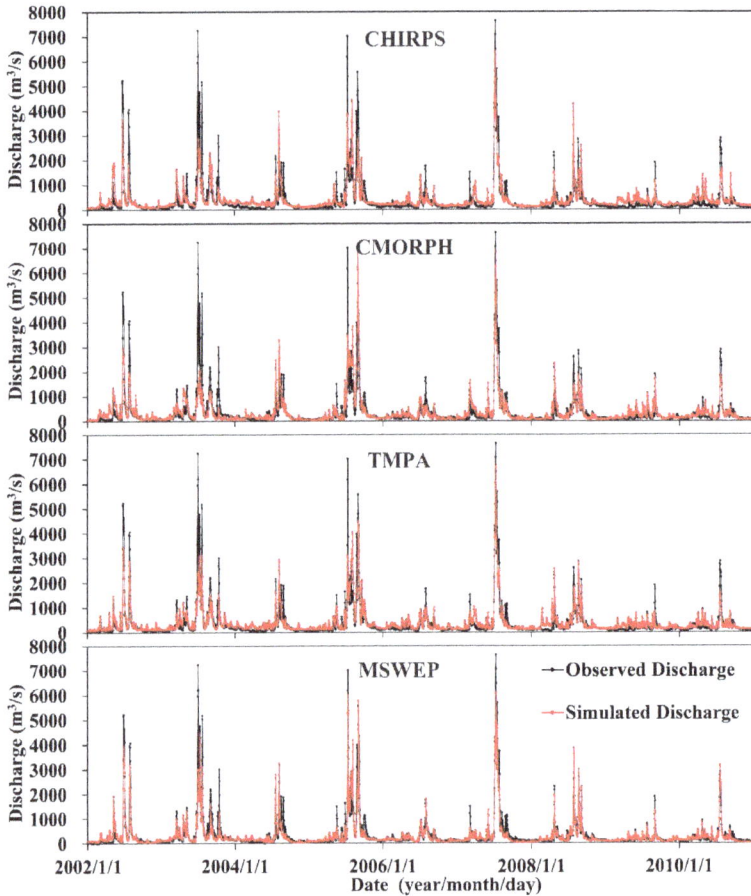

Figure 8. The comparison between the observed and simulated discharge of CHIRPS, CMORPH, TMAP, and MSWEP at the Wangjiaba Station under Scenario II.

The simulated streamflow during the flood season (from June to September) under two scenarios was also analyzed to further investigate the performance of the four QPEs (Table 5). Except for TMPA (RE = 10.0%), all simulations tended to underestimate the streamflow in the flood season under Scenario I, especially CMORPH and MSWEP (RE = −17.3% and −21.2%, respectively), which is largely due to the significant negative RE of CMORPH and MSWEP. The simulation results under Scenario II improved compared with the simulations under Scenario I, which is consistent with the results for the whole period, reflected by an increasing NSCE and more acceptable RE. Generally, MSWEP presents the best performance among the four QPEs during the flood season, with a desirable NSCE of 0.81 and 0.84 under the two scenarios, respectively. TMPA takes the second place with an acceptable NSCE of 0.68 and 0.76, respectively. CHIRPS and CMORPH generally perform poorly, with NSCEs not exceeding 0.65.

The HWCC forced simulation (Figure 7) and simulated streamflow are driven by the four QPEs under Scenario II (Figure 8) show an underestimation at the high streamflow and an overestimation at the low streamflow. The exceedance probability plots (Figure 9) are presented to further validate the performance of the four QPEs products at different exceedance probabilities. When the exceedance frequencies are less than 5.0%, all simulations underestimate the streamflow. Generally, HWCC yields the least underestimation, followed by MSWEP in this frequency interval. When the exceedance frequencies increase and exceed 5.0%, the streamflow simulated by CHIRPS, CMORPH, and TMPA gradually approaches and then exceeds the observed streamflow. The higher accuracy is concentrated in a small frequency range from 5.0 to 10.0%. HWCC forced simulation gradually exceeds the observed streamflow when the exceedance frequency is about 12.0% and presents its best performance with an exceedance frequency range from 10.0 to 17.5%. MSWEP performs best when the exceedance frequency is beyond 20%, but the streamflow simulated by MSWEP consistently underestimates the observed streamflow until the exceedance frequency is above 25%.

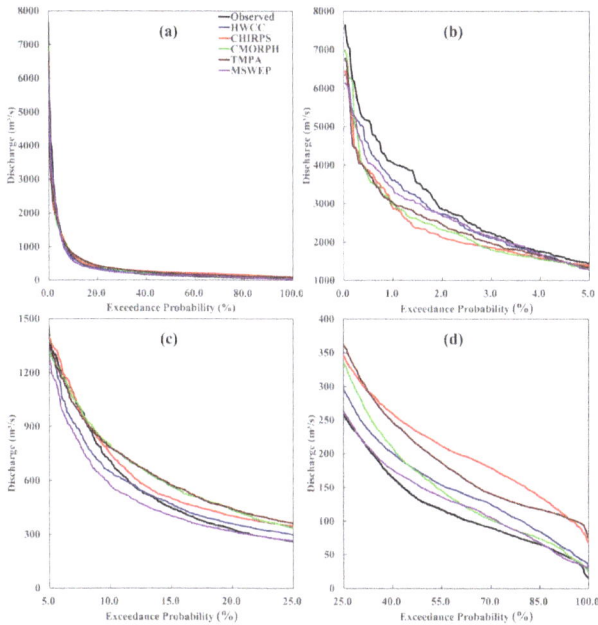

Figure 9. The exceedance probabilities of the daily streamflow during the whole period under Scenario II: (**a**) all frequencies; (**b**) frequencies below 5.0%; (**c**) frequencies ranging from 5.0 to 25.0%; (**d**) frequencies exceeding 25.0%.

Table 5. Similar to Table 4, but for the flood season during 2002–2010.

Precipitation Products	Scenario I		Scenario II	
	RE (%)	NSCE	RE (%)	NSCE
HWCC	−14.2	0.94		
CHIRPS	−3.3	0.60	−10.1	0.65
CMORPH	−17.3	0.58	−8.1	0.65
TMPA	10.0	0.68	−8.4	0.76
MSWEP	−21.2	0.81	−13.4	0.84

4.3. Evaluation of the Soil Moisture Simulations under Two Parameterization Scenarios

In situ soil moisture data from January 2008 to December 2010 were adopted to further evaluate the hydrological utility of different precipitation products. The box plots of the evaluation metrics for soil moisture at depths of 0–10 cm and 0–40 cm under two scenarios are presented in Figure 10.

The HWCC and QPEs forced simulations significantly underestimate the soil moisture of the two soil depths under Scenario I. In addition, there is almost no difference between the REs of simulated soil moisture with different forcing precipitation, which indicates that the REs of the simulated soil moisture has little to do with forcing precipitation. Overall, the soil moisture simulated by HWCC with the least ubRMSE (the average ubRMSE is 0.056 and 0.043 m³/m³ at depths of 0–10 cm and 0–40 cm, respectively) and MSWEP is subordinate with a slightly higher ubRMSE of 0.057 and 0.045 m³/m³, respectively. CMORPH and CHIRPS are the worst performers, especially at a depth of 0–10 cm. Compared with the results for the depth of 0–10 cm, the ubRMSE of soil moisture simulated at a depth of 0–40 cm is significantly reduced. Furthermore, the difference in the ubRMSE of simulated soil moisture at a depth of 0–40 cm with different forcing precipitation products is relatively small. The CC of simulated and observed soil moisture at a depth of 0–40 cm is slightly better than that at 0–10 cm, except for MSWEP. HWCC performs best with the highest CC (the average CC is 0.53 and 0.55 at depths of 0–10 cm and 0–40 cm, respectively), followed by MSWEP (both are 0.50). Similarly, CHIRPS and CMORPH exhibit worse performances with lower CC compared with MSWEP and TMPA. With respect to the ubRMSE and CC, improving the precision of precipitation can effectively improve the precision of simulated soil moisture, especially at a depth of 0–10 cm, which indicates that the results of the simulated soil moisture at a depth of 0–10 cm are more sensitive to the accuracy of the forcing precipitation than that at a depth of 0–40 cm.

Unlike the results of simulated streamflow, the accuracy of simulated soil moisture under Scenario II seems to not notably improve compared with the simulations under Scenario I. Specifically, both the ubRMSE and CC of simulated soil moisture under Scenario II are slightly inferior to the results of Scenario I at a depth of 0–10 cm. Regarding the simulated results at a depth of 0–40 cm, both the ubRMSE and CC of simulated soil moisture under Scenario II are slightly lower than the results of Scenario I, except for the CC forced by MSWEP and CMORPH. Generally, MSWEP shows the best performance among the four QPEs in terms of simulated soil moisture under Scenario II; TMPA takes the second place, CHIRPS and CMORPH generally perform poorly, which is consistent with the results under Scenario I.

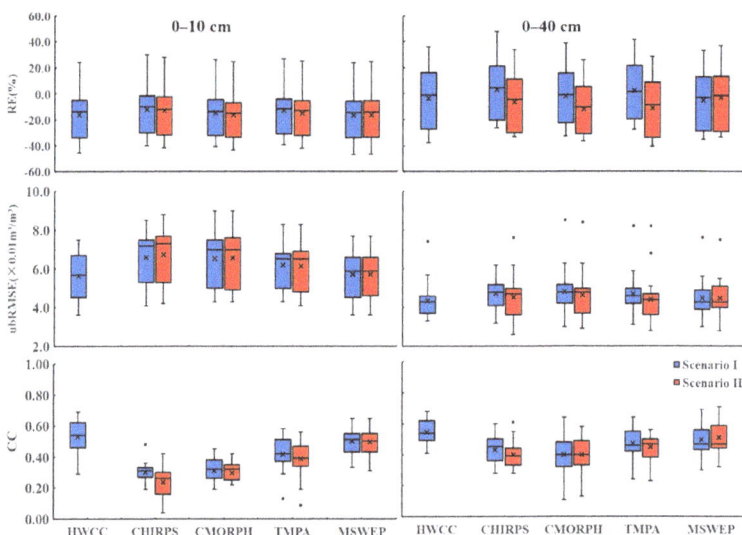

Figure 10. The box plots of the evaluation metrics for the soil moisture at depths of 0–10 cm and 0–40 cm under two scenarios.

5. Discussion

The IR data measured from geostationary satellites and microwave (MV) data measured from low earth orbiting satellites are the main data sources for SPPs. Generally, SPPs are from IR data with fine temporal resolutions but with less accuracy. The MV data provide direct and accurate precipitation estimates, however, at the cost of coarse temporal resolution [41,43]. Generally, CHIRPS performs unsatisfactorily compared with TMPA and CMORPH, which is probably due to the fact that CHIRPS is mainly based on IR data, while CMORPH and TMPA combined IR and MV data. The main difference between CMORPH and TMPA is the gauge adjustment algorithm adopted in these two datasets. To develop CMORPH, the probability density function (PDF) matching against the CPC unified daily gauge analysis was used to adjust the biases. Monthly GPCC rain gauge analyses and inverse-error-variance weighting were used for the TMPA to adjust the biases. However, TMPA generally outperforms CMORPH, suggesting that the PDF matching adopted in CMOPRH is not superior to the monthly gauge adjustment algorithm used in TMPA [19], probably because PDF matching does not work well with undetected precipitation. The limitation of PDF matching mentioned above may also be the reason CMORPH performs worst (with the largest FBI and FAR and lowest ETS) in terms of the statistical scores for no rain events, which further leads to CMORPH being dominated by a significant negative RE.

MSWEP has the best performance compared with the other three QPEs, except for the significant negative RE. Tong et al. [32] pointed out that SPPs are most accurate during summer and at lower latitudes, while the atmospheric model reanalysis datasets perform better than the SPPs during winter and at higher latitudes. The optimal merging of the gauge, satellite, and reanalysis precipitation estimates complements the advantages of the three different data sources. Moreover, the finer spatial resolution of MSWEP compared with the other three QPEs provides a unique advantage over the other QPEs. The merged precipitation products MSWEP not only performs better than the other three QPEs but also, to some extent, makes up for the lack of accuracy of the single precipitation product varying depending on the seasons (in terms of CC showed in Table 3). However, MSWEP still has larger uncertainties in mountain areas, the same as SPPs.

The streamflow modeling results under Scenario I reveal that the RE of simulated streamflow mainly depends on RE of precipitation estimates. Due to the highly nonlinear rainfall–runoff response, any overestimation/underestimation of precipitation estimates can be transformed into a larger overestimation/underestimation in the simulated streamflow [7,32,44]. Specifically, the grid-averaged RE of precipitation estimates from CHIRPS and TMPA is 7.5% and 10.6%, respectively, resulting in a 14.4% and 21.0% overestimation of the streamflow simulations from 2002 to 2010. Similarly, the grid–averaged RE of precipitation estimates from CMORPH and MSWEP is −2.5% and −6.0%, respectively, resulting in an 8.2% and 17.2% underestimation of streamflow simulations from 2002 to 2010. However, the impact of the RE of precipitation estimates on the simulated streamflow is diminished under Scenario II (the RE of simulated streamflow forced by CHIRPS, CMORPH, TMPA, and MSWEP is 10.7%, 1.9%, 10.6% and −7.4%, respectively), mainly because the model parameters may change according to the precipitation input to match the streamflow. Thus, the model parameters recalibrated under Scenario II are more suitable for the corresponding precipitation products for streamflow simulation. Among the hydrological parameters of the VIC model, the shape of the variable infiltration capacity curve (B) and the depth of the three soil layers (d1, d2, and d3) are the most influential factors on streamflow and soil moisture simulation. An increase of B tends to enhance streamflow production. Soil thickness mainly controls the maximum moisture storage capacity. The thicker soil depths have higher moisture storage capacities, thus, more evaporation loss and less streamflow production. Three base flow parameters (Dsmax, Ds, and Ws) determine how quickly the water storage in the third layer evacuates and are generally less sensitive than the parameters B and the thicker soil depths [45]. The MSWEP precipitation estimate decreases d2 after recalibration compared with the model parameters under Scenario I (Table 6) to retain less soil moisture and enhance streamflow production. The model parameters recalibrated by MSWEP are mostly close to the results of HWCC, which further indicates that MSWEP has satisfactory accuracy. The recalibrated results of CHIRPS have a smaller B and thicker soil thickness with the potential purpose to offset the positive RE of precipitation estimates. The CMORPH and TMPA precipitation estimate increases B and soil thickness tends to yield more streamflow and retain more soil moisture. However, the soil thickness recalibrated by TMPA increases significantly, which means that more precipitation is stored in the soil, further reducing the RE of the simulated streamflow. Although the increase of B recalibrated by CMORPH can compensate for the negative RE of precipitation estimates, the decrease of the seasonal peak discharge caused by the increase of the soil thickness may be one of the main reasons for the worst performance of CMORPH during the validation period under Scenario II.

Table 6. The optimal model parameters of the Variable Infiltration Capacity (VIC) model under two scenarios.

Parameters	Unit	Scenario I	Scenario II			
			CHIRPS	CMORPH	TMPA	MSWEP
B		0.1	0.06	0.18	0.14	0.1
Ds		0.998	0.092	0.068	0.035	0.530
Dsmax	mm	0.5	10.0	9.0	13.0	0.5
Ws		0.99	0.99	0.70	0.99	0.72
d1	m	0.1	0.1	0.1	0.1	0.1
d2	m	0.50	0.60	0.58	1.2	0.42
d3	m	0.11	0.37	0.15	0.81	0.11

The model structure, forcing data, and parameters are the main factors affecting the accuracy of soil moisture simulation [46,47], which has been proved by the results of this study. There are little differences between the RE of the simulated soil moisture with different forcing precipitation products, which indicates that the existing systematic errors of the VIC model are probably the main causes of the RE of simulated soil moisture. However, the dynamic change tendency of the simulated soil moisture driven by precipitation with higher precision is more consistent with the observed results (with lower ubRMSE and higher CC). This demonstrates that rainfall forcing data are one of the important factors influencing the accuracy of simulated soil moisture. In this study, the model parameters are calibrated in terms of the highest NSCE between the simulated and observed streamflow at the watershed outlet, without considering the simulation results for other hydrological variables (such as soil moisture). Therefore, the parameters determined by this method may not reasonably reflect the actual conditions of the study area. Comparatively speaking, the parameters calibrated using a relatively dense network of rain gauge stations (Scenario I) can be considered as the best possible approximation of watershed hydrological features [44]. In Scenario II, these parameters were recalibrated with individual QPEs, which takes into account the potential impacts of input uncertainty on the streamflow simulations [19]. Thus, with the recalibrated model parameters, the streamflow simulations are likely to produce more accurate results than the case under Scenario I. However, the recalibrated model parameters may not reflect watershed hydrological features well, which results in the improvements of the simulated soil moisture under Scenario II being insignificant or even worse than the results under Scenario I, indicating that the model parameters may be the major factors influencing the accuracy of the simulated soil moisture.

The evaluation of the hydrological simulation capability in this study partly indicates that the merged precipitation product, MSWEP, has great potential to be a reliable dataset for conducting long-term hydrological studies compared with the three SPPs. However, the simulated results using HWCC are closer to observations than that using MSWEP, which may be due to two reasons: (1) the layout of rain gauge stations used in this study takes into consideration the factors of rainstorm and flood formation, so, it can reflect the precipitation characteristics of the study area well and it is suitable for hydrological simulation; and (2) the satellite-based precipitation estimates are inevitably subject to errors resulting from indirect measurements although rain gauge information is incorporated to reduce biases, but due to the insufficient gauge observations in the bias–corrected or merging process, the uncertainty in the satellite-based precipitation estimate has not been fully solved.

6. Conclusions

This study provides a comprehensive assessment of the newly merged precipitation product MSWEP and three typical SPPs based on rain gauge observations and the VIC hydrologic model in the upper Huaihe River basin. The primary conclusions can be summarized as follows:

(1) All four QPEs are subject to significant errors, although the spatial patterns of average annual precipitation are generally consistent with the rain gauge observations. Specifically, CHIRPS and TMPA generally overestimate precipitation; however, CMORPH and MSWEP generally underestimate precipitation. Overall, MSWEP performs best, TMPA takes the second place, and CHIRPS exhibits a relatively poor performance. The contribution of the random error generally larger than the systematic error of the four QPEs and the spatial distributions of the two error components are closely related to the topography. Simultaneously, all QPEs are prone to underestimate the precipitation in regions daily precipitation with larger temporal variability (higher standard deviation) and the precipitation detection capabilities of QPEs decrease with increasing precipitation magnitude.

(2) Under Scenario I (model parameters calibrated by rain gauge observations), similar to the results of the statistical evaluation, CHIRPS and TMPA forced simulations tend to overestimate streamflow, while serious underestimations are observed in the streamflow simulations of both CMORPH and MSWEP, which indicates that the RE of simulated streamflow is greatly affected by the RE of the forcing precipitation products. MSWEP outperforms the other three QPEs (NSCE lower than 0.70) with the highest NSCE (NSCE = 0.87 and 0.74 for the calibration and validation period, respectively), implying that the MSWEP products could better reveal the precipitation spatial pattern.

(3) Under scenario II (model parameters separately recalibrated using the QPEs), all simulations are improved compared with Scenario I. Similarly, MSWEP exhibits the best performance, followed by TMPA. CHIRPS and CMORPH forced simulation performed poorly. The influence of the RE of the QPEs on simulated streamflow may be mitigated by recalibrating the parameters with the corresponding QPEs. Furthermore, all QPE forced simulations underestimate the observed streamflow with exceedance probabilities less than 5.0% whilst they overestimate the observed streamflow with exceedance probabilities of more than 30.0%. HWCC performs best with exceedance probabilities that are less than 5.0%, followed by MSWEP, while MSWEP performs best with exceedance probabilities of more than 20.0%.

(4) Under the two scenarios, the difference of RE of soil moisture simulated by different precipitation products is insignificant, implying that the existing systematic error of the VIC model is probably the main cause of the RE of simulated soil moisture. However, with increasing accuracy of the forcing precipitation products, the dynamic changes of the simulated soil moisture become more consistent with the measured results (lower ubRMSE and higher CC), especially for the results at a depth of 0–10 cm. Generally, the soil moisture simulated by different precipitation products at a depth of 0–40 cm is slightly better than the results for the depth of 0–10 cm. Unlike the simulated discharge, the accuracy of the simulated soil moisture under Scenario II is even worse than the results under Scenario I, which indicates that the model parameters are the main factors influencing the accuracy of the simulated soil moisture, in addition to the forcing data.

In summary, the merged precipitation product MSWEP presents a satisfactory performance both in the statistical evaluation and hydrological simulation in the study area, indicating the large potential to substitute rain gauge observations. However, there are still some limitations in the MSWEP product, for example, the MSWEP product has great uncertainty in mountainous regions the same as SPPs.

Future studies are needed to further contrastive evaluation of the applicability of MSWEP and relevant SPPs over different climate zones, and then improve the data fusion scheme according to the error characteristics of MSWEP, to obtain more accurate precipitation estimates.

Author Contributions: All authors contributed extensively to the work presented in this paper. Z.W. designed the framework of this study and wrote the paper. Z.X. conducted the analysis and wrote the paper. F.W. contributed to the writing of the paper. H.H. provided constructive comments and revised the paper. J.Z. collected observation data and revised the paper. X.W. and Z.L. revised the paper.

Funding: This research was funded by the National Key Research and Development Project (No. 2017YFC1502403), the National Science Foundation of China (Grant Nos. 51779071, 51579065), the Fundamental Research Funds for the Central Universities (Grant No. 2017B10514), Central Public Welfare Operations for Basic Scientific Research of Research Institute (Grant No. HKY-JBYW-2017-12) and Water Science and Technology Project of Jiangsu Province (2017007).

Conflicts of Interest: The authors declare no conflict of interest.

References

1. Beck, H.E.; Vergopolan, N.; Pan, M.; Levizzani, V.; van Dijk, A.I.J.M.; Weedon, G.P.; Brocca, L.; Pappenberger, F.; Huffman, G.J.; Wood, E.F. Global-scale evaluation of 22 precipitation datasets using gauge observations and hydrological modeling. *Hydrol. Earth Syst. Sci.* **2017**, *21*, 6201–6217. [CrossRef]
2. Liu, X.; Yang, T.; Hsu, K.; Liu, C.; Sorooshian, S. Evaluating the streamflow simulation capability of persiann-cdr daily rainfall products in two river basins on the tibetan plateau. *Hydrol. Earth Syst. Sci.* **2017**, *21*, 169–181. [CrossRef]
3. Sorooshian, S.; AghaKouchak, A.; Arkin, P.; Eylander, J.; Foufoula-Georgiou, E.; Harmon, R.; Hendrickx, J.M.H.; Imam, B.; Kuligowski, R.; Skahill, B.; et al. Advanced concepts on remote sensing of precipitation at multiple scales. *Bull. Am. Meteorol. Soc.* **2011**, *92*, 1353–1357. [CrossRef]
4. Prakash, S.; Mitra, A.K.; AghaKouchak, A.; Pai, D.S. Error characterization of trmm multisatellite precipitation analysis (tmpa-3b42) products over india for different seasons. *J. Hydrol.* **2015**, *529*, 1302–1312. [CrossRef]
5. Xue, X.; Hong, Y.; Limaye, A.S.; Gourley, J.J.; Huffman, G.J.; Khan, S.I.; Dorji, C.; Chen, S. Statistical and hydrological evaluation of trmm-based multi-satellite precipitation analysis over the wangchu basin of bhutan: Are the latest satellite precipitation products 3b42v7 ready for use in ungauged basins? *J. Hydrol.* **2013**, *499*, 91–99. [CrossRef]
6. Duethmann, D.; Zimmer, J.; Gafurov, A.; Guentner, A.; Kriegel, D.; Merz, B.; Vorogushyn, S. Evaluation of areal precipitation estimates based on downscaled reanalysis and station data by hydrological modelling. *Hydrol. Earth Syst. Sci.* **2013**, *17*, 2415–2434. [CrossRef]
7. Zhu, Q.; Xuan, W.; Liu, L.; Xu, Y.-P. Evaluation and hydrological application of precipitation estimates derived from persiann-cdr, trmm 3b42v7, and ncep-cfsr over humid regions in china. *Hydrol. Process.* **2016**, *30*, 3061–3083. [CrossRef]
8. Funk, C.; Peterson, P.; Landsfeld, M.; Pedreros, D.; Verdin, J.; Shukla, S.; Husak, G.; Rowland, J.; Harrison, L.; Hoell, A.; et al. The climate hazards infrared precipitation with stations—A new environmental record for monitoring extremes. *Sci. Data* **2015**, *2*, 150066. [CrossRef] [PubMed]
9. Joyce, R.J.; Janowiak, J.E.; Arkin, P.A.; Xie, P.P. Cmorph: A method that produces global precipitation estimates from passive microwave and infrared data at high spatial and temporal resolution. *J. Hydrometeorol.* **2004**, *5*, 487–503. [CrossRef]
10. Huffman, G.J.; Adler, R.F.; Bolvin, D.T.; Gu, G.; Nelkin, E.J.; Bowman, K.P.; Hong, Y.; Stocker, E.F.; Wolff, D.B. The trmm multisatellite precipitation analysis (tmpa): Quasi-global, multiyear, combined-sensor precipitation estimates at fine scales. *J. Hydrometeorol.* **2007**, *8*, 38–55. [CrossRef]
11. Ushio, T.; Sasashige, K.; Kubota, T.; Shige, S.; Okamoto, K.; Aonashi, K.; Inoue, T.; Takahashi, N.; Iguchi, T.; Kachi, M.; et al. A kalman filter approach to the global satellite mapping of precipitation (gsmap) from combined passive microwave and infrared radiometric data. *J. Meteorol. Soc. Jpn.* **2009**, *87A*, 137–151. [CrossRef]

12. Huffman, G.J.; Bolvin, D.T.; Braithwaite, D.; Hsu, K.; Joyce, R.; Kidd, C.; Sorooshian, S.; Xie, P.; Yoo, S.H. Developing the integrated multi-satellite retrievals for gpm (imerg). In Proceedings of the EGU General Assembly Conference, Vienna, Austria, 22–27 April 2012.

13. Beria, H.; Nanda, T.; Bisht, D.S.; Chatterjee, C. Does the gpm mission improve the systematic error component in satellite rainfall estimates over trmm? An evaluation at a pan-india scale. *Hydrol. Earth Syst. Sci.* **2017**, *21*, 6117–6134. [CrossRef]

14. Wang, Z.; Zhong, R.; Lai, C.; Chen, J. Evaluation of the gpm imerg satellite-based precipitation products and the hydrological utility. *Atmosp. Res.* **2017**, *196*, 151–163. [CrossRef]

15. Prakash, S.; Mitra, A.K.; Aghakouchak, A.; Liu, Z.; Norouzi, H.; Pai, D.S. A preliminary assessment of gpm-based multi-satellite precipitation estimates over a monsoon dominated region. *J. Hydrol.* **2016**, *556*, 865–876. [CrossRef]

16. Zhao, H.; Yang, B.; Yang, S.; Huang, Y.; Dong, G.; Bai, J.; Wang, Z. Systematical estimation of gpm-based global satellite mapping of precipitation products over china. *Atmosp. Res.* **2018**, *201*, 206–217. [CrossRef]

17. Zubieta, R.; Getirana, A.; Espinoza, J.C.; Lavado, W. Impacts of satellite-based precipitation datasets on rainfall-runoff modeling of the western amazon basin of peru and ecuador. *J. Hydrol.* **2015**, *528*, 599–612. [CrossRef]

18. Zulkafli, Z.; Buytaert, W.; Onof, C.; Manz, B.; Tarnavsky, E.; Lavado, W.; Guyot, J.-L. A comparative performance analysis of trmm 3b42 (tmpa) versions 6 and 7 for hydrological applications over andean-amazon river basins. *J. Hydrometeorol.* **2014**, *15*, 581–592. [CrossRef]

19. Sun, R.; Yuan, H.; Liu, X.; Jiang, X. Evaluation of the latest satellite-gauge precipitation products and their hydrologic applications over the huaihe river basin. *J. Hydrol.* **2016**, *536*, 302–319. [CrossRef]

20. Gao, Z.; Long, D.; Tang, G.; Zeng, C.; Huang, J.; Hong, Y. Assessing the potential of satellite-based precipitation estimates for flood frequency analysis in ungauged or poorly gauged tributaries of China's yangtze river basin. *J. Hydrol.* **2017**, *550*, 478–496. [CrossRef]

21. Sahoo, A.K.; Sheffield, J.; Pan, M.; Wood, E.F. Evaluation of the tropical rainfall measuring mission multi-satellite precipitation analysis (tmpa) for assessment of large-scale meteorological drought. *Remote Sens. Environ.* **2015**, *159*, 181–193. [CrossRef]

22. Bayissa, Y.; Tadesse, T.; Demisse, G.; Shiferaw, A. Evaluation of satellite-based rainfall estimates and application to monitor meteorological drought for the upper blue nile basin, ethiopia. *Remote Sens.* **2017**, *9*, 669. [CrossRef]

23. Tote, C.; Patricio, D.; Boogaard, H.; van der Wijngaart, R.; Tarnavsky, E.; Funk, C. Evaluation of satellite rainfall estimates for drought and flood monitoring in mozambique. *Remote Sens.* **2015**, *7*, 1758–1776. [CrossRef]

24. Yang, N.; Zhang, K.; Hong, Y.; Zhao, Q.; Huang, Q.; Xu, Y.; Xue, X.; Chen, S. Evaluation of the trmm multisatellite precipitation analysis and its applicability in supporting reservoir operation and water resources management in hanjiang basin, China. *J. Hydrol.* **2017**, *549*, 313–325. [CrossRef]

25. Maggioni, V.; Meyers, P.C.; Robinson, M.D. A review of merged high-resolution satellite precipitation product accuracy during the tropical rainfall measuring mission (trmm) era. *J. Hydrometeorol.* **2016**, *17*, 1101–1117. [CrossRef]

26. Zhao, H.; Yang, S.; You, S.; Huang, Y.; Wang, Q.; Zhou, Q. Comprehensive evaluation of two successive v3 and v4 imerg final run precipitation products over mainland China. *Remote Sens.* **2018**, *10*, 34. [CrossRef]

27. He, Y.; Zhang, Y.; Kuligowski, R.; Cifelli, R.; Kitzmiller, D. Incorporating satellite precipitation estimates into a radar-gauge multi-sensor precipitation estimation algorithm. *Remote Sens.* **2018**, *10*, 106. [CrossRef]

28. Beck, H.E.; van Dijk, A.I.J.M.; Levizzani, V.; Schellekens, J.; Miralles, D.G.; Martens, B.; de Roo, A. Mswep: 3-hourly 0.25 degrees global gridded precipitation (1979–2015) by merging gauge, satellite, and reanalysis data. *Hydrol. Earth Syst. Sci.* **2017**, *21*, 589–615. [CrossRef]

29. Boudevillain, B.; Delrieu, G.; Wijbrans, A.; Confoland, A. A high-resolution rainfall re-analysis based on radar-raingauge merging in the cevennes-vivarais region, France. *J. Hydrol.* **2016**, *541*, 14–23. [CrossRef]

30. Ciabatta, L.; Brocca, L.; Massari, C.; Moramarco, T.; Gabellani, S.; Puca, S.; Wagner, W. Rainfall-runoff modelling by using sm2rain-derived and state-of-the-art satellite rainfall products over italy. *Int. J. Appl. Earth Obs. Geoinform.* **2016**, *48*, 163–173. [CrossRef]

31. Massari, C.; Camici, S.; Ciabatta, L.; Brocca, L. Exploiting satellite-based surface soil moisture for flood forecasting in the mediterranean area: State update versus rainfall correction. *Remote Sens.* **2018**, *10*, 292. [CrossRef]

32. Tong, K.; Su, F.; Yang, D.; Hao, Z. Evaluation of satellite precipitation retrievals and their potential utilities in hydrologic modeling over the tibetan plateau. *J. Hydrol.* **2014**, *519*, 423–437. [CrossRef]

33. Wu, Z.Y.; Lu, G.H.; Wen, L.; Lin, C.A. Reconstructing and analyzing china's fifty-nine year (1951–2009) drought history using hydrological model simulation. *Hydrol. Earth Syst. Sci.* **2011**, *15*, 2881–2894. [CrossRef]

34. Wu, J.; Lu, G.; Wu, Z. Flood forecasts based on multi-model ensemble precipitation forecasting using a coupled atmospheric-hydrological modeling system. *Nat. Hazards* **2014**, *74*, 325–340. [CrossRef]

35. Liang, X.; Wood, E.F.; Lettenmaier, D.P. Surface soil moisture parameterization of the vic-2l model: Evaluation and modification. *Glob. Planet. Chang.* **1996**, *13*, 195–206. [CrossRef]

36. Hansen, M.C.; Defries, R.S.; Townshend, J.R.G.; Sohlberg, R. Global land cover classification at 1km spatial resolution using a classification tree approach. *Int. J. Remote Sens.* **2000**, *21*, 1331–1364. [CrossRef]

37. Reynolds, C.A.; Jackson, T.J.; Rawls, W.J. Estimating soil water-holding capacities by linking the food and agriculture organization soil map of the world with global pedon databases and continuous pedotransfer functions. *Water Resour. Res.* **2000**, *36*, 3653–3662. [CrossRef]

38. Rosenbrock, H.H. An automatic method for finding the greatest or least value of a function. *Comput. J.* **1960**, *3*, 175–184. [CrossRef]

39. Sun, Y.; Huang, S.; Ma, J.; Li, J.; Li, X.; Wang, H.; Chen, S.; Zang, W. Preliminary evaluation of the smap radiometer soil moisture product over china using in situ data. *Remote Sens.* **2017**, *9*, 292. [CrossRef]

40. Bai, L.; Shi, C.; Li, L.; Yang, Y.; Wu, J. Accuracy of chirps satellite-rainfall products over mainland China. *Remote Sens.* **2018**, *10*, 362. [CrossRef]

41. Duan, Z.; Liu, J.; Tuo, Y.; Chiogna, G.; Disse, M. Evaluation of eight high spatial resolution gridded precipitation products in adige basin (Italy) at multiple temporal and spatial scales. *Sci. Total Environ.* **2016**, *573*, 1536–1553. [CrossRef] [PubMed]

42. Alijanian, M.; Rakhshandehroo, G.R.; Mishra, A.K.; Dehghani, M. Evaluation of satellite rainfall climatology using cmorph, persiann-cdr, persiann, trmm, mswep over Iran. *Int. J. Climatol.* **2017**, *37*, 4896–4914. [CrossRef]

43. Serrat-Capdevila, A.; Merino, M.; Valdes, J.B.; Durcik, M. Evaluation of the performance of three satellite precipitation products over Africa. *Remote Sens.* **2016**, *8*, 836. [CrossRef]

44. Yuan, F.; Zhang, L.; Win, K.W.W.; Ren, L.; Zhao, C.; Zhu, Y.; Jiang, S.; Liu, Y. Assessment of gpm and trmm multi-satellite precipitation products in streamflow simulations in a data-sparse mountainous watershed in Myanmar. *Remote Sens.* **2017**, *9*, 302. [CrossRef]

45. Zhang, L.; Su, F.; Yang, D.; Hao, Z.; Tong, K. Discharge regime and simulation for the upstream of major rivers over tibetan plateau. *J. Geophys. Res. Atmosp* **2013**, *118*, 8500–8518. [CrossRef]

46. Mo, K.C.; Lettenmaier, D.P. Objective drought classification using multiple land surface models. *J. Hydrometeorol.* **2014**, *15*, 990–1010. [CrossRef]

47. Tramblay, Y.; Bouvier, C.; Ayral, P.A.; Marchandise, A. Impact of rainfall spatial distribution on rainfall-runoff modelling efficiency and initial soil moisture conditions estimation. *Nat. Hazards Earth Syst. Sci.* **2011**, *11*, 157–170. [CrossRef]

remote sensing

MDPI

Article

Impact of Radiance Data Assimilation on the Prediction of Heavy Rainfall in RMAPS: A Case Study

Yanhui Xie [1], Jiancheng Shi [2,3], Shuiyong Fan [1,*], Min Chen [1], Youjun Dou [1] and Dabin Ji [2]

[1] Institute of Urban Meteorology, China Meteorological Administration, Beijing 100089, China; lyxieyanhui@163.com (Y.X.); mchen@ium.cn (M.C.); yjdou@ium.cn (Y.D.)

[2] State Key Laboratory of Remote Sensing Science, Institute of Remote Sensing and Digital Earth, Chinese Academy of Sciences, Beijing 100101, China; shijc@radi.ac.cn (J.S.); jidb@radi.ac.cn (D.J.)

[3] Joint Center for Global Change Studies (JCGCS), Beijing 100875, China

* Correspondence: syfan@ium.cn; Tel.: +86-10-6840-0742

Received: 20 July 2018; Accepted: 23 August 2018; Published: 30 August 2018

Abstract: Herein, a case study on the impact of assimilating satellite radiance observation data into the rapid-refresh multi-scale analysis and prediction system (RMAPS) is presented. This case study targeted the 48 h period from 19–20 July 2016, which was characterized by the passage of a low pressure system that produced heavy rainfall over North China. Two experiments were performed and 24 h forecasts were produced every 3 h. The results indicated that the forecast prior to the satellite radiance data assimilation could not accurately predict heavy rainfall events over Beijing and the surrounding area. The assimilation of satellite radiance data from the advanced microwave sounding unit-A (AMSU-A) and microwave humidity sounding (MHS) improved the skills of the quantitative precipitation forecast to a certain extent. In comparison with the control experiment that only assimilated conventional observations, the experiment with the integrated satellite radiance data improved the rainfall forecast accuracy for 6 h accumulated precipitation after about 6 h, especially for rainfall amounts that were greater than 25 mm. The average rainfall score was improved by 14.2% for the 25 mm threshold and by 35.8% for 50 mm of rainfall. The results also indicated a positive impact of assimilating satellite radiances, which was primarily reflected by the improved performance of quantitative precipitation forecasting and higher spatial correlation in the forecast range of 6–12 h. Satellite radiance observations provided certain valuable information that was related to the temperature profile, which increased the scope of the prediction of heavy rainfall and led to an improvement in the rainfall scoring in the RMAPS. The inclusion of satellite radiance observations was found to have a small but beneficial impact on the prediction of heavy rainfall events as it relates to our case study conditions. These findings suggest that the assimilation of satellite radiance data in the RMAPS can provide an overall improvement in heavy rainfall forecasting.

Keywords: heavy rainfall prediction; satellite radiance; data assimilation; RMAPS

1. Introduction

Heavy rainfall events represent a significant area of concern among the scientific community due to their dramatic social, economic, and ecological impacts. Local heavy rainfall is generally a short-range mesoscale weather process, which makes it difficult to accurately predict the evolution and development of these mesoscale weather systems that lead to heavy rainfall events. Although over the last several decades significant improvements have been made in short-range forecasting, the accurate prediction and quantification of heavy rainfall events still remains a challenge [1]. Local heavy rainfall forecasting is greatly dependent on the accuracy of the initial conditions that are considered in numerical weather prediction (NWP) models [2]. With the development of remote sensing technologies, satellite observations have played a significant role in the improvement of numerical forecasting techniques by providing a more accurate estimation of the initial conditions. Thus, the assimilation of satellite observations into the operational NWP system has emerged as an important method for improving quantitative precipitation forecasting.

There are two basic approaches for assimilating information that has been obtained from satellites into a data assimilation system [3]. During the initial phase of development, the primary method of satellite data assimilation is to incorporate conventional retrieval from satellite measurements into the system. This method is one possible and effective way, especially for variables that are related to precipitation, such as cloud-top pressure, rain rate, and total column water vapor. Assimilating these retrieved data provides valuable insight into cloud-affected observations in the system, which can improve the hydrometeor process and cloud properties in the analysis and forecast atmosphere fields [4]. However, certain additional prior information and restrictive assumptions are generally required in the conversion process [5]. They can also bring some retrieval problems, which have been discussed previously [4–8]. The second approach is to assimilate the satellite radiances directly into the data assimilation system. This approach requires a non-linear observation operator to transform model variables into radiances, where the observation operator is expressed mathematically by a forward radiative transfer equation that is used to calculate radiance from model-state vertical profiles. Currently, the assimilation of satellite radiances is largely performed under clear-sky conditions, especially for infrared sounder data. Direct assimilation of cloud-affected radiances is very difficult due to limited understanding of the vertical structure of cloud parameters and uncertainties that are associated with nonlinearity of the moist physics process [4,9]. Wang et al. compared the assimilations of radiance and three-layered precipitable water from the Advanced Himawari Imager (AHI) [10], with results indicating a similar or comparable overall impact on precipitation forecasting. Theoretically, satellite radiance assimilation is superior to the retrieval assimilation [3,7] because the errors of the radiance observations are much simpler and more justified in direct radiance assimilation than in retrieval assimilation.

Radiances from various satellite instruments have been assimilated directly in most NWP models using variational data assimilation schemes [5,11–14]. These satellite radiances have shown a significant improvement in the forecasting accuracy of numerical models with better representations of mesoscale features [15,16]. A number of studies have demonstrated the value of satellite radiance observations in precipitation forecasting. Xu et al. assessed the impact of radiance observations from the Advanced Television and Infrared Observation Satellite Operational Vertical Sounder (ATOVS) on precipitation forecasts in Southwest Asia, with results revealing a reduction in the precipitation forecast errors for most locations within the 24 h hindcasts [17]. Zou and Qin reported that the assimilation of the imager channel radiances from Geostationary Operational Environmental Satellites-11 and -12 (GOES-11 and GOES-12) resulted in a significant positive impact on coastal quantitative precipitation forecasting (QPF) that was performed near the northern region of the Gulf of Mexico [18,19]. Zou also studied the impact of assimilating satellite microwave humidity sounding data with a newly added cloud detection algorithm for a coastal precipitation event [20], wherein the precipitation threat scores were increased by more than 50% after 3–6 h of model forecasting for 3 h rainfall thresholds exceeding 1.0 mm. Singh et al. analyzed the impact of assimilating radiances that were obtained from the Indian National Satellite-3D (INSAT-3D) on short-range weather forecasts [21], where the results demonstrated the ability of temperature and water vapor sensitive radiances to improve not only the temperature and moisture fields, but also the wind fields. Sagita et al. examined the influence of the satellite radiance data assimilation on the prediction of two days of heavy rainfall in the Java region [22]. The results indicated that the satellite radiance data assimilation displayed a higher QPF accuracy than that of surface observation data assimilation. Wang et al. [23] investigated the impact of Himawari-8's AHI on severe rainstorm, which occurred over North China during 18–21 July 2016, and the assimilation of AHI radiances from water vapor channels was found to have a clear positive impact on rainfall forecasting accuracy for the first 6 h of lead time. Previous studies have reported the positive impact of satellite radiance data assimilation on the accuracy of NWP models, especially for quantitative precipitation forecasts.

This study aimed to examine the impact of the ingestion and assimilation of satellite radiance data along with conventional observations on the accuracy of heavy rainfall simulations in the rapid-refresh multi-scale analysis and prediction system (RMAPS). The rainfall event that was the subject of this study occurred over North China during 19–20 July 2016. It was one of the heaviest rainfalls in the past 60 years and caused major damage in and around Beijing [23]. The upper-level low pressure center slowly moved eastward, while the southwesterly flow intensified under the influence of peripheral flow around a subtropical high in the middle and lower levels, which produced persistent precipitation over North China. We focused on the influence of assimilating radiances data from the Advanced Microwave Sounding Unit-A (AMSU-A) and Microwave Humidity Sounding (MHS) on RMAPS heavy rainfall forecasting. The rest of this paper is organized as follows: Section 2 describes the RMAPS, satellite radiance observations, the observed data, and strategy for verification that was used in this study; Section 3 details the experimental design and the quality control of the satellite radiance data; the study results are presented and discussed in Section 4, and the conclusions are presented in Section 5.

2. Materials and Methods

2.1. Model Description

RMAPS is a rapid-refresh multi-scale analysis and prediction system that is based on version 3.8.1 of the Weather Research and Forecasting (WRF) model and WRF Data Assimilation (WRFDA) [12,24]. It was developed by the Institute of Urban Meteorology from the China Meteorological Administration in collaboration with the National Center for Atmospheric Research (NCAR) and has run operationally at the Beijing Meteorological Bureau since 2015. The RMAPS configuration includes Thompson double moment microphysics, ACM2 PBL, RRTMG short and long wave schemes, and 50 vertical computational layers. The land use data that was used in the RMAPS have been reprocessed, which has a higher accuracy and finer classification for urban areas. For the data assimilation system, considerable conventional and radar observations have been included. The background error covariance is domain-dependent, which has been generated based on forecasts over a period of one month in the summer using the NMC (National Meteorological Center) method [25].

Two nested domains are configured in the RMAPS, as shown in Figure 1. The 9-km horizontal resolution domain covers all of China, with 649 × 500 grid points. The black box represents a 3 km horizontal resolution domain that is primarily centered in North China, with 550 × 424 grid points. The 3-dimension variational (3DVar) technique was used for the analysis component in the RMAPS. Most conventional observation types are assimilated in both domains, including SYNOP (conventional grounded-based), METAR (airport ground-based), SHIP (ship observation), BUOY (oceanographic buoy), AMDAR (aircraft), RAOB (sounding), PILOT (pilot balloon system), and GPSZTD (ground-based GPS zenith total delay) observations. In addition to conventional data, radial velocity and reflectivity from radar observations were also assimilated in the 3 km horizontal resolution domain. Surface and atmospheric fields as initial conditions from the European Centre for Medium-Range Weather Forecasts were introduced once in a day in the RMAPS at 1800 UTC and 6 h spin-up run to 0000 UTC. From 0000 UTC, the RMAPS was able to produce 3 h updated analyses and 0–24 h forecasts serving the North China and Beijing area.

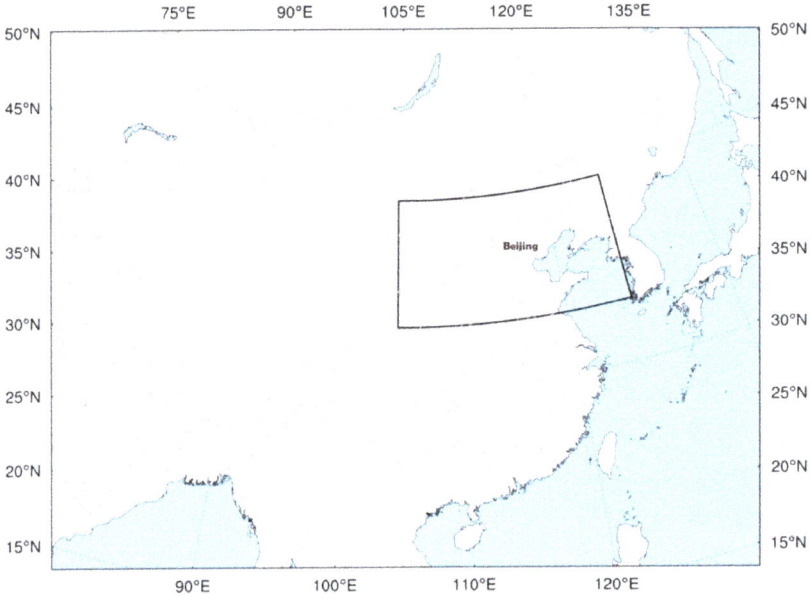

Figure 1. Two nested domains in the RMAPS (Rapid-Refresh Multi-Scale Analysis and Prediction system). The outside box is a 9 km horizontal resolution domain, the inside black box represents a 3-km horizontal resolution domain that is centered in Beijing.

2.2. Satellite Radiance Observations

The satellite radiance observations that were used in this study were obtained from AMSU-A and MHS instruments that were mounted on polar orbiting satellites NOAA18/19 and METOP-A/B. Both the instruments can provide passive measurements of the radiation that was emitted from the earth's surface and throughout the atmosphere. AMSU-A has an instantaneous field-of-view of 3.3° at half-power points with a nominal spatial resolution of 48 km at nadir. It has 15 measurement channels, including 4 channels (1, 2, 3, and 15) that measure in the "window" spectral regions, which receive energy primarily from the surface and the boundary layer. Remaining channels are "temperature sounding" channels, which can be used to derive atmospheric temperature profiles [26]. MHS has an instantaneous field-of-view of 1.1° at the half-power points, and the spatial resolution at nadir is nominally 16 km. It has a total of 5 channels, of which 2 channels (1 and 2) measure in the "window" spectral regions. Channels 3, 4, and 5 are "humidity sounding" channels which provide atmospheric humidity profiles.

Additionally, the satellite radiance data that were used in this study were taken from NCAR, which underwent substantial preprocessing by the National Environmental Satellite, Data, and Information Service before being made available to users, including statistical limb adjustment and surface emissivity corrections in the microwave channels.

2.3. Verification Data

Two types of precipitation data were used for verification in this study. First, ground observations from automatic weather stations (AWS) were used to provide a reliable surface precipitation amount with high temporal frequency using rain gauges [27–29]. To match the gridded forecasts to point observation locations, the forecast value at P was assigned the value at the nearest grid point for comparison. Another type of precipitation data was derived from the Climate Prediction Center

Morphing (CMORPH) products, which uses global precipitation estimates of satellite retrieved precipitation that is merged with hourly precipitation from ground AWS in China. The CMORPH precipitation is the gridded data with $0.1° \times 0.1°$ horizontal resolution. Several studies have evaluated the quality and accuracy of CMORPH precipitation [27,30,31], and it has been found that it can capture the precipitation process very well with a more reasonable precipitation amount and spatial distribution. In this study, the CMORPH data were interpolated to the forecast grid for comparison through simple inverse distance weighting.

Additionally, conventional vertical observations were also used for comparison and evaluation. In the vertical, if forecasts and observations were at the same vertical level, then they were paired as is. If any discrepancy existed between the vertical levels, then the forecasts were interpolated to the level of the observation in natural log of pressure coordinates.

2.4. Verification Strategy

In this study, several standard verification statistics were used for evaluating the heavy rainfall forecasts, including bias score (BIAS), critical success index (CSI, or Threat Score), equitable threat score (ETS, or Gilbert Skill Score), probability of detection (POD), and false alarm ration (FAR).

Consider a set of forecasts that can only have two alternatives of yes and no. The relationship between the forecasts and the observed events can be described through a four-cell contingency table (Table 1). a (hits) is the number of positive (yes) forecasts that correspond to an occurrence of the observed event; b (false alarms) is the number of positive forecasts that were not accompanied by an observed event; c (miss) is the number of negative (no) forecasts that had an occurrence of the observed event; and d is the number of negative forecasts that did not have any associated events. Then, BIAS indicates whether the forecast system had a tendency to under-forecast (BIAS < 1) or over-forecast (BIAS > 1) events. CSI can be considered as the accuracy when correct negatives have been removed from consideration. ETS measures the fraction of the observed and/or forecast events that were correctly predicted, adjusted for hits that were associated with random chance. POD is the percent of events that were correctly forecast. The FAR index explains the inaccuracy of the forecast to exclude non-event cases. These statistical criteria, as listed in Table 2, mainly investigate the consistency of the precipitation forecast based on observations [28,32,33].

Table 1. Four-cell contingency table.

		Observed Events	
		yes	no
Forecasts	yes	a	b
	no	c	d

Table 2. Statistical performance measures used in evaluation and comparison.

Index	Formula	Range	Perfect Value
Bias (BIAS)	$\frac{a+b}{a+c}$	$0\sim+\infty$	1
Critical Success Index (CSI)	$\frac{a}{a+b+c}$	$0\sim1$	1
Equitable Threat Score (ETS)	$\frac{a-a_{ref}}{a-a_{ref}+b+c}$	$-1/3\sim1$	1
Probability of Detection (POD)	$\frac{a}{a+c}$	$0\sim1$	1
False Alarm Ration (FAR)	$\frac{b}{a+b}$	$0\sim1$	0

a_{ref} : $(a+b)(a+c)/N$, where $N = a+b+c+d$.

In addition, the Pearson correlation ρ_p and mean error (ME) were also employed to quantitatively evaluate the agreement of the spatial distributions between the forecasts and observations [28,34], where X_i and Y_i represent the ith data points from forecasts and observations, respectively, and $\overline{X}$ and $\overline{Y}$ are the means of the X_i and Y_i fields, respectively.

$$\rho_p = \frac{\sum(X_i - \overline{X})(Y_i - \overline{Y})}{\sqrt{\sum(X_i - \overline{X})^2}\sqrt{\sum(Y_i - \overline{Y})^2}},\tag{1}$$

$$\mathrm{ME} = \frac{1}{N}\sum_{i=1}^{N}(X_i - Y_i),\tag{2}$$

3. Assimilation Experiments

3.1. Experimental Design

This work is based on the operational 3D-Var settings of the RMAPS. To evaluate the impact of assimilating satellite radiance observations for the severe rainfall event over North China during 19–20 July 2016, two parallel experiments were configured. The first experiment ("CTRL") assimilated conventional observations in domain 1 with a horizontal resolution of 9 km. Observations that were taken within ±1.5 h of each analysis time were assimilated every 3 h. All of the observations were assumed to be valid at the time of analysis for each experiment. In addition to the conventional data, radial velocity and reflectivity from radar observations were also assimilated in domain 2 with a horizontal resolution of 3 km. The second experiment ("DA_RAD") assimilated the same conventional observations as CTRL in domain 1 along with the satellite radiance data that were obtained from AMSU-A and MHS. The radiance data were assimilated in RMAPS through the community radiative transfer model (CRTM) that was incorporated within the WRFDA [24,35]. Satellite radiance observations within ±1.5 h of each analysis time were also taken for assimilation every 3 h in the system. For both experiments, two partially-cycling analysis/forecast runs were conducted at an interval of 24 h. Each partially-cycling run began at 1800 UTC, which was 6 h earlier than the synoptic times (0000 UTC next day), and then 3 h analysis updates and 24 h cycling forecasts were obtained using the previous cycle's characteristics as the background.

3.2. Data Quality Control

One of the most important aspects of this type of data assimilation is the monitoring of the observation quality in which the data are being assimilated. Various errors that are associated with satellite radiance data need to be treated correctly, such as cloud and precipitation simulation errors, systematic bias, and random errors of observations arising from the errors in numerical models and instruments. In this study, a series of quality control procedures were performed based on those that were available for assimilation systems that use satellite radiance data, including observation error statistics, limb adjustment, cloud detection, data thinning (with 120 km), and variational bias correction.

Figure 2 shows the brightness temperatures of channel 6 in AMSU-A and channel 3 in MHS from the satellite NOAA18 at the analysis time of 0000 UTC 20 July 2016. The coverage of all of the observations before quality control is shown in Figure 2a,b, which includes dense data. After all of the quality control procedures were performed, the number of retained observations was 751 for AMSU-A and 697 for MHS, as shown in Figure 2c,d, respectively. The proportions of data that were used which were really ingested in the data assimilation system was greatly reduced.

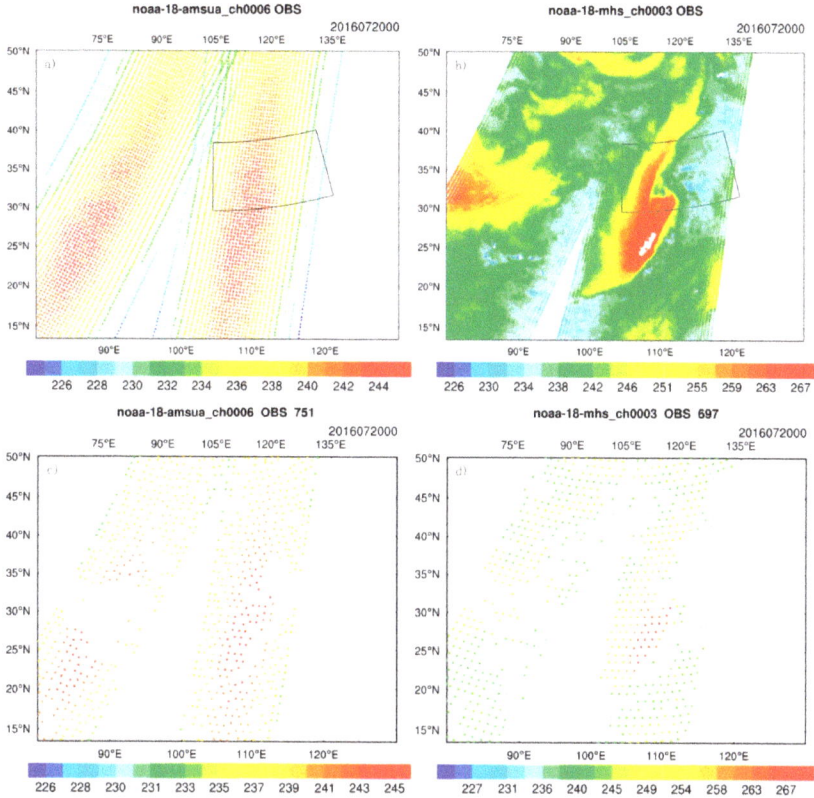

Figure 2. The brightness temperatures (K) of channel 6 in AMSU-A (Advanced Microwave Sounding Unit-A) and channel 3 in MHS (Microwave Humidity Sounding) from the satellite NOAA18 at the analysis time of 0000 UTC 20 July 2016. (**a,c**) show the coverages of channel 6 in AMSU-A before and after quality control, and (**b,d**) show the coverages of channel 3 in MHS before and after quality control.

4. Results and Discussion

In this section, the results of assimilation experiments using RMAPS are described for the heavy rainfall that occurred over North China during 19–20 July 2016. The description of the figures and the discussions of the results are included in this section.

4.1. Results

Radiance bias correction is crucial for properly assimilating satellite radiance data. Figure 3 shows the scatter plots of the observed (OBS) versus CRTM-calculated brightness temperatures for channel 6 of AMSU-A and channel 3 of MHS from NOAA18 at the analysis time 0000 UTC 20 July 2016. It can be seen that the root-mean-square error (RMSE) of background (BAK) with bias correction (BC) decreased from 1.312 to 0.291 (by about 78%) in comparison with the bias before correction (no BC) for channel 6 of AMSU-A. The RMSE of analysis (ANA) with radiance data assimilation was further reduced to 0.225 and the standard deviation (STDV) was also reduced from 0.290 to 0.225. The BAK RMSE with BC for channel 3 of MHS slightly increased. In comparison with the statistics of BAK after BC, the RMSE and STDV of ANA were reduced by approximately 55.6% and 53.5%, respectively. The correlation coefficients between the simulated brightness temperatures and the observations were both more than 0.995 for BAK and ANA of AMSU-A, which were higher than the corresponding values of MHS.

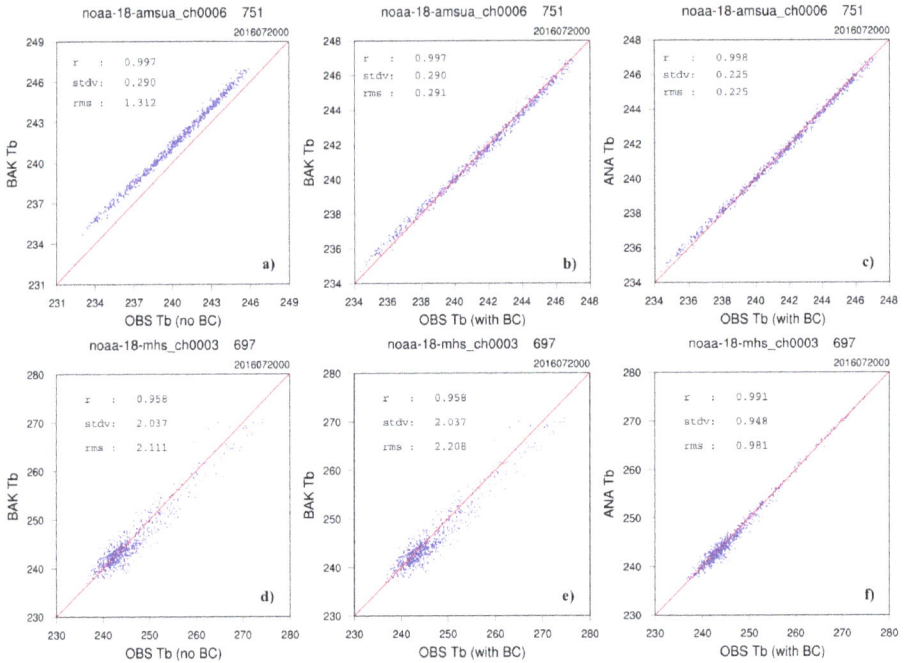

Figure 3. The scatter plots of the observed (OBS) versus CRTM-calculated brightness temperatures (K) for channel 6 of AMSU-A (the first line) and channel 3 of MHS (the second line) from NOAA18 at the analysis time 0000 UTC 20 July 2016. (**a,d**): BAK (background) versus OBS without bias correction (no BC), (**b,e**): BAK versus OBS with bias correction (with BC), and (**c,f**): ANA (analysis) versus OBS with bias correction (with BC).

Figure 4 shows the vertical profiles of average bias (BIAS) and root-mean-square error (RMSE) at the analysis time 0000 UTC against conventional observations for temperature (TMP), specific humidity (SPFH), geopotential height (HGT), zonal (UGRD), and meridional (VGRD) wind. For temperature, the BIAS of the DA_RAD was smaller than that of the CTRL between 300 hPa and 600 hPa, however it increased in the upper and lower levels. For humidity, the BIAS and RMSE of DA_RAD were not reduced in comparison with the CTRL and even increased in the middle levels. For geopotential height, DA_RAD obtained a smaller BIAS in the upper and lower levels. For wind fields, the assimilation of satellite radiances had a little impact, with only a slight decrease found for BIAS at 700 hPa and at 300 hPa for RMSE. The assimilation of radiances reduced the bias of temperature in the middle levels, however it increased the bias and RMSE of the moisture fields. Radiance assimilation also showed an indirect positive impact on the geopotential height fields through multivariate correlations, however it showed little impact on wind fields.

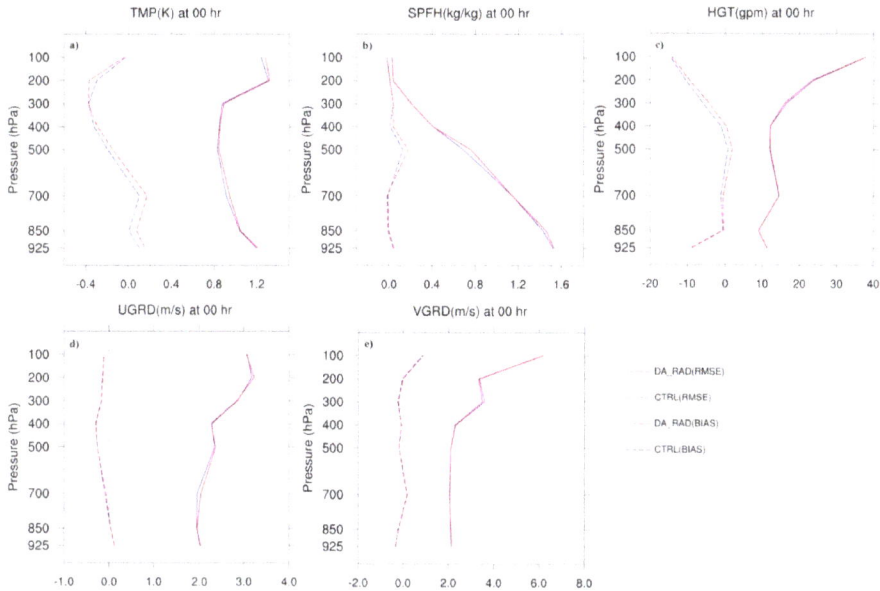

Figure 4. The vertical profiles of average BIAS and RMSE (Root-Mean-Square Error) at analysis time 0000 UTC against conventional observations for (**a**) temperature (TMP), (**b**) specific humidity (SPFH), (**c**) geopotential height (HGT), (**d**) zonal (UGRD), and (**e**) meridional (VGRD) wind.

Rainfall observations from ground stations were used to verify the rainfall forecasts. The area of rainfall forecast verification was the domain of 3-km horizontal resolution over North China, as shown in Figure 1. The rainfall scores were aggregated over sixteen 24 h forecasts during the experimental period. Figure 5 shows the CSI scores of the rainfall forecasts for 6 h of accumulated precipitation with six different thresholds, i.e., 0.1, 1, 5, 10, 25, and 50 mm. Below the threshold of 25 mm after 6 h, the rainfall scores of the experiment DA_RAD were higher than that for the CTRL. In comparison with the CTRL, the average CSI score has been improved by 14.2% in DA_RAD for the threshold of 25 mm after 6 h. For the threshold of 50 mm, the average rainfall score of the DA_RAD displayed an improvement of 35.8% for the first 18 h forecast range, with a substantial increase appearing over the 12–18 h forecast range. In addition, the CSI scores of the DA_RAD were better than the CTRL during the 6–18 h period for all of the thresholds. Figure 6 gives the corresponding BIAS scores for the six thresholds, and the BIASs of DA_RAD were all above 1.0 but below the threshold of 10 mm for the first 6 h, which is indicative of an over-prediction. In comparison with the CTRL, the BIASs of the DA_RAD were closer to 1.0 for the threshold of 25 mm, which indicates an improvement for the under-prediction in the CTRL forecast. For the threshold of 50 mm, the BIAS value displayed an obvious increase over the 12–18 h period in the DA_RAD, which was much closer to 1.0 than the CTRL. The ETS, POD, and FAR indices provided more details for the accumulated rainfall over the 6–12 h of the forecasts, as shown in Figure 7. The results suggest that the ETS of the DA_RAD was higher than that of the CTRL for all of the thresholds. As the PODs indicated, the fraction of events that are correctly forecast in the DA_RAD was greater for all of the thresholds except for 10 mm in comparison with the CTRL. Meanwhile, the fraction of false alarms in the DA_RAD was less than that of the CTRL, as the FARs indicated.

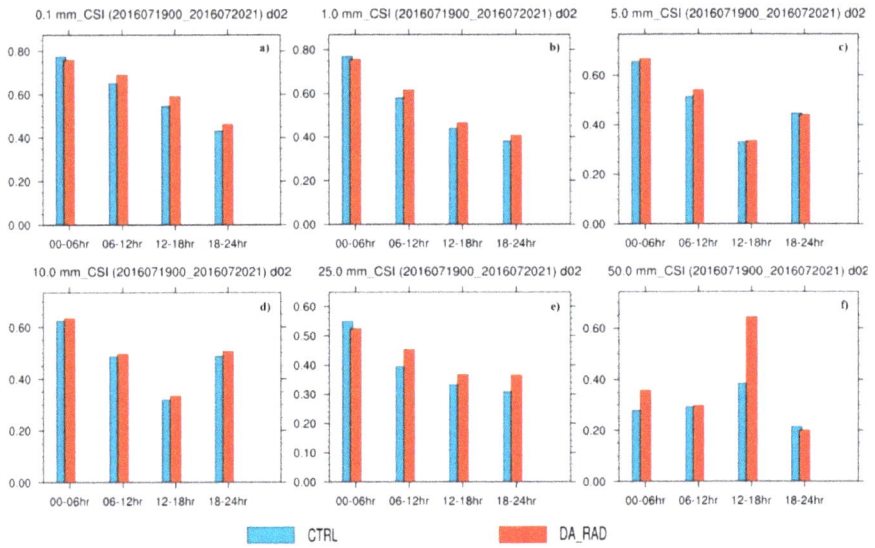

Figure 5. The CSI scores of the rainfall forecasts for 6 h of accumulated precipitation for the six different thresholds of (**a**) 0.1 mm, (**b**) 1 mm, (**c**) 5 mm, (**d**) 10 mm, (**e**) 25 mm, and (**f**) 50 mm.

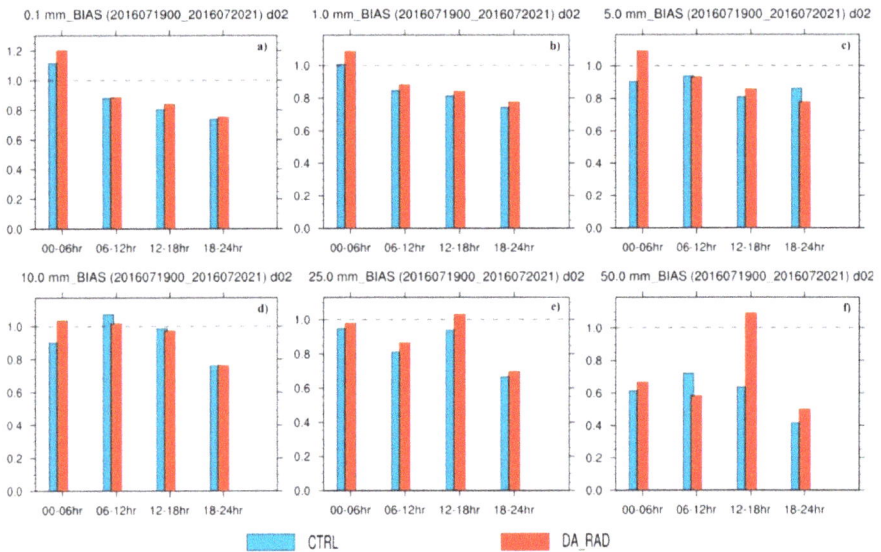

Figure 6. The BIAS scores of the rainfall forecasts for 6 h of accumulated precipitation for the six different thresholds of (**a**) 0.1 mm, (**b**) 1 mm, (**c**) 5 mm, (**d**) 10 mm, (**e**) 25 mm, and (**f**) 50 mm.

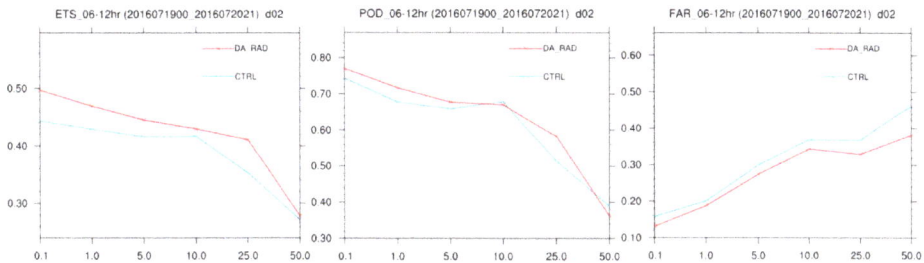

Figure 7. The ETS, POD, and FAR statistical indices for the accumulated rainfall over the 6–12 h period of the forecasts.

Figure 8 presents the rainfall forecast scores CSI and BIAS for hourly accumulated precipitation with thresholds of 0.1, 5, and 10 mm. The CSI scores of the experiment DA_RAD were slightly higher than the CTRL between 6 and 16 h for the threshold of 0.1 mm. The rainfall in the CTRL was underestimated between 8 and 16 h, which was improved in the DA_RAD. For the threshold of 5 mm, there was an improvement in the CSI scores of the DA_RAD between 6 and 12 h in comparison with that of the CTR. For the 10 mm threshold, the rainfall scores of the DA_RAD were better than the CTRL between 8 and 14 h. The BIAS score differences between the two experiments were small before 12 h for the 5 and 10 mm thresholds. There was a positive impact of assimilating satellite radiances, reflected mainly on the rainfall forecast during the 6–12 h period. Based on the precipitation products of CMORPH, Figure 9 gives the spatial correlation of hourly precipitation between the forecasts and observations from 0000 UTC to 2400 UTC on 20 July 2016. It can be seen that the spatial correlation of precipitation from the DA_RAD forecast was higher than that of the CTRL most of the time after 6 h, especially over the period of 6–12 h.

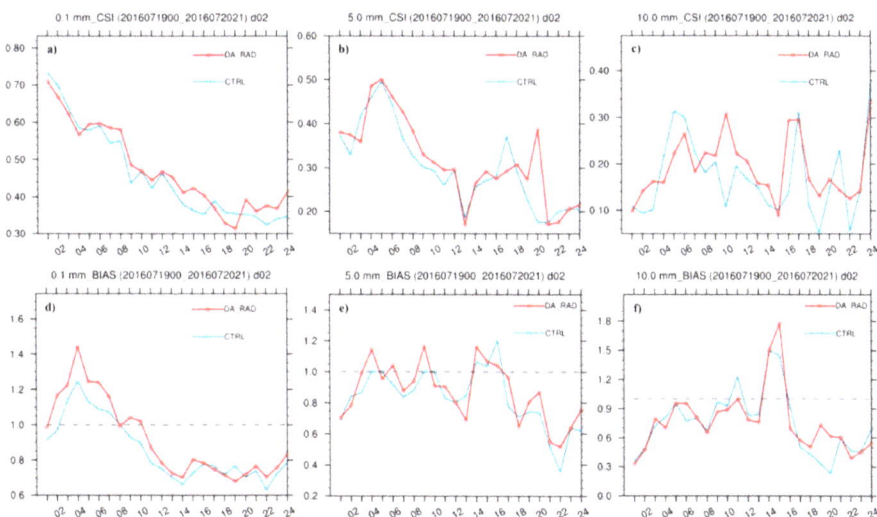

Figure 8. The CSI (the first line) and BIAS (the second line) scores for hourly accumulated precipitation for the thresholds of (**a,d**) 0.1 mm, (**b,e**) 5 mm, and (**c,f**) 10 mm.

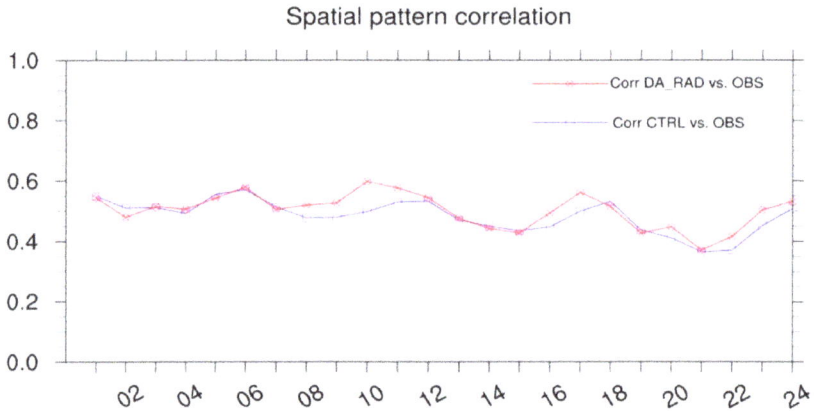

Figure 9. Spatial correlation (Corr) of hourly precipitation between forecasts and observations (OBS) from 0000 UTC to 2400 UTC 20 July 2016 over the domain of 3 km horizontal resolution.

4.2. Discussion

The results suggest that assimilating satellite radiance observations improved the rainfall score to a certain extent for the 19–20 July 2016 severe rainfall event over North China. For the 6 h accumulated precipitation, the rainfall scores of the DA_RAD experiment were better than the CTRL below the threshold of 25 mm after a 6 h run. For the threshold of 50 mm, the average rainfall score of the DA_RAD was substantially improved by 35.8% for the first 18 h, with most of that improvement associated with the period from 12 to 18 h. In addition, a positive impact of assimilating satellite radiance data on the rainfall scores was primarily found in the forecast range from 6 to 12 h.

To further investigate and explain the results of forecast rainfall scores, Figure 10 gives the spatial distribution patterns of 6 h accumulated precipitation on 20 July 2016. The first line displays the location distributions and 6 h accumulated rainfall observations of the ground stations, which are used for the score skills. The spatial distributions of 6 h of accumulated precipitation from the CTRL and the DA_RAD forecasts are presented in the second and third lines, respectively. The heavy rainfall that was based on observations was primarily located in the south and southeast of Beijing for the first 6 h, whereas the high rainfall from the CTRL forecast primarily appeared southeast of Beijing. The spatial distribution pattern from the DA_RAD forecast was similar to the CTRL, however the scope of rainfall above 10 mm and the rainfall intensity south of Beijing increased. From 6–12 h, the high rainfall observations were concentrated in and to the east of Beijing, whereas the center of strong rainfall moved east of Beijing for both CTRL and DA_RAD. In comparison with CTRL, the scope of rainfall between 20 and 50 mm in the forecast of DA_RAD increased and slightly decreased above 50 mm. It was consistent with the BIAS scores that were presented in Figure 6. For 12–18 h, the center of heavy rainfall was located toward the northeast of Beijing for both the CTRL and the DA_RAD forecasts, which displayed a good agreement with the rainfall observations.

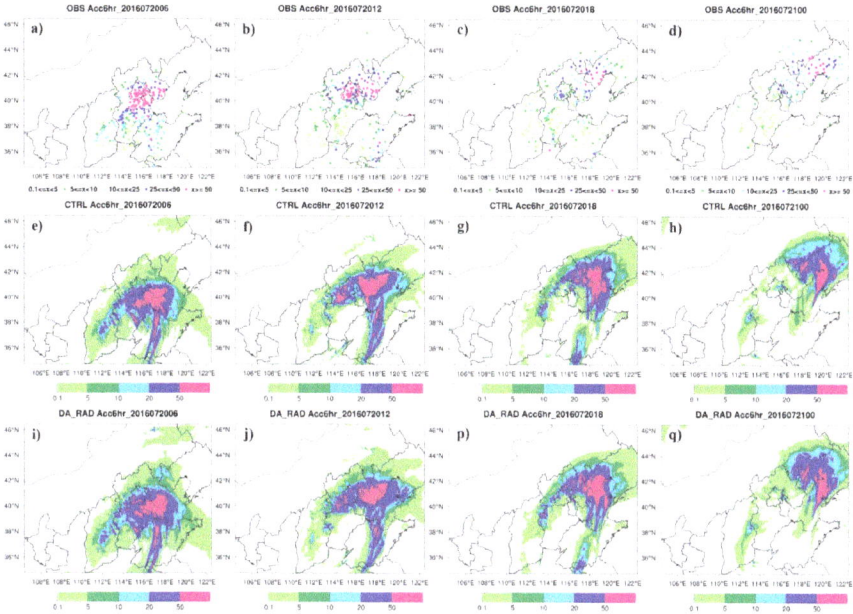

Figure 10. The spatial distribution patterns of 6 h of accumulated precipitation on 20 July 2016. (**a–d**) are for observations at 06, 12, 18, and 24 h; (**e–h**) are for the CTRL forecast at 06, 12, 18, and 24 h; and (**i,j**), (**p,q**) are for the DA_RAD forecast at 06, 12, 18, and 24 h.

The forecast from the DA_RAD increased the scope of rainfall above 50 mm which hardly covered all of the corresponding observations, leading to an obvious increase in the CSI score in comparison with that of the CTRL. The heavy rainfall continued to move toward east and weakened after 18 h in both CTRL and DA_RAD experiments, which covered parts of the sea. During this period, the rainfall observations above 50 mm were mainly located in the northeast and the rainfall score was greatly influenced by the spatial distribution and the number of observation sites. Although satellite radiance data were shown to increase the heavy rainfall forecast score over a relatively data-rich area, the spatial distribution pattern and the location of strong rainfall changed slightly.

Figure 11 shows the spatial distribution patterns of 24 h of accumulated precipitation on 20 July 2016 for observations and forecasts of the CTRL and the DA_RAD. An obvious difference in the location of heavy rainfall was found between forecasts and observations. The rainfall observations above 100 mm mainly covered Beijing and the area toward the northeast. For the CTRL experiment, the heavy rainfall above 100 mm was located in the east and northeast of Beijing, which did not cover the Beijing area. Assimilation of satellite radiance data in the DA_RAD increased the scope of heavy rainfall, however the spatial distribution pattern of heavy rainfall was similar to that of the CTRL. Based on the precipitation products of CMORPH, Figure 12 gives the spatial distribution of the ME index between the forecasts and observations for hourly precipitation over the domain of 3-k horizontal resolution. ME spatial variability shows that the forecasts from DA_RAD and CTRL were both under-forecasted during the rainfall event, however the under-forecasted area in DA_RAD was smaller than that in the CTRL.

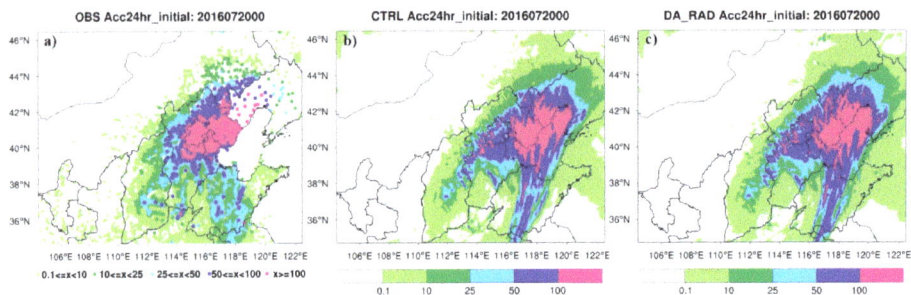

Figure 11. The spatial distribution patterns of 24 h of accumulated precipitation on 20 July 2016. (**a**) is for observations, (**b**) is for the CTRL forecast, and (**c**) is for the DA_RAD forecast.

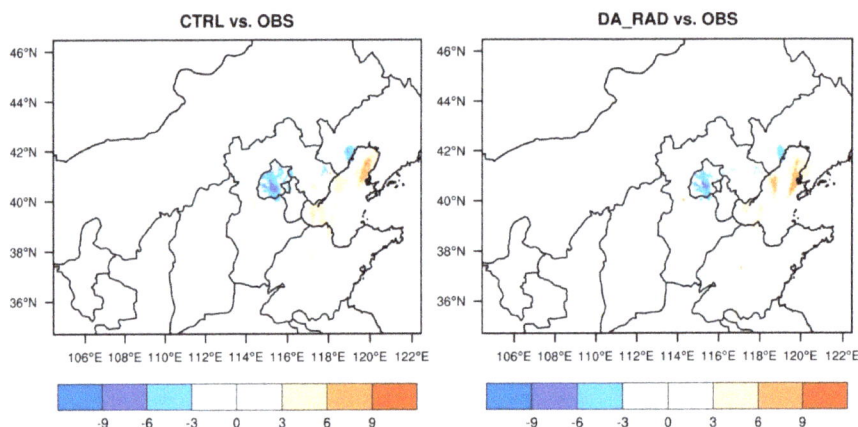

Figure 12. The spatial distribution of the ME (Mean Error) index between forecasts and observations for hourly precipitation over the domain of 3 km horizontal resolution.

The wind, humidity, and geopotential height fields for 500 and 700 hPa at 0000 UTC on 20 July 2016 are shown in Figure 13. The atmospheric analysis fields from the European Centre for Medium-Range Weather Forecasts (EC) have been used for comparison. For 500 hPa, a low-pressure system in the southwest–northeast direction can be found from the atmospheric fields of EC. There was a similar low-pressure system in both the CTRL and the DA_RAD, however the direction changed a small amount and the humidity in the center of the low pressure increased, which directly influenced the subsequent evolution of the low-pressure system. Assimilation of satellite radiance data also increased the humidity in the eastern part of the low-pressure system in comparison with that of CTRL. For 700 hPa, the center of low pressure was located towards the south of Beijing in the EC's atmospheric fields. In the CTRL forecast, the center of the low pressure was towards the southwest of Beijing and the high humidity was mainly concentrated in the northeast of the low-pressure system, leading to heavy rainfall towards the east and northeast of Beijing during the next 24 h. Similarly, the humidity in the east of the low pressure increased with satellite radiance data, however there was no change in the position of the low-pressure center. Consequently, the forecast from the DA_RAD experiment had a similar spatial distribution pattern of heavy rainfall to that of the CTRL, however it had a positive impact on the rainfall scores to some extent.

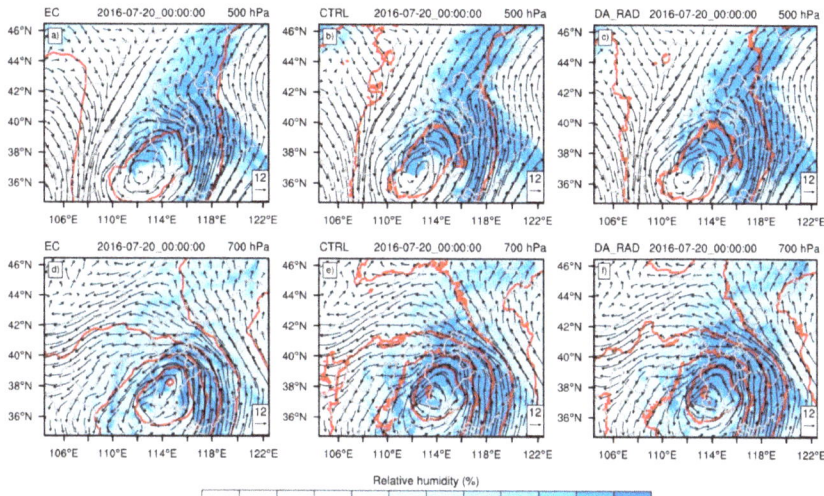

Figure 13. The fields of wind, humidity, and geopotential height for 500 hPa (the first line) and 700 hPa (the second line) at 0000 UTC 20 July 2016. (**a,d**) are from EC, (**b,e**) are from the CTRL forecast, and (**c,f**) are from the DA_RAD forecast.

5. Conclusions

RMAPS is an operational NWP based on the WRF and WRFDA model with special configurations mainly for North China. In this study, the impacts of assimilating satellite radiance observations on the accuracy of heavy rainfall forecasts were evaluated using RMAPS. The rainfall event that was selected for this study occurred in North China over the 19–20 July 2016. Two assimilation experiments were conducted and 24 h forecasts were produced every 3 h. The CTRL experiment, which assimilated only conventional data, had the same setting as the operational running of the RMAPS. The DA_RAD experiment assimilated the satellite radiance data from the AMSU-A and MHS along with the conventional observations in the RMAPS. The accuracy and performance of quantitative precipitation forecasts from the two experiments were compared and investigated against the observations. The main conclusions are summarized as follows:

(1) Assimilating satellite radiance data in the RMAPS reduced the temperature bias in the middle levels for the 19–20 July 2016 rainfall event that occurred over North China. It also improved the forecast accuracy of 6 h of accumulated precipitation after 6 h of forecasting, especially for the threshold of above 25 mm. The average CSI rainfall score -was improved by 14.2% for the 25 mm threshold and by 35.8% for the 50 mm threshold.

(2) Satellite radiances assimilation had a positive impact on QPF in the forecast range of 6–12 h. For hourly precipitation, satellite radiances increased the CSI scores below the 10 mm threshold during the period of 6–12 h. For the 6 h accumulated precipitation, the correct forecast rate increased with satellite radiances that were associated with higher ETS and POD indices, and the false alarm ratio decreased with low FAR values.

(3) Satellite radiance assimilation in the RMAPS increased the spatial distribution correlation of heavy rainfall after 6 h of forecasting, especially for the period from 6 to 12 h. Over Beijing, the under-forecasted area decreased with satellite radiances, which was associated with the better ME performance index.

(4) The spatial distribution of heavy rainfall from the CTRL and the DA_RAD followed a similar pattern, where the center of high rainfall shifted to the northeast. In comparison with the EC's

atmospheric fields, a shift of the low pressure center can also be found at different pressure levels at analysis time, which directly influenced the subsequent development of the heavy rainfall system. Satellite radiances from AMSU-A and MHS provided some valuable information for the temperature profile that increased the scope of the heavy rainfall, leading to an improvement of the rainfall scores. Nevertheless, satellite radiance observations did not improve the shift in the low pressure center's position, resulting in a similar spatial distribution pattern of heavy rainfall in both the CTRL and the DA_RAD.

Based on the results of this case study, satellite radiance data assimilation in the system of RMAPS had a positive impact on the quantitative precipitation forecast accuracy for heavy rainfall, especially from 6 to 12 h. Further research on the forecasting of precipitation is essential to effectively evaluate the impact of assimilating radiance data in the RMAPS, which will be a primary focus in our future work.

Author Contributions: Y.X. wrote the paper; J.S. supervised the study and reviewed the manuscript; S.F. supervised and reviewed the manuscript; M.C., Y.D., and D.J. reviewed the manuscript.

Funding: This research was funded by Beijing Science & Technology Commission (Grant No. Z161100001116098).

Acknowledgments: This study is supported by the Beijing Science & Technology Commission (Grant No. Z161100001116098), the National Key Research and Development Plan (Grant No. 2016YFE0117300), the Key Research Program of Frontier Sciences, CAS (QYZDY-SSW-DQC011), the National Key Research, and the Development Program of China (Grant No. 2017YFB0504105).

Conflicts of Interest: The authors declare no conflict of interest.

References

1. Peng, S.Q.; Zou, X. Impact on short-range precipitation forecasts from assimilation of ground-based GPS zenith total delay and rain gauge precipitation observations. *J. Meteorol. Soc. Jpn.* **2004**, *82*, 491–506. [CrossRef]

2. Chandrasekar, A. The impact of assimilation of AMSU data for the prediction of a tropical cyclone over India using a mesoscale model. *Int. J. Remote Sens.* **2006**, *27*, 4621–4653.

3. Xu, J.; Rugg, S.; Horner, M.; Byerle, L. Application of ATOVS Radiance with ARW WRF/GSI Data Assimilation System in the Prediction of Hurricane Katrina. *Open Atmos. Sci. J.* **2009**, *3*, 13–28. [CrossRef]

4. Li, J.; Wang, P.; Han, H.; Li, J.; Zheng, J. On the assimilation of satellite sounder data in cloudy skies in numerical weather prediction models. *J. Meteorol. Res.* **2016**, *30*, 169–182. [CrossRef]

5. Derber, J.C.; Wu, W.S. The use of TOVS cloud-cleared radiances in the NCEP SSI analysis system. *Mon. Weather Rev.* **1998**, *126*, 2287–2299. [CrossRef]

6. Mcnally, A.P.; Derber, J.C.; Wu, W.; Katz, B.B. The use of TOVS level-lb radiances in the NCEP SSI analysis system. *Q. J. R. Meteorol. Soc.* **2000**, *126*, 689–724. [CrossRef]

7. Baker, N.L.; Hogan, T.F.; Campbell, W.F.; Pauley, R.L.; Swadley, S.D. *The Impact of AMSU-A Radiance Assimilation in the US Navy's Operational Global Atmospheric Prediction System (NOGAPS)*; Naval Research Lab Monterey CA Marine Meteorology Div: Monterey, CA, USA, 2005.

8. Rodgers, C.D. *Inverse Methods for Atmospheric Sounding: Theory and Practice*; World Scientific: Singapore, 2000.

9. Errico, R.M.; Bauer, P.; Mahfouf, J.F. Issues regarding the assimilation of cloud and precipitation data. *J. Atmos. Sci.* **2009**, *65*, 3785–3798. [CrossRef]

10. Wang, P.; Li, J.; Lu, B.; Schmit, T.J.; Lu, J.; Lee, Y.-K.; Li, J.; Liu, Z. Impact of moisture information from advaned Himawari imger measurements on heavy precipitation forecasts in a regional NWP model. *J. Geophys. Res. Atmos.* **2018**, *123*, 6022–6038. [CrossRef]

11. McNally, A.P.; Watts, P.D.; Smith, J.A.; Engelen, R.; Kelly, G.A.; Thépaut, J.N.; Matricardi, M. The assimilation of AIRS radiance data at ECMWF. *Q. J. R. Meteorol. Soc.* **2006**, *132*, 935–957. [CrossRef]

12. Barker, D.; Huang, X.Y.; Liu, Z.; Auligné, T.; Zhang, X.; Rugg, S.; Ajjaji, R.; Bourgeois, A.; Bray, J.; Chen, Y.; et al. The weather research and forecasting model's community variational/ensemble data assimilation system: WRFDA. *Bull. Am. Meteorol. Soc.* **2012**, *93*, 831–843. [CrossRef]

13. Buehner, M.; Caya, A.; Carrieres, T.; Pogson, L. Assimilation of SSMIS and ASCAT data and the replacement of highly uncertain estimates in the Environment Canada Regional Ice Prediction System. *Q. J. R. Meteorol. Soc.* **2016**, *142*, 562–573. [CrossRef]

14. Kazumori, M. Satellite radiance assimilation in the JMA operational mesoscale 4DVAR system. *Mon. Weather Rev.* **2014**, *142*, 1361–1381. [CrossRef]

15. Bauer, P.; Geer, A.J.; Lopez, P.; Salmond, D. Direct 4D-Var assimilation of all-sky radiance. Part I: Implementation. *Q. J. R. Meteorol. Soc.* **2010**, *136*, 1868–1885. [CrossRef]

16. Liu, Z.Q.; Schwartz, C.S.; Snyder, C.; Ha, S.Y. Impact of assimilating AMSU-A radiances on forecasts of 2008 Atlantic tropical cyclones initialized with a limited-area ensemble kalman filter. *Mon. Weather Rev.* **2012**, *140*, 4017–4034. [CrossRef]

17. Xu, J.J.; Powell, A.M. Dynamical downscaling precipitation over Southwest Asia: Impacts of radiance data assimilation on the forecasts of the WRF-ARW model. *Atmos. Res.* **2012**, *111*, 90–103. [CrossRef]

18. Zou, X.; Qin, Z.; Weng, F. Improved coastal precipitation forecasts with direct assimilation of GOES-11/12 imager radiances. *Mon. Weather Rev.* **2011**, *139*, 3711–3729. [CrossRef]

19. Qin, Z.; Zou, X.; Weng, F. Evaluating added benefits of assimilating GOES Imager radiance data in GSI for coastal QPFs. *Mon. Weather Rev.* **2013**, *141*, 75–92. [CrossRef]

20. Zou, X.; Qin, Z.; Weng, F. Improved Quantitative Precipitation Forecasts by MHS Radiance Data Assimilation with a Newly Added Cloud Detection Algorithm. *Mon. Weather Rev.* **2013**, *141*, 3203–3221. [CrossRef]

21. Singh, R.; Ojha, S.P.; Kishtawal, C.M.; Pal, P.K.; Kiran Kumar, A.S. Impact of the assimilation of INSAT-3D radiances on short-range weather forecasts. *Q. J. R. Meteorol. Soc.* **2016**, *142*, 120–131. [CrossRef]

22. Sagita, N.; Hidayati, R.; Hidayat, R.; Gustari, I. Satellite radiance data assimilation for rainfall prediction in Java Region. *IOP Conf. Ser. Earth Environ. Sci.* **2017**, *54*. [CrossRef]

23. Wang, Y.; Liu, Z.; Yang, S.; Min, J.; Chen, L.; Chen, Y.; Zhang, T. Added value of assimilating Himawari-8 AHI water vapor radiances on analyses and forecasts for "7.19" severe storm over north China. *J. Geophys. Res. Atmos.* **2018**, *123*, 3374–3394. [CrossRef]

24. Skamarock, W.C.; Klemp, J.B.; Dudhia, J.; Gill, D.O.; Barker, D.M.; Duda, M.G.; Huang, X.Y.; Wang, W.; Powers, J.G. *A Description of the Advanced Research WRF Version 3*; NCAR Technical Note NCAR/TN/u2013475; Mesoscale and Microscale Meteorology Division: Boulder, CO, USA, 2008.

25. Parrish, D.F.; Derber, J.C. The National Meteorological Center's spectral statistical interpo-lation analysis system. *Mon. Weather Rev.* **1992**, *120*, 1747–1763. [CrossRef]

26. Amstrup, B. *Impact of ATOVS AMSU-A Radiance Data in the DMI-HIRLAM 3D-Var Analysis and Forecasting System*; Danish Meteorological Institute: Copenhagen, Denmark, 2001.

27. Xie, P.; Arkin, P.A. Analyses of global monthly precipitation using gauge observations, satellite estimates, and numerical model predictions. *J. Clim.* **1996**, *9*, 840–858. [CrossRef]

28. Omranian, E.; Sharif, H.O.; Tavakoly, A.A. How well can Global Precipitation Measurement (GPM) capture Hurricanes? Case study: Hurricane Harvey. *Remote Sens.* **2018**, *10*, 1150. [CrossRef]

29. Omranian, E.; Sharif, H.O. Evaluation of the Global Precipitation Measurement (GPM) satellite rainfall products over the Lower Colorado River Basin, Texas. *JAWRA J. Am. Water Resour. Assoc.* **2018**. [CrossRef]

30. Joyce, R.J.; Janowiak, J.E.; Arkin, P.A.; Xie, P.P. CMORPH: A method that produces global precipitation estimates from passive microwave and infrared data at high spatial and temporal resolution. *J. Hydrometeorol.* **2004**, *5*, 487–503. [CrossRef]

31. Shen, Y.; Pan, Y.; Yu, J.J.; Zhao, P.; Zhou, Z.J. Quality assessment of hourly merged precipitation product over China. *Trans. Atmos. Sci.* **2013**, *36*, 37–46.

32. Casati, B.; Wilson, L.J.; Stephenson, D.B.; Nurmi, P.; Ghelli, A.; Pocernich, M.; Damrath, U.; Ebert, E.E.; Brown, B.G.; Mason, S. Forecast verification: Current status and future directions. *Meteorol. Appl.* **2008**, *15*, 3–18. [CrossRef]

33. Schaffer, J.T. The critical success index as an indicator of warning skill. *Weather Forecast.* **1990**, *5*, 570–575. [CrossRef]

34. Walsh, K.; Mcgregor, J. An assessment of simulations of climate variability over Australia with a limited area model. *Int. J. Climatol.* **1997**, *17*, 201–223. [CrossRef]

35. Weng, F.; Han, Y.; Delst, P.V.; Liu, Q.; Kleespies, T.; Yan, B.; Marshall, J.L. *JCSDA Community Radiative Transfer Model (CRTM)—Version 1*; NOAA Technical Report NESDIS 122; American Geophysical Union: Washington, DC, USA, 2005.

remote sensing

MDPI

Article

Evaluation and Intercomparison of High-Resolution Satellite Precipitation Estimates—GPM, TRMM, and CMORPH in the Tianshan Mountain Area

Chi Zhang [1], Xi Chen [2], Hua Shao [2], Shuying Chen [2], Tong Liu [3], Chunbo Chen [2], Qian Ding [1] and Haoyang Du [4,*]

[1] Shandong Provincial Key Laboratory of Water and Soil Conservation and Environmental Protection, College of Resources and Environment, Linyi University, Linyi 276000, China; zc@ms.xjb.ac.cn (C.Z.); dingq2017@163.com(Q.D.)

[2] State Key Laboratory of Desert and Oasis Ecology, Xinjiang Institute of Ecology and Geography, Chinese Academy of Sciences, Urumqi 830011, China; chenxi@ms.xjb.ac.cn (X.C.); shaohua@ms.xjb.ac.cn (H.S.); chenshuying16@mails.ucas.ac.cn (S.C.); ccb_8586@ms.xjb.ac.cn (C.C.)

[3] College of Life Science, Shihezi University, Shihezi 832000, China; betula@126.com

[4] Jiangsu Provincial Key Laboratory of Geographic Information Science and Technology, Nanjing University, Nanjing 210093, China

* Correspondence: duhaoyang15@mails.ucas.ac.cn; Tel.: +86-131-5030-3289

Received: 10 September 2018; Accepted: 22 September 2018; Published: 25 September 2018

Abstract: With high resolution and wide coverage, satellite precipitation products like Global Precipitation Measurement (GPM) could support hydrological/ecological research in the Tianshan Mountains, where the spatial heterogeneity of precipitation is high, but where rain gauges are sparse and unevenly distributed. Based on observations from 46 stations from 2014–2015, we evaluated the accuracies of three satellite precipitation products: GPM, Tropical Rainfall Measurement Mission (TRMM) 3B42, and the Climate Prediction Center morphing technique (CMORPH), in the Tianshan Mountains. The satellite estimates significantly correlated with the observations. They showed a northwest–southeast precipitation gradient that reflected the effects of large-scale circulations and a characteristic seasonal precipitation gradient that matched the observed regional precipitation pattern. With the highest correlation (R = 0.51), the lowest error (RMSE = 0.85 mm/day), and the smallest bias (1.27%), GPM outperformed TRMM and CMORPH in estimating daily precipitation. It performed the best at both regional and sub-regional scales and in low and mid-elevations. GPM had relatively balanced performances across all seasons, while CMORPH had significant biases in summer (46.43%) and winter (−22.93%), and TRMM performed extremely poorly in spring (R = 0.31; RMSE = 1.15 mm/day; bias = −20.29%). GPM also performed the best in detecting precipitation events, especially light and moderate precipitation, possibly due to the newly added Ka-band and high-frequency microwave channels. It successfully detected 62.09% of the precipitation events that exceeded 0.5 mm/day. However, its ability to estimate severe rainfall has not been improved as expected. Like other satellite products, GPM had the highest RMSE and bias in summer, suggesting limitations in its way of representing small-scale precipitation systems and isolated deep convection. It also underestimated the precipitation in high-elevation regions by 16%, suggesting the difficulties of capturing the orographic enhancement of rainfall associated with cap clouds and feeder–seeder cloud interactions over ridges. These findings suggest that GPM may outperform its predecessors in the mid-/high-latitude dryland, but not the tropical mountainous areas. With the advantage of high resolution and improved accuracy, the GPM creates new opportunities for understanding the precipitation pattern across the complex terrains of the Tianshan Mountains, and it could improve hydrological/ecological research in the area.

Keywords: satellite precipitation product; Tianshan Mountains; GPM; TRMM; CMORPH

1. Introduction

Precipitation is a key meteorological variable and a major climate change indicator, directly affecting the energy and water exchanges between the biosphere and atmosphere. Traditional climate studies relied heavily on field observations [1], however, it is difficult to develop high-resolution spatial dataset of precipitation based on field observations [2]. Compared to air temperature, precipitation has more complex spatiotemporal patterns, which can only be accounted for with a dense network of rain gauge stations that are usually unavailable in a study area, particularly in the remote mountains in developing countries [3,4]. Such data issues have seriously hindered ecological and hydrological studies in the Tianshan Mountains area, which is known as "the water tower" of Central Asia [5], the largest dryland and one of the most climate sensitive ecosystems in the world [6]. The Tianshan Mountains are characterized by a large contrast in elevation, from 154 m below sea level to 7439 m above sea level (asl), which creates highly heterogeneous precipitation that varies from <100 mm/yr in the low mountain deserts to >900 mm/yr in the windward slopes of the high mountain ranges [7,8]. However, there is only a sparse and unevenly distributed network of meteorological stations across this complex geography [9]. Many of the stations stopped functioning with the collapse of the former Soviet Union in the early 1990s [10], and they further worsened the data availability [11]. As the result, widely used spatially interpolated precipitation datasets like the Climate Research Unit (CRU) are not reliable for hydrological/ecological research in both the Tianshan Mountains and the Central Asia dryland [9,12].

The development of satellite remotely sensed datasets has provided an opportunity to retrieve the spatiotemporal pattern of precipitation with high resolution in remote areas that have had few field observations [13]. These precipitation products are generated by combining thermal infrared reflection (IR), passive microwave (PM), and precipitation radar (PR) data from various satellite sensors. Over the four decades, multiple generations of satellite precipitation projects have been launched, e.g., the Precipitation Estimation from Remotely Sensed Information using Artificial Neural Network (PERSIANN) [14], the Naval Research Laboratory Global Blended-Statistical Precipitation Analysis (NRLgeo) [15], the Climate Prediction Center Morphing Technology (CMORPH) [16], the Tropical Rainfall Measurement Mission (TRMM) Multi-sensor Precipitation Analysis (TMPA) [17], and the Global Precipitation Measurement (GPM) project [18]. With a high spatial resolution of 0.25° and a long temporal period (1998–2015), the TMPA has been widely applied in ecological/hydrological researches [17,19,20]. With the highest spatial resolution of ~0.1° among the satellite precipitation datasets, the CMORPH and GPM products also have a huge potential for research in mountainous areas [16,18]. In particular, the Integrated Multi-Satellite Retrievals for GPM (IMERG) products have inherited the strengths of previous satellite precipitation projections. In addition, the GPM Core Observatory is equipped with a multi-channel GPM Microwave Imager (GMI), an expanded Ku/Ka-band at 13.6 GHz/35.5 GHz, and an upgraded dual-frequency precipitation radar (DPR). These on-board sensors are more sensitive to light rainfall and snowfall than their predecessors [18].

Satellite precipitation estimates, however, are subject to large uncertainties due to their indirect nature [21–23]. As uncertainties can emanate from various elements, including the retrieval algorithm, thermal radiance, and cloud noise in these satellite precipitation products, strict and comprehensive evaluations are necessary [21,24]. Launched in 2014, the GPM mission is still in the early stages of the development and evaluation cycle. Duan et al. [25] and Huffman et al. [26,27] provided preliminary comparisons between the IMERG and TMPA monthly precipitation products. Their studies showed the two products are similar over land. Liu (2016) [28] made further comparisons on a global scale and found that IMERG performed better over land with high precipitation. Studies in India also indicated a notable advantage of IMERG over TMPA in terms of heavy monsoon rainfall detection [29]. The improvement was attributed to the newly added DPR [30]. However, another study in tropical

Asia showed that IMERG performed worse than TMPA in heavy rain detection, according to 75% of the gauge stations [31]. The evaluation study in the US mid-Atlantic region also indicated that IMERG tended to underestimate heavy rain with considerable random errors [32]. Albeit at low frequency, heavy precipitating events have a significant hydrological impact, leading to extreme floods and landslides in mountainous areas [25] and would be considerable affection on the global climate models [30]. Therefore, it is important to evaluate and improve the capacities of satellite instruments in detecting and estimating heavy rainfall.

After 2016, there were more studies evaluating the performances of IMERG in comparison with other satellite precipitation products in Xinjiang, China [33,34], where the major part of Tianshan is located, and in the Tibetan Plateau, a region adjacent to Tianshan [35–37]. Both Chen and Li (2016) [33] and Lu et al. (2018) [34] found that in comparison with TMPA, IMERG significantly improved the estimation accuracy of precipitation over the Xinjiang region. They attributed the improvement to the upgraded DPR and PM sensors, which were supposed to increase GPM's sensitivity to light precipitation [33]. However, the study by Wei et al. (2018) [37] showed that IMERG performed worse than TRMM 3B42 in typical arid/semi-arid regions of China, indicating that its sensitivity to light precipitation might not have been as improved as expected. In addition, the precipitation in the Tianshan Mountains is highly influenced by topography and convective systems, and satellite precipitation products have been found to perform relatively poorly in regions with strong orographic effects and complex convective systems [33,38,39]. Although previous studies have confirmed the suitability of satellite precipitation products over complex mountain terrains in Asia [40,41], their performance in the Tianshan Mountains are still unclear. In general, previous studies have shown that the IMERG as well as other satellite-retrieved precipitation products have not been very reliable in the Xinjiang region [33]. A comprehensive evaluation of the performances of satellite precipitation products in the Tianshan Mountains could provide valuable feedback for developers to improve the related retrieval algorithm, and for data users to assess their usefulness in ecological/hydrological research in Central Asia.

In this study, the accuracies of the estimated precipitation in the Tianshan Mountains from three satellite products: IMERG, TRMM 3B42, and CMORPH were evaluated through comparisons with observations from 46 stations. The study period was set to April 2014–March 2015, which was the overlapping time period of the three satellite missions (see Section 2.2.1.). While this short time period limits the representativeness of the temporal pattern in weather, it is still possible to evaluate whether the satellite products can reflect the spatial pattern and seasonal variation of precipitation. This study aims to improve our understanding of the suitability and uncertainty of satellite precipitation products in the Tianshan Mountains, and evaluates whether the upgrades in GPM actually helped to enhance its capacity in capturing light precipitation and solid precipitation in the Tianshan Mountains of the Central Asia dryland.

2. Materials and Methods

2.1. Study Region

Straddling the border between China and Kyrgyzstan, the Tianshan Mountains ($39°30'–45°45'$N, $74°10'–96°15'$E) are the largest mountain system in Central Asia [42]. Due to the issue of rain-gauge data availability, only the part of the Tianshan Mountains that are located in China was investigated in this study (Figure 1). Stretching about 1700 km from west–southwest to east–northeast and with a central width of about 350 km, the study area (3.5×10^5 km^2) accounts for two thirds of the whole Tianshan Mountain area. The mountains have a rough terrain, with elevation ranging from -154 m to 7439 m. The tallest peaks in the Tianshan area are a central cluster of mountains forming a knot, from which ridges extend along the boundaries between China, Kyrgyzstan, and Kazakhstan.

Sitting in the center of Central Asia, the Tianshan Mountains are the source of major rivers and lakes in the dryland including the Tarim River, the Syr Darya, and the Ili River (Figure 1).

The region's atmospheric circulation is controlled by moist westerly Atlantic air masses and cold northerly/northwesterly inflows. The seasonal pattern of precipitation is controlled by the seasonal dynamic of the northern jet stream and the southwesterly cyclones from the Arabian Sea, which results in strong precipitation in spring [10]. Topography also plays a vital role in the formation of distinct local climates that vary from <100 mm/yr in the low-mountain deserts in the eastern Tianshan Mountains to >900 mm/yr in the windward slopes of high ranges in the western Tianshan Mountains. Following the climate gradient, the mountain vegetation/land cover changes from low-mountain deserts/dry grasslands, to mid-mountain forests, to alpine meadow/dwarf shrubs, before rising into glaciers in higher elevations.

Based on the monthly mean precipitation data from 163 ground stations, the study area was further divided into sub-regions with similar climates using the K-nearest neighbor (KNN) method [43]. To find the optimal spatial constraint parameters, we tried a series of K values ranging between 2 and 10. The number of clusters was automatically decided by the clustering analysis tool. This was done in the analysis by finding the most effective number of clusters. The effectiveness was quantified by the Calinski–Harabasz pseudo F-statistic that reflects the similarity and difference between the group, and a larger F means a better clustering result:

$$F = \frac{\frac{R^2}{n_C - 1}}{\frac{1 - R^2}{n - n_c}}$$

where $R^2 = \frac{BGD - WGS}{BGD}$; WGS is a reflection of within-cluster similarity, and BGD reflects the between-cluster difference:

$$BGD = \sum_{i=1}^{n_c} \sum_{j=1}^{n_i} \sum_{k=1}^{n_v} (v_{ij}^k - v_j^k)^2$$

$$WGS = \sum_{i=1}^{n_c} \sum_{j=1}^{n_i} \sum_{k=1}^{n_v} (v_{ij}^k - v_i^k)^2$$

In the above equations,

n = the number of the objects to be regionalized.

n_i = the number of the objects in cluster i.

n_c = the number of the cluster.

n_v = the number of the variables used to cluster objects.

v_{ij}^k = the value of the kth variable of the jth object in the ith cluster.

v_j^k = the mean value of the kth variable.

v_j^k = the mean value of the kth variable in cluster i.

The highest coefficient of determination R^2 = 0.79 and the largest F = 81.18 were reached when K = 8 and the category is set to 3. Therefore, a value of K = 8 was used for the spatial constraint condition of KNN, and the clustering analysis yielded three climate sub-regions in the Tianshan Mountain area, each of which was coherent in precipitation variation and seasonal circulation pattern: the southern and western Tianshan Mountains (SW TS) that are bounded to the south by the Tarim Basin and links up with the Pamir Mountains in the southwest; the northern and eastern Tianshan Mountains (NE TS), which are bounded to the north by the Junggar Basin and meets the Altai Mountains in the east, and the central Tianshan Mountains (CN TS) that includes the Ili River Valley and the central mountain clusters [41] (Figure 1). The sub-regions identified in this study agreed well with the three main climatic sub-regions identified by Sorg et al. (2012) [44]: (1) a moist sub-region bounded by the outer ranges of Tianshan from the northwest and the inner ranges of Tianshan from the southwest, corresponding to CN TS, (2) the inner ranges of Tianshan and its arid southern slope, corresponding to SW TS, and (3) the eastern ranges of Tianshan that accounted for >70% of the NE TS area in this study.

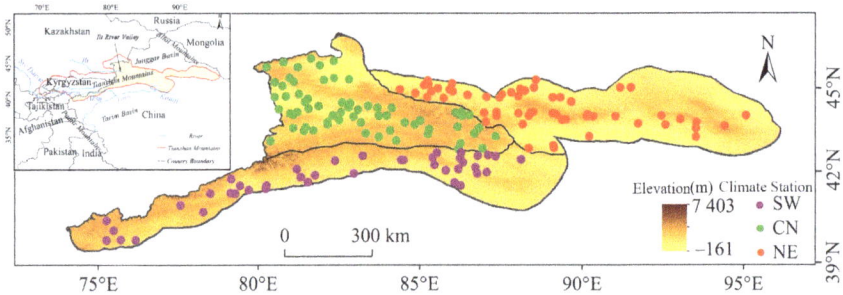

Figure 1. The study area and distribution of precipitation monitoring stations.

2.2. Data

2.2.1. Satellite Precipitation Products

The satellite precipitation products used in this study include the GPM IMERG, TRMM 3B42, and CMORPH (Table 1). CMORPH was developed by the National Oceanic and Atmospheric Administration Climate Prediction Center (NOAA-CPC). It estimates precipitation from low orbiter satellite passive microwave observations, compensated by high-resolution IR imagery [16]. The dataset had 30-min temporal and 0.07° spatial resolutions. Despite its high resolutions, the CMORPH may not perform as well as the TRMM precipitation product in Central Asia and the nearby Tibetan Plateau [45,46]. This study used the version 7 TRMM data (3B42V7), which had 3 h temporal and 0.25° spatial resolution. The TRMM product combined multiple satellite observations, including (the first) space-based precipitation radar, IR and PM sensor data, and precipitation observations from the Global Precipitation Climatology Centre (GPCC) [47]. The TRMM project was discontinued in April 2015, and it was succeeded by the GPM mission, an international constellation of satellites that consisted of a GPM core observatory satellite and 10 partner satellites. There were several major improvements, including a Ka-band (35.5 GHz) to the DPR and newly added high-frequency channels (165.6 and 183.3 GHz) to the GMI [30]. IMERG was the level-3 product of the GPM mission [26]. With 30-min temporal and 0.1° spatial resolutions, it inherited the strengths of previous satellite precipitation projections, including CMORPH, TRMM, and PERSIANN [27]. Considering that the level-3 GPM IMERG final-run products started in March 2014 while the TRMM was discontinued by April 2015, the study period was confined to the 12 months between April 2014 and March 2015.

Table 1. The satellite precipitation products used in this study.

	Spatial/Temporal Resolutions	Temporal Extent	Coverage	Data Sources
GPM IMERG	0.1°/0.5 h	2014–present	60°N–60°N	[48]
TRMM 3B42V7	0.25°/3 h	1998–2015	50°N–50°N	[49]
CMORPH	0.07°/0.5 h	2002–present	60°N–60°N	[50]

2.2.2. Rain Gauge Data

Hourly and daily precipitation data were obtained from the Xinjiang Meteorological Network (XMN, [51], last visited on 15 August 2018), which is part of China's National Meteorological Network [35,52]. The XMN included 1934 meteorological stations, the majority of which commenced after 2012. Only 163 of the 1934 stations fell within the study area and were used in this study. The monthly mean precipitation data from these stations were used to divide the study area into sub-regions (see Section 2.1). Among the 163 stations, 64 were distributed in CN TS, 58 in the NE TS, and 41 in the SW TS (Figure 1). This dataset underwent strict quality controls, including a spatial consistency check, an internal consistency check, and an extreme value check [53]. In particular, stations with 20% of missing observations in any 30-day period or with missing observations for more than five consecutive days were excluded. Otherwise, the missing values in the precipitation records were filled by linear interpolation. If an interpolated daily precipitation fell below the observed minimum daily precipitation within the nearest 90 days, it was assumed that there was no precipitation on that day and the daily precipitation was set to zero. Only the 46 stations that had daily precipitation records throughout the study period (i.e., April 2014–March 2015) were used in the evaluation study of the three satellite precipitation products. We found that less than 5% of the stations could be used in the GPCC [54] for adjustments of the TRMM/GPM products in the study area. In addition, both the TRMM- and GPM-integrated GPCC data at a monthly scale, while this study used daily data [27]. Therefore, the dependence between the gauge- and satellite-based data should not be a big issue. To our knowledge, this is the best observational dataset of precipitation in the Tianshan Mountain area.

2.3. Methodology

2.3.1. Spatial Downscaling Method

Mountain precipitation is under the influence of orographic effects, and it has high spatial variation. To warrant an accurate evaluation, it is necessary to downscale the grid-level satellite precipitation estimates to the site level at each gauge station [9]. The traditional bilinear interpolation or direct data extraction approach [55,56], however, ignores the orographic effect, and this could lead to unwanted effects from spreading convective precipitation during summer [57]. To avoid these problems, this study adopted the precipitation–topography partial least squares (PTPLS) downscaling method, which uses the partial least squares method to estimate local topographical influences on precipitation [9]. Motivated by previous studies, we used elevation, slope, and aspect ratio, as well as latitude and longitude to estimate the effects of orography and geography on precipitation in the Tianshan Mountains [58,59]. The following procedures were conducted to downscale the gridded satellite precipitation estimates to match the site-level observations from the gauge stations:

(1) For a gauge station (S), we identified the grid cell in which that station fell in the spatial dataset of a satellite precipitation product. This grid cell was noted as A, and the precipitation at station S and at the grid representing A were noted as P_S and P_A, respectively;

(2) We identified 25 adjacent grid cells centered around station S in the spatial dataset satellite of precipitation, and extracted their terrain and geographic information, i.e., elevation (X_1), aspect ratio (X_2), slope (X_3), latitude (X_4), and longitude (X_5) from the digital elevation model of the study region (GTOPO30) [60].

(3) Based on the information, the local terrain effects on precipitation were estimated at site S. Let X be the matrix of five terrain and geographic factors at each of the 25 grid cells:

$$X = \begin{pmatrix} x_{1,1} & \cdots & x_{1,5} \\ \vdots & \vdots & \vdots \\ x_{25,1} & \cdots & x_{25,5} \end{pmatrix} = \begin{pmatrix} X_1 & \cdots & X_5 \end{pmatrix}.$$

where X_i (i = 1, 2, ... , 5) are column vectors; and let P be the matrix of precipitation of these 25 grid cells:

$$P^T = \begin{pmatrix} p_1 & \cdots & p_{25} \end{pmatrix},$$

where P^T was the transpose of P. Both X and P were normalized. To estimate the effects of X_i on local precipitation, matrix $W^{(1)}$ was constructed as:

$$W^{(1)} = \frac{1}{\sqrt{\sum_{i=1}^{5} Cov^2(P^T, X)}} \begin{pmatrix} Cov(P^T, X_1) \\ \vdots \\ Cov(P^T, X_5) \end{pmatrix},$$

where $Cov(P^T, X)$ is the covariance of the time series P^T and X. Applying $W^{(1)}$, we obtained the first order estimate of the terrain/geographic effects on local precipitation $T^{(1)} = X*W^{(1)}$, and the precipitation was estimated by $P = r_1 T^{(1)} + P^{(1)}$, where $r_1 = \frac{PT^{(1)}}{\|T^{(1)}\|}$ and $P^{(1)}$ was the residual vector of P.

Repeating this process by treating $P^{(1)}$ as P, we obtained the second order estimate of the local terrain/geographic effect, with $P^{(1)} = r_2 T^{(2)} + P^{(2)}$ with $r_2 = \frac{PT^{(2)}}{\|T^{(2)}\|}$ and $P^{(2)}$ being the residual vector of $P^{(1)}$. Repeating this process n times, we obtained the number of estimates $\{P^{(1)}, P^{(2)}, \ldots, P^{(n)}\}$, taking into account the portions of the terrain/geographic influences on local precipitation. Among the estimates, we used the least square method to obtain the precipitation for grid cell A, P'_A, from the most relevant $P^{(i)}$ $(1 \le i \le n)$.

P'_A was the downscaled satellite estimates of precipitation at station S. The value was then compared to the observations at station S.

2.3.2. Evaluation Metrics

The evaluations were conducted based on daily precipitation datasets, which were aggregated from the sub-daily data. Following Ma et al. (2016) [35], three quantitative statistical metrics were used to quantify the accuracy or discrepancy between the rain gauge observations (OBS) and satellite precipitation estimates: (1) the Pearson linear correlation coefficient (R) that measured the strength and direction of the linear association between OBS and satellite estimates; (2) the root-mean-square error (RMSE) that measured the averaged magnitude of the deviation that a satellite product will have from the OBS; and (3) the relative bias (PB) that measured any persistent bias in satellite estimates to either underestimate or overestimate OBS:

$$\text{PB} = \frac{\sum_{i=1}^{n}(Pe - Po)}{\sum_{i=1}^{n} Po} \times 100\% \tag{1}$$

$$\text{R} = \frac{Cov(Pe - Po)}{\delta e \delta o} \tag{2}$$

$$\text{RMSE} = \sqrt{\frac{\sum_{i=1}^{n}(Pe - Po)^2}{n}} \tag{3}$$

where Po and Pe are the OBS and satellite estimated precipitations, respectively; n is the sample size; δ is standard deviation; and cov() is covariance. The significance of R at the 95% level is tested.

In addition, to better check the appearance possibility of rainfall events from satellite products, the false alarm ratio (FAR), probability of detection (POD), frequency bias index (FBI), and equitable threat score (ETS) were calculated [61]. An estimate has a high accuracy when POD, ETS, and FBI approach 1, and FAR approaches 0. POD measures the ability of the satellite product to correctly detect rainfall, where the best score is '1' and the worst score is '0'. FBI of less (or greater) than 1 measures under (or over) forecast frequency. FAR measures how often a satellite product incorrectly reports rainfall events when no rain has occurred, and the score value can range from '0' to '1', where '1' is the worst score and '0' is the best score. ETS penalizes false alarms and misses equally, and it was designed to account for hits that would occur purely randomly. The parameter has a value of 0 for no skill and 1 for perfect correspondence. The detailed information of the evaluation indices can be found in Tian et al. (2009) [61] and Kenawy et al. (2015) [62].

$$FBI = \frac{H + F}{H + M} \tag{4}$$

$$POD = \frac{H}{H + M} \tag{5}$$

$$FAR = \frac{F}{H + F} \tag{6}$$

$$\text{ETS} = \frac{H - Hs}{H + M + F - Hs}; \text{ where } Hs = \frac{(H + M)(H + F)}{Total} \tag{7}$$

where M represents the observed precipitation events that are missed by the satellite products, while H is the correctly detected precipitation events, and F represents the precipitation events that are falsely reported by the satellite but not observed by the rain gauge. Hs indicates random hits. These parameters were calculated according to a precipitation threshold and a contingency table between the satellite estimates and OBS (Table 2). To measure the performances of the satellite estimates for different ranges of precipitation intensity, multiple evaluations were conducted with eight levels of threshold values: 0.5 mm/day, 1 mm/day, 2 mm/day, 4 mm/day, 8 mm/day, 10 mm/day, 15 mm/day, and 20 mm/day.

Table 2. Contingency table between the rain gauge observations and satellite estimates.

	Observations ≥ Threshold	Observations < Threshold	Total
Estimates ≥ Threshold	H	F	Estimated yes
Estimates < Threshold	M	Z #	Estimated no
Total	Observed yes	Observed no	total

Z stands for correct estimation of no precipitation.

3. Results

3.1. Accuracy in Describing Regional Precipitation

The observed precipitation ranged from 4.8 ± 5.7 mm in December to 31.3 ± 27.8 mm in June, with an annual total of 176.7 ± 113.6 mm during the study period. The highest sub-regional precipitation (297.9 ± 79.0 mm/yr) was observed in CN TS, while the lowest (129.6 ± 85.1 mm/yr) was found in SW TS. Our evaluations showed that the daily precipitation estimates from three satellite products were significantly correlated with the OBS ($p < 0.05$). The GPM product had the highest correlation of 0.53 ± 0.25, about 17% higher than the correlations of the other two satellite products (Figure 2a). The GPM also had the lowest RMSE (0.85 ± 0.54 mm/day) and the smallest bias (PB = 1.27 ± 30.41%) among the three satellite products, according to the observed precipitation, and the CMORPH data had the highest RMSE (1.15 ± 0.69 mm/day) and the highest bias (PB = 8.34 ± 42.21%) (Figure 2b,c). While the CMORPH overestimated precipitation, the TRMM underestimated precipitation by 5.00 ± 34.43% (Figure 2c). Overall, the GPM performed the best while the CMORPH performed the worst in estimating precipitation in the Tianshan Mountain area.

All satellite products correctly reflected the general pattern of the observed intra-annual variations in precipitation at the gauge stations during the study period, which peaked in June 2014 and gradually declined until reaching its lowest point in December 2014 (Figure 3). In particular, the GPM product closely followed the fluctuations of observed precipitation from month to month throughout the year, while the CMORPH product showed a false renunciation of precipitation in August and the TRMM product failed to show the decline of precipitation from April to May. Of the four seasons, the satellite estimates had the highest correlations with the OBS in autumn (SON), with the highest R value (0.63 ± 0.25) found in the GPM (Figure 2a). In the winter season (DJF), the satellite estimates had the lowest correlation with OBS, but also the smallest error (RMSE) (Figure 2a,b). Again, the GPM had the smallest RMSE (0.53 ± 0.31 mm/day) among the satellite products in winter (Figure 2b). The satellite estimates had relatively high RMSE in summer (JJA). Moreover, the CMORPH overestimated the summer precipitation significantly (PB = 46.43 ± 35.33%). In addition, the CMORPH significantly underestimated the winter precipitation (PB = −22.93 ± 21.20%). The TRMM also seriously underestimated the precipitation (PB = −20.29 ± 39.10%) in spring (MAM). In fact, the TRMM performed the poorest in spring of all of the satellite products, by all the quantitative metrics. Overall, the GPM performed better than the other two satellite products in describing seasonal precipitations, especially in the autumn. Although it overestimated summer precipitation by 12.17 ± 34.71% and underestimated winter precipitation by 11.82 ± 27.69%, the magnitudes of the biases were much lower than that of the CMORPH. Compared to the CMORPH estimates, which showed significant biases in summer and winter, and the TRMM, which performed extremely poorly in spring, the GPM had relatively balanced performances across all seasons. It should be noted that Figure 3 only reflects the average precipitation of the rain gauge stations and it does not represent the pattern of intra-annual variations of precipitation across the Tianshan Mountain area. Actually, the seasonality of precipitation varied among the sub-regions (as shown in the following section).

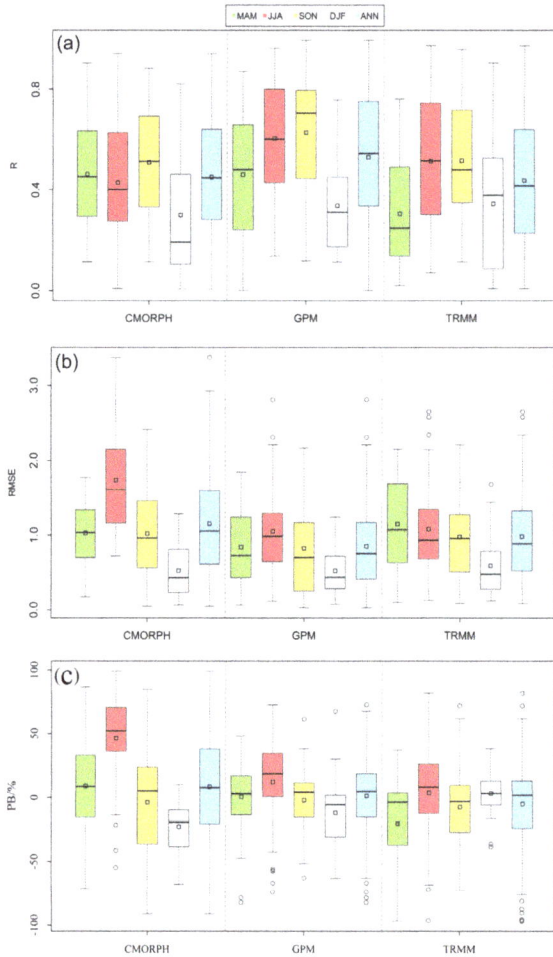

Figure 2. Results of different accuracy estimators for each satellite-based product in different seasons. The statistical metrics are (**a**) Pearson linear correlation coefficient, (**b**) root-mean-square error, and (**c**) relative bias. The solid black line represents the median value, the square represents the average value; from top to bottom, the four horizontal lines are the upper edge line, the upper quartile, the lower quartile, and the lower edge, respectively, and the empty black dots represent outliers.

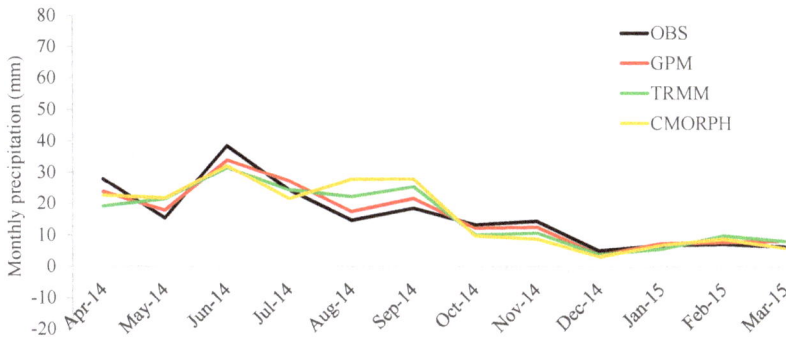

Figure 3. Comparison of the monthly precipitation patterns from the Global Precipitation Measurement (GPM), Tropical Rainfall Measurement Mission (TRMM), and Climate Prediction Center morphing technique (CMORPH) products against the OBS from April 2014 to March 2015, averaged for the 46 stations. Shaded area shows 1 standard errors (SE) of the observations. The monthly precipitation values (averaged for the 46 stations) from all satellite products fell within 1 SE of the OBS.

3.2. Accuracy in Describing Spatial Distribution of Precipitation

The satellite products correctly reflected the spatial pattern of the precipitation in the Tianshan Mountain area, where the precipitation decreased from the north to the south and from the west to the east (Figure 4). Although the overall pattern of precipitation was stable throughout the year, the precipitation gradient was stronger in spring and autumn than in the summer season. Our analysis showed a gradient in seasonality of precipitation across the study area, with precipitation peaking in late spring/early summer in the northwest (the CN TS and the northern slope of the Tianshan Mountains), and summer/early autumn in the southern (SW TS) and the eastern ranges of the Tianshan Mountains. Compared with the gauge observations, the satellite products tended to underestimate the spatial gradient in annual precipitation. The precipitation in northeastern Tianshan was overestimated, especially by the COMORPH product (Figure 4b,k,n), while the precipitation in the Ili River Valley was underestimated by the TRMM product (Figure 4a,d,m).

Among the three sub-regions, the satellite products performed relatively poorly in the CN TS, where the smallest R values and the largest RMSEs were found. The CMORPH tended to overestimate precipitation in the SW TS (Figure 5c), especially in the upper reaches of the Tarim River and Konqi River (Figures 1 and 4b,h,k). The GPM IMERG performed better than the other two satellite products in all sub-regions by all quantitative metrics (Figure 5a–c). Among the three sub-regions, the GPM had the highest R value (0.62 ± 0.14) and lowest RMSE (3.77 ± 1.63 mm/day) in the NE TS; however, it overestimated the precipitation in the NE TS by $8.96 \pm 32.67\%$ and underestimated precipitation in the CN TS by $-4.25 \pm 37.95\%$.

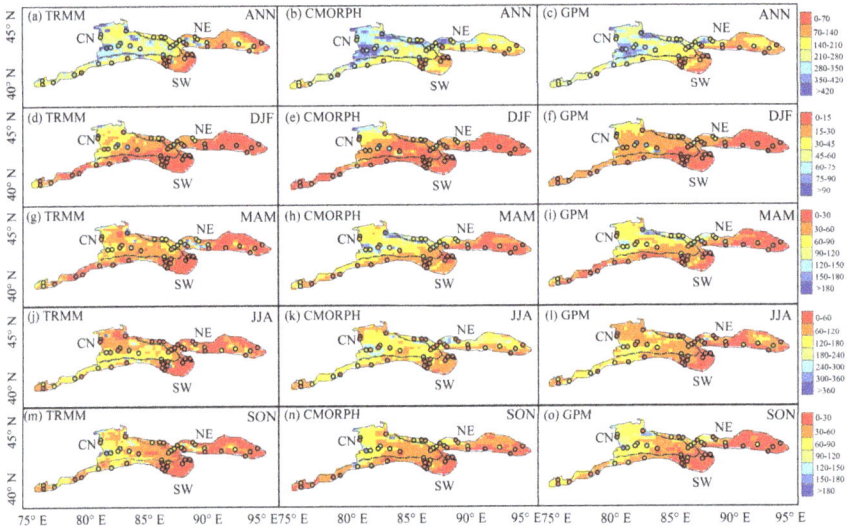

Figure 4. Spatial patterns of the observed precipitation (mm) by the gauge stations (circles) and the satellite estimated precipitations estimated by the TRMM (**a,d,g,j,m**), CMORPH (**b,e,h,k,n**), and GPM (**c,f,i,l,o**) at annual and seasonal scales. ANN: annual (**a–c**), DJF: winter (**d–f**), MAM: spring (**g–i**), JJA: summer (**j–l**), SON: autumn (**m–o**). CN: Central Tianshan area, NE: northeastern Tianshan area, SW: southwestern Tianshan.

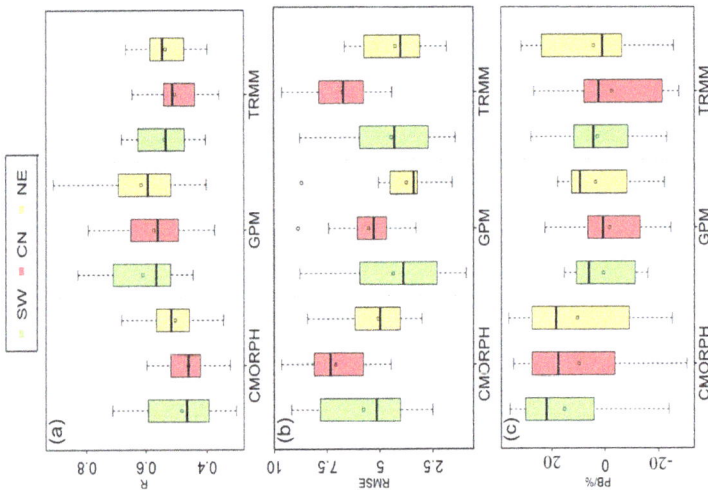

Figure 5. Results of the different accuracy estimators for each satellite-based product in different sub-regions. The statistical metrics are (**a**) Pearson linear correlation coefficient, (**b**) root-mean-square error, and (**c**) relative bias. In the box plot, the solid black line represents the median value, the square represents the average value, and the four horizontal lines from top to bottom are the upper edge line, upper quartile, lower quartile, and lower edge, respectively, and the hollow black point represents the outliers.

3.3. Elevation Impact Analyses

The elevation impacts were further analyzed by comparing the variations of the accuracies among the low-mountain (<1250 m asl.), mid-mountain (1250–2800 m asl.) and high-mountain (>2800 m asl.) ranges that were defined according to Jenks Natural Breaks [63]. All satellite products had the highest correlation with OBS in the mid-mountain areas, with the highest R value (≈0.71) found in the GPM (Figure 6). In addition, both the GPM and CMORPH had low RMSE in the mid-mountain areas, and both the GPM and TRMM had a small bias (PB ≈ 2%) in the mid-mountain areas. Although the GPM performed much better than the other two satellite products in the low-mountain areas, its correlation with the OBS in the high-mountain areas was relatively low. Notably, the CMORPH product consistently overestimated the precipitation, particularly in the low-mountain areas (PB = 29.05%). In comparison, the GPM and TRMM underestimated precipitation by 13–16% in the high-mountain areas.

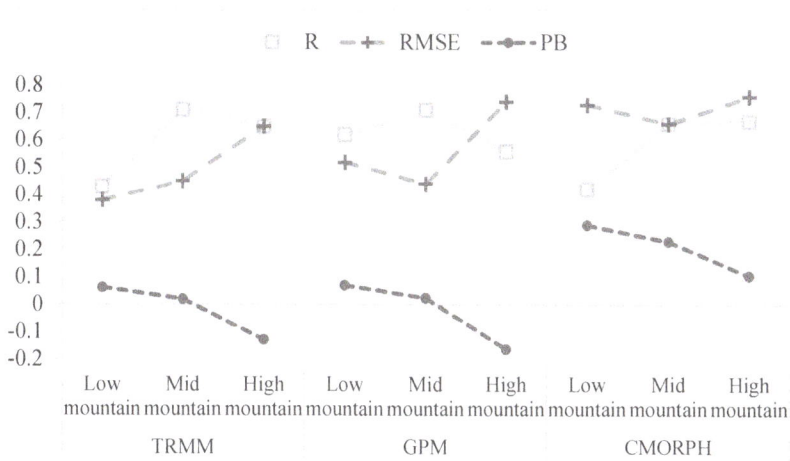

Figure 6. Comparison of the precisions of three satellite-based products at different elevation zones.

3.4. Contingency Statistics

To measure the algorithm performance for different precipitation rates, it is useful to plot the categorical scores as a function of an increasing precipitation threshold, as shown in Figure 7. According to the scores of FAR, ETS, and FBI, the GPM performed better than the other two satellite products, especially for precipitation exceeding 1–10 mm/day (Figure 7a,b,d). It also had the best performance in detecting light precipitation. It successfully detected 62.09% of the precipitation events that exceeded 0.5 mm/day (Figure 7c). The performance of the TRMM was consistently poorer than the other two satellite products, according to the scores of ETS and POD, especially for precipitation exceeding 8–15 mm/day, indicating its relatively low skill in detecting precipitation, and its tendency to miss precipitation events. The CMORPH strongly overestimated the frequency of precipitation, except for extremely heavy rains (>20 mm/day) (Figure 7a). Its FBI was as high as 3.10 for precipitation exceeding 2 mm/day. According to the scores of ETS, POD, and FAR, the performances of the satellite estimates generally declined as the thresholds of precipitation events narrowed to heavier precipitations (Figure 7b–d). Notably, all satellite estimates had high false alarm ratios (FAR > 50%).

The GPM overestimated the frequency of light precipitation (FBI > 1.5 when thresholds ∈ {0.5, 1, 2}) and underestimated the frequency of heavy precipitation (FBI < 0.5 when thresholds ∈ {15, 20}). Nevertheless, it performed well in capturing light precipitation, achieving the maximum detection skill for precipitation exceeding 0.5–2 mm/day (the ETS ranged from 0.23 to 0.24) (Figure 7b), and the lowest false alarm ratios for precipitation exceeding 1 mm/day to 4 mm/day (Figure 7d).

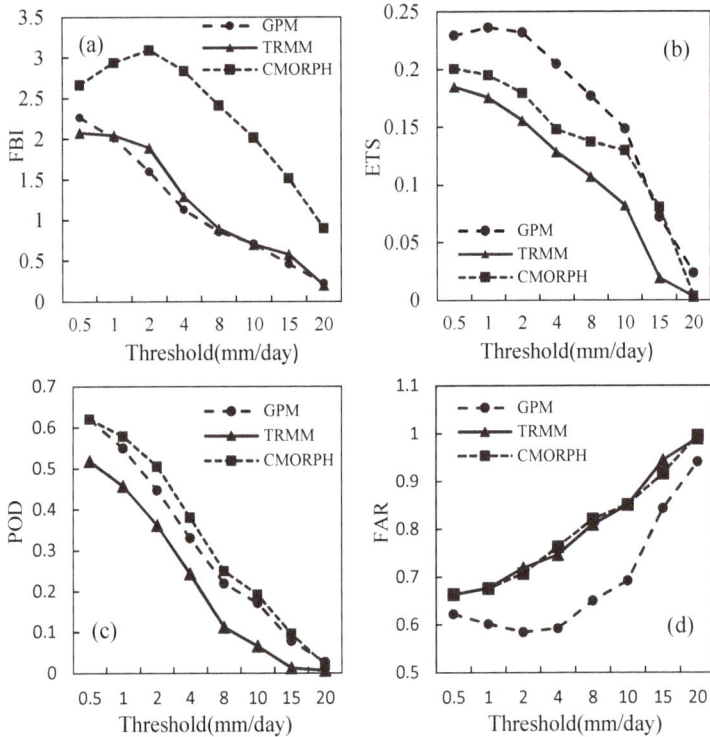

Figure 7. (**a**) frequency bias index (FBI), (**b**) equitable threat score (ETS), (**c**) probability of detection (POD), and (**d**) false alarm ratio (FAR) values for the different satellite-based products.

4. Discussion

Previous studies have suggested that satellite precipitation products like the GPM IMERG and TRMM TEMPA are unreliable in the Xinjiang Province of China [33], particularly under the influence of complex terrain [35]. Our evaluations, however, showed that the daily precipitation estimates from the three satellite products were significantly correlated with the OBS ($p < 0.05$) in the Tianshan Mountains, Xinjiang, China. The spatial patterns of the satellite precipitation products also agreed well with the observed precipitation gradient, which decreased from the north to the south, and from the west to the east (Figure 4). This pattern reflects the fact that the Tianshan Mountains as well, as Central Asia receives its moisture primarily from the westerly flow of the atmosphere. The large-scale circulation and the mountain barrier effect create a distinct continentality gradient with decreasing precipitation rates from northwest to southeast [64]. The north–south precipitation gradient, however, was weakened in the summer (Figure 4j,k,l) when monsoonal circulations from the Pacific and Indian Oceans also brought in southerly moisture fluxes [44,65,66]. In fact, the Tianshan Mountains themselves play an important role in enhancing the East Asian summer monsoon [44,65]. In addition, model simulations suggested that the Tianshan Mountains could enhance the precipitation seasonality gradient across the Central Asia dryland by blocking western winter precipitation and increasing eastern summer precipitation [66]. As a result, in Central Asia's western deserts (Kyzyl Kum and Kara-Kum), precipitation peaks in the winter and spring, while in the deserts to the east of the Tianshan

Mountains (Taklimakan and Gobi Deserts), precipitation peaks in the summer [67]. This pattern was confirmed by our satellite precipitation datasets, which showed that precipitation peaked in late spring/early summer in the northwestern Tianshan Mountains and summer/early autumn in the southern and eastern Tianshan Mountains (Figure 4). Our analysis showed that the satellite precipitation products not only correctly reflected the effects of large-scale circulations on the spatial and seasonal patterns of precipitation in the study area, but could also provide additional evidence for the orographic effects of the Tianshan Mountains on the observed precipitation seasonality gradient across the Central Asia dryland.

According to our evaluation in the Tianshan Mountain area, the GPM IMERG outperformed the TRMM 3B42V7 and CMORPH products in estimating accumulated precipitation at all spatial scales and all elevation ranges, except for the alpine region. Although previous evaluations in China indicated that the GPM products might perform worse than the TRMM products in the winter season [68,69], our study showed that the IMERG outperformed the other two satellite products in all seasons and had more balanced seasonal performances. The GPM product also had the best performance in detecting precipitation in the Tianshan Mountains, particularly in capturing light and moderate precipitation (i.e., <15 mm/day). Similarly, studies in the adjacent Tibetan Plateau showed that the GPM IMERG had appreciably better correlations, lower errors, and smaller FAR than its predecessor the TRMM 3B42V7 [35,36]. Like our study, these studies also found an improved detecting ability for light rainfall events by GPM. This improvement is particularly helpful in retrieving precipitation in the low-mountain drylands of Central Asia. The improvements could be attributed to the GPM's newly added Ka-band (35.5 GHz), which enhanced its ability to capture light precipitation [30].

It should be noted that the findings in this study do not indicate that the GPM products will outperform their predecessors in all mountainous areas. A study in a tropical mountainous watershed found no significant improvements from the IMERG in comparison with the TRMM 3B42V7 [31]. In fact, the GPM product performed worse than the TRMM 3B42V7 in heavy rain detection according to 75% of the gauge stations according to the study by Yuan et al. (2017) [31]. Although studies in India indicated a notable advantage of the IMERG over the TMPA in terms of heavy monsoon rainfall detection [70], our study did not find any obvious improvement in the GPM's ability to detect severe rainfall in the Tianshan Mountains (Figure 7) [30]. Previous studies showed that the PR attenuation correction (i.e., surface reference technique and the Hitschfeld and Bordan method applied to correct for atmospheric attenuation) for the GPM tended to underestimate convective rain, particularly for heavy rain accumulations [25,71]. Such weakness limits the GPM's usefulness in tropical mountainous areas as heavy precipitating events have significant hydrologic impacts on extreme floods and landslides in these regions. However, our study and several other studies [33,40,41] indicated that the GPM could significantly improve the estimation accuracy of precipitation over the mountainous areas in high- and mid-latitude regions, possibly because the high-frequency channels (165.6 and 183.3 GHz) added to the GMI improved the GPM's ability to detect solid precipitation and gave it the capability to provide the probability of the liquid phase for all grid boxes in the IMERG data.

The GPM, like the TRMM, underestimated precipitation by 13–16% at high-elevation regions in the Tianshan Mountains (Figures 2 and 6). Recent studies in the adjacent Tibetan Plateau also found that the GPM and TRMM underestimated precipitation at high-elevation regions [35,36]. Studies in the Great Smoky Mountains, USA indicated that the precipitation radar in the TRMM and GPM tended to underestimate low level orographic enhancement of rainfall associated with fog, cap clouds, and cloud to cloud feeder–seeder interactions over ridges [25]. Previous studies indicated that mountain precipitation associated with small-scale systems and isolated deep convection tended to be underestimated by the GPM and TRMM, which could be attributed to non-uniform beam-filling effects, due to the spatial averaging of reflectivity at their PR resolution [25,72]. In addition, the passive microwave algorithms depend primarily on scattering by ice, but the orographic rains might not produce much ice aloft, thus resulting in the underestimation of precipitation at high altitudes [73]. These uncertainties in satellite products and a shortage of high-elevation gauge stations limit our ability to assess the spatiotemporal pattern of precipitation in the alpine Tianshan Mountains, where prominent climate change is threatening the sustainability of the major glaciers in Central Asia [44,74].

Despite the uncertainties, satellite precipitation estimates are very valuable for Tianshan Mountain studies, where meteorological stations are sparse and mainly distributed in the low mountain dryland, causing biased estimation of precipitation by spatial interpolation of rain gauge data [11,13]. For example, the widely used CRU data underestimated the precipitation of Tianshan Mountain by about 34% [75]. For this reason, ecological and hydrological studies in Central Asia, including Tianshan Mountains, had to rely on climate reanalysis products (CRP; e.g., the NCEP-CFSR data [76]) as precipitation inputs [77]. Our study showed that the GPM product not only has much higher spatial resolution (5 km vs. >40 km in CRP) but also higher accuracy in estimating mountain precipitation in Central Asia than the CRP data (e.g., R_{GPM} = 0.65 vs. $R_{CRP} \leq$ 0.42; $RMSE_{GPM}$ = 0.85 mm/day vs. $RMSE_{CRP} \geq$ 1.03 mm/day; PB_{GPM} = 1.27% vs. $PB_{CRP} \geq$ 36%, where GPM and CRP denote the GPM and climate reanalysis products respectively) [9]. It is noteworthy that the GPM product performed especially well in the mid-mountain area (Figure 6), where the highest ecosystem biomass/productivity in Central Asia were found [77]. In addition, the major rivers/lakes in the Central Asia dryland mainly relied on the water yields from the mid-mountain area of Tianshan [42]. Therefore, the ecological and hydrological studies in the Tianshan Mountain and Central Asia areas can be significantly benefit by using the new generation of satellite precipitation products—the GPM IMERG.

5. Conclusions

Estimates by the satellite precipitation products significantly correlated with gauge observations. They showed a northwest–southeast precipitation gradient that reflected the effects of large-scale circulations and showed characteristic seasonal precipitation patterns among different sub-regions that matched the observed precipitation seasonality gradient in the study region. Among the high-resolution satellite products evaluated in this study, the GPM IMERG outperformed the TRMM and CMORPH in estimating the accumulated precipitation at all temporal (daily and seasonal) and spatial (regional and sub-regional) scales, and in low and mid-elevation regions. The GPM had relatively balanced performances across all seasons, while the CMORPH had significant biases in summer and winter, and the TRMM performed extremely poorly in spring. In addition, the GPM had the best performance in detecting precipitation events, especially for light and moderate precipitation, possibly due to the newly added Ka-band and high-frequency microwave channels. However, the dual-frequency precipitation radar did not significantly improve the GPM's ability to estimate severe rainfall, as expected. The satellite product had the highest RMSE and bias in summer when small-scale convective rain was common in Tianshan Mountains. These findings suggest that GPM may outperform its predecessors in the mid- or high-latitude dryland areas, but not in the tropical mountainous areas. While this short time period in this study limits the representativeness of the temporal pattern in weather, previous studies with longer time periods generally support these findings.

The improved accuracy in satellite precipitation data creates new opportunities for understanding the spatial variation of precipitation across complex terrains, and it provides alternatives for estimating rainfall in areas where field observations are inadequate or unreliable. The GPM IMERG has relatively high accuracy and shows more spatially detailed information when compared with its predecessors, which will extend the application of satellite precipitation data from a global or national scale to a regional scale, especially in the mountainous areas in mid or high-latitudes. Previous evaluation studies based on hourly data suggested that the GPM IMERG showed appreciably better correlations and lower errors than the TRMM 3B42V7 for all assessment indicators [35]. Moreover, the higher temporal resolution (30 min) of the GPM IMERG in comparison with the TRMM products (3 h) can help to eliminate the anomalous values in the TRMM-estimated precipitation and could improve the accuracy of hydrological modeling [33].

Author Contributions: Conceptualization, C.Z. and X.C.; Data curation, C.C. and Q.D.; Formal analysis, C.Z. and H.D.; Funding acquisition, C.Z. and X.C.; Investigation, C.Z. and T.L.; Methodology, H.S. and T.L.; Resources, S.C. and Q.D.; Software, S.C. and C.C.; Supervision, C.Z.; Validation, H.S. and H.D.; Visualization, H.D.; Writing—original draft, C.Z. and H.D.; Writing—review & editing, C.Z. and H.D.

Funding: This project was funded by the Strategic Priority Research Program of Chinese Academy of Sciences, Grant No. XDA2006030201 and Chi Zhang is supported by the Taishan Scholars Program of Shandong, China, Grant No. ts201712071.

Conflicts of Interest: The authors declare no conflict of interest. The funders had no role in the design of the study; in the collection, analyses, or interpretation of data; in the writing of the manuscript, and in the decision to publish the results.

References

1. Wheater, H.S.; Isham, V.S.; Cox, D.R.; Chandler, R.E.; Kakou, A.; Northrop, P.J.; Oh, L.; Onof, C.; Rodriguez-Iturbe, I. Spatial-temporal rainfall fields: Modelling and statistical aspects. *Hydrol. Earth Syst. Sci.* **2000**, *4*, 581–601. [CrossRef]

2. Wilheit, T.T. Some comments on passive microwave measurement of rain. *Bull. Am. Meteorol. Soc.* **1986**, *67*, 1226–1232. [CrossRef]

3. Boe, J.; Terray, L.; Cassou, C.; Najac, J. Uncertainties in european summer precipitation changes: Role of large scale circulation. *Clim. Dyn.* **2009**, *33*, 265–276. [CrossRef]

4. Chung, C.T.; Power, S.B.; Arblaster, J.M.; Rashid, H.A.; Roff, G.L. Nonlinear precipitation response to el niño and global warming in the indo-pacific. *Clim. Dyn.* **2014**, *42*, 1837–1856. [CrossRef]

5. Siegfried, T.; Bernauer, T.; Guiennet, R.; Sellars, S.; Robertson, A.W.; Mankin, J.; Bauer-Gottwein, P.; Yakovlev, A. Will climate change exacerbate water stress in Central Asia? *Clim. Chang.* **2012**, *112*, 881–899. [CrossRef]

6. Seddon, A.W.R.; Macias-Fauria, M.; Long, P.R.; Benz, D.; Willis, K.J. Sensitivity of global terrestrial ecosystems to climate variability. *Nature* **2016**, *531*, 229–232. [CrossRef] [PubMed]

7. Bohner, J. General climatic controls and topoclimatic variations in central and high Asia. *Boreas* **2006**, *35*, 279–295. [CrossRef]

8. Domrös, M.; Peng, G. *The Climate of China*; Springer Science & Business Media: Berlin/Heidelberg, Germany, 2012.

9. Hu, Z.Y.; Hu, Q.; Zhang, C.; Chen, X.; Li, Q.X. Evaluation of reanalysis, spatially interpolated and satellite remotely sensed precipitation data sets in Central Asia. *J. Geophys. Res. Atmos.* **2016**, *121*, 5648–5663. [CrossRef]

10. Schiemann, R.; Luthi, D.; Vidale, P.L.; Schar, C. The precipitation climate of Central Asia—Intercomparison of observational and numerical data sources in a remote semiarid region. *Int. J. Climatol.* **2008**, *28*, 295–314. [CrossRef]

11. Lioubimtseva, E.; Cole, R. Uncertainties of climate change in arid environments of Central Asia. *Rev. Fish. Sci.* **2006**, *14*, 29–49. [CrossRef]

12. Mitchell, T.D.; Jones, P.D. An improved method of constructing a database of monthly climate observations and associated high-resolution grids. *Int. J. Climatol.* **2005**, *25*, 693–712. [CrossRef]

13. Michaelides, S.; Levizzani, V.; Anagnostou, E.; Bauer, P.; Kasparis, T.; Lane, J. Precipitation: Measurement, remote sensing, climatology and modeling. *Atmos. Res.* **2009**, *94*, 512–533. [CrossRef]

14. Hsu, K.-L.; Gao, X.; Sorooshian, S.; Gupta, H.V. Precipitation estimation from remotely sensed information using artificial neural networks. *J. Appl. Meteorol.* **1997**, *36*, 1176–1190. [CrossRef]

15. Turk, F.J.; Miller, S.D. Toward improved characterization of remotely sensed precipitation regimes with modis/amsr-e blended data techniques. *IEEE Trans. Geosci. Remote Sens.* **2005**, *43*, 1059–1069. [CrossRef]

16. Joyce, R.J.; Janowiak, J.E.; Arkin, P.A.; Xie, P.P. CMORPH: A method that produces global precipitation estimates from passive microwave and infrared data at high spatial and temporal resolution. *J. Hydrometeorol.* **2004**, *5*, 487–503. [CrossRef]

17. Kummerow, C.; Barnes, W.; Kozu, T.; Shiue, J.; Simpson, J. The tropical rainfall measuring mission (TRMM) sensor package. *J. Atmos. Ocean. Technol.* **1998**, *15*, 809–817. [CrossRef]

18. Hou, A.Y.; Kakar, R.K.; Neeck, S.; Azarbarzin, A.A.; Kummerow, C.D.; Kojima, M.; Oki, R.; Nakamura, K.; Iguchi, T. The global precipitation measurement mission. *Bull. Am. Meteorol. Soc.* **2014**, *95*, 701–722. [CrossRef]

19. Su, F.G.; Gao, H.L.; Huffman, G.J.; Lettenmaier, D.P. Potential utility of the real-time TMPA-rt precipitation estimates in streamflow prediction. *J. Hydrometeorol.* **2011**, *12*, 444–455. [CrossRef]

20. Xue, X.W.; Hong, Y.; Limaye, A.S.; Gourley, J.J.; Huffman, G.J.; Khan, S.I.; Dorji, C.; Chen, S. Statistical and hydrological evaluation of TRMM-based multi-satellite precipitation analysis over the wangchu basin of bhutan: Are the latest satellite precipitation products 3B42V7 ready for use in ungauged basins? *J. Hydrol.* **2013**, *499*, 91–99. [CrossRef]

21. Yang, Y.F.; Luo, Y. Evaluating the performance of remote sensing precipitation products CMORPH, persiann, and TMPA, in the arid region of northwest China. *Theor. Appl. Climatol.* **2014**, *118*, 429–445. [CrossRef]

22. Bellerby, T.J. Satellite rainfall uncertainty estimation using an artificial neural network. *J. Hydrometeorol.* **2007**, *8*, 1397–1412. [CrossRef]

23. AghaKouchak, A.; Nasrollahi, N.; Habib, E. Accounting for uncertainties of the TRMM satellite estimates. *Remote Sens.* **2009**, *1*, 606–619. [CrossRef]

24. Hossain, F.; Anagnostou, E.N.; Bagtzoglou, A.C. On latin hypercube sampling for efficient uncertainty estimation of satellite rainfall observations in flood prediction. *Comput. Geosci.* **2006**, *32*, 776–792. [CrossRef]

25. Duan, Y.; Wilson, A.M.; Barros, A.P. Scoping a field experiment: Error diagnostics of TRMM precipitation radar estimates in complex terrain as a basis for iphex2014. *Hydrol. Earth Syst. Sci.* **2015**, *19*, 1501–1520. [CrossRef]

26. Huffman, G.J.; Bolvin, D.T.; Nelkin, E.J. Integrated multi-satellite retrievals for GPM (IMERG) technical documentation. *NASA/GSFC Code* **2015**, *612*, 47.

27. Huffman, G.J.; Bolvin, D.T.; Braithwaite, D.; Hsu, K.; Joyce, R.; Xie, P.; Yoo, S.-H. Nasa global precipitation measurement (GPM) integrated multi-satellite retrievals for GPM (IMERG). *Algorithm Theor. Basis Doc. Version* **2015**, *4*, 30.

28. Liu, Z. Comparison of integrated multisatellite retrievals for GPM (IMERG) and TRMM multisatellite precipitation analysis (TMPA) monthly precipitation products: Initial results. *J. Hydrometeorol.* **2016**, *17*, 777–790. [CrossRef]

29. Prakash, S.; Mitra, A.K.; AghaKouchak, A.; Liu, Z.; Norouzi, H.; Pai, D.S. A preliminary assessment of GPM-based multi-satellite precipitation estimates over a monsoon dominated region. *J. Hydrol.* **2018**, *556*, 865–876. [CrossRef]

30. Liu, C.T.; Zipser, E.J. The global distribution of largest, deepest, and most intense precipitation systems. *Geophys. Res. Lett.* **2015**, *42*, 3591–3595. [CrossRef]

31. Yuan, F.; Zhang, L.; Win, K.W.W.; Ren, L.; Zhao, C.; Zhu, Y.; Jiang, S.; Liu, Y. Assessment of GPM and TRMM multi-satellite precipitation products in streamflow simulations in a data-sparse mountainous watershed in myanmar. *Remote Sens.* **2017**, *9*, 302. [CrossRef]

32. Tan, J.; Petersen, W.A.; Tokay, A. A novel approach to identify sources of errors in IMERG for GPM ground validation. *J. Hydrometeorol.* **2016**, *17*, 2477–2491. [CrossRef]

33. Chen, F.R.; Li, X. Evaluation of IMERG and TRMM 3B43 monthly precipitation products over mainland China. *Remote Sens.* **2016**, *8*, 18. [CrossRef]

34. Lu, X.; Wei, M.; Tang, G.; Zhang, Y. Evaluation and correction of the TRMM 3B43v7 and GPM 3IMERGM satellite precipitation products by use of ground-based data over Xinjiang, China. *Environ. Earth Sci.* **2018**, *77*, 209. [CrossRef]

35. Ma, Y.Z.; Tang, G.Q.; Long, D.; Yong, B.; Zhong, L.Z.; Wan, W.; Hong, Y. Similarity and error intercomparison of the GPM and its predecessor-TRMM multisatellite precipitation analysis using the best available hourly gauge network over the Tibetan Plateau. *Remote Sens.* **2016**, *8*, 17. [CrossRef]

36. Xu, R.; Tian, F.Q.; Yang, L.; Hu, H.C.; Lu, H.; Hou, A.Z. Ground validation of GPM IMERG and TRMM 3B42V7 rainfall products over southern Tibetan Plateau based on a high-density rain gauge network. *J. Geophys. Res. Atmos.* **2017**, *122*, 910–924. [CrossRef]

37. Wei, G.; Lü, H.; Crow, W.T.; Zhu, Y.; Wang, J.; Su, J. Evaluation of satellite-based precipitation products from IMERG V04A and V03D, CMORPH and TMPA with gauged rainfall in three climatologic zones in China. *Remote Sens.* **2017**, *10*, 30. [CrossRef]

38. Qin, Y.; Chen, Z.; Shen, Y.; Zhang, S.; Shi, R. Evaluation of satellite rainfall estimates over the Chinese mainland. *Remote Sens.* **2014**, *6*, 11649–11672. [CrossRef]

39. Zhao, T.; Yatagai, A. Evaluation of TRMM 3B42 product using a new gauge-based analysis of daily precipitation over China. *Int. J. Climatol.* **2014**, *34*, 2749–2762. [CrossRef]

40. Kim, J.H.; Ou, M.L.; Park, J.D.; Morris, K.R.; Schwaller, M.R.; Wolff, D.B. Global precipitation measurement (GPM) ground validation (GV) prototype in the korean peninsula. *J. Atmos. Ocean. Technol.* **2014**, *31*, 1902–1921. [CrossRef]

41. Golian, S.; Moazami, S.; Kirstetter, P.E.; Hong, Y. Evaluating the performance of merged multi-satellite precipitation products over a complex terrain. *Water Resour. Manag.* **2015**, *29*, 4885–4901. [CrossRef]

42. Chen, X. *Retrieval and Analysis of Evapotranspiration in Central Areas of Asia*; China Meteorological News Pres: Beijing, China, 2012; p. 8.

43. Duque, J.C.; Ramos, R.; Suriñach, J. Supervised regionalization methods: A survey. *Int. Reg. Sci. Rev.* **2007**, *30*, 195–220. [CrossRef]

44. Sorg, A.; Bolch, T.; Stoffel, M.; Solomina, O.; Beniston, M. Climate change impacts on glaciers and runoff in Tien Shan (Central Asia). *Nat. Clim. Chang.* **2012**, *2*, 725–731. [CrossRef]

45. Tong, K.; Su, F.G.; Yang, D.Q.; Hao, Z.C. Evaluation of satellite precipitation retrievals and their potential utilities in hydrologic modeling over the Tibetan Plateau. *J. Hydrol.* **2014**, *519*, 423–437. [CrossRef]

46. Guo, H.; Chen, S.; Bao, A.M.; Hu, J.J.; Gebregiorgis, A.S.; Xue, X.W.; Zhang, X.H. Inter-comparison of high-resolution satellite precipitation products over Central Asia. *Remote Sens.* **2015**, *7*, 7181–7211. [CrossRef]

47. Huffman, G.; Adler, R.; Bolvin, D.; Nelkin, E. *The TRMM Multi-Satellite Precipitation Analysis (TMPA). Chapter in Satellite Applications for Surface Hydrology*; Hossain, F., Gebremichael, M., Eds.; Springer Verlag: Berlin/Heidelberg, Germany, 2009. Available online: ftp://meso.gsfc.nasa.gov/agnes/huffman/papers/TMPA_hydro_rev.pdf (accessed on 12 January 2017).

48. GPM IMERG. Available online: https://pmm.nasa.gov/data-access/downloads/GPM (accessed on 12 January 2017).

49. TRMM 3B42V7. Available online: https://pmm.nasa.gov/data-access/downloads/TRMM (accessed on 15 January 2017).

50. CMORPH. Available online: https://rda.ucar.edu/datasets/ds502.0/ (accessed on 15 January 2017).

51. Daily Precipitation Data. Available online: http://222.82.235.66/RadarDoc/WeatherService.aspx (accessed on 15 January 2017).

52. Shen, Y.; Xiong, A.Y.; Wang, Y.; Xie, P.P. Performance of high-resolution satellite precipitation products over China. *J. Geophys. Res. Atmos.* **2010**, *115*, 1–17. [CrossRef]

53. Ren, Z.; Zhao, P.; Zhang, Q.; Zhang, Z.; Cao, L.; Yang, Y.; Zou, F.; Zhao, Y.; Zhao, H.; Chen, Z. Quality control procedures for hourly precipitation data from automatic weather stations in China. *Meteorol. Mon.* **2010**, *36*, 123–132.

54. GPCC. Available online: https://kunden.dwd.de/GPCC/Visualizer (accessed on 17 January 2017).

55. Blacutt, L.A.; Herdies, D.L.; de Goncalves, L.G.G.; Vila, D.A.; Andrade, M. Precipitation comparison for the CFSR, MERRA, TRMM3B42 and combined scheme datasets in Bolivia. *Atmos. Res.* **2015**, *163*, 117–131. [CrossRef]

56. Gao, Y.C.; Liu, M.F. Evaluation of high-resolution satellite precipitation products using rain gauge observations over the Tibetan Plateau. *Hydrol. Earth Syst. Sci.* **2013**, *17*, 837–849. [CrossRef]

57. Accadia, C.; Mariani, S.; Casaioli, M.; Lavagnini, A.; Speranza, A. Sensitivity of precipitation forecast skill scores to bilinear interpolation and a simple nearest-neighbor average method on high-resolution verification grids. *Weather Forecast.* **2003**, *18*, 918–932. [CrossRef]

58. Daly, C.; Neilson, R.P.; Phillips, D.L. A statistical topographic model for mapping climatological precipitation over mountainous terrain. *J. Appl. Meteorol.* **1994**, *33*, 140–158. [CrossRef]

59. Basist, A.; Bell, G.D.; Meentemeyer, V. Statistical relationships between topography and precipitation patterns. *J. Clim.* **1994**, *7*, 1305–1315. [CrossRef]

60. GTOPO30. Available online: http://eros.usgs.gov (accessed on 17 January 2017).

61. Tian, Y.D.; Peters-Lidard, C.D.; Eylander, J.B.; Joyce, R.J.; Huffman, G.J.; Adler, R.F.; Hsu, K.L.; Turk, F.J.; Garcia, M.; Zeng, J. Component analysis of errors in satellite-based precipitation estimates. *J. Geophys. Res. Atmos.* **2009**, *114*, 1–15. [CrossRef]

62. El Kenawy, A.M.; Lopez-Moreno, J.I.; McCabe, M.F.; Vicente-Serrano, S.M. Evaluation of the TMPA-3B42 precipitation product using a high-density rain gauge network over complex terrain in northeastern Iberia. *Glob. Planet. Chang.* **2015**, *133*, 188–200. [CrossRef]

63. Jenks, G.F.; Caspall, F.C. Error on choroplethic maps: Definition, measurement, reduction. *Ann. Assoc. Am. Geogr.* **1971**, *61*, 217–244. [CrossRef]

64. Aizen, V.B.; Aizen, E.M.; Melack, J.M.; Dozier, J. Climatic and hydrologic changes in the Tien Shan, Central Asia. *J. Clim.* **1997**, *10*, 1393–1404. [CrossRef]

65. Tang, H.; Micheels, A.; Eronen, J.T.; Ahrens, B.; Fortelius, M. Asynchronous responses of east Asian and indian summer monsoons to mountain uplift shown by regional climate modelling experiments. *Clim. Dyn.* **2013**, *40*, 1531–1549. [CrossRef]

66. Baldwin, J.; Vecchi, G. Influence of the tian shan on arid extratropical Asia. *J. Clim.* **2016**, *29*, 5741–5762. [CrossRef]

67. Bothe, O.; Fraedrich, K.; Zhu, X. Precipitation climate of Central Asia and the large-scale atmospheric circulation. *Theor. Appl. Climatol.* **2012**, *108*, 345–354. [CrossRef]

68. Guo, H.; Chen, S.; Bao, A.; Behrangi, A.; Hong, Y.; Ndayisaba, F.; Hu, J.; Stepanian, P.M. Early assessment of integrated multi-satellite retrievals for global precipitation measurement over China. *Atmos. Res.* **2016**, *176*, 121–133. [CrossRef]

69. Tang, G.; Ma, Y.; Long, D.; Zhong, L.; Hong, Y. Evaluation of GPM day-1 IMERG and TMPA version-7 legacy products over mainland China at multiple spatiotemporal scales. *J. Hydrol.* **2016**, *533*, 152–167. [CrossRef]

70. Prakash, S.; Mitra, A.K.; Pai, D.; AghaKouchak, A. From TRMM to GPM: How well can heavy rainfall be detected from space? *Adv. Water Resour.* **2016**, *88*, 1–7. [CrossRef]

71. Liao, L.; Meneghini, R. Validation of TRMM precipitation radar through comparison of its multiyear measurements with ground-based radar. *J. Appl. Meteorol. Climatol.* **2009**, *48*, 804–817. [CrossRef]

72. Durden, S.L.; Haddad, Z.S.; Kitiyakara, A.; Li, F.K. Effects of nonuniform beam filling on rainfall retrieval for the TRMM precipitation radar. *J. Atmos. Ocean. Technol.* **1998**, *15*, 635–646. [CrossRef]

73. Dinku, T.; Chidzambwa, S.; Ceccato, P.; Connor, S.J.; Ropelewski, C.F. Validation of high-resolution satellite rainfall products over complex terrain. *Int. J. Remote Sens.* **2008**, *29*, 4097–4110. [CrossRef]

74. Sorg, A.; Huss, M.; Rohrer, M.; Stoffel, M. The days of plenty might soon be over in glacierized Central Asian catchments. *Environ. Res. Lett.* **2014**, *9*, 104018. [CrossRef]

75. Yin, G.; Chen, X.; Tiyip, T.; Shao, H.; Bai, L.; Hu, Z.; Zhang, C.; Xu, T. A comparison study between site-extrapolation-based and regional climate model-simulated climate datasets. *Geogr. Res.* **2015**, *34*, 631–643.

76. Saha, S.; Moorthi, S.; Pan, H.-L.; Wu, X.; Wang, J.; Nadiga, S.; Tripp, P.; Kistler, R.; Woollen, J.; Behringer, D. The NCEP climate forecast system reanalysis. *Bull. Am. Meteorol. Soc.* **2010**, *91*, 1015–1058. [CrossRef]

77. Zhang, C.; Ren, W. Complex climatic and CO_2 controls on net primary productivity of temperate dryland ecosystems over Central Asia during 1980–2014. *J. Geophys. Res. Biogeosci.* **2017**, *122*, 2356–2374. [CrossRef]

remote sensing

MDPI

Article

Tropical Cyclone Rainfall Estimates from FY-3B MWRI Brightness Temperatures Using the WS Algorithm

Ruanyu Zhang [1,2], Zhenzhan Wang [1,*] and Kyle A. Hilburn [3]

[1]   Key Laboratory of Microwave Remote Sensing, National Space Science Center, Chinese Academy of Sciences, Beijing 100190, China; ruanyu_zhang@163.com (R.Z.); wangzhenzhan@mirslab.cn (Z.W.)

[2]   University of Chinese Academy of Sciences, Beijing 100049, China

[3]   Cooperative Institute for Research in the Atmosphere, Colorado State University, Fort Collins, CO 80523, USA; kyle.hilburn@colostate.edu

*   Correspondence: wangzhenzhan@mirslab.cn; Tel.: +86-10-6258-6454

Received: 24 September 2018; Accepted: 5 November 2018; Published: 8 November 2018

Abstract: A rainfall retrieval algorithm for tropical cyclones (TCs) using 18.7 and 36.5 GHz of vertically and horizontally polarized brightness temperatures (Tbs) from the Microwave Radiation Imager (MWRI) is presented. The beamfilling effect is corrected based on ratios of the retrieved liquid water absorption and theoretical Mie absorption coefficients at 18.7 and 36.5 GHz. To assess the performance of this algorithm, MWRI measurements are matched with the National Snow and Ice Data Center (NSIDC) precipitation for six TCs. The comparison between MWRI and NSIDC rain rates is relatively encouraging, with a mean bias of -0.14 mm/h and an overall root-mean-square error (RMSE) of 1.99 mm/h. A comparison of pixel-to-pixel retrievals shows that MWRI retrievals are constrained to reasonable levels for most rain categories, with a minimum error of -1.1% in the 10–15 mm/h category; however, with maximum errors around -22% at the lowest (0–0.5 mm/h) and highest rain rates (25–30 mm/h). Additionally, Advanced Microwave Scanning Radiometer for Earth Observing System (AMSR-E) Tbs are applied to retrieve rain rates to assess the sensitivity of this algorithm, with a mean bias and RMSE of 0.90 mm/h and 3.11 mm/h, respectively. For the case study of TC Maon (2011), MWRI retrievals underestimate rain rates over 6 mm/h and overestimate rain rates below 6 mm/h compared with Precipitation Radar (PR) observations on storm scales. The Tropical Rainfall Measuring Mission (TRMM) Microwave Imager (TMI) rainfall data provided by the Remote Sensing Systems (RSS) are applied to assess the representation of mesoscale structures in intense TCs, and they show good consistency with MWRI retrievals.

Keywords: tropical cyclone; rain rate; precipitation; remote sensing; radiometer; retrieval algorithm

1. Introduction

Tropical cyclones (TCs) are high-impact meteorological phenomena accompanied by high winds, torrential precipitation, and storm surges, and occur in nearly every ocean basin in the Northern Hemisphere every year, which not only has a dramatic impact on coastal and inland regions, but also brings catastrophic impacts to human life, property, and ecology [1–3]. Hence, improving TC forecasts is an important field of study. Tropical precipitating cloud systems are one of the most important elements in the global climate system [4]. Globally, precipitation measurements did not become available until the era of satellite Earth observations which employ infrared and microwave radiometric techniques [5]. The advantage of microwave frequency is that microwaves penetrate or "see through" clouds with little attenuation and give an uninterrupted observation down to the sea surface during the day and night [6]. This is a distinct advantage over space-borne infrared and optical measurements that are obstructed by clouds due to their shorter wavelengths than microwaves [7]. Hence, passive microwave remote sensing plays a key role in providing vital information for understanding and studying global weather and climate changes. Since most of the large range of TCs often occurs over oceans, TC rain rates are difficult to quantify based on ground-based instruments such as rain gauges and radars. Even if coastal precipitation measurements are detected by rain gauges, they do not adequately detect and measure the structure and rain rate of TCs over oceans or their surrounding environment. The surface radar datasets have a limited spatial extent [8]. Consequently, satellite-derived measurements are frequently used to detect Earth system processes and parameters, especially to monitor the tracks of TCs. Although satellite observations have relatively large random errors at small scales, their global nature makes them suitable for addressing potential changes in global precipitation extremes [9].

Passive microwave radiometers onboard low-Earth orbiting satellites are able to quickly cover the globe, providing approximately two overpasses per day at a given location [10]. Radiometers have been widely applied to global observations and measurements of Earth system processes and parameters, especially monitoring TCs' development and characterizing mesoscale structures. The first passive microwave instrument in orbit dates back to the mid-1960s [11]. The first Special Sensor Microwave Imager (SSM/I) aboard the Defense Meteorological Satellite Program (DMSP) was launched successfully in 1987 [12], making it possible to provide measurements of atmospheric moisture and global surface precipitation [13]. The Microwave Imager (TMI) and Precipitation Radar (PR) onboard the Tropical Rainfall Measuring Mission (TRMM) satellite began operating in 1997 to provide detailed satellite measurements between 40°N and 40°S [14]. Based on the success of TRMM, the Global Precipitation Measurement (GPM) core instruments Microwave Imager (GMI) and dual-frequency Precipitation Radar (DPR) provide improved capabilities of precipitation measurements [15]. Blended products of global rainfall measurements are available at 30 min of temporal resolution across the globe [16]. The Advanced Microwave Scanning Radiometer for Earth Observing System (AMSR-E) was on board the National Aeronautics and Space Administration (NASA) Aqua satellite working from 2002 to 2011 [17]. The Microwave Radiation Imager (MWRI), the first space-borne microwave sensor in China, began operating in 2010 [18].

The concept, using satellite microwave observations to retrieve accurate oceanic precipitation rates, was first proposed by [19]. With the rapid development of satellite-based instruments, a variety of models and algorithms for retrieving precipitation rates based on microwave remote sensing have been developed and have evolved to extract information from passive microwave radiometer measurements [8]. These precipitation retrieval algorithms are generally classified into two schemes: empirical and physically based schemes. The empirical models are generally based on the simple empirical relationship between the brightness temperature (Tb) observed by radiometers and rain rates detected by conventional instruments, such as scatter methods [20,21], and polarization-corrected temperature (PCT) [22,23]. However, empirical approaches are difficult to extend to global regions, while physical methods generally provide more precipitation parameters, such as vertical profiles of precipitation and hydrometeor types [8,24]. Over the ocean, physical methods have been developed to retrieve rain rates for SSM/I, TMI, GMI, AMSR, AMSR-E, and the Advanced Microwave Scanning Radiometer 2 (AMSR2) [13,25–29], using the relationship between Tbs and rain rate. However, there are few studies based on MWRI observations [30,31].

Over the past 20 years, TC studies based on radiometers have mainly focused on three aspects: TC best-track databases [32], meteorological characteristics of TCs [33], and rainfall retrievals of TCs [24,34]. In surface rainfall retrieval algorithms, however, deficiencies have been shown when compared with estimates of radar and TMI precipitation observations, with an underestimation of inner-core and high rain rate regions of TCs [34]. MWRI Tbs observations have not yet been thoroughly utilized in detecting surface precipitation rates in TCs, while the rainfall and hydrometeor profiles have been retrieved based on the Goddard profiling (GPROF [35]) algorithm [30]. Therefore, the purpose of this paper is to find a suitable methodology for MWRI observations to estimate the surface rain intensities of TCs over oceans. For accurate radiometer retrievals of surface rain rates, methods combining different frequencies have been used. A physical method is developed to retrieve surface rain rate using vertically and horizontally polarized Tbs provided by the SSM/I, AMSR and AMSR-E instruments [7,13,25], which ignores precipitation scattering effects and only considers the emission effects on measurements. However, these algorithms mainly focus on global precipitation estimates. In an attempt to demonstrate the rainfall retrieval capability of TCs utilizing measurements of microwave radiometers, it is necessary to do more studies on MWRI observations.

This paper advances the effort to estimate surface rain rates of TCs based on the Wentz and Spencer (hereafter, WS) physical algorithm [13] using FY-3B MWRI 18.7 and 36.5 GHz vertically and horizontally polarized Tb measurements. A brief description of MWRI channel characteristics is given in Section 2. Section 3 introduces all the datasets used in this paper, including MWRI Tbs, AMSR-E Tbs, the Global Forecast System (GFS) environmental product, the National Snow and Ice Data Center (NSIDC) rainfall product, and the TRMM rainfall products. The algorithm scheme for surface rainfall retrievals using MWRI Tbs is described in Section 4. Detailed results of the MWRI WS algorithm are provided and evaluated in Section 5, including the case study of TC Maon (2011). Section 5 also presents mesoscale structure comparisons of select TCs. Section 6 provides a discussion of the results. Finally, Section 7 draws conclusions based on our results.

2. Instrument Description

The Feng-Yun 3B (FY-3B), a Chinese second-generation afternoon-configured polar-orbiting satellite, travels 14.17 orbits over 24 h and crosses the equator in ascending modes at 1:40 P.M. local time with a revisit period of 6 days [36]. The MWRI instrument on board the FY-3B satellite conically scans the Earth with a swath width of 1400 km and a viewing angle of 45° [18]. It is a total power passive radiometer that has five frequencies at 10.65, 18.7, 23.8, 36.5, and 89.0 GHz, and each frequency has vertical (V) and horizontal (H) polarizations.

The channel configuration and viewing geometry of MWRI are very similar to the existing microwave imagers, such as AMSR-E, SSM/I and TMI, which provide comparable measurements and products as MWRI retrievals. The instrument performance requirements of MWRI are provided in Table 1. Since MWRI's launch, the sensor has been operating with high stability, providing continuous observations. MWRI collects observations under different weather conditions during the day and night to monitor total precipitation water (TPW), cloud liquid water, surface rain, snow–water equivalent, sea surface temperature (SST), wind speed, and sea ice constant. These parameters are helpful for improving the accuracy of weather forecasting and for TC monitoring.

Table 1. The main specifications of the FY-3B Microwave Radiation Imager (MWRI) [36].

Frequency (GHz)	10.65	18.7	23.8	36.5	89.0
Polarization	V/H	V/H	V/H	V/H	V/H
Resolution (km)	51×85	30×50	27×45	18×30	9×15
Sensitivity (K)	0.5	0.5	0.5	0.5	0.8
Calibration error (K)	1.5	1.5	1.5	1.5	1.5
Swath width			1400 km		
Scan mode			Conical scanning		
Scan cycle			1.8 s		
Viewing angle			45°		
Samples			254		
Sampling interval			2.08 ms		

3. Datasets and Processing

This study employs the MWRI Tbs, the AMSR-E Tbs, the NSIDC rain rates, the PR surface rain rates, and the Remote Sensing Systems (RSS) TMI rainfall product to explore the accuracy and mesoscale structures of MWRI rainfall retrievals for six selected TCs. The domain is limited to the western North Pacific Ocean (5°N–50°N, 100°E–175°E) to form a geographically well-understood test bed. The GFS is used to provide environmental parameters, such as SST, TPW, and wind speed for this rainfall retrieval algorithm. The NSIDC rainfall product and the TRMM precipitation data serve as the evaluation reference and a priori surface rain rates. Additionally, the AMSR-E Tb dataset provided by the NSIDC is used to retrieve rain rates for the same selected TCs which will be compared with MWRI retrievals.

3.1. Brightness Temperature

The Tbs used in the WS algorithm are the FY-3B level-1 MWRI products provided by the National Satellite Meteorological Center Feng Yun Satellite Data Center (http://satellite.nsmc.org.cn/Portal-Site/default.aspx). Yang et al. [18] developed a unique calibration system designed with a main reflector viewing both cold and hot calibration targets for the MWRI system. The calibration targets and the Earth scene are all viewed by the same main reflector, and the calibration is proven to work well. As shown in Table 1, MWRI has five frequencies varying from 10.65 to 89 GHz, and the absolute calibration errors are 1.5 K for all frequencies. The details of the calibration approach and accuracy evaluation are described in [18]. During each scan, each channel consists of different locations and resolution. The sensitivity of Tbs at different channels varies with atmospheric and surface geophysical parameters [36]. MWRI includes a 6.9 GHz frequency, and its poor spatial resolution will limit the usefulness for rainfall retrievals [29]. Therefore, the 18.7 and 36.5 GHz Tbs (hereafter, 19 and 37 GHz) are chosen for rainfall retrievals in the MWRI WS algorithm, which are more sensitive to precipitation.

The AMSR-E instrument is a conical scanning passive microwave radiometer, which provides measurements for 12 channels (6.9 V/H, 10.79 V/H, 18.79 V/H, 23.89 V/H, 36.59 V/H, 89.09 V/H GHz) over the globe [17]. Because MWRI and AMSR-E have the same local ascending/descending times, similar channels, and view geometries, AMSR-E Tbs are collected to evaluate the sensitivity of the MWRI WS algorithm. The AMSR-E Tbs are provided by the AMSR-E/Aqua L2A Global Swath Spatially-Resampled Tbs (Version 3), which are downloaded from the NSIDC website (http://nsidc.org/data/AE_L2A).

3.2. GFS Environmental Product

To provide input parameters for this algorithm, the GFS forecast data are used as an a priori dataset produced by the National Centers for Environmental Prediction (NCEP) (https://www.ncd-c.noaa.gov/data-access/model-data/model-datasets/global-forcast-system-gfs). The entire globe is covered by the GFS at a horizontal resolution of 1 degree in latitude and longitude. It is particularly suitable for the algorithm requirements and provides high-precision environmental parameters, especially wind speed, TPW, and SST measurements at the time of satellite overpasses.

3.3. NSIDC Rainfall Product

The AMSR-E/Aqua Level-2B precipitation product (Version 3) provided by the NSIDC (https://nsidc.org/data/AE_Rain) is used as the reference dataset and applied for evaluating the accuracy of the retrieved rain rate. In each satellite orbit, the AMSR-E/Aqua Level-2B precipitation provides high-quality instantaneous surface rain rates over land and oceans with a 5 km spatial resolution along the scan and 10 km spatial resolution along the track. The precipitation dataset is generated from the GPROF2010 (Version 2) [28] using the AMSR-E Level-2A Tbs (Version 3). The GPROF2010 is a Bayesian retrieval scheme that is developed for TRMM TMI by Kummerow et al. [28]. The GPROF rainfall results compare well with atoll gauge data over oceans and the Global Precipitation Climatology Centre (GPCC) gauge network over land [26]. Therefore, the GPROF has been finalized for AMSR-E. Wolff and Fisher [37] have assessed the performance of 4-year (2003–2006) AMSR-E GPROF retrievals using ground-based rain estimates from TRMM ground validation sites, which shows the AMSR-E estimates exhibit the highest correlation (0.83) and a low error (−5.3%) over oceans. The AMSR-E retrievals have also been investigated in correspondence with the radar-based convective and stratiform rain indexing [38]. Due to the similar local time of ascending nodes (LTAN) of AMSR-E and MWRI (13:30 and 13:40, respectively), the NSIDC rainfall product is a good candidate for the comparison with MWRI rainfall retrievals.

3.4. TRMM Precipitation Data

The TRMM satellite was launched on 27 November 1997, carrying five instruments, two of which are the TMI and PR. The TMI is a multi-channel, dual polarized, conical scanning passive microwave radiometer capable of quantifying not only surface rain rate but also SST, sea surface wind speed, columnar water vapor, and cloud liquid water [14]. In this study, the TMI rainfall data provided by RSS (http://www.remss.com/missions/tmi/) are used as a reference dataset to present the mesoscale of selected TCs. The TMI orbital data mapped to 0.25° grid covering the globe and divided into two maps based on ascending and descending passes.

The PR instrument is a Ku-band radar operating at 13.8 GHz with a field of view (FOV) of approximately 5 km. The Ku-band radar has a swath of approximately 250 km, and it provides vertical sampling with a vertical resolution of 250 m from ground level up to 20 km [39]. The PR offers vertical profiles of precipitation in global tropic regions based on the time delay of precipitation-backscattered return power [40]. The standard TRMM PR 2A25 product (Version 7) obtained from the NASA Precipitation Measurement Missions (PPS) (https://pmm.nasa.gov/) provides the attenuation-corrected radar reflectivity profile, freezing level estimate, and near-surface rain rate, which is used to evaluate MWRI rainfall retrievals. The evaluation analysis of near-surface rain rate estimates from PR 2A25 product, Japanese high-temporal-resolution rain gauge dataset, and the National Oceanic and Atmospheric Administration/National Severe Storms Laboratory (NOAA/NSSL)'s ground-based radar dataset shows that the near-surface rain rate in Version 7 is generally reliable and has been improved largely compared with PR 2A25 Version 6 [41,42]. The detailed information of validation is described in [41,42].

3.5. Data Matchup

To evaluate the rain retrievals of TCs, MWRI measurements coincident with the NISDC rainfall product are collected. This paper mainly focuses on the TCs occurring over the western North Pacific Ocean, which generally affects the spatial distribution of regional precipitation in China [43]. MWRI observations for six selected TCs during 2011 are collected: Songda in May, Maon in July, Mufia in August, Talas in August, Sonca in September, and Nesat in September. The TC dataset, termed as best-track data, consists of the best estimates of storm position and intensity at 6-h intervals only. The best-track data are a result of postseason reanalysis of a storm's position and intensity from all available data, such as the ship, surface, and satellite observations [32]. Figure 1 depicts the corresponding intensity and best-track analysis data of the six selected TCs, which are obtained from the Joint Typhoon Warning Center (JTWC) (http://www.metoc.navy.mil/jtwc/jtwc.html?best-tracks).

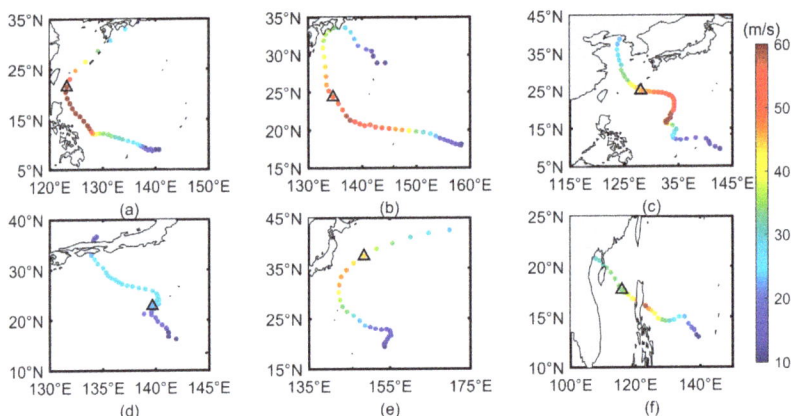

Figure 1. Corresponding intensity and best-track locations of the six selected tropical cyclones (TCs): (**a**) Songda, 20 to 29 May; (**b**) Maon, 11 to 2 July; (**c**) Mufia, 25 July to 8 August; (**d**) Talas, 25 August to 4 September; (**e**) Sonca, 14 to 20 September; (**f**) Nesat, 23 to 30 September. The triangle area in each TC is the target area for analysis.

The time of matching MWRI observations and NSIDC rain rates is within 30 min. Within the 30-min time interval between these two products, a TC does not move a relatively large distance and its overall distribution of precipitation does not undergo significant changes. All the NSIDC points within a 5 km radius of each MWRI pixel are selected for data matchup, in which the closest point with this MWRI pixel will be picked and used as the best-matched point. The NSIDC rainfall product is based on AMSR-E rainfall retrievals developed by the GPROF2010 algorithm [28], and it performs at a high resolution, approximately 5 km × 10 km, which is around the size of the AMSR-E 89 GHz FOV. Therefore, the NSIDC rainfall product is resampled to make it comparable to MWRI retrievals at the 37 GHz footprint. Using the nearest neighbor interpolation, all NSIDC rainfall points within a radius of 23 km (around the size of MWRI 37 GHz, 18 km × 30 km) are selected and then averaged with a weighting inversely proportional to distance. The total number of matched retrieved rain rates and NSIDC rainfall data is 101,879, which is used for rain retrieval intercomparison. Table 2 shows the statistics of the six selected TCs in the datasets, which includes the TC's name, date, the maximum wind speed, and the matched points of each TC.

Table 2. Statistics of the six selected TCs: TC's name, MWRI observation date, and the matched point number of MWRI and Advanced Microwave Scanning Radiometer for Earth Observing System (AMSR-E) observations.

TC's Name	Date	Matched Point Number
Songda	17:25, 27 May 2011	18,263
Maon	03:53, 17 July 2011	19,234
Mufia	17:21, 4 August 2011	15,153
Talas	04:19, 27 August 2011	14,095
Sonca	16:07, 19 September 2011	21,852
Nesat	05:53, 28 September 2011	13,282
Total number of matched pixels: 101,879		

Since the TRMM PR FOV is approximately 5 km, the spatial and temporal intervals for the collocation between MWRI rain retrievals and TRMM PR rain rates are same as the method of NSIDC product. However, there is only one TC coincident with both MWRI and PR observations due to the narrow swath of PR and the different equatorial crossing times of PR and MWRI.

4. Methodology

4.1. Microwave Radiation

The Tbs observed by a satellite radiometer depend on a number of sea surface and atmospheric geophysical variables, as well as measurement geometry, which has been used to estimate various environmental parameters, such as precipitation rate, SST, and wind speed [13,25,27,28,44,45]. Tbs at the top of atmosphere measured by a satellite radiometer are expressed as the sum of upwelling atmospheric radiation (T_{UP}), the sea surface direct emission attenuated by the intervening atmosphere, the down-welling atmospheric radiation (T_{DN}) reflected upward by sea surface, and the reflected cosmic background radiation (T_{CO}) [46]. The basic form of the radiation transfer equation for the microwave band is expressed as follows:

$$T_b = T_{UP} + \tau[ET_S + (1 - E)(\Omega T_{DN} + \tau T_{CO}], \tag{1}$$

where E is the sea surface emissivity, T_s is the SST, and τ is the transmittance through the atmosphere. The Ω accounts for non-specular reflection from the rough sea surface. The cosmic background radiation temperature T_{CO} is approximately 2.7 K. Figure 2 shows the Tbs received by a radiometer.

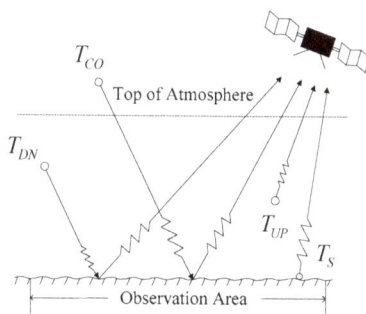

Figure 2. The Tbs measured by a radiometer.

4.2. Retrieval Algorithm

Previous studies have developed rainfall retrieval algorithms for SSM/I [13], TMI [26], and AMSR-E [29]. The study of the WS algorithm focused on 19 and 37 GHz H and V polarization data from SSM/I to retrieve global rain rates. An improved algorithm, the unified microwave ocean retrieval algorithm (UMORA) developed by [25], has been applied to SSM/I, TMI, GMI, AMSR-E, AMSR2, and WindSat. However, there are few studies focusing on rainfall retrievals inside TCs based on MWRI Tbs. Consequently, based on the WS algorithm, an algorithm for retrieving rain rates inside TCs by using MWRI 19 and 37 GHz Tbs is proposed. Since the upwelling radiation at higher frequencies is strongly affected by ice scattering, the retrieval uses lower frequencies where radiation is absorbed and re-emitted by liquid hydrometeors to obtain the information of the column-integrated liquid water [47]. High rain intensities not only change the atmospheric attenuation but also influence the sea surface dielectric properties in a complicated manner [48]. The emissivity over the ocean is approximately from 0.4 to 0.5, and this low surface emissivity ensures a strong contrast between a radiometrically cold and warm background and the precipitation-related atmospheric signature [21,26]. Since the background radiometric signal over the ocean is low, the additional emissions from the precipitation and polarization signature are utilized to identify rain intensity using low-frequency channels (10 to 37 GHz) [10].

One of the key steps of rainfall retrieval is relating the Tbs to rain rates based on Equation (1). Generally, three processes should be taken into account: absorption, emission, and scattering. The atmospheric absorption includes three components: water vapor, oxygen, and liquid water in the form of clouds or rain rates [46]. Raindrops increase the attenuation of the atmosphere at high radar frequencies [48], and it should be noted that the scattering induced by raindrops can be ignored for frequencies below 10 GHz when the rain intensity is less than 12 mm/h [49]. Because Equation (1) obscures the essential physics of rain retrieval, a simplified form of Equation (1) is necessary. Two assumptions are made: (1) we ignore the cosmic microwave background and small effects of non-specular reflection; and (2) we assume that the ocean–atmosphere system is isothermal with the effective temperature (T_E). In this algorithm, dual-polarization Tb measurements are used to separate emission signals from scattering signals. Based on the radiation transfer equation in Equation (1), a highly simplified model for Tb is obtained following [13]:

$$T_B = T_E(\tau)\left(1 - \tau^2 \rho_P\right),\tag{2}$$

where ρ_P is the reflectivity of the sea surface of different polarizations, P (V and H). Despite its simplicity, Equation (2) is still a good approximation to the Tbs model function. A two-way transmittance is given by

$$\tau^2 = \frac{T_{BV} - T_{BH}}{\rho_H T_{BV} - \rho_V T_{BH}}.\tag{3}$$

Using dual-polarization observations from MWRI 19 or 37 GHz Tbs, the T_E term is eliminated and τ_{19}^2 or τ_{37}^2 is separated from T_E. Hence, the accurate liquid water absorption A_{L19} and A_{L37} are obtained through the two-way transmittance τ_{19}^2 and τ_{37}^2, respectively. Additionally, the total atmosphere transmittance is given by

$$\tau = \exp[-sec\theta(A_O + A_V + A_L)],\tag{4}$$

where A_O, A_V, and A_L are oxygen, water vapor, and liquid water absorption, respectively. θ is the incident angle of the MWRI instrument. A_O and A_V are given by [46] as functions of effective air temperature and columnar water vapor. The footprint-averaged attenuation A_L is

$$A_L = \int A' P(A')\, dA',\tag{5}$$

where $P(A')$ is the probability distribution function for attenuation within the footprint. Then, the footprint-averaged two-way transmittance is given by

$$\tau_L^2 = \int exp(-2A' sec\theta) P(A') dA'.\tag{6}$$

If the beamfilling were uniform, $P(A')$ would be the delta function, and integrating (6) yields

$$\tau_L = \exp(-sec\theta \cdot A_L),\tag{7}$$

and this is inverted to provide the total liquid water (cloud and rain water) absorption A_L:

$$A_L = -\ln(\tau_L)/sec\theta.\tag{8}$$

Because the nonlinear relationship between rain rate and Tb occurs when averaging over the radiometer footprint, there is a systematic underestimation of total liquid water absorption, which is referred to as the "beamfilling effect". The WS algorithm shows that if the sub-pixel absorption follows an exponential distribution, the relationship between the absorption A_L within the radiometer footprint and the true mean value $\hat{A}_L$ is given by

$$\hat{A}_L = \frac{e^{2A_L\beta^2 sec\theta} - 1}{2\beta^2 sec\theta},$$

(9)

where A_L and $\hat{A}_L$ represent liquid water absorption before and after beamfilling correction, respectively. β is the normalized RMSE variation for the cloud and rain water absorption A_L, which is related to the variability of liquid water in footprint and is independent of frequency.

The ratio A_{L37}/A_{L19} is less than that predicted by the theoretical Mie absorption ratio when the beamfilling effect is significant. When $A_{L37}/A_{L19} \geq \hat{A}_{L37}/\hat{A}_{L19}$, the beamfilling effect is not significant and no beamfilling correction is performed ($\beta = 0$). The theoretical Mie absorption ratio varies from 3.5 for light absorption to 2.8 for heavy absorption values. The beamfilling correction is made by finding a β value that yields the ratio $\hat{A}_{L37}/\hat{A}_{L19}$ close to the given Mie theory ratio with a monotonically increasing value of β from 0 to 1. The maximum value of the exponent $2A_{L37}\beta^2 \sec\theta$ in Equation (9) is capped at 3.0. Additionally, the maximum values of $\hat{A}_{L19}$ and $\hat{A}_{L37}$ are not allowed to exceed 1.2, which correspond to observations of the heavy rain rate that the 37 and 19 GHz Tbs have saturated. The effect of these limits is to place a rain rate as the upper bound of the algorithm's retrieval ability. As described in Wentz and Spencer [13], because the retrieved rain rate is about 25 mm/h when $\hat{A}_{L19}$ reaching the value of 1.2 for a rain column height of 3 km, the WS algorithm's retrieval ability is limited to 25 mm/h. For MWRI WS retrievals, the limit of retrieved rain rates is approximately 27 mm/h, as shown in Section 5.1.

Figure 3 shows the 37 GHz liquid water absorption versus 19 GHz absorption values, comparing with theoretical Mie absorption (black line), before (red line) and after (blue line) beamfilling correction. The curves are generated based on 101,879 MWRI observations. The $\hat{A}_{L19}$ and $\hat{A}_{L37}$ values curve closely follows the theoretical curve until the value of $A_{L19} \approx 0.4$. Above this value, the maximum value restriction of $\hat{A}_{L37}$ becomes important and the curve value asymptotically approaches 1.2. The value of $\hat{A}_{L37}$ is a constant (1.2) for the high absorption value.

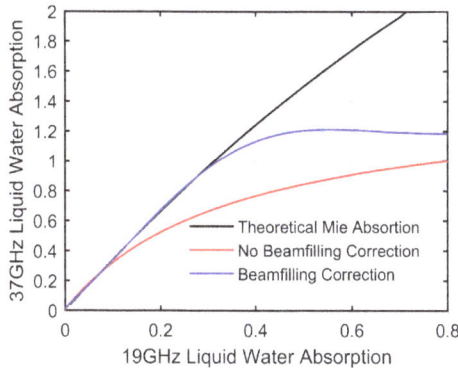

Figure 3. The 37 and 19 GHz liquid water absorption before (red line) and after (blue line) beamfilling correction compared with the theoretical Mie absorption (black line).

Finally, based on the assumption that surface rain rates equal vertically averaged rain rates, the basic Equations (10) and (11) are used to infer rain rate from liquid water attenuation ($\hat{A}_{L19}$ and $\hat{A}_{L37}$, after beamfilling correction), which are described in [7]:

$$\hat{A}_{L19} = 0.0556[1 - 0.0288(T_L - 283)]L + 0.0113[1 + 0.004(T_L - 283)]HR^{1.0636},$$

(10)

$$\hat{A}_{L37} = 0.2027[1 - 0.0261(T_L - 283)]L + 0.0425[1 - 0.002(T_L - 283)]HR^{0.9546}, \tag{11}$$

where L is columnar cloud water and H is the height of the rain column, which are factored out by [13]. Assuming the rain cloud temperature is the mean temperature of the surface and freezing level: $T_L = (T_S + 273)/2$.

5. Results

Because there are no in-situ observations to validate the retrievals for any TC case over oceans, we provide an intercomparison of the MWRI retrievals with other available remotely sensed datasets (see Section 3). Our evaluation consists of four separate parts. The first activity is to compare the MWRI rainfall retrieval to the NSIDC rain rate to see how they agree instantaneously using matched points from six selected TCs and find reasons for disagreements. The second is to examine the sensitivity of this algorithm by comparing AMSR-E Tbs rain retrievals with MWRI retrievals and see how well they match. The third part is to compare MWRI rain retrievals against TRMM PR rain rates. The final part focuses on the capability of MWRI retrievals in presenting the mesoscale structure of selected TCs compared with RSS TMI rainfall data.

5.1. Comparison with the NSIDC rainfall product

This comparison is established with 101,879 matched points of MWRI measurements and NSIDC rain rates from the six TCs. Figure 4 compares MWRI retrieved rain rates to the NSIDC rainfall data provided by the AMSR-E GPROF2010 algorithm.

Generally, MWRI rain retrievals are consistent with NSIDC rain rates in that they reproduce the main vortex structures of these selected TCs. In TCs, the rainfall outside the eyewall is primarily associated with a series of rainbands, which have a spiral geometry rather than the quasi-circular geometry of the eyewall [50]. Figure 4a,b show that the inner-core rainbands of TC Songda between MWRI and NSIDC data are consistent in rainfall distribution. However, the distant environmental regions near Taiwan island of these two figures show some differences. The eyes of TC Maon (Figure 4c,d) and TC Mufia (Figure 4e,f) are free of precipitation and are surrounded by eyewalls that contain intense rain rates and high wind speeds. The distant rainband near the inner-core region of TC Maon in Figure 4c shows consistency with that of the NSIDC product. The inner-core rainband around the eye of TC Mufia is consistent with the rainfall structure provided by NSIDC. TC Talas (Figure 4g,h) and TC Nesat (Figure 4k,l) have larger inner-core rainbands than other selected TCs, and the MWRI rainfall distributions of these two TCs show a good match with that of NSIDC in both inner-core and distant environmental regions. In TC Sonca (Figure 4i,j), there is no obvious eye in the TC vortex structure, and these two figures show the same distant rainbands in the northeastern quadrant and consistency in inner-core rainbands in the south quadrant. The MWRI retrievals show reasonable consistency with the NSIDC rainfall data in both inner-core and distant environment rainband distributions. The further statistical analysis of the MWRI retrieval is shown in Figures 5–7 and Table 3.

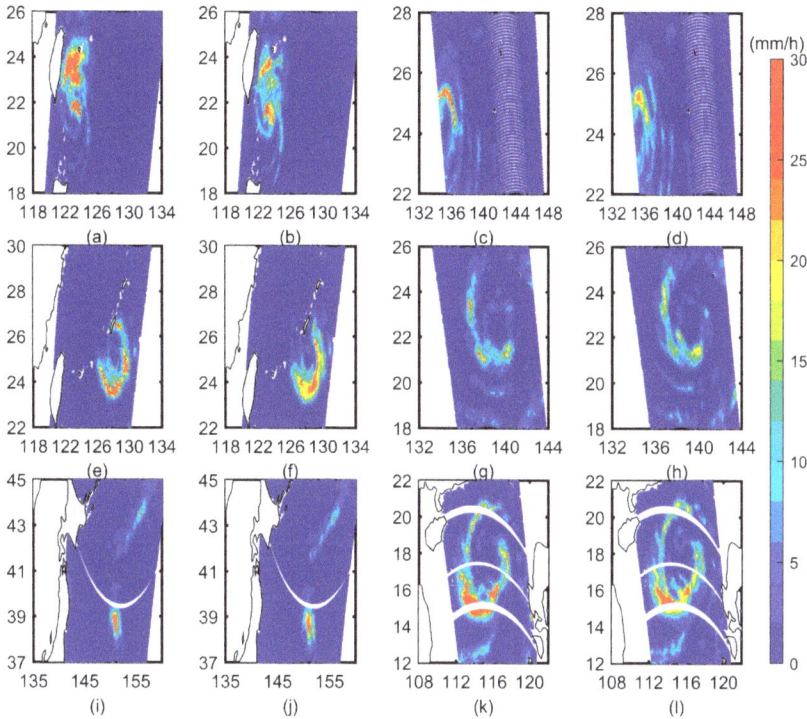

Figure 4. MWRI WS retrieved rain rate of six TCs versus the rain products provided by National Snow and Ice Data Center (NSIDC) data: (**a**) MWRI and (**b**) NSIDC Songda; (**c**) MWRI and (**d**) NSIDC Maon; (**e**) MWRI and (**f**) NSIDC Mufia; (**g**) MWRI and (**h**) NSIDC Talas; (**i**) MWRI and (**j**) NSIDC Sonca; (**k**) MWRI and (**l**) NSIDC Nesat. The blank spots in (**a,b,e,f**) are due to the missing sea surface temperature (SST), wind speed, water vapor data, land pixels, or no MWRI observations. The blank stripes in (**i–l**) are due to no MWRI observations.

Figure 5 shows pixel-to-pixel comparisons between MWRI rain retrievals and NSIDC rain rates for each TC. Two basic statistical measures—namely, the total bias and RMSE—are used to assess the performance of retrievals for each case. The negative total biases (title in Figure 5) are found in TCs, including Maon, Mufia, Talas, and Nesat, and they show that MWRI retrievals slightly underestimate rain intensities compared with the NSIDC product. Even if all the NSIDC rain rates (AMSR-E GPROF2010 product) within the MWRI 37 GHz FOV have been averaged before these comparisons, they still have higher rainfall resolution and therefore higher rain rates compared with MWRI retrievals. Figure 5 shows an obvious limited maximum rain rate (approximately 27 mm/h) at high intensities and shows underestimates at high rain rates for each case. As mentioned in Section 4, the limiting value for the maximum retrieved rain rate is mainly due to the saturation of frequencies. The retrievals of TC Songda (Figure 5a) and Sonca (Figure 5e) give positive total biases for 0.10 mm/h and 0.09 mm/h, respectively, which show a slight overestimation compared with NSIDC rain rates. Generally, the MWRI-retrieved rain rates show good agreement with the NSIDC product.

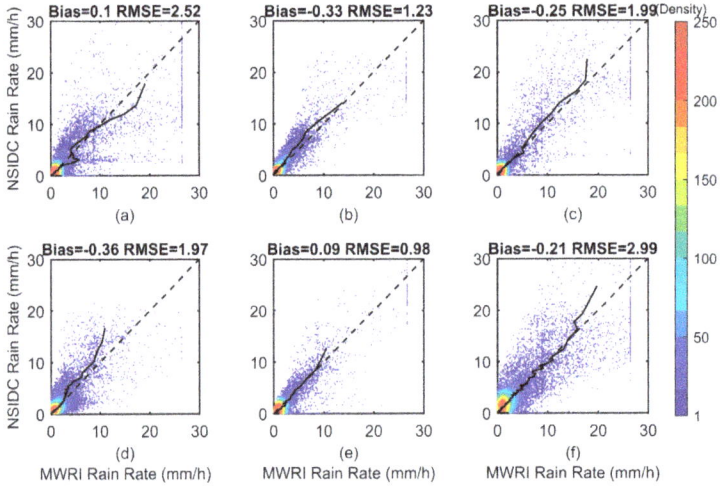

Figure 5. The pixel-to-pixel comparisons between the MWRI-retrieved rain rate and NSIDC rainfall product for each selected TC: (**a**) Songda; (**b**) Maon; (**c**) Mufia; (**d**) Talas; (**e**) Sonca; (**f**) Nesat. The mean of MWRI retrievals (solid line) and 1:1 line (dashed line) are presented in each plot.

The occurrences of MWRI surface rain rates versus NSIDC rain rates are presented in Figure 6. The histograms of these six TCs do not differ much in the low rain rate distribution of each TC but change significantly at the high rain intensity distribution. In particular, the number of the maximum MWRI-retrieved rain rates (approximately 27 mm/h) in Songda (Figure 6a), Mufia (Figure 6c), and Nesat (Figure 6f) are both over 100 compared with other cases. It can be seen that the performance of MWRI retrievals is relatively consistent with the rainfall products in histograms.

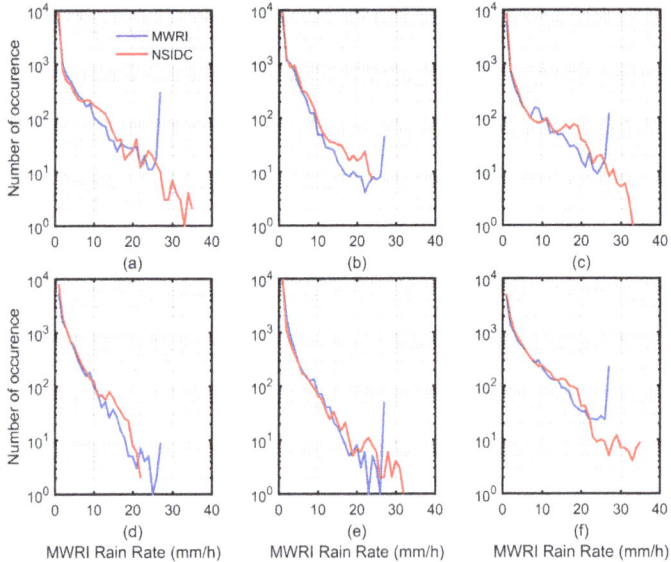

Figure 6. Histograms of FY-3B MWRI retrievals versus precipitation products provided by NSIDC for six selected TCs: (**a**) Songda; (**b**) Maon; (**c**) Mufia; (**d**) Talas; (**e**) Sonca; (**f**) Nesat.

To assess the total rain intensity distribution between the MWRI retrieved rain rate and the NSIDC rain rate for all cases, the total pixel-to-pixel scatter plot and statistic evaluation are displayed in Figure 7 and Table 3. Figure 7 exhibits relatively consistent performance between the MWRI retrievals and NSIDC rainfall product. The scatter in this figure can be attributed to the differences in the sampling of these two instruments and differences between the retrieval algorithms. The mean rain rates of MWRI retrievals (dashed line) compared with the one-to-one line are depicted, and this shows that the MWRI WS algorithm tends to underestimate the rain rates of 5–0 mm/h and over 15 mm/h.

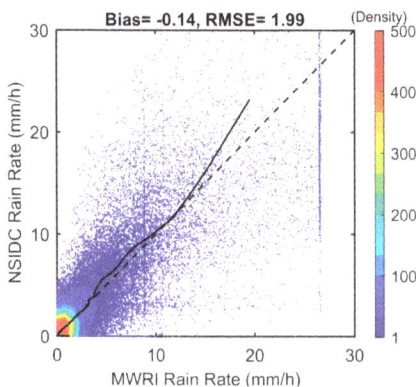

Figure 7. The pixel-to-pixel comparisons between the MWRI-retrieved rain rate and NSIDC rainfall product for all six TCs. The mean of MWRI retrieval line (solid line) and 1:1 line (dashed line) are shown.

To analyze the statistical quality of MWRI retrievals, the bias, RMSE, and bias percentage values between the MWRI rain rates and NSIDC rain rates for all TCs are shown in Table 3. The overall mean bias and RMSE of the MWRI retrievals for all six TCs are −0.14 mm/h and 1.99 mm/h, respectively, which shows an encouraging accuracy compared to the NSIDC rain intensity. Biases of different rainfall categories between these two rain products are almost negative, except the 1.0–.0 mm/h and 2.0–3.0 mm/h rainfall categories, which is due to a few factors. The NSIDC rain rate estimates a higher rain rate than that of MWRI because the AMSR-E GPROF2010 algorithm has a smaller FOV (5 km × 10 km) compared with that of MWRI WS algorithm (18 km × 30 km). In the GPROF2010 scheme, different combinations of all AMSR-E channel frequencies are used, while there are only two low frequencies applied to the MWRI WS algorithm. Consequently, the fundamental observation–rainfall relationship will be different.

The MWRI retrievals show reasonable levels for most rain categories because the absolute value of bias percentage in each category is lower than 15% except for the 0.0–0.5 mm/h and 25–30 mm/h categories. The 0.0–0.5 mm/h category has a bias percentage of −21.4%. The 10–15 mm/h category holds a bias percentage of −1.1% and performs well regarding the retrieved results with NSIDC rain rates. The mean rain rate of MWRI retrievals is about 8.7% lower than that of NSIDC rain rates at the 15–25 mm/h category, which shows a good match with NSIDC rain rates. The bias percentage of the 25–30 mm/h category is −22.8% which shows an underestimation of MWRI retrieval over the WS algorithm's limited rain rate. Saturation in the MWRI WS algorithm limits the maximum rain intensity.

Table 3. Quantitative assessment of MWRI retrievals compared with NSIDC rain rates for the six selected TCs combined. The rain interval, the number of pixels, mean rain rates, bias (MWRI–NSIDC), bias percentage ((MWRI - NSIDC)/ NSIDC × 100%), and total RMSE are listed. The unit is mm/h except for bias percentage.

Rain Interval	Pixel Number	MWRI Retrievals	NSIDC Rain Rate	Bias	Bias Percentage
0.0–0.5	55,172	0.11	0.14	−0.03	−21.4%
0.5–1.0	10,794	0.65	0.71	−0.06	−8.4%
1.0–2.0	7711	1.46	1.43	0.03	2.1%
2.0–3.0	4555	2.57	2.48	0.09	3.6%
3.0–4.0	3500	3.35	3.47	−0.12	−3.4%
4.0–6.0	4076	4.18	4.87	−0.69	−14.1%
6.0–8.0	2445	6.17	6.94	−0.77	−11.1%
8.0–10.0	1671	8.34	8.95	−0.71	−7.9%
10.0–15.0	2356	12.07	12.21	−0.14	−1.1%
15.0–25.0	1629	16.80	18.40	−1.60	−8.7%
25.0–30.0	237	21.00	27.20	−6.20	−22.8%
Total bias: −0.14			Total RMSE: 1.99		

5.2. Comparison with AMSR-E Tbs Retrievals

Because the MWRI and AMSR-E instruments have the same LTAN and viewing geometries, the collocations of these two instruments are close. To assess the application of the MWRI WS algorithm to other sensors, the AMSR-E Tbs provided by the NSIDC are used as input data for this algorithm to get rainfall retrievals utilizing the matched points between MWRI, AMSR-E observations, and NSIDC rain rates for the six TCs. Since the retrieved results are derived from the same algorithm and parameters, the comparison between MWRI and AMSR-E retrievals is not a rigorous test, but rather a consistency check.

Table 4 shows the rain retrievals of MWRI and AMSR-E measurements compared with the NSIDC rain rates, respectively. The overall mean RMSE of rain rates between MWRI retrievals and NSIDC rain rates is much lower than that of AMSR-E observations (1.99 mm/h and 3.11 mm/h, respectively). The bias of AMSR-E retrievals is 0.90 mm/h, which is much higher than that of MWRI. Each individual case has a similar result regarding the total bias and RMSE. The RMSE of TC Nesat in MWRI retrievals is 2.99 mm/h, which is the highest RMSE in MWRI-retrieved results and much lower than that of AMSR-E (4.49 mm/h). The minimum RMSE of the MWRI-retrieved rain rates of 0.98 mm/h occurred in TC Sonca, which is nearly half of AMSR-E (2.00 mm/h). In addition, all the biases of AMSR-E retrievals are positive, which shows an overestimation of rain retrievals.

Table 4. Statistical errors of MWRI and AMSR-E retrievals against the NSIDC product (mm/h).

TC's Name	MWRI		AMSR-E	
	Bias	RMSE	Bias	RMSE
Songda	0.01	2.52	1.11	3.63
Maon	−0.33	1.23	0.51	2.48
Mufia	−0.25	1.99	0.73	3.22
Talas	−0.36	1.97	0.85	2.84
Sonca	0.09	0.98	0.65	2.00
Nesat	−0.21	2.99	1.74	4.49
Total	−0.14	1.99	0.90	3.11

The overestimation of the AMSR-E-retrieved rain rates for the six TCs is mainly attributed to two factors. The first reason is due to the vertical Tbs (Figure 8a) of AMSR-E from the six TCs being about 5 K higher than that of MWRI, while the horizontal Tbs (Figure 8b) of both instruments are nearly the same. Secondly, the MWRI algorithm implemented in this paper does not explicitly account for the resolution of AMSR-E footprint. The beamfilling correction works well for MWRI resolution (Table 1), but it assumes more spatial variability when it is actually present in a smaller footprint of AMSR-E. Therefore, the beamfilling correction tends to overcorrect AMSR-E rainfall retrievals, similar to the behavior shown in [25]. Because it neglects the resolution dependence in the beamfilling correction during the rain retrieval algorithm, the AMSR-E instrument with higher retrieval resolution has higher biases than that of MWRI.

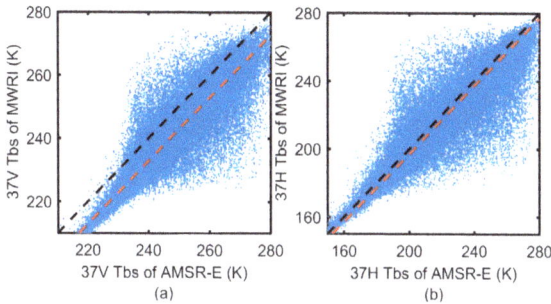

Figure 8. The Tb biases of AMSR-E and MWRI sensors (red dashed line), and 1:1 line (black dashed line) are depicted: (**a**) vertical; (**b**) horizontal.

5.3. Case Study: Comparison with TRMM PR data for Maon

So far, we have assessed the MWRI rain rates against other passive microwave retrievals. In this section, we compare MWRI rain rates against active microwave retrievals. Using the matchup method described in Section 3, we only get one TC (TC Maon) to make a comparison between MWRI retrieved rain rates and TRMM PR observations due to the narrow swath of PR. The MWRI and PR observation time difference of TC Maon is approximately 19 min.

Figures 9 and 10 present the comparison of the MWRI-retrieved rain rates and NSIDC rainfall product with PR precipitation for TC Maon, respectively, and show the differences between these products organized with storm scales. MWRI retrievals and NSIDC rainfall data are higher in the eyewall and PR rain rates are higher in distant rainbands. The performance of MWRI retrievals and NSIDC rain rates for TC Maon shows good agreement with PR rain intensities on the storm scale. Regional biases can occur because of the differences in these retrieval algorithms. The PR algorithm determines rain rates based on the relationship between reflectivity and rain rate [39], while the WS algorithm retrieves rain rates utilizing the relationship between liquid water absorption and rain rate [13], and the AMSR-E GPROF2010 algorithm is a Bayesian retrieval scheme mainly based on the priori oceanic database created for AMSR-E [28].

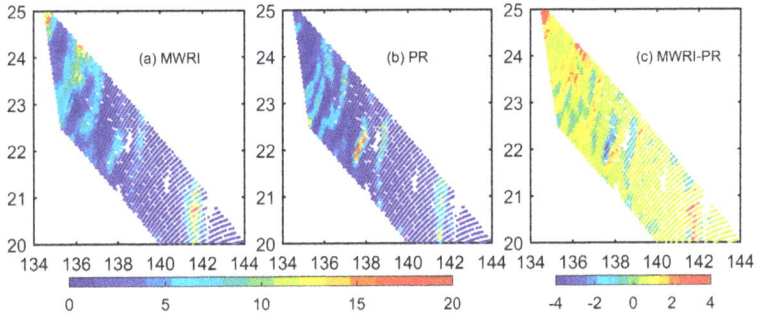

Figure 9. The rain rate (mm/h) comparison of TC Maon: (**a**) the MWRI retrieved rain rates; (**b**) TRMM Precipitation Radar (PR) rainfall product; (**c**) the MWRIPR rainfall difference for Maon. The blank spots in figures are the missing MWRI retrieval pixels.

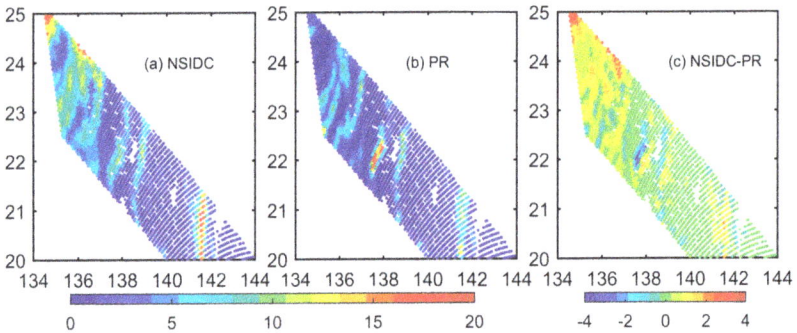

Figure 10. As in Figure 9, but for NSIDC rain rate (mm/h): (**a**) the NSIDC rain rates; (**b**) TRMM PR rainfall product; (**c**) the NSIDC–PR rainfall difference for Maon.

Table 5 shows the bias and RMSE values of MWRI retrievals and NSIDC rain rates with PR rain rates for different rain categories, respectively. For MWRI retrievals, the maximum error of −8.79 mm/h is reached at the rain category of 10–25 mm/h, and the biases for rain categories below (over) 6 mm/h are positive (negative). The NSIDC rain rates show an underestimation at rain rates over 8 mm/h and overestimation below 8 mm/h.

The overestimate of the MWRI-retrieved rain rates lower than 6 mm/h is likely due to overcorrecting the beamfilling effect for MWRI. Additionally, the PR algorithm described in detail by [39] uses an a priori drop size distribution for light and moderate rain rates (<5 mm/h), while it corrects the attenuation of the radar beam for the heavy rainfall [26]. Both factors can introduce errors to PR products and introduce biases when comparing with MWRI and NSIDC rain intensities. The underestimation of MWRI and NSIDC rain rates is mainly ascribed to the larger footprint than that of PR (5 km × 5 km). This can be expected as a radiometer is not as capable a rain sensor as the PR, even if is averaged to the same spatial resolution. While PR shows low sensitivity to low rain rates, it is sensitive to high rain intensities and gives good measurements of heavy rainfall. The MWRI WS algorithm determines heavy rain intensities using the relationship between the liquid water absorption (FOV averaged to MWRI 37 GHz) and rain rates, which leads to a lower rate of heavy rain rate than that of PR. For NSIDC, the underestimation of high rain intensity is mainly due to the fewer rainfall profiles in the oceanic database, which determines the retrieval accuracy of the GPROF2010 algorithm [28].

Table 5. The pixel number, PR average rain rate, averaged rain rate MWRI and NSIDC, and bias (MWRI-PR and NSIDC-PR) of MWRI and NSIDC rain rates with PR rain rates, respectively, for each rain interval, and the total bias and RMSE of MWRI and NSIDC rain rates compared with PR, respectively. The unit is mm/h except for the pixel number.

Rain Interval	Pixel Number	PR Rain Rate	MWRI		NSIDC	
			Rain Rate	Bias	Rain Rate	Bias
0.0–0.5	797	0.18	0.73	0.55	0.99	0.81
0.5–1.0	338	0.74	1.91	1.17	2.34	1.60
1.0–2.0	420	1.47	2.89	1.42	3.69	2.22
2.0–3.0	248	2.50	3.76	1.26	4.56	2.06
3.0–4.0	201	3.47	4.90	1.43	5.78	2.31
4.0–6.0	295	4.89	5.25	0.34	6.26	1.37
6.0–8.0	53	7.42	6.27	−1.15	8.43	1.01
8.0–10.0	17	9.48	5.73	−3.75	9.20	−0.28
10.0–25.0	27	13.76	4.97	−8.79	7.34	−6.42
			Total bias:	0.74		1.41
			Total RMSE:	2.40		2.64

5.4. Mesoscale Structure of the Selected TCs

To check whether the saturation of the MWRI WS algorithm could impact the structure of TCs, the RSS TMI rainfall product is selected and used as evaluation data. To assess the capability of MWRI retrievals in presenting the mesoscale structure of intense TCs, the resulting rain fields of the MWRI-retrieved rain rates and the TMI rainfall data are shown in Figures 11 and 12, respectively. They not only offer a series of well-defined characteristic structures, but also a large range of rain intensities. These two figures are differently color-coded because the TMI rainfall data are gridded data, which have lower rain rates than MWRI retrievals. This section is not intended to compare the rainfall deviation between MWRI retrievals and TMI gridded rainfall data, but to check whether the TC structures of MWRI retrievals match TMI rain rates.

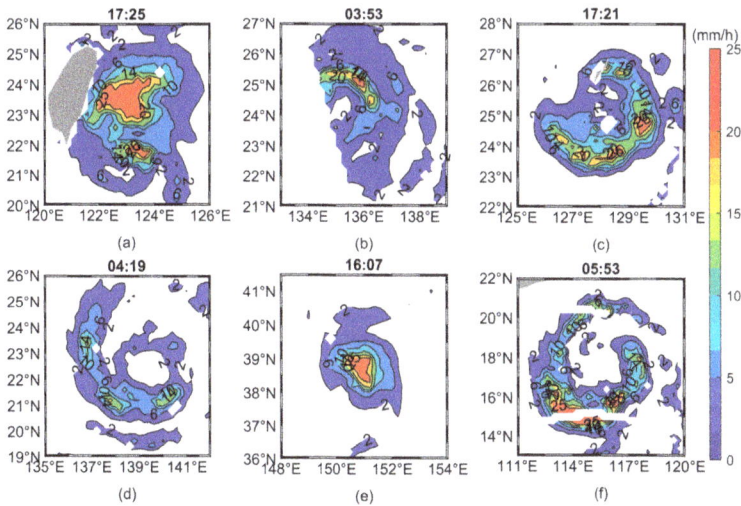

Figure 11. MWRI retrieval counters for six selected track records: (**a**) Sonda; (**b**) Maon; (**c**) Mufia; (**d**) Talas; (**e**) Sonca; (**f**) Nesat. The blank stripes in (**f**) are due to the missing MWRI observations. The title in each TC is the start scan time of MWRI observation in the plotted area.

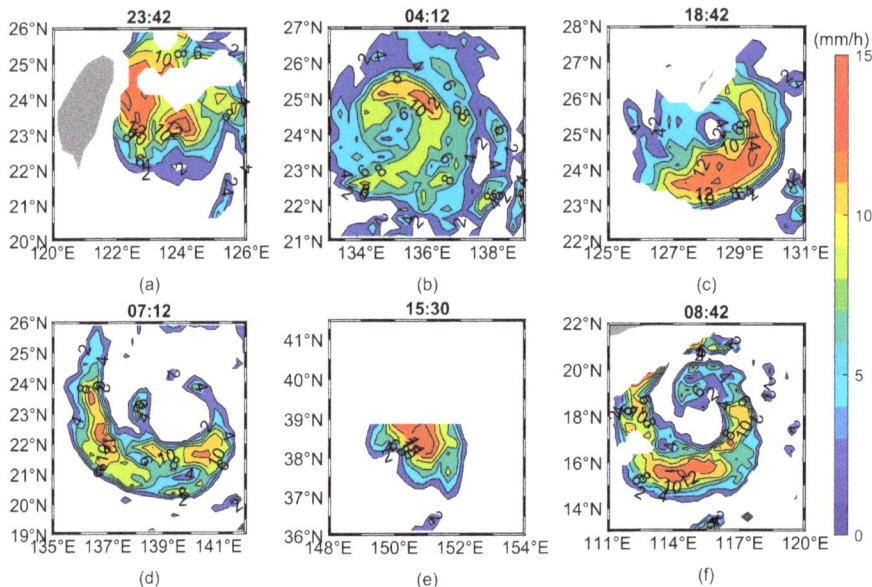

Figure 12. Remote Sensing Systems (RSS) TMI rainfall counters: (**a**) Sonda; (**b**) Maon; (**c**) Mufia; (**d**) Talas; (**e**) Sonca; (**f**) Nesat. The blank spots in (**a**), (**c**), (**e**), and (**f**) are due to the missing SST, wind speed, water vapor data, land pixels, or no TMI observations. The title in each TC is the start scan time of TMI gridded data in the plotted area.

The performance of these two precipitation products shows that MWRI retrievals represent clear mesoscale structures of the selected TCs. Figure 11a shows the retrieved rainfall structure of Songda from MWRI: a principal rainband near eyewall with 20 mm/h contour surrounds, and a broader distant rainband near Taiwan island. Even though the time difference of MWRI and TMI observations is approximately 6 h, TMI (Figure 12a) shows a similar rainband structure, which is partially obscured by more aggressive land masking of the Yaeyama and Miyako Islands. TC Maon (Figure 11b) indicates a clearly visible TC eye, which is free of precipitation and surrounded by an intense eyewall. The 20 mm/h rainfall contour of Maon corresponds to an eyewall region, and similarly, the maximum contour (12 mm/h) of TMI rain intensities (Figure 12b) occurs in the north-eastern quadrant. TMI shows a broader area of moderate rain rates in the distant rainband (the south-eastern quadrant). These two figures (Figures 11b and 12b) have the minimum time difference (~19 min), and TC Maon does not move a large distance. The structures of Mufia from MWRI and TMI (Figures 11c and 12c) are in relatively good agreement within a 1.5-h time difference. The inner-core structure of Mufia from MWRI (Figure 11c) is clear, with a 10 mm/h rainfall contour corresponding to the vortex region and high rain rates of 10 to 20 mm/h that extend to the most of Mufia structure, particularly in the southeast quadrant.

The rain field structure of Talas from MWRI (Figure 11d) has a larger TC scale; however, its retrieved rain rates are lower than other cases. TMI (Figure 12d) exhibits three relative maxima in the western, south-western, and south-eastern parts of the storm, in agreement with MWRI. The TC Sonca from MWRI (Figure 11e) within a 30-min time difference is the smallest storm among all the cases, and its rainfall distribution is extremely compact with low rain rates between 0 and 10 mm/h extending over the inner-core structure of the TC system, and TMI has the same compact structure (Figure 12e). The structure of TC Nesat from MWRI (Figure 11f) displays the main vortex zone with 25 mm/h contour corresponding to the TC eyewall. Despite the missing MWRI scans, the mainly high rainband shows a broad principal rainband of Nesat. TMI has a similar area of high rain rates along the southern portion of the storm (Figure 12f). The results show that the saturation of MWRI WS algorithm does not impact the structure of TCs, and MWRI retrievals present clear mesoscale structures of intense TCs.

6. Discussion

Our results have shown that a rain retrieval algorithm closely based on the WS algorithm, applied to MWRI measurements, provides reasonable rain rates for TCs and well represents the TC mesoscale structures. The WS algorithm is a physical scheme based on dual-polarization passive microwave measurements to provide a two-way transmittance through the atmosphere. The advantage of this simple strategy is that the parameterizations are more tightly constrained [25]. However, this retrieval algorithm applied to MWRI still shows some problems. The MWRI retrievals were low relative to the NSIDC dataset in high intensities. However, positive mean differences were found when the same algorithm was applied to AMSR-E measurements. The retrievals did not match from different sensors because the WS algorithm did not include the resolutions of different sensors in the beamfilling correction. The unified microwave ocean retrieval algorithm (UMORA) algorithm developed by Hilburn and Wentz [25] is an improvement of the WS algorithm in that it corrects this problem. For practical applications of the new rain retrieval algorithm, it is planned to apply the UMORA algorithm to MWRI measurements for TCs.

Although the MWRI retrievals for six selected TCs shown in Figure 7 provides consistent performance with the NSIDC rainfall product from the GPROF2010 algorithm, there are some differences due to fundamental algorithm differences. GPROF retrieves a higher surface rain rate compared with MWRI because of differences in microphysical assumptions. In the GPROF algorithm, the vertical rain water profile is more physically constrained, while the WS algorithm assumes a constant vertical precipitation profile and thus obtains a surface rain rate that is equal to the column-average rain rate. Another issue is that it does not explicitly account for the spatial resolution of the satellite observations. Hilburn and Wentz [25] found that the probability distribution function $P(A')$ (see Equation (5)) changes systematically with the footprint size. The WS algorithm assumed that this distribution function works well for SSM/I resolution (37 GHz, 28 km × 37 km); however, it assumes more spatial variability than in the smaller AMSR-E footprint (37 GHz, 8 km × 14 km). Thus, the beamfilling overcorrected AMSR-E. As shown in Table 4, the rain retrievals from the higher-resolution sensor (AMSR-E) are biased higher than MWRI retrievals if the resolution dependence in the beamfilling correction is neglected. While the MWRI instrument has a similar resolution (37 GHz, 18 km × 30 km) to SSM/I, it works better in the WS algorithm than the AMSR-E instrument.

7. Conclusions

With the development of passive microwave radiometers, retrieving rain rates inside TCs has become possible. The goal of this study is to develop an algorithm for estimating surface rain rates of TCs using 19 and 37 GHz V and H polarized Tbs from FY-3B MWRI. Based on the algorithm derived by Wentz and Spencer [13], an ocean-only rainfall retrieval algorithm for the MWRI has been developed. Since the nonlinear relationship between rain rate and Tb occurs when averaging over the radiometer footprint, the beamfilling correction is considered by comparing ratios of liquid water absorption with the theoretical Mie absorption between 37 and 19 GHz.

MWRI WS retrievals are in reasonable agreement with other datasets. The MWRI measurements for six selected TCs and rain rates provided by the NSIDC are used to develop and verify the accuracy of this retrieval algorithm. The results are encouraging, with an overall mean bias and RMSE of −0.14 mm/h and 1.99 mm/h, respectively. The histograms of these two products for six TCs show encouraging agreement and do not differ much at the low rain rate distribution of each TC; however, they differ more at high rain intensity distribution. A comparison of instantaneous pixel-to-pixel retrievals shows that the MWRI retrieval shows reasonable agreement for most rain categories with absolute bias percentages within 15%, except for the lowest (0–0.5 mm/h) and highest (25–30 mm/h) rain rates (−21.4% and −22.8%, respectively). The maximum rain rate value (~27 mm/h) of MWRI retrievals is limited due to the frequency saturation.

To assess the sensitivity of the MWRI WS algorithm to the sensor configuration, the algorithm was also applied to AMSR-E Tbs. The overall mean bias and RMSE are 0.90 mm/h and 3.11 mm/h, respectively, which are higher than that of MWRI retrievals. The AMSR-E WS retrievals overestimate rain rates compared with the NSIDC rain rate. A case study for TC Maon (2011) showed that MWRI WS retrievals agree well with PR rain rates on storm scales; however, MWRI rain rates overestimate rain rates below 6 mm/h and underestimate over 6 mm/h relative to PR rain rates. Additionally, the RSS TMI rainfall product is used to assess the capability of the MWRI WS algorithm in representing the mesoscale structure of rain rates in intense TCs. The performance of MWRI retrievals shows the clear mesoscale structures of the selected TCs. Despite the remaining uncertainties in the MWRI WS rainfall retrieval, the reasonable intercomparison of datasets indicates that the retrieved rain rate can be used for the TCs' surface precipitation and mesoscale structure studies. The MWRI sensor performs well and provides good quality Tbs.

Author Contributions: Data curation, R.Z.; formal analysis, R.Z. and K.A.H.; funding acquisition, Z.W.; investigation, R.Z.; methodology, R.Z. and K.A.H.; project administration and resources, Z.W.; software, R.Z. and K.A.H.; supervision, Z.W.; validation and visualization, R.Z.; writing—original draft preparation, R.Z.; writing—review & editing, R.Z., Z.W., and K.A.H.

funding: This research was funded by the National Natural Science Foundation of China grant number 61501433.

Acknowledgments: The authors would like to thank the National Satellite Meteorological Center Feng Yun Satellite Data Center, the National Centers for Environmental Prediction, the National Snow and Ice Data Center, the NASA Precipitation Measurement Missions team, and the Remote Sensing Systems for providing the satellite datasets and the JTWC team for the provision of best-track data for TCs.

Conflicts of Interest: The authors declare no conflict of interest. The funding sponsors had no role in the design of the study; in the collection, analyses, or interpretation of data; in the writing of the manuscript, and in the decision to publish the results.

References

1. Gray, W.M. Global view of origin of tropical disturbances and storms. *Mon. Weather Rev.* **1968**, *96*, 669–700. [CrossRef]
2. Cutter, S.L.; Emrich, C.T. Moral hazard, social catastrophe: The changing face of vulnerability along the hurricane coasts. *Ann. Am. Acad. Political Soc. Sci.* **2006**, *604*, 102–112. [CrossRef]
3. Zick, S.E.; Matyas, C.J. A shape metric methodology for studying the evolving geometries of synoptic-scale precipitation patterns in tropical cyclones. *Ann. Am. Assoc. Geogr.* **2016**, *106*, 1217–1235. [CrossRef]
4. Shige, S.; Takayabu, Y.N.; Tao, W.-K.; Johnson, D.E. Spectral retrieval of latent heating profiles from TRMM PR data. Part I: Development of a model-based algorithm. *J. Appl. Meteorol.* **2004**, *43*, 1095–1113. [CrossRef]
5. Atlas, D.; Thiele, O.W. Precipitation Measurements from Space. In Proceedings of the IEEE International Geoscience and Remote Sensing Symposium, NASA/Goddard Space Flight Center, Greenbelt MD, USA, 1 October 1981; p. 441.
6. Milman, A.S.; Wilheit, T.T. Sea surface temperatures from the scanning multichannel microwave radiometer on Nimbus 7. *J. Geophys. Res.* **1985**, *90*, 11631. [CrossRef]
7. Wentz, F.J.; Meissner, T. *AMSR Ocean Algorithm Theoretical Basis Document, Version 2*; Remote Sensing Systems: Santa Rosa, CA, USA, November 2000.
8. Kidd, C.; Matsui, T.; Chern, J.; Mohr, K.; Kummerow, C.D.; Randel, D.L. Global precipitation estimates from cross-track passive microwave observations using a physically based retrieval scheme. *J. Hydrometeorol.* **2016**, *17*, 383–400. [CrossRef]
9. Petković, V.; Kummerow, C.D.; Randel, D.L.; Pierce, J.R.; Kodros, J.K. Improving the quality of heavy precipitation estimates from satellite passive microwave rainfall retrievals. *J. Hydrometeorol.* **2018**, *19*, 69–85. [CrossRef]
10. Kidd, C.; Levizzani, V. Status of satellite precipitation retrievals. *Hydrol. Earth Syst. Sci.* **2011**, *15*, 1109–1116. [CrossRef]

11. Basharinov, A.; Yegorov, S.; Gurvich, A.; Oboukhov, A. Some results of microwave sounding of the atmosphere and ocean from the satellite Cosmos 243. *Space Res.* **1971**, *11*, 593–600.

12. Hollinger, J.P.; Peirce, J.L.; Poe, G.A. SSM/I instrument evaluation. *IEEE Trans. Geosci. Remote Sens.* **1990**, *28*, 781–790. [CrossRef]

13. Wentz, F.J.; Spencer, R.W. SSM/I rain retrievals within a unified all-weather ocean algorithm. *J. Atmos. Sci.* **1998**, *55*, 1613–1627. [CrossRef]

14. Kummerow, C.D.; Barnes, W.; Kozu, T.; Shiue, J.; Simpson, J. The Tropical Rainfall Measuring Mission (TRMM) sensor package. *J. Atmos. Ocean. Technol.* **1998**, *15*, 809–817. [CrossRef]

15. Hou, A.Y.; Kakar, R.K.; Neeck, S.; Azarbarzin, A.A.; Kummerow, C.D.; Kojima, M.; Oki, R.; Nakamura, K.; Iguchi, T. The Global Precipitation Measurement Mission. *Bull. Am. Meteorol. Soc.* **2014**, *95*, 701–722. [CrossRef]

16. Huffman, G.J.; Bolvin, D.T.; Braithwaite, D.; Hsu, K.; Joyce, R.; Kidd, C.; Nelkin, E.J.; Xie, P. *NASA Global Precipitation Measurement Integrated Multi-Satellite Retrievals for GPM (IMERG) Algorithm Theoretical Basis Document (ATBD) Version 4.5*; National Aeronautics and Space Administration: Washington, DC, USA, November 2015.

17. Kawanishi, T.; Sezai, T.; Ito, Y.; Imaoka, K.; Takeshima, T.; Ishido, Y.; Shibata, A.; Miura, M.; Inahata, H.; Spencer, R.W. The Advanced Microwave Scanning Radiometer for the Earth Observing System (AMSR-E), NASDA's contribution to the EOS for global energy and water cycle studies. *IEEE Trans. Geosci. Remote Sens.* **2003**, *41*, 184–194. [CrossRef]

18. Yang, H.; Weng, F.; Lv, L.; Lu, N.; Liu, G.; Bai, M.; Qian, Q.; He, J.; Xu, H. The FengYun-3 Microwave Radiation Imager on-orbit verification. *IEEE Trans. Geosci. Remote Sens.* **2011**, *49*, 4552–4560. [CrossRef]

19. Buettner, K.J.K. Regenortung vom wettersatelliten mit hilfe von zentimeterwellen (Rain localization from a weather satellite via centimeter waves). *Naturwiss* **1963**, *50*, 591–592. [CrossRef]

20. Grody, N.C. Classification of snow cover and precipitation using the Special Sensor Microwave Imager. *J. Geophys. Res. Atmos.* **1991**, *96*, 7423–7435. [CrossRef]

21. Ferraro, R.R.; Marks, G.F. The development of SSM/I rain-rate retrieval algorithms using ground-based radar measurements. *J. Atmos. Ocean. Technol.* **1995**, *12*, 755–770. [CrossRef]

22. Spencer, R.W.; Goodman, H.M.; Hood, R.E. Precipitation retrieval over land and ocean with the SSM/I: Identification and characteristics of the scattering signal. *J. Atmos. Ocean. Technol.* **1989**, *6*, 254–273. [CrossRef]

23. Kidd, C. On rainfall retrieval using polarization-corrected temperatures. *Int. J. Remote Sens.* **1998**, *19*, 981–996. [CrossRef]

24. Brown, P.J.; Kummerow, C.D.; Randel, D.L. Hurricane GPROF: An optimized ocean microwave rainfall retrieval for tropical cyclones. *J. Atmos. Ocean. Technol.* **2016**, *33*, 1539–1556. [CrossRef]

25. Hilburn, K.A.; Wentz, F.J. Intercalibrated passive microwave rain products from the Unified Microwave Ocean Retrieval Algorithm (UMORA). *J. Appl. Metrorol. Climatol.* **2008**, *47*, 778–794. [CrossRef]

26. Kummerow, C.D.; Hong, Y.; Olson, W.S.; Yang, S.; Adler, R.F.; McCollum, J.; Ferraro, R.; Petty, G.; Shin, D.B.; Wilheit, T.T. The evolution of the Goddard profiling algorithm (GPROF) for rainfall estimation from passive microwave sensors. *J. Appl. Meteorol.* **2001**, *40*, 1801–1820. [CrossRef]

27. Kummerow, C.D.; Randel, D.L.; Kulie, M.; Wang, N.Y.; Ferraro, R.; Munchak, S.J.; Petkovic, V. The evolution of the Goddard Profiling Algorithm to a fully parametric scheme. *J. Atmos. Ocean. Technol.* **2015**, *32*, 2265–2280. [CrossRef]

28. Kummerow, C.D.; Ringerud, S.; Crook, J.; Randel, D.; Berg, W. An observationally generated a priori database for microwave rainfall retrievals. *J. Atmos. Ocean. Technol.* **2011**, *28*, 113–130. [CrossRef]

29. Wilheit, T.; Kummerow, C.D.; Ferraro, R. Rainfall algorithms for AMSR-E. *IEEE Trans. Geosci. Remote Sens.* **2003**, *41*, 204–214. [CrossRef]

30. Li, X.; Zhao, F. Characteristics of precipitating clouds in typhoon Ma-on from MWRI and TMI observations. In Proceedings of the Remote Sensing, Environment and Transportation Engineering (RSETE), Nanjing, China, 1–3 June 2012; pp. 1–4.

31. Zhang, R.; Wang, Z.; Zhang, L.; Li, Y. Rainfall retrieval of tropical cyclones using FY-3B microwave radiation imager (MWRI). In Proceedings of the Geoscience and Remote Sensing Symposium (IGARSS), Fort Worth, TX, USA, 23–28 July 2017; pp. 550–553.

32. Knapp, K.R.; Kruk, M.C.; Levinson, D.H.; Diamond, H.J.; Neumann, C.J. The international best track archive for climate stewardship (IBTrACS) unifying tropical cyclone data. *Bull. Am. Meteorol. Soc.* **2010**, *91*, 363–376. [CrossRef]

33. Liu, C.; Zipser, E.J.; Cecil, D.J.; Nesbitt, S.W.; Sherwood, S. A cloud and precipitation feature database from nine years of TRMM observations. *J. Appl. Metrorol. Climatol.* **2008**, *47*, 2712–2728. [CrossRef]

34. Viltard, N.; Burlaud, C.; Kummerow, C.D. Rain retrieval from TMI brightness temperature measurements using a TRMM PR–based database. *J. Appl. Metrorol. Climatol.* **2006**, *45*, 455–466. [CrossRef]

35. Kummerow, C.D.; Olson, W.S.; Giglio, L. A simplified scheme for obtaining precipitation and vertical hydrometeor profiles from passive microwave sensors. *IEEE Trans. Geosci. Remote Sens.* **1996**, *34*, 1213–1232. [CrossRef]

36. Yang, H.; Zou, X.; Li, X.; You, R. Environmental data records from FengYun-3B Microwave Radiation Imager. *IEEE Trans. Geosci. Remote Sens.* **2012**, *50*, 4986–4993. [CrossRef]

37. Wolff, D.B.; Fisher, B.L. Assessing the relative performance of microwave-based satellite rain-rate retrievals using TRMM ground validation data. *J. Appl. Metrorol. Climatol.* **2009**, *48*, 1069–1099. [CrossRef]

38. Islam, T.; Rico-Ramirez, M.A.; Han, D.; Srivastava, P.K. Using S-band dual polarized radar for convective/stratiform rain indexing and the correspondence with AMSR-E GSFC profiling algorithm. *Adv. Space Res.* **2012**, *50*, 1383–1390. [CrossRef]

39. Iguchi, T.; Kozu, T.; Meneghini, R.; Awaka, J.; Okamoto, K.I. Rain-profiling algorithm for the TRMM precipitation radar. *J. Appl. Meteorol.* **2000**, *39*, 2038–2052. [CrossRef]

40. Shige, S.; Takayabu, Y.N.; Tao, W.K.; Shie, C.L. Spectral retrieval of latent heating profiles from TRMM PR data. Part II: Algorithm improvement and heating estimates over Tropical Ocean regions. *J. Appl. Metrorol. Climatol.* **2007**, *46*, 1098–1124. [CrossRef]

41. Seto, S.; Iguchi, T.; Utsumi, N.; Kiguchi, M.; Oki, T. Evaluation of extreme rain estimates in the TRMM/PR standard product version 7 using high-temporal-resolution rain gauge datasets over Japan. *SOLA* **2013**, *9*, 98–101. [CrossRef]

42. Kirstetter, P.-E.; Hong, Y.; Gourley, J.J.; Schwaller, M.; Petersen, W.; Zhang, J. Comparison of TRMM 2A25 products, version 6 and version 7, with NOAA/NSSL ground radar–based National Mosaic QPE. *J. Hydrometeorol.* **2013**, *14*, 661–669. [CrossRef]

43. Tu, J.Y.; Chou, C.; Chu, P.S. The abrupt shift of typhoon activity in the vicinity of taiwan and its association with western North Pacific-East Asian climate change. *J. Clim.* **2009**, *22*, 3617–3628. [CrossRef]

44. Shibata, A. A wind speed retrieval algorithm by combining 6 and 10 GHz data from Advanced Microwave Scanning Radiometer: Wind speed inside hurricanes. *J. Oceanogr.* **2006**, *62*, 351–359. [CrossRef]

45. Yin, X.; Wang, Z.; Song, Q.; Huang, Y.; Zhang, R. Estimate of ocean wind vectors inside tropical cyclones from polarimetric radiometer. *IEEE J. Sel. Top. Appl. Earth Obs. Remote Sens.* **2017**, *10*, 1701–1714. [CrossRef]

46. Wentz, F.J. A well-calibrated ocean algorithm for Special Sensor Microwave/Imager. *J. Geophys. Res. Oceans* **1997**, *102*, 8703–8718. [CrossRef]

47. Petković, V.; Kummerow, C.D. Understanding the sources of satellite passive microwave rainfall retrieval systematic errors over land. *J. Appl. Metrorol. Climatol.* **2017**, *56*, 597–614. [CrossRef]

48. Meissner, T.; Wentz, F.J. Wind-vector retrievals under rain with passive satellite microwave radiometers. *IEEE Trans. Geosci. Remote Sens.* **2009**, *47*, 3065–3083. [CrossRef]

49. Ulaby, F.T. *Microwave Remote Sensing Active and Passive-Volume III: From Theory to Applications*; Artech House: Dedham, MA, USA, 1986.

50. Houze, R.A., Jr. Clouds in tropical cyclones. *Mon. Weather Rev.* **2010**, *138*, 293–344. [CrossRef]

remote sensing

MDPI

Article

NWP-Based Adjustment of IMERG Precipitation for Flood-Inducing Complex Terrain Storms: Evaluation over CONUS

Xinxuan Zhang [1,*], Emmanouil N. Anagnostou [1] and Craig S. Schwartz [2]

[1] Department of Civil and Environmental Engineering, University of Connecticut, Storrs, CT 06269, USA; emmanouil.anagnostou@uconn.edu
[2] National Center for Atmospheric Research, Boulder, CO 80301, USA; schwartz@ucar.edu
*   Correspondence: xinxuan.zhang@uconn.edu; Tel.: +1-860-617-9015

Received: 20 February 2018; Accepted: 15 April 2018; Published: 21 April 2018

Abstract: This paper evaluates the use of precipitation forecasts from a numerical weather prediction (NWP) model for near-real-time satellite precipitation adjustment based on 81 flood-inducing heavy precipitation events in seven mountainous regions over the conterminous United States. The study is facilitated by the National Center for Atmospheric Research (NCAR) real-time ensemble forecasts (called model), the Integrated Multi-satellitE Retrievals for GPM (IMERG) near-real-time precipitation product (called raw IMERG) and the Stage IV multi-radar/multi-sensor precipitation product (called Stage IV) used as a reference. We evaluated four precipitation datasets (the model forecasts, raw IMERG, gauge-adjusted IMERG and model-adjusted IMERG) through comparisons against Stage IV at six-hourly and event length scales. The raw IMERG product consistently underestimated heavy precipitation in all study regions, while the domain average rainfall magnitudes exhibited by the model were fairly accurate. The model exhibited error in the locations of intense precipitation over inland regions, however, while the IMERG product generally showed correct spatial precipitation patterns. Overall, the model-adjusted IMERG product performed best over inland regions by taking advantage of the more accurate rainfall magnitude from NWP and the spatial distribution from IMERG. In coastal regions, although model-based adjustment effectively improved the performance of the raw IMERG product, the model forecast performed even better. The IMERG product could benefit from gauge-based adjustment, as well, but the improvement from model-based adjustment was consistently more significant.

Keywords: GPM; IMERG; satellite precipitation adjustment; numerical weather prediction; heavy precipitation; flood-inducing storm; complex terrain

---

1. Introduction

Accurate measurement of precipitation is a prerequisite for understanding related hydrologic processes. The fact that precipitation is highly discontinuous in space and time presents challenges for obtaining accurate spatio-temporal quantification of precipitation, especially over topographically-complex regions, due to the variability and uncertainty introduced by orographic effects [1,2]. Generally, observed gridded precipitation datasets can be generated by three approaches: gauge data interpolation, surface radar network and satellite-based observation.

The accuracy of gauge interpolation depends largely on gauge density and measurement quality. Gauge locations are not homogeneously distributed; there are more gauges at low elevations and in densely-populated areas relative to mountainous terrain because of the higher costs of gauge installation and maintenance over complex topography. Moreover, since gauge networks around the world are operated by different countries, the observations are less accessible due to different data-sharing policies. Hence, gauge-based gridded precipitation datasets usually have coarse temporal and spatial resolutions; most global products have monthly or daily time scales and 0.25° to 2.5° spatial resolutions [3–6].

However, for meso-scale studies of such as extreme rainfall events and related floods, precipitation products with higher spatial and temporal resolution are required. Although surface radar networks provide fine resolution products, the data quality is limited in complex terrain due to severe beam blocking and strong ground clutter [7–9]. In addition, considering the expensive operating and maintenance costs, spatial coverages of radar networks are very limited especially in mountainous or less populated regions.

Besides surface observations, techniques of satellite-based measurements have developed rapidly over the past 30+ years [10]. As a result, a variety of satellite-based precipitation products is now available with quasi-global coverage, including the Tropical Rainfall Measuring Mission (TRMM) near-Real-Time Multi-satellite Precipitation product (3B42RT) [11], the National Oceanic and Atmospheric Administration (NOAA) Climate Prediction Center (CPC) morphing technique (CMORPH) [12], the Precipitation Estimation from Remotely Sensed Information Using Artificial Neural Networks (PERSIANN) [13], the Global Satellite Mapping of Precipitation Microwave-IR Combined Product (GSMaP) produced by the Earth Observation Research Center (EORC) of the Japan Aerospace Exploration Agency (JAXA) [14,15] and products of Global Precipitation Measurement (GPM) Integrated Multi-satellitE Retrievals for GPM (IMERG) [16]. Many studies indicate that these satellite products tend to underestimate heavy precipitation over mountainous regions [17–22].

Apart from single source datasets, products with combined data sources are available, as well. Typically, gauge observations are incorporated into raw radar or satellite products to improve accuracy [23–25]. In fact, most satellite products mentioned above have their gauge-adjusted counterparts [11,16,26–28]. In general, gauge-adjusted satellite products are released weeks to months after the observation time because of the delay of high quality gauge datasets, and the accuracy largely depends on the spatio-temporal representativeness of the gauge networks. However, over mountainous regions, which usually have sparsely-distributed gauge networks and temporally coarser gauge observations, there are great uncertainties about the performance of gauge-adjusted satellite precipitation products [20,22].

To address the aforementioned disadvantages of gauge-based adjustment, Zhang et al. [29] developed a numerical model-based technique for satellite precipitation adjustment. This technique is designed specifically for heavy precipitation events over topographically-complex regions, where the raw satellite products considerably underestimate heavy precipitation [20,30] and can remedy the negative bias without gauge data input. In addition, the model-adjusted product can be generated in near-real-time, while gauge-adjusted products are only available several months later. Previous studies [22,31–33] have successfully applied this technique to the raw CMORPH and GSMaP products with model simulations for severe storms over the Alps, Andes, Appalachians, Rockies and mountains in Taiwan.

In this paper, we apply the evaluation of Zhang et al.'s [29] model-adjustment technique to the latest near-real-time satellite precipitation product (IMERG) by incorporating an ensemble precipitation forecast dataset produced by NCAR [34]. The study focuses on a large number of flood-inducing storms that occurred in mountainous areas over the conterminous United States (CONUS) and is unique relative to past studies in that it contrasts model-adjustment performance characteristics across different complex terrain domains (coastal vs. inland) and uses model precipitation forecasts for near-real-time adjustment of satellite precipitation datasets.

Section 2 provides information about the study regions and precipitation datasets, while Section 3 explains the methodology of model-based satellite adjustment and data evaluation. Section 4 presents results and a brief discussion of findings from previous publications related to this topic. Section 5 presents conclusions and thoughts for future study.

2. Study Regions and Datasets

2.1. Study Regions

The selection criteria for study regions considered both terrain complexity and annual precipitation amounts. We picked study regions from major mountain ranges in the CONUS: the Appalachians, Rocky Mountains, Olympic Mountains, Pacific Coast Ranges, Cascade Range and Sierra Nevada. Each region was composed of multiple counties with complex terrain and relatively high annual precipitation. Four of the regions (Figure 1, Regions (a), (b), (c), and (d)) were along the Pacific coastline, where the climate is heavily influenced by the ocean and characterized by wet winters and dry summers. The other three (Figure 1, Regions (e), (f), and (g)) were inland regions with continental climates. Regional elevation maps are shown in Figure 2.

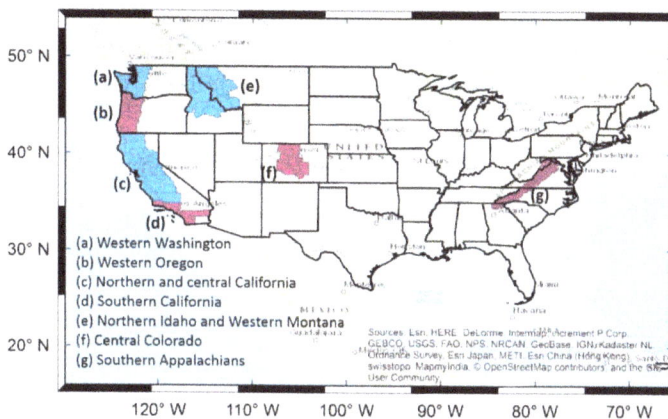

Figure 1. Location of the seven study regions over the conterminous United States (CONUS).

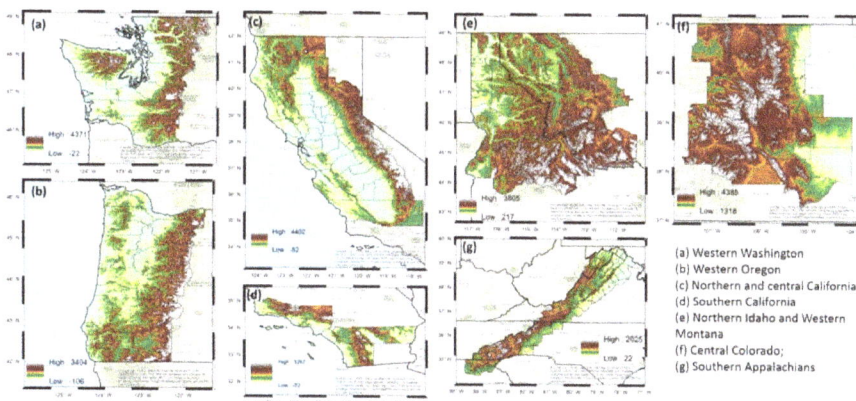

Figure 2. Map of terrain elevation for the seven study regions (meters). DEM data are from the USGS Shuttle Radar Topography Mission (SRTM, https://lta.cr.usgs.gov/SRTM). Data are available at https://dds.cr.usgs.gov/srtm/version2_1/SRTM3/North_America/.

The Western Washington region (Figure 2a) covers the Olympic Mountains and the windward side of the North Cascade Range. The terrain has elevations ranging from 500 to 1500 m a.s.l. and several volcanoes reaching significantly higher altitudes than the rest of the mountains. This region is characterized by an oceanic climate with mild temperatures in all seasons. It has relatively dry summers, and most precipitation occurs in winter, spring and fall. The annual average precipitation varies roughly from 1500 to 3300 mm in higher altitude areas. On the western slopes of the Olympic Mountains, annual precipitation can exceed 4000 mm, which makes this region the rainiest in the CONUS.

The Western Oregon region (Figure 2b) is composed of the Pacific Coast Ranges, the windward side of the Central Cascade Range, the Northern Klamath Mountains and Willamette Valley. The elevations of most mountainous areas range from approximately 500 to 1500 m a.s.l. Like Western Washington, Western Oregon has an oceanic climate, with very wet winters and dry summers. The overall annual average precipitation in Western Oregon's complex topography ranges between 1200 and 3000 mm, which is slightly lower than in Western Washington.

The study region of Northern and Central California (Figure 2c) covers the Southern Klamath Mountains, the Coast Ranges, the windward side of the Southern Cascade Range, the Sierra Nevada and the Great Valley. This region is characterized by extremely steep topographic gradients from the valleys to the mountains. The elevation ranges from approximately 400 to 2300 m a.s.l., while a small portion of the Sierra Nevada exceeds 3000 m a.s.l. Most of the precipitation occurs in mountainous areas. The northwestern part of this region has annual average precipitation between 1300 and 3000 mm, while other mountains in the region have less, ranging from 800 to 2300 mm.

The Southern California study region (Figure 2d) is smaller than the others. It includes the Peninsular Ranges and part of the Transverse Ranges. Elevations in most mountainous areas range from 200 to 2000 m a.s.l. Like the above three coastal regions, this region is under maritime influence, but the climate is much drier and hotter. Annual average precipitation ranges from approximately 200 to 700 mm.

The study region of Northern Idaho and Western Montana (Figure 2e) is an inland area covering part of the Middle Rocky Mountains. Elevation gradually increases from north to south and ranges roughly from 1000 to over 3800 m a.s.l. where Borah Peak is located. This region is dominated by a continental and subarctic climate with annual precipitation ranging from 500 to 1400 mm.

The Central Colorado region (Figure 2f) is located in the Southern Rocky Mountains. It is between the Continental Divide and western boundary of the Colorado Plains and includes Colorado's most populated area (Front Range). The topography of most of this region is very complex, with elevations ranging between 1500 and 4300 m a.s.l. Similar to the Idaho and Montana region, Central Colorado has a continental or subarctic climate, but it has less precipitation. Annual average precipitation ranges between 350 and 800 mm.

The third inland region is located in the Southern Appalachians (Figure 2g). Specifically, it covers all of the Blue Ridge Mountains and part of the Ridge-and-Valley Appalachians, which are two physiographic provinces of the larger Appalachian range. Elevations of the mountainous areas range from approximately 500 to 1700 m a.s.l. Although it has a humid subtropical and temperate oceanic climate, we still count it as an inland region in this research because it includes no coastal area. Annual average precipitation ranges from 1000 to 2500 mm with no significant seasonal differences.

2.2. Precipitation Datasets

2.2.1. Satellite-Retrieved Product

The IMERG precipitation product is available at 0.1°/30-min resolution with quasi-global coverage (60°N–60°S). The IMERG algorithm merges all available satellite microwave precipitation estimates, the microwave-calibrated infrared (IR) satellite estimates, gauge analyses and other precipitation estimators from the TRMM and GPM eras [16]. In the GPM era (starting in 2014), IMERG is considered a more comprehensive precipitation product than those of the TRMM era (CMORPH, TRMM Multi-satellite Precipitation Analysis (TMPA), PERSIANN, GSMaP). In the consideration of observation data latencies, IMERG runs twice to provide quick estimates in near-real-time, which are the early run (~4-h latency, ftp://jsimpson.pps.eosdis.nasa.gov/data/imerg/early/) and late run (~12-h latency, ftp://jsimpson.pps.eosdis.nasa.gov/data/imerg/late/). After about 2.5 months, the IMERG final run (ftp://arthurhou.pps.eosdis.nasa.gov/gpmdata) provides a research-level product that is generated with more available satellite-based data and gauge data adjustment.

Since the research goal was to conduct near-real-time IMERG correction solely by the numerical weather prediction (NWP) model, an NWP-based adjustment was applied to the IMERG Version 5B Late run (~12-h latency) estimates (herein called raw IMERG-L). To compare the two adjustment methods—NWP-based and gauge-based—we also included the IMERG Version 5B Final run (~2.5-month latency) gauge-adjusted estimates (herein called gauge-adjusted IMERG-F) in the error analyses discussed in this paper.

2.2.2. Numerical Weather Prediction

We extracted precipitation forecasts from NCAR's experimental real-time ensemble prediction system [34] (https://rda.ucar.edu/datasets/ds300.0/), which is a 10-member ensemble prediction system that produces daily 48-h forecasts with 3-km horizontal grid spacing over the CONUS using the Weather Research and Forecasting (WRF) model [35]. The ensemble data have been available since April 2015. All ensemble members share the same physics and dynamics, which makes them all equally likely to represent the "true" atmospheric state.

Schwartz et al. [34] evaluated the NCAR ensemble precipitation over the central and eastern CONUS and showed that the ensemble generally produced reasonable amplitudes of precipitation from the viewpoint of multi-month accumulation (7 April to 5 July 2015), while analyses of hourly precipitation rates revealed over-prediction at higher rates ($\geq$5.0 mm/h) and under-prediction at lower rates. While Schwartz et al. [34] evaluated precipitation over an area with relatively lower elevation than the study domains in this research, Gowan et al. [36] found the NCAR ensemble performed well at high-altitude sites in the western United States. These collective results indicate that the NCAR ensemble precipitation was potentially suitable for conducting model-based correction on the underestimation of the raw IMERG-L product for heavy precipitation events in the case study areas.

2.2.3. NCEP Stage IV Product

For the reference precipitation data, we used the NCEP Stage IV precipitation dataset [23] (https://data.eol.ucar.edu/dataset/21.093), which is a multi-sensor (radar and gauge) product available over CONUS at approximately 4.7-km horizontal grid spacing. The final product is mosaicked by observations from twelve National Weather Service (NWS) River Forecast Centers (RFCs). Stage IV data are available in hourly, six-hourly and 24-hourly temporal resolutions. The six-hourly and 24-hourly products cover the entire CONUS, while the hourly product is not available in some of the western coastal states, where four of the coastal case study regions are located. Moreover, the hourly Stage IV product does not always include manual quality control from every RFC [37], but the six-hourly product does. Therefore, this study used the six-hourly product to evaluate the various precipitation estimates.

We note that Stage IV has less accuracy over the western mountainous states [37], where radar coverage is relatively sparse [38], and the corresponding RFCs use a unique rainfall processing algorithm named Mountain Mapper [39]. Nelson et al. [37] indicated underestimation in western RFC's analysis data, which is due to the application of the Mountain Mapper algorithm. To quantify the underestimation, they showed bias ratios of Stage IV versus gauge data based on an 11-year period (2002 to 2012). The three western RFCs—Northwest, California-Nevada and Colorado basin—have seasonal bias ratios in the range of 0.78 to 0.82, 0.68 to 1.18 and 0.78 to 0.87, respectively. Consequently, the results of this paper may be affected by the Stage IV accuracy in western regions. Nevertheless, Stage IV is still used as reference data in this study because it is widely considered as the best gridded precipitation dataset over the CONUS.

3. Methodology

3.1. Event Selection

The precipitation events we selected occurred between May 2015 and December 2016 (a 20-month period). Since our research focused solely on flood-inducing storms, we first collected precipitation-caused flood reports from the NOAA Storm Events Database [40] for that period for each study region. We then identified precipitation events associated with these flood reports and eliminated coastal flood reports from the study because, precipitation aside, coastal flooding usually depends greatly on storm surge. We found a total of 523 precipitation-induced flood and flash flood reports for the seven study regions and study period, associated with 81 heavy precipitation events. The event lengths vary from 24 to 120 h. Although the NCAR ensemble provides daily 48-h forecasts, we only use the first 24-h forecasts for each event. In other words, events longer than 24 h employ the forecasts from two or more different model runs.

3.2. IMERG Adjustment

Before applying the model-based adjustment to raw IMERG-L, the hourly NCAR ensemble precipitation forecasts were summed to produce 6-hourly accumulations and upscaled from the original (3-km) grid to the 0.1° IMERG grid. We performed the remapping procedure by assigning model grid centers to each IMERG grid box. The average value of the NCAR model grid cells collocated with a particular IMERG grid box represented the remapped model value on the IMERG grid. We temporally aggregated the model and IMERG values at 6-hourly precipitation rates for consistency with the Stage IV temporal resolution.

We adjusted raw IMERG-L precipitation values by matching the raw IMERG-L precipitation quantiles with the model quantiles using a power-law function. Specifically, we computed precipitation quantile values from all non-zero, 6-hourly precipitation rates of each dataset. To simplify the calculation, we used only 5%, 10%, 15%, ..., 95% quantile values in the data fitting equation shown below,

$$Y = a \times X^b, \tag{1}$$

where X and Y represent the precipitation quantile values of raw IMERG-L and the model, respectively. We estimated the parameters a and b by the least squares method. The adjustment was done at the event scale, meaning a and b varied for each precipitation event. The quantile-quantile plot between raw IMERG-L and the mean of NCAR ensemble model products (Figure 3a) shows that the raw IMERG-L product has lower quantiles than the NCAR model in all study regions except Western Oregon. Further error analysis in Section 4.1 will show that the raw IMERG-L product underestimates over Western Oregon in terms of occurrence comparison at different precipitation thresholds, which is due to the fact that the NCAR model estimates precipitation more frequently than the satellite product. Figure 3b is a scatter plot of the values of parameters a and b for events in all regions. The parameter values from events occurring in the same region tend to be grouped together. Moreover, events in inland regions tend to have lower a values and higher b values, while events in coastal regions tend to have higher a values and lower b values. This indicates that the model-based adjustment performs differently for coastal and inland regions.

After data fitting, we applied the a and b values back to all raw IMERG-L precipitation rates by Equation 1 again to produce model-adjusted IMERG-L data. The adjustment can change the precipitation magnitude of the raw IMERG-L data, but cannot change the spatial pattern. Note that model precipitation is a 10-member ensemble dataset. Each member was used in the adjustment separately. Eventually, a new product was generated: model-adjusted IMERG-L ensemble precipitation.

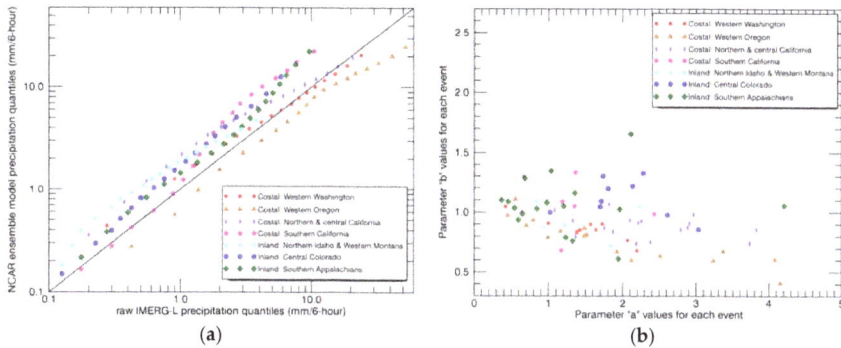

Figure 3. (**a**) Quantile-quantile plot of the raw IMERG-L (Late run) product vs. the NCAR model product; (**b**) scatter plot of the values of parameters *a* and *b* for events in all regions.

3.3. Data Evaluation

We evaluated precipitation estimates for each event at $0.1°$ horizontal grid spacing, meaning we remapped Stage IV reference precipitation onto the IMERG grid by the same procedure we used for model remapping. We compared four estimators (model ensemble, raw IMERG-L, gauge-adjusted IMERG-F and model-adjusted IMERG-L ensemble precipitation) against the Stage IV reference precipitation at 6-hourly/$0.1°$ grid spacing. The results shown below are based on median values of the model ensemble and the model-adjusted IMERG-L ensemble.

We quantitatively analyzed the 6-hourly precipitation rates using the Bias Ratio of frequency (BS), Heidke Skill Score (HSS) [41] and Critical Success Index (CSI). These error metrics are derived from the 2×2 contingency table representing the following four occurrence conditions:

A is counted when estimator $\geq$ threshold and Stage IV $\geq$ threshold (hits);
B is counted when estimator $\geq$ threshold and Stage IV < threshold (false alarms);
C is counted when estimator < threshold and Stage IV $\geq$ threshold (misses);
D is counted when estimator < threshold and Stage IV < threshold (correct rejections).

Each score is calculated at three thresholds. Threshold values varied for each region, depending on local precipitation intensity. The equation for BS is:

$$BS = \frac{A + B}{A + C}, \tag{2}$$

which shows an estimator's bias for an entire study domain and a whole event, meaning BS is affected by the overall estimation, but not the exact location and timing of rainfall. It has a perfect value of 1, with below or above 1 representing under- or over-estimation, respectively.

The following equations are used to estimate the HSS:

$$PC = \frac{A + D}{A + B + C + D}, \tag{3a}$$

$$F = \frac{(A + B)(A + C) + (B + D)(C + D)}{(A + B + C + D)^2}, \tag{3b}$$

$$HSS = \frac{PC - F}{1 - F} = \frac{2(A \times D - B \times C)}{(A + C)(C + D) + (A + B)(B + D)}, \tag{3c}$$

PC is the percentage of correct estimates, and F is the fraction of correct estimates expected by chance. Finally, HSS is defined as the percentage of correct estimates that has been adjusted by the number expected to be correct by chance. Spatial mismatches in rainfall patterns would affect HSS performance and result in lower values. HSS values range from $-\infty$ to 1, with 1 indicating a perfect set of estimation and negative values indicating that the given estimation has fewer hits (H) than a random estimation.

Finally, the CSI score is defined as:

$$CSI = \frac{A}{A + B + C},$$ (4)

CSI measures the fraction of precipitation rates that were correctly estimated. It examines the accuracy of the estimator without considering the correct rejections (D). CSI is sensitive to hits and penalized for misses and false alarms, so it is a function of Probability Of Detection (POD) and False Alarm Ratio (FAR). Unlike HSS, CSI is not affected by spatial mismatches in the rainfall patterns of different products, which means it can provide an evaluation for overall precipitation occurrences. The range of CSI values is from 0 to 1, with 1 as the perfect value.

To examine the performance of model-based IMERG adjustment in different topographic and climatic conditions, we classified the study regions into two groups: Pacific coastal regions and inland regions. Then, we analyzed the domain average event total precipitation estimation performance by the Pearson correlation coefficient (CORR) and normalized root-mean-square-error (NRMSE),

$$NRMSE = \frac{\sqrt{\frac{1}{n}\sum_{i=1}^{n}\left(\left(E_i - \frac{1}{n}\sum_{i=1}^{n}E_i\right) - \left(S_i - \frac{1}{n}\sum_{i=1}^{n}S_i\right)\right)^2}}{\frac{1}{n}\sum_{i=1}^{n}S_i},$$ (5)

where n is number of events in each group and E and S are precipitation of the estimator and Stage IV, respectively. *NRMSE* measures the random component of error after removing the bias. CORR reveals the similarity of each estimator to Stage IV data.

4. Results and Discussion

4.1. Comparisons of Precipitation Rates

The seven study regions are discussed individually in this section. Figure 4 shows the error statistics of the six-hourly precipitation rate in the Western Washington region. The BS, HSS and CSI scores of all ten precipitation events occurring in this region are shown in boxplots, with three different rain rate thresholds for all estimators.

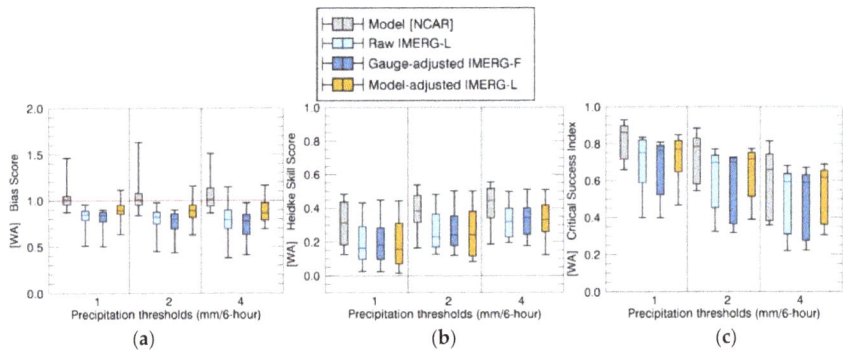

Figure 4. Error statistics of precipitation rate in Western Washington Region. (**a**) Bias Ratio of frequency (BS), (**b**) Heidke Skill Score (HSS) and (**c**) Critical Success Index (CSI).

As the BS plot shows (Figure 4a), the raw IMERG-L product tended to underestimate the number of occurrences at all three precipitation thresholds, while the model data tended toward slight overestimation. The gauge-based adjustment did not improve IMERG performance; in fact, it enhanced the underestimation. In contrast, the model-adjusted IMERG-L product showed considerable improvement, with BS values much closer to one for all precipitation thresholds.

HSS values (Figure 4b) showed that model forecasts exhibited higher scores than the IMERG estimates for all thresholds, and the raw IMERG-L score was relatively low. The performances of the gauge-adjusted IMERG-F and model-adjusted IMERG-L products were similar to that of the raw IMERG-L. This result indicated that neither IMERG adjustment could reduce the random component of the error.

The CSI values (Figure 4c) for all estimators decreased as the rainfall threshold increased. Although the CSI values of the raw IMERG-L and gauge-adjusted IMERG-F products were relatively low, the model-adjusted IMERG-L product produced higher values, indicating that the model-based adjustment effectively increased the percentage of correct estimates. Overall, the model-adjusted IMERG-L performed better than all three IMERG-related products, although the NCAR model forecast provided an even better estimation.

Figure 5 shows the results of the analysis of 16 heavy precipitation events in the Western Oregon region. The raw IMERG-L exhibited underestimation (Figure 5a). While the model-based adjustment effectively moderated the negative biases, the gauge-based adjustment had no impact on the raw IMERG-L product. In fact, the gauge-adjusted IMERG-F showed no improvement for any error metric (BS, HSS or CSI). Meanwhile, the model forecast was consistently superior to the three IMERG products at lower precipitation thresholds (2 and 4 mm/6 h) for all error metrics, while at the high threshold (8 mm/6 h), the performance of the model-adjusted IMERG-L was comparable to that of the model forecast.

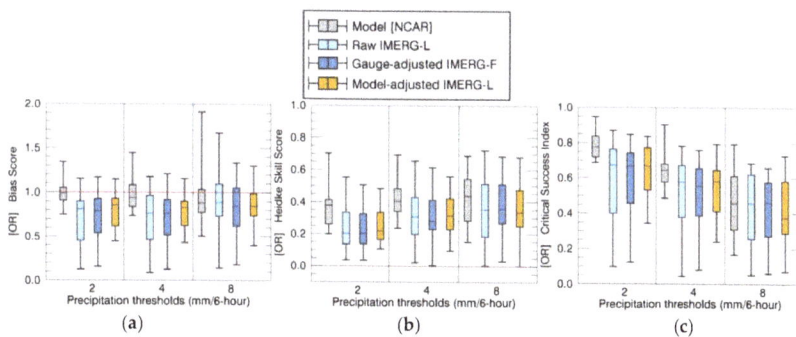

Figure 5. Error statistics of precipitation rate in Western Oregon Region. (**a**) Bias Ratio of frequency (BS), (**b**) Heidke Skill Score (HSS) and (**c**) Critical Success Index (CSI).

Results for the North and Central California region (Figure 6) were based on 17 heavy precipitation events. BS values of the raw IMERG-L product continued to exhibit severe underestimation (Figure 6a). Unlike in the Washington and Oregon regions, the gauge-based adjustment in this region did improve the precipitation estimates. Meanwhile, the model-based adjustment performed even better, especially at the 8 mm/6 h threshold, where the median BS value was very close to one. The advantage of model-based adjustment could also be found in the HSS and CSI (Figure 6b,c). Overall, the model-adjusted IMERG-L had not only higher HSS and CSI median values, but also narrower score value ranges than the other two IMERG products. Still, although the model-adjusted IMERG-L proved superior to the other IMERG products, the model forecast showed the overall best performance for all error metrics in this region.

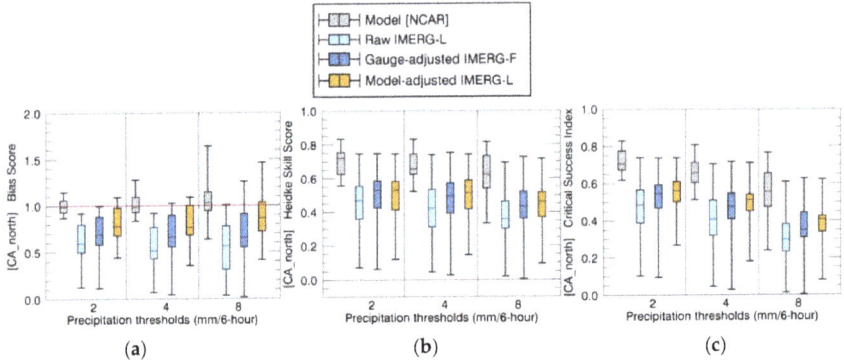

Figure 6. Error statistics of precipitation rate in Northern and Central California Region. (**a**) Bias Ratio of frequency (BS), (**b**) Heidke Skill Score (HSS) and (**c**) Critical Success Index (CSI).

Southern California (Figure 7) is the last coastal study region discussed here. Given the relatively dry climatic conditions of this area, only seven flood-inducing precipitation events were identified. Similar to the above three coastal regions, the raw IMERG-L product was shown to be the least accurate estimator, with apparent underestimations (Figure 7a), and the model forecast performed the best for all error metrics. Nevertheless, unique to this region was the slightly better performance of the gauge-based adjustment relative to that of the model-based adjustment for BS, HSS and CSI, even though both adjustments appeared to be more accurate than the raw IMERG-L. Moreover, the comparison of BS plots across all four coastal regions showed Southern California with the greatest raw IMERG-L underestimation, and the two adjustment methods were unable to improve IMERG to a reasonable level.

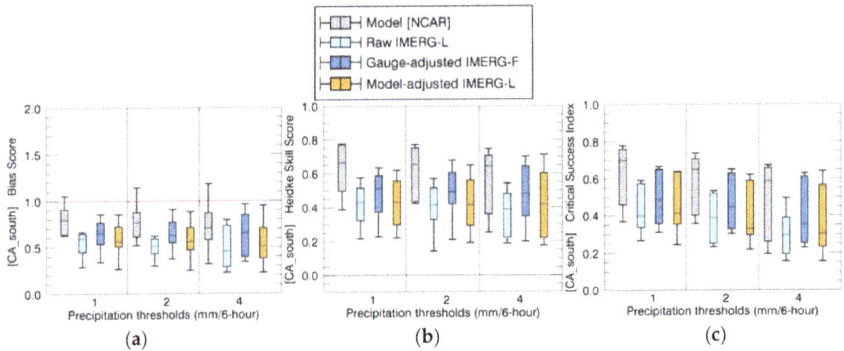

Figure 7. Error statistics of precipitation rate in Southern California Region. (**a**) Bias Ratio of frequency (BS), (**b**) Heidke Skill Score (HSS) and (**c**) Critical Success Index (CSI).

With regards to inland areas, five flood-inducing precipitation events were included in the error metrics for the Northern Idaho and Western Montana region (Figure 8). The raw IMERG-L product exhibited underestimation, and the model forecast had BS values mostly around one (Figure 8a). At the 1 and 2 mm/6 h thresholds, the gauge-based adjustment shrank the BS range, but the median BS values remained similar, indicating no improvement, while the model-adjusted IMERG-L product exhibited substantial improvement. At the 4 mm/6 h threshold, the gauge-adjusted IMERG-F increased BS values, but still had a wide value range, while model-adjusted IMERG-L exhibited underestimation, but with a narrower value range. HSS and CSI plots (Figure 8b,c) showed similar error characteristics in the comparison among the four estimators. Basically, the two adjusted IMERG products performed comparably to the model forecast at high thresholds and were less accurate at lower thresholds.

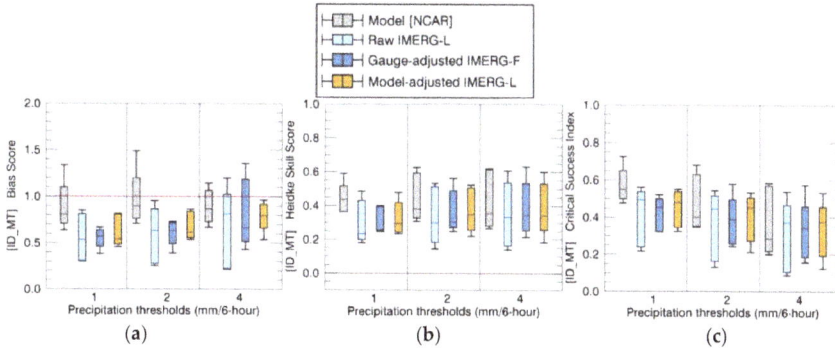

Figure 8. Error statistics of precipitation rate in Northern Idaho and Western Montana Region. (**a**) Bias Ratio of frequency (BS), (**b**) Heidke Skill Score (HSS) and (**c**) Critical Success Index (CSI).

The Central Colorado region had nine precipitation events included in the analysis (Figure 9). As in all the above regions, the raw IMERG-L showed severe underestimation for all rain rate thresholds (Figure 9a). Gauge-based adjustment had very limited impact on the IMERG product; thus, the error scores showed no substantial improvement. Comparison of the BS, HSS and CSI metrics indicated that the model-adjusted IMERG-L product was superior to any of the other estimators, including the model forecast. In fact, the model forecast in this region had a general trend of overestimation and relatively low performance in terms of HSS and CSI values.

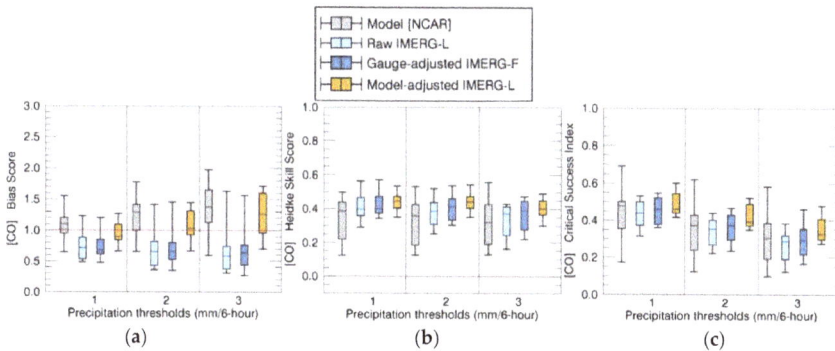

Figure 9. Error statistics of precipitation rate in Central Colorado Region. (**a**) Bias Ratio of frequency (BS), (**b**) Heidke Skill Score (HSS) and (**c**) Critical Success Index (CSI).

Seventeen precipitation events were analyzed in the Southern Appalachians region (Figure 10), with results similar to those for the Colorado region; the raw IMERG-L tended to underestimate for all rain rate thresholds, and the model-adjusted product had the best HSS and CSI (Figure 10b,c). The BS of the model forecast was comparable to that of the model-adjusted IMERG-L (Figure 10a). A possible explanation for the disagreement between the HSS/CSI and BS scores is that the model successfully predicted the domain average rainfall intensity, but with the wrong spatial patterns.

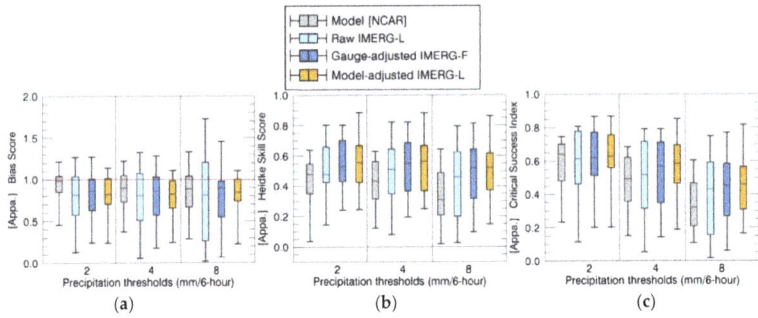

Figure 10. Error statistics of precipitation rate in Southern Appalachians Regions. (**a**) Bias Ratio of frequency (BS), (**b**) Heidke Skill Score (HSS) and (**c**) Critical Success Index (CSI).

Overall, for all study regions, the raw IMERG-L product exhibited the poorest performance of rain rate estimation. Both model- and gauge-based adjustments reflected the effectiveness of IMERG correction. Table 1 compares the two adjustments over coastal regions in terms of the number of events that are improved, exhibit no change or worsened by the adjustments. The model-based adjustment outperforms the gauge-based adjustment with BS and CSI, but not with HSS. Over inland regions (Table 2), IMERG-L benefited more from model-based adjustment than from gauge-based adjustment for all statistical scores. The best-performing product varied under different topographic and climatic conditions. In coastal regions, the model forecast was superior to the two adjusted IMERG (IMERG-F and model-adjusted IMERG-L) products, especially for lower rain rates, while the performance of these products was better than or comparable to that of the model forecast over inland regions.

Table 1. The impact of model- and gauge-based adjustments on 6-hourly precipitation over coastal regions.

Coastal Regions (50 Events)	Model-Based Adjustment			Gauge-Based Adjustment		
Percentage of Event (%)	BS	HSS	CSI	BS	HSS	CSI
Improved	64	52	60	62	58	50
Neutral	8	28	26	4	30	32
Worsened	28	20	14	34	12	18

Table 2. The impact of model- and gauge-based adjustments on 6-hourly precipitation over inland regions.

Inland Regions (31 Events)	Model-Based Adjustment			Gauge-Based Adjustment		
Percentage of Event (%)	BS	HSS	CSI	BS	HSS	CSI
Improved	74	58	58	61	52	52
Neutral	3	35	32	10	32	29
Worsened	23	6	10	29	16	19

4.2. Comparisons of Event Total Precipitation

To illustrate the spatial rainfall distribution of each dataset, we show event total precipitation maps for three events (Figure 11). First, the typical characteristics of each product in coastal regions are shown by a 42-h event that occurred in Northern and Central California, beginning on 15 October 2016, at 18:00 UTC (Figure 11, top row). Taking the Stage IV product as a reference, the model forecast captured all major rain bands in this area with reasonable magnitude, which supports the finding that the model had the best performance in rain rate error metrics over coastal regions. Although the raw IMERG-L product had severe underestimation, it accurately captured the northwestern rain band. Meanwhile, the rain band in the Sierra Nevada was not captured correctly.

The model-based adjustment was effective in dealing with the underestimation of precipitation over the northwestern corner, but it could not improve the estimation for Sierra Nevada because the adjustment is sensitive only to magnitude correction. The gauge-adjusted IMERG-F product did not show enough improvement, either.

The second event occurred in central Colorado on 7 May 2015, at 18:00 UTC, and lasted for three days (Figure 11, middle row). The model prediction was fairly accurate with regard to the overall rainfall magnitude, but the most intense precipitation was erroneously located at the northeastern corner of the domain, while Stage IV showed intense rain over the southeastern part. In contrast, although the raw IMERG-L product largely underestimated the rainfall magnitude, it captured the correct location of the intense rain. After the model-based adjustment, the IMERG-L product achieved the best estimation of the four estimators.

The model precipitation location issue arose in the Southern Appalachians, as well (Figure 11, bottom row). This was a two-day event starting on 29 September 2016, at 00:00 UTC. As with the Colorado event, the model predicted rainfall intensity well from the domain average perspective, but with the wrong spatial distribution of precipitation. The raw IMERG-L product showed severe underestimation again, but with correct spatial distribution of precipitation. The model-adjusted IMERG-L product had the best performance, taking advantage of rainfall intensity from the model and spatial distribution from the raw IMERG-L.

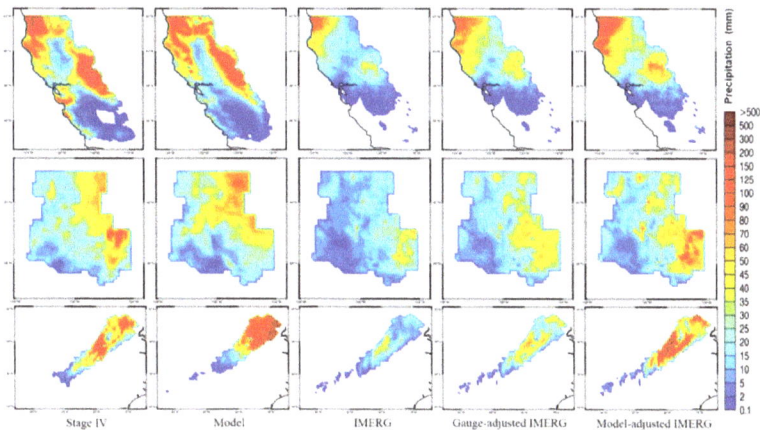

Figure 11. Event total precipitation maps for selected storms. Top panel: event in Northern and Central California (start at 15 October 2016 18:00 UTC, 42-h length). Middle panel: event in Central Colorado (start at 7 May 2015 18:00 UTC, 84-h length). Bottom panel: event in Southern Appalachians (start at 29 September 2016 00:00 UTC, 48-h length).

Statistics for the domain-average event-total precipitation validated the findings from the rain maps (Table 3). We classified the mountainous heavy precipitation events into two groups: coastal region (50 events) and inland region (31 events). Then, we calculated the CORR and NRMSE with respect to reference Stage IV data. The raw IMERG-L product exhibited the lowest CORR and the highest NRMSE for both the coastal and inland regions. The two IMERG adjustment methods effectively improved the pure satellite product. Comparison between the two methods showed that the model-based adjustment was always superior to the gauge-based adjustment, with the exception of coastal region CORR, for which the performance was the same. Taking model forecast data into account, the model performed better than the model-adjusted IMERG-L product for coastal region events. For inland region events, however, the model-adjusted IMERG-L was more accurate than the model itself. This may be due to the fact that the model tended to erroneously place intense precipitation over inland regions, while the raw IMERG-L product was more likely to have difficulty capturing rain band structures over coastal regions, but generally showed correct spatial distributions over inland regions.

Table 3. Statistics of domain average event total precipitation. CORR, Pearson correlation coefficient.

	Coastal Regions (50 Events)		Inland Regions (31 Events)	
	CORR	NRMSE	CORR	NRMSE
Model	0.981	0.19	0.844	0.323
Raw IMERG-L	0.916	0.444	0.609	0.629
Gauge-adjusted IMERG-F	0.961	0.277	0.754	0.377
Model-adjusted IMERG-L	0.961	0.255	0.863	0.28

4.3. Discussion: Comparison to Previous Studies

The technique of model-based satellite adjustment was first introduced by Zhang et al. [29] and demonstrated on five heavy precipitation events over the European Alps and Massif Central. The raw CMORPH product largely underestimated high rain rates, while the WRF simulations provided reasonable overall rainfall magnitudes. Results based on the limited storms from that study showed that the technique efficiently reduced the underestimation of high rain rates, thus providing an improved product.

After the successful first attempt in middle-latitude regions, a more comprehensive study examined the technique in three tropical mountainous regions (Colombian Andes, Peruvian Andes and Taiwan) from 81 storm cases [33]. The raw and gauge-adjusted CMORPH and GSMaP products were involved. As expected, raw CMORPH and GSMaP exhibited severe underestimation in all regions, and the bias was more significant at higher rain rates. Improvements of gauge-based adjustment were shown to be limited, possibly due to the sparse gauge network incorporated in the satellite adjustment. Meanwhile, WRF model-adjusted products outperformed either the gauge-based adjustments or the WRF model itself. The adjustments for higher rain rates were more effective than low rain rates.

Aside from the above mid- and low-latitude applications, there were two studies focusing on CONUS mountain ranges. One of them examined six extreme events induced by hurricane landfalls in the Southern Appalachians [32]. Again, raw CMORPH underestimated all events. Improvements were comparable between WRF model- and gauge-adjusted products. In order to evaluate the impact of satellite adjustment on flood simulations, a hydrological model ran for 20 basins over the study region and showed considerable improvements on the runoff outputs simulated by adjusted CMORPH products.

Nikolopoulos et al. [31] focused on a single extreme rainfall event that occurred in September 2013, in Colorado. Model forecasts produced by the Regional Atmospheric Modeling System and Integrated Community Limited Area Modeling System (RAMS-ICLAMS) were utilized to adjust raw CMORPH, TRMM 3B42RT and weather radar (Multi-Radar Multi-Sensor (MRMS)) estimates. The adjustments were applied by two different procedures: (i) mean field bias and (ii) the adjustment technique herein. Both procedures provided improvements to raw satellite and radar products, with the latter one performing better in terms of random error and correlation.

These previous studies have focused on products during the TRMM era. In the GPM era, the new IMERG products are expected to be extensively used in many applications. In addition, an important gap in past studies was the use of NWP analysis (with the exception of [31]) vs. forecasts that is needed for applications with near-real-time IMERG products. The current study used the NCAR real-time ensemble forecasts for this purpose and demonstrated improvements based on a large number of flash flood events over complex terrain areas in the CONUS. We are encouraged by the results that the model-based adjustment technique can provide improvements to the state-of-the-art IMERG products, especially by the fact that the model-adjusted product outperformed the gauge-adjusted one, which is consistent with the findings of previous studies applied to CMORPH and GSMaP across global mountainous areas.

Combining our studies so far, the technique of model-based satellite precipitation adjustment has been examined over mountainous areas around the world with different terrain complexity and climatic conditions. Results show that the model-adjusted products outperform, or at least are comparable to, the gauge-adjusted products for all high-resolution satellite datasets examined. Moreover, the model-based adjustment requires no gauge network and much less processing time. The results are promising for future applications of model-based satellite precipitation adjustment over mountainous areas, especially for areas lacking ground observations. To successfully apply the technique, there are two prerequisites: (i) the raw satellite data capture the relative spatial and temporal variabilities of precipitation (i.e., no significant surface contamination effects on satellite precipitation detection); and (ii) the model provides relatively accurate precipitation outputs in terms of overall magnitude (not necessarily location).

5. Conclusions

The primary objectives of this study were, first, to examine the feasibility of an ensemble model-based IMERG adjustment technique and, second, to compare the performances of the model-adjusted IMERG-L, a gauge-adjusted IMERG-F and the model itself. Major conclusions are summarized below.

The raw IMERG-L product consistently underestimated heavy precipitation in all study regions over the CONUS, while the rainfall magnitudes exhibited in the NCAR real-time model forecast were fairly accurate. From the perspective of spatial distribution, the raw IMERG-L product was more likely to have difficulty capturing rain band structures over coastal regions, but generally showed correct spatial distributions over inland regions. On the other hand, the model tended to erroneously place intense precipitation over inland regions.

In general, the model-based adjustment could successfully increase the raw IMERG-L precipitation magnitude without changing its spatial pattern and, ultimately, provided a more accurate product than the raw product. While the IMERG-F product could benefit from gauge-based adjustment, as well, the improvement from model-based adjustment was consistently more significant, except in the Southern California region. Comparison between the model forecast and the model-adjusted IMERG-L product showed that the former performed even better than the latter for coastal region events. For inland events, however, the model-adjusted IMERG-L was more accurate than the model itself.

As described in the IMERG technical document [16], the final gauge-adjusted IMERG-F product usually has a 2.5-month latency before it is publicly available. On the other hand, the model-adjusted IMERG-L precipitation can be produced concurrently with the raw IMERG-L, as it requires no gauge observations and is based on NWP model forecasts. Moreover, given that the model-adjusted IMERG-L product performs consistently better than its gauge-adjusted counterpart, it is safe to conclude that model-based adjustment is a feasible technique to improve the quality of the near-real-time IMERG-L product for mountainous heavy precipitation events.

Since the precipitation events in this research were all flood-inducing storms, future studies can focus on the hydrological processes of related flood events. The three IMERG products analyzed here can be used to force a hydrological model and simulate runoff for corresponding basins to evaluate error propagation.

Acknowledgments: Glen Romine, Kate Fossell and Ryan Sobash are thanked for their efforts to produce the NCAR ensemble forecasts. NCAR is sponsored by the National Science Foundation. Xinxuan Zhang was funded by the Eversource Energy Center at the University of Connecticut.

Author Contributions: Xinxuan Zhang and Emmanouil N. Anagnostou conceived of and designed the model-based satellite precipitation adjustment. Xinxuan Zhang performed the adjustment and analyzed the data. Craig S. Schwartz contributed the NCAR ensemble forecast data. Xinxuan Zhang wrote the paper. Emmanouil Anagnostou and Craig Schwartz contributed to the writing of the paper.

Conflicts of Interest: The authors declare no conflict of interest. The founding sponsors had no role in the design of the study; in the collection, analyses or interpretation of data; in the writing of the manuscript; nor in the decision to publish the results.

References

1. Roe, G.H. Orographic precipitation. *Annu. Rev. Earth Planet. Sci.* **2005**, *33*, 645–671. [CrossRef]
2. Houze, R.A. Orographic effects on precipitating clouds. *Rev. Geophys.* **2012**, *50*. [CrossRef]
3. Becker, A.; Finger, P.; Meyer-Christoffer, A.; Rudolf, B.; Schamm, K.; Schneider, U.; Ziese, M. A description of the global land-surface precipitation data products of the Global Precipitation Climatology Centre with sample applications including centennial (trend) analysis from 1901–present. *Earth Syst. Sci. Data* **2013**, *5*, 71–99. [CrossRef]
4. Schamm, K.; Ziese, M.; Becker, A.; Finger, P.; Meyer-Christoffer, A.; Schneider, U.; Schröder, M.; Stender, P. Global gridded precipitation over land: A description of the new GPCC First Guess Daily product. *Earth Syst. Sci. Data* **2014**, *6*, 49–60. [CrossRef]
5. Haylock, M.R.; Hofstra, N.; Klein Tank, A.M.G.; Klok, E.J.; Jones, P.D.; New, M. A European daily high-resolution gridded data set of surface temperature and precipitation for 1950–2006. *J. Geophys. Res.* **2008**, *113*. [CrossRef]
6. Yatagai, A.; Arakawa, O.; Kamiguchi, K.; Kawamoto, H.; Nodzu, M.I.; Hamada, A. A 44-year daily gridded precipitation dataset for Asia based on a dense network of rain gauges. *Sola* **2009**, *5*, 137–140. [CrossRef]
7. Krajewski, W.F.; Smith, J.A. Radar hydrology: Rainfall estimation. *Adv. Water Resour.* **2002**, *25*, 1387–1394. [CrossRef]
8. Germann, U.; Galli, G.; Boscacci, M.; Bolliger, M. Radar precipitation measurement in a mountainous region. *Q. J. R. Meteorol. Soc.* **2006**, *132*, 1669–1692. [CrossRef]
9. Villarini, G.; Krajewski, W.F. Review of the different sources of uncertainty in single polarization radar-based estimates of rainfall. *Surv. Geophys.* **2010**, *31*, 107–129. [CrossRef]
10. Kidd, C.; Levizzani, V. Status of satellite precipitation retrievals. *Hydrol. Earth Syst. Sci.* **2011**, *15*, 1109–1116. [CrossRef]
11. Huffman, G.J.; Bolvin, D.T.; Nelkin, E.J.; Wolff, D.B.; Adler, R.F.; Gu, G.; Hong, Y.; Bowman, K.P.; Stocker, E.F. The TRMM multisatellite precipitation analysis (TMPA): Quasi-global, multiyear, combined-sensor precipitation estimates at fine scales. *J. Hydrometeorol.* **2007**, *8*, 38–55. [CrossRef]
12. Joyce, R.J.; Janowiak, J.E.; Arkin, P.A.; Xie, P. CMORPH: A method that produces global precipitation estimates from passive microwave and infrared data at high spatial and temporal resolution. *J. Hydrometeorol.* **2004**, *5*, 487–503. [CrossRef]
13. Sorooshian, S.; Hsu, K.L.; Gao, X.; Gupta, H.V.; Imam, B.; Braithwaite, D. Evaluation of PERSIANN system satellite–based estimates of tropical rainfall. *Bull. Am. Meteorol. Soc.* **2000**, *81*, 2035–2046. [CrossRef]
14. Kubota, T.; Shige, S.; Hashizume, H.; Aonashi, K.; Takahashi, N.; Seto, S.; Takayabu, Y.N.; Ushio, T.; Nakagawa, K.; Iwanami, K. Global precipitation map using satellite-borne microwave radiometers by the GSMaP project: Production and validation. *IEEE Trans. Geosci. Remote Sens.* **2007**, *45*, 2259–2275. [CrossRef]
15. Ushio, T.; Sasashige, K.; Kubota, T.; Shige, S.; Okamoto, K.; Aonashi, K.; Inoue, T.; Takahashi, N.; Iguchi, T.; Kachi, M. A Kalman filter approach to the Global Satellite Mapping of Precipitation (GSMaP) from combined passive microwave and infrared radiometric data. *J. Meteorol. Soc. Jpn. Ser. II* **2009**, *87*, 137–151. [CrossRef]
16. Huffman, G.J.; Bolvin, D.T.; Braithwaite, D.; Hsu, K.; Joyce, R.; Kidd, C.; Nelkin, E.J.; Sorooshian, S.; Tan, J.; Xie, P. NASA global precipitation measurement (GPM) integrated multi-satellite retrievals for GPM (IMERG). In *Algorithm Theoretical Basis Document, version 5.1*; National Aeronautics and Space Administration: Washington, DC, USA, 2017.
17. Hirpa, F.A.; Gebremichael, M.; Hopson, T. Evaluation of high-resolution satellite precipitation products over very complex terrain in Ethiopia. *J. Appl. Meteorol. Climatol.* **2010**, *49*, 1044–1051. [CrossRef]
18. Gao, Y.C.; Liu, M. Evaluation of high-resolution satellite precipitation products using rain gauge observations over the Tibetan Plateau. *Hydrol. Earth Syst. Sci.* **2013**, *17*, 837. [CrossRef]
19. Stampoulis, D.; Anagnostou, E.N.; Nikolopoulos, E.I. Assessment of high-resolution satellite-based rainfall estimates over the Mediterranean during heavy precipitation events. *J. Hydrometeorol.* **2013**, *14*, 1500–1514. [CrossRef]
20. Derin, Y.; Anagnostou, E.; Berne, A.; Borga, M.; Boudevillain, B.; Buytaert, W.; Chang, C.-H.; Delrieu, G.; Hong, Y.; Hsu, Y.C. Multiregional satellite precipitation products evaluation over complex terrain. *J. Hydrometeorol.* **2016**, *17*, 1817–1836. [CrossRef]
21. Maggioni, V.; Meyers, P.C.; Robinson, M.D. A review of merged high-resolution satellite precipitation product accuracy during the Tropical Rainfall Measuring Mission (TRMM) era. *J. Hydrometeorol.* **2016**, *17*, 1101–1117. [CrossRef]

22. Beck, H.E.; Vergopolan, N.; Pan, M.; Levizzani, V.; van Dijk, A.I.; Weedon, G.P.; Brocca, L.; Pappenberger, F.; Huffman, G.J.; Wood, E.F. Global-scale evaluation of 22 precipitation datasets using gauge observations and hydrological modeling. *Hydrol. Earth Syst. Sci.* **2017**, *21*, 6201. [CrossRef]

23. Lin, Y.; Mitchell, K.E. The NCEP stage II/IV hourly precipitation analyses: Development and applications. In Proceedings of the 19th Conf. Hydrology, San Diego, CA, USA, 9–13 January 2005.

24. Sinclair, S.; Pegram, G. Combining radar and rain gauge rainfall estimates using conditional merging. *Atmos. Sci. Lett.* **2005**, *6*, 19–22. [CrossRef]

25. Goudenhoofdt, E.; Delobbe, L. Evaluation of radar-gauge merging methods for quantitative precipitation estimates. *Hydrol. Earth Syst. Sci.* **2009**, *13*, 195–203. [CrossRef]

26. Mega, T.; Ushio, T.; Kubota, T.; Kachi, M.; Aonashi, K.; Shige, S. Gauge adjusted global satellite mapping of precipitation (GSMaP_Gauge). In Proceedings of the 2014 XXXIth URSI General Assembly and Scientific Symposium (URSI GASS), Beijing, China, 16–23 August 2014; pp. 1–4. [CrossRef]

27. Xie, P.; Yoo, S.-H.; Joyce, R.; Yarosh, Y. Bias-corrected CMORPH: A 13-year analysis of high-resolution global precipitation. *Geophys. Res. Abstr.* **2011**, *13*, Abstract EGU2011-1809.

28. Xie, P.; Joyce, R.; Wu, S.; Yoo, S.-H.; Yarosh, Y.; Sun, F.; Lin, R. Reprocessed, bias-corrected CMORPH global high-resolution precipitation estimates from 1998. *J. Hydrometeorol.* **2017**, *18*, 1617–1641. [CrossRef]

29. Zhang, X.; Anagnostou, E.N.; Frediani, M.; Solomos, S.; Kallos, G. Using NWP simulations in satellite rainfall estimation of heavy precipitation events over mountainous areas. *J. Hydrometeorol.* **2013**, *14*, 1844–1858. [CrossRef]

30. Scofield, R.A.; Kuligowski, R.J. Status and outlook of operational satellite precipitation algorithms for extreme-precipitation events. *Weather Forecast.* **2003**, *18*, 1037–1051. [CrossRef]

31. Nikolopoulos, E.I.; Bartsotas, N.S.; Anagnostou, E.N.; Kallos, G. Using high-resolution numerical weather forecasts to improve remotely sensed rainfall estimates: The case of the 2013 Colorado flash flood. *J. Hydrometeorol.* **2015**, *16*, 1742–1751. [CrossRef]

32. Zhang, X.; Anagnostou, E.N.; Vergara, H. Hydrologic Evaluation of NWP-Adjusted CMORPH Estimates of Hurricane-Induced Precipitation in the Southern Appalachians. *J. Hydrometeorol.* **2016**, *17*, 1087–1099. [CrossRef]

33. Zhang, X.; Anagnostou, E.N. Evaluation of Numerical Weather Model-based Satellite Precipitation Adjustment in Tropical Mountainous Regions. *J. Hydrometeorol.* **2018**. under review.

34. Schwartz, C.S.; Romine, G.S.; Sobash, R.A.; Fossell, K.R.; Weisman, M.L. NCAR's experimental real-time convection-allowing ensemble prediction system. *Weather Forecast.* **2015**, *30*, 1645–1654. [CrossRef]

35. Skamarock, W.C.; Klemp, J.B.; Dudhia, J.; Gill, D.O.; Barker, D.M.; Duda, M.G.; Huang, X.; Wang, W.; Powers, J.G. *A Description of the Advanced Research WRF, version 3*; NCAR Technical Note; NCAR/TN-475+STR; National Center for Atmospheric Research: Boulder, CO, USA, 2008.

36. Gowan, T.M.; Steenburgh, W.J.; Schwartz, C.S. Validation of mountain precipitation forecasts from the convection-permitting NCAR Ensemble and operational forecast systems over the Western United States. *Weather Forecast.* **2018**, in press. [CrossRef]

37. Nelson, B.R.; Prat, O.P.; Seo, D.-J.; Habib, E. Assessment and implications of NCEP stage IV quantitative precipitation estimates for product intercomparisons. *Weather Forecast.* **2016**, *31*, 371–394. [CrossRef]

38. Maddox, R.A.; Zhang, J.; Gourley, J.J.; Howard, K.W. Weather radar coverage over the contiguous United States. *Weather Forecast.* **2002**, *17*, 927–934. [CrossRef]

39. Hou, D.; Charles, M.; Luo, Y.; Toth, Z.; Zhu, Y.; Krzysztofowicz, R.; Lin, Y.; Xie, P.; Seo, D.-J.; Pena, M. Climatology-calibrated precipitation analysis at fine scales: Statistical adjustment of stage IV toward CPC gauge-based analysis. *J. Hydrometeorol.* **2014**, *15*, 2542–2557. [CrossRef]

40. NOAA Storm Events Database. Available online: https://www.ncdc.noaa.gov/stormevents/ (accessed on 20 February 2018).

41. Heidke, P. Berechnung des Erfolges und der Güte der Windstärkevorhersagen im Sturmwarnungsdienst. *Geogr. Ann.* **1926**, *8*, 301–349. [CrossRef]

remote sensing

MDPI

Article

Evaluation and Hydrologic Validation of Three Satellite-Based Precipitation Products in the Upper Catchment of the Red River Basin, China

Yueyuan Zhang [1], Yungang Li [1,*], Xuan Ji [1], Xian Luo [1] and Xue Li [2]

1   Yunnan Key Laboratory of International Rivers and Transboundary Eco-Security, Yunnan University, Kunming 650091, China; zhangyueyuan1@163.com (Y.Z.); jixuan@ynu.edu.cn (X.J.); luoxian@ynu.edu.cn (X.L.)
2   School of Geography and Planning, Sun Yat-sen University, Guangzhou 510275, China; lixue33333z@126.com (X.L.)
*   Correspondence: ygli@ynu.edu.cn; Tel.: +86-871-65034577

Received: 15 October 2018; Accepted: 21 November 2018; Published: 24 November 2018

Abstract: Satellite-based precipitation products (SPPs) provide alternative precipitation estimates that are especially useful for sparsely gauged and ungauged basins. However, high climate variability and extreme topography pose a challenge. In such regions, rigorous validation is necessary when using SPPs for hydrological applications. We evaluated the accuracy of three recent SPPs over the upper catchment of the Red River Basin, which is a mountain gorge region of southwest China that experiences a subtropical monsoon climate. The SPPs included the Tropical Rainfall Measuring Mission (TRMM) 3B42 V7 product, the Climate Prediction Center (CPC) Morphing Algorithm (CMORPH), the Bias-corrected product (CMORPH_CRT), and the Precipitation Estimation from Remotely Sensed Information using Artificial Neural Networks (PERSIANN) Climate Data Record (PERSIANN_CDR) products. SPPs were compared with gauge rainfall from 1998 to 2010 at multiple temporal (daily, monthly) and spatial scales (grid, basin). The TRMM 3B42 product showed the best consistency with gauge observations, followed by CMORPH_CRT, and then PERSIANN_CDR. All three SPPs performed poorly when detecting the frequency of non-rain and light rain events (<1 mm); furthermore, they tended to overestimate moderate rainfall (1–25 mm) and underestimate heavy and hard rainfall (>25 mm). GR (Génie Rural) hydrological models were used to evaluate the utility of the three SPPs for daily and monthly streamflow simulation. Under Scenario I (gauge-calibrated parameters), CMORPH_CRT presented the best consistency with observed daily (*Nash–Sutcliffe efficiency coefficient, or NSE* = 0.73) and monthly (*NSE* = 0.82) streamflow. Under Scenario II (individual-calibrated parameters), SPP-driven simulations yielded satisfactory performances (*NSE* >0.63 for daily, *NSE* >0.79 for monthly); among them, TRMM 3B42 and CMORPH_CRT performed better than PERSIANN_CDR. SPP-forced simulations underestimated high flow (18.1–28.0%) and overestimated low flow (18.9–49.4%). TRMM 3B42 and CMORPH_CRT show potential for use in hydrological applications over poorly gauged and inaccessible transboundary river basins of Southwest China, particularly for monthly time intervals suitable for water resource management.

Keywords: TRMM 3B42 V7; CMORPH_CRT; PERSIANN_CDR; GR models; hydrological simulation; Red River Basin

---

1. Introduction

Precipitation is one of the most important water balance components of the global water cycle, and has great variability across different spatial and temporal scales [1,2]. The accurate observation or estimation of precipitation has important theoretical and practical significance for flood warnings, drought monitoring, and water resource management [3,4]. Gauge observations provide relatively accurate point-based measurements of precipitation [5]; however, owing to significant precipitation heterogeneity across a variety of spatiotemporal scales, rain gauge observations only represent local conditions, and can result in potential errors when interpolated to larger scales, especially in mountainous areas with complex terrain [6]. Moreover, the spatial distribution of rain gauges is extremely uneven, with sparse gauges in remote areas, less developed regions, or areas with complex terrain [7]. Therefore, in situ gauge data usually cannot meet the requirements of applications that depend on high spatial–temporal resolution precipitation data (e.g., hydrological simulations [5]).

Satellite-based precipitation estimates have the advantage of adequate temporal resolution and fine spatial resolution with wide coverage, enabling accurate precipitation estimates in data-scarce or ungauged regions [5,8–10]. A number of satellite-based precipitation products (SPPs) are currently available, including the Global Precipitation Climatology Project (GPCP) [11], the Precipitation Estimation from Remotely Sensed Information using Artificial Neural Networks (PERSIANN) [12], the Climate Prediction Center (CPC) morphing algorithm (CMORPH) [13], the Tropical Rainfall Measuring Mission (TRMM) Multi-satellites Precipitation Analysis (TMPA) [14], the Global Precipitation Measurement (GPM) [3], and the global gridded precipitation dataset Multi-Source Weighted-Ensemble Precipitation (MSWEP) [15]. With wide coverage and high spatial–temporal resolution (mostly finer than $0.25° \times 0.25°$ at spatial and three-hour temporal scales), SPPs have been extensively applied in many fields, including hydrological simulation [16–18], extreme event analysis [19–21], and water resource management [22,23].

Since SPPs are based on an indirect approach that utilizes sensors, the results inevitably contain uncertainties caused by measurement errors, sampling, retrieval algorithms, and bias correction processes [17,24,25]. Furthermore, the error characteristics change depending on the climate region, season, altitude, and other factors [10,26]. In general, quantitative statistical and hydrological modelling evaluations are effective tools that are used to evaluate the precision of SPPs [4,17]. The former focuses on the comparison and evaluation of SPPs against gauge data or ground-based radar estimates. By this principle, temporal characteristics and spatial distributions of SPPs are not only investigated, but can also be quantitatively analyzed; however, the scale discrepancy problem remains when using rain gauge data for validation. The latter evaluates SPPs based on their predictive ability of streamflow rate in a hydrological modeling framework; precipitation products are evaluated at the watershed scale with respect to a specific application [27].

Over the past decades, numerous studies have improved our understanding of SPP performance at global and regional scales [8,28–31]. For example, TMPA products were validated in various parts of the world [32–36], and those results showed that TMPA products perform reasonably well over most regions. Following the successful TMPA, the Integrated Multisatellite Retrievals for GPM (IMERG), which incorporates observations from several satellites, offers improvements over the TMPA in quality and spatiotemporal resolution of precipitation data [3]. A range of studies comparing the TMPA and GPM products for the United States [37], Brazil [38], Africa [39], Iran [40], India [41], Pakistan [42], Malaysia [43], Singapore [44], and China [45] indicated that GPM is superior to TMPA products. In China, evaluation and validation using hydrologic simulations have been explored over many basins, including the Ganzi River basin [46], the upper Yangtze River and upper Yellow River basins over the Tibetan Plateau [47], the Huaihe River basin of eastern China [18], the Beijiang and Dongjiang River basins of southern China [5], the Luanhe River basin of northern China [4], the Tiaoxi watershed, which is part of the southern catchment of Taihu Lake in southeastern China [48], the Lancang River basin of southwest China [49], and the Huifa River basin of northeast China [50]. These studies all highlighted the great potential of different SPPs in hydrologic simulations, although SPPs have variable accuracy and distinct hydrological performance in different basins.

The Red River is an important transboundary river in Southeast Asia. Precipitation distribution is significantly uneven across the basin due to the complex terrain and subtropical monsoon climate [51].

About 85% of the annual total precipitation falls during the summer season [52]. Consequently, the Red River has an irregular flow regime. The high variability of river discharge in space and time leads to substantial challenges related to flooding and water stress, particularly in the Red River delta, which is a densely populated area of great importance to Vietnam for its agricultural productivity and economic activity [35]. The upstream region in China has a mean annual flow of 48.3 billion m^3, which contributes 37% of total flow of the Red River (131.4 billion m^3) [53]. The transboundary water resource is virtual for agriculture irrigation, hydropower, and ecosystem services. However, the rain gauge network has a low density (around 300 km^2 per rain gauge in Yunnan province, China), spatially uneven distribution, and is insufficient over mountainous areas. The scarcity and mismatch of the precipitation observations from upstream and downstream countries make it imperative to use SPPs in hydrological modeling, drought monitoring, and water resources management. Unfortunately, there is little work focusing on the evaluation of SPPs and their hydrological applicability over the Red River Basin in China.

This study aimed to assess the performance of three latest SPPs over the upper catchment of the Red River Basin for the time period 1998–2010. The SPPs included TRMM 3B42 V7, CMORPH_CRT (CMORPH Bias-corrected product), and PERSIANN_CDR (PERSIANN Climate Data Record). The main objectives were to: (1) statistically evaluate the quality of the three SPPs through comparison with rain gauge observations; and (2) comprehensively explore and compare the capability of these three SPPs in streamflow simulations using GR (Génie Rural) hydrological models at daily and monthly scales. This study will improve our understanding of the reliability of the three latest SPPs, and provide a reference for their applications in hydrological simulation and transboundary water resource management in the Red River Basin.

2. Materials and Methods

2.1. Study Area

The Red River originates in a mountainous area of Yunnan Province, China; it flows 1200 km to the southeast, and ends in the Gulf of Tonkin, in the South China Sea [54,55]. The Red River Basin drains an area of 156,451 km^2, of which 50.3% is in Vietnam, 48.8% is in China, and 0.9% is in Laos [52].The upper catchment of the Red River Basin (URRB) refers to the catchment north of the China–Vietnam border (Figure 1). The catchment covers an approximate area of 33,614 km^2. The elevation of the catchment ranges from 76 m to 3123 m above sea level, and decreases from the northwest to the southeast [51]. It is characterized by a subtropical monsoon climate [52], with annual average temperatures of 14.8–23.8°C. The annual average precipitation over the URRB from 1998 to 2010 was about 1044 mm, ranging from 772 mm to 1276 mm; approximately 85% of the annual precipitation is concentrated in the rainy season (May to October) [52]. The climate confers the typical hydrologic regime characterized by large runoff during the summer and low runoff during the winter [56]. The annual average discharge at Manhao Station for the period 1998–2010 was 282 m^3/s.

2.2. Datasets

Daily observed discharge data for the period 1998–2010 at the Manhao hydrological station was obtained from the Hydrological Year Book and the Hydrological Bureau of Yunnan Province (HBYP). Daily precipitation data from 25 rain gauges were provided by the Meteorological Agency of Yunnan Province (MAYP) (Figure 1). It is noteworthy that these rain gauges are independent from the Global Precipitation Climatology Centre (GPCC) gridded gauge-analysis precipitation dataset. Daily meteorological observations at 21 stations for the same period were collected from the China Meteorological Administration (CMA), including mean air temperature, mean relative humidity, mean wind speed at 10-m height, and hours of bright sunshine. The quality of the above datasets has been checked by the HBYP, MAYP, and CMA. We also performed routine quality assessment including statistical tests, visual data plots, and histograms, to ensure that there were no missing or erroneous records. The descriptive statistics of precipitation and discharge during 1998–2010 are provided in the supplementary material (Table S1 and Table S2). The daily potential evapotranspiration for each station was estimated using the Penman–Monteith equation, as recommended by the Food and Agriculture Organization of the United Nations (FAO) [57]. The FAO Penman–Monteith method is provided in the supplementary material (S1). Areal average rainfall and potential evapotranspiration over the catchment were produced by using the Thiessen polygon approach [58].

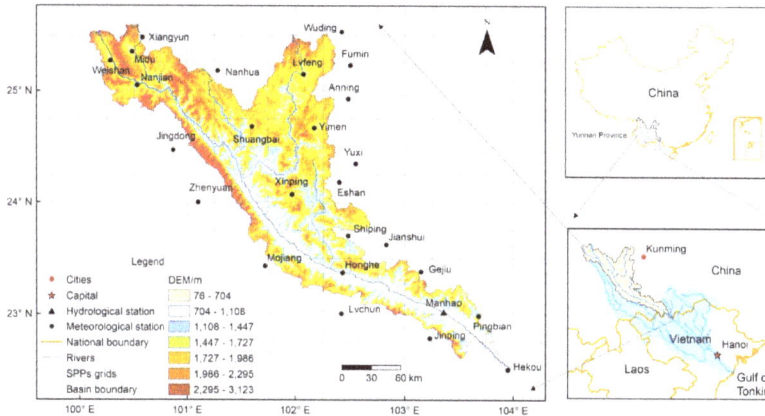

Figure 1. Location of the upper catchment of the Red River Basin (URRB) and distribution of meteorological and hydrological stations.

Three SPPs (TRMM 3B42 V7, CMORPH_CRT, and PERSIANN_CDR) were considered. The TRMM 3B42 V7 product is one of the TMPA products, which were designed based on a wide variety of satellite datasets and are supplied by the National Aeronautics and Space Administration (NASA) [14]. This product provides precipitation at a spatial resolution of 0.25° and a three-hour temporal resolution; it has quasi-global coverage of 50°N–50°S from 1998 to the present, combining information from calibrated passive microwave (PMW) and infrared (IR) data. The 3B42 V7 product was adjusted using monthly rain gauge precipitation data from the GPCC [59]. Here, the 3B42 V7 with a daily temporal resolution and a 0.25° spatial resolution from 1998 to 2010 was employed. Daily precipitation was obtained by the accumulation of three-hour precipitation data.

NOAA's (National Oceanic and Atmospheric Administration) CPC CMORPH contains global satellite-based precipitation generated by integrated PMW and IR data [13]. The latest CMORPH V1.0 product includes a raw satellite-only precipitation product (CMORPH_RAW), CMORPH_CRT, and a satellite-gauge blended product (CMORPH_BLD), covering 60°S–60°N and 180°W–180°E. The CMORPH_CRT product is generated by adjusting the CMORPH_RAW product against the CPC unified daily gauge analysis over land, and the pentad GPCP over ocean using the probability density function (PDF) matching bias correction method [60]. Three combinations of spatial–temporal resolutions are available: eight km and 30 min, 0.25° and 3 h, and 0.25° and daily. Here, the CMORPH_CRT product with 0.25° and daily spatial–temporal resolution for the period 1998–2010 was used.

The original PERSIANN is one of the popular global precipitation estimations for estimating historical precipitation from March 2000 to present; it was developed by combining PMW observations and IR data. The latest PERSIANN_CDR product used the archive of the GridSat-B1 IR data [61] as input to the trained PERSIANN model; then, the biases in the PERSIANN estimated precipitation were adjusted using GPCP 2.5° monthly data version 2.2 [12,27]. Since no PMW is used in the PERSIANN_CDR product, the PERSIANN model parameters were pretrained using National Centers for Environmental Prediction (NCEP) stage-IV hourly precipitation data. Currently, this version of PERSIANN_CDR provides daily precipitation estimates at a spatial resolution of 0.25° for quasi-global coverage (60°N–60°S) from 1983 to the present. In this study, a subset of data for the period 1998–2010 was employed.

2.3. Methods

2.3.1. Evaluation Indices

Comparisons between the three SPPs and rain gauge data were performed both on grid and basin scales. For the grid scale, the SPPs precipitation at the grid boxes with rain gauge stations are extracted and compared with the corresponding rain gauge precipitation. For the basin scale, spatially averaged pixel values of the SPPs precipitation were compared with the areal-averaged precipitation from rain gauge stations using the Thiessen polygon approach [58].

Several widely used statistical indices were adopted to quantify the performance of the three SPPs against rain gauge observations, including Spearman's Rank correlation coefficient (*CC*) [62], root mean squared error (*RMSE*), mean absolute error (*MAE*), and relative bias (*Bias*). In addition, the probability of detection (*POD*), frequency of hit (*FOH*), false alarm ratio (*FAR*), critical success index (*CSI*), and the Heidke skill score (*HSS*) indices were calculated to evaluate the precipitation detection ability of the three SPPs [63]. *POD* provides the fraction of precipitation events that the satellite products detect among all the actual precipitation events; *FOH* measures how often the satellite products detect rainfall when there is actually rainfall; *FAR* measures the fraction of unreal events among all the events that the satellite products detected; *CSI* represents the overall fraction of precipitation events correctly detected by the satellite products; and *HSS* measures the accuracy of the estimates accounting for matches due to random chance. Furthermore, *CC*, *Bias*, and the Nash–Sutcliffe efficiency coefficient (*NSE*) [64] were employed to evaluate the performance of the hydrological model in streamflow simulations. *NSE* describes the prediction skill of the simulated streamflow as compared to the observed. The formulas to calculate the indices mentioned above are listed in Table 1.

Table 1. Indices used to evaluate the performance of satellite precipitation estimates. *CC*: correlation coefficient, *RMSE*: root mean squared error, *MAE*: mean absolute error, *Bias*: relative bias, *POD*: probability of detection, *FOH*: frequency of hit, *FAR*: false alarm ratio, *CSI*: critical success index, *HSS*: Heidke skill score, *NSE*: Nash–Sutcliffe efficiency coefficient.

Statistical Metric	Unit	Equation	Range	Perfect Value	Reference
CC	-	$CC = 1 - \dfrac{6\sum_{i=1}^{n}[X(i)-Y(i)]^2}{n(n^2-1)}$	−1 to 1	1	[62]
RMSE	mm	$RMSE = \sqrt{\dfrac{\sum_{i=1}^{n}(S_i-G_i)^2}{n}}$	0 to ∞	0	[16]
MAE	mm	$MAE = \dfrac{\sum_{i=1}^{n}|S_i-G_i|}{n}$	0 to ∞	0	[16]
Bias	%	$Bias = \dfrac{\sum_{i=1}^{n}(S_i-G_i)}{\sum_{i=1}^{n}G_i} \times 100\%$	−∞ to ∞	0	[16]
POD	-	$POD = \dfrac{N_{11}}{N_{11}+N_{01}}$	0 to 1	1	[63]
FOH	-	$FOH = \dfrac{N_{11}}{N_{11}+N_{10}}$	0 to 1	1	[63]
FAR	-	$FAR = \dfrac{N_{10}}{N_{11}+N_{10}}$	0 to 1	0	[63]
CSI	-	$CSI = \dfrac{N_{11}}{N_{11}+N_{01}+N_{10}}$	0 to 1	1	[63]
HSS	-	$HSS = \dfrac{2(N_{11}-N_{00}-N_{10}N_{01})}{(N_{11}+N_{01})(N_{01}+N_{00})+(N_{11}+N_{10})(N_{10}+N_{00})}$	0 to 1	1	[63]
NSE	-	$NSE = 1 - \dfrac{\sum_{i=1}^{n}(Q_{oi}-Q_{si})^2}{\sum_{i=1}^{n}(Q_{oi}-\overline{Q_o})^2}$	−∞ to 1	1	[64]

Notation: n, number of samples; S_i, satellite precipitation; G_i, gauged observation; $X(i) = $ rank$[S(i)]$; $Y(i) = $ rank$[G(i)]$; Q_{si}, simulated streamflow; Q_{oi}, observed streamflow; $\overline{Q_o}$, observed mean streamflow. N_{11}, Satellite is > 0 and gauge is > 0; N_{10}, Satellite is > 0 and gauge equals 0; N_{01}, Satellite equals 0 and gauge is > 0; N_{00}, Satellite equals 0 and gauge equals 0.

2.3.2. GR Hydrological Models

The GR hydrological models were developed by Irstea, which is a national applied research institute of France; they are lumped rain–runoff models that can be applied at various time steps, ranging from hourly to annual [65,66]. In this study, only the daily (GR4J) and monthly (GR2M) models were employed.

(1) GR4J model

The GR4J model was originally developed and tested on 429 different catchments in France, the United States of America (USA), Brazil, and the Côte d'Ivoire [67]. GR4J simulates runoff via two functions. First, a production function accounts for precipitation (P) and potential evapotranspiration (PET), and determines the effective precipitation that contributes to flow and supplies the production reservoir. Second, a routing function calculates runoff at the catchment outlet. The quantity of water feeding the routing part of the model comprises the percolation that is added to the remaining fraction of water. This flow is divided into two components according to a fixed split: 90% is routed by a unit hydrograph, UH1, and then a non-linear routing store, and the remaining 10% are routed by a single unit hydrograph, UH2. The purpose of the unit hydrographs is to account for differences in runoff delays between the two conceptual reservoirs. GR4J requires the calibration of four free parameters (Table 2). The median values and approximate 80% confidence intervals for the four parameters are provided in Table 2, which were obtained using a large variety of catchments [68].

Table 2. Median values and approximate 80% confidence intervals of four model parameters.

Parameter	Median Value	80% Confidence Interval
x1: maximum capacity of the production store (mm)	350	100–1200
x2: groundwater exchange coefficient (mm)	0	−5 to 3
x3: maximum capacity of the routing store (mm)	90	20–300
x4: time base of unit hydrograph UH1 (days)	1.7	1.1–2.9

(2) GR2M Model

The GR2M has two parameters and has been shown to have the best performance among several similar models when using a benchmark test consisting of 410 basins around the world [69,70]. The production function of the GR2M model has strong similarities with the daily version, but uses a simplified routing scheme [66]. This model is characterized by two functions: (1) a function of production that revolves around a reservoir ground of a maximum capacity (x1), which is the first parameter to be wedged (transferring a percolation of reservoir ground is ensured by the dependent feature of the stock status 'S'), and (2) a transfer function represented by a quadratic draining reservoir with a capacity fixed at 60 mm. This reservoir is modified by an underground exchange, whose coefficient (x2) is the second parameter to optimize [71].

(3) Hydrological Simulation Scenarios

In this study, both models were warmed up for one year (1998), and then split into calibration (1999–2004) and validation (2005–2010) periods. Model calibration was achieved with the aid of the default algorithm provided in the airGR package developed at Irstea [66]. Two parameterization scenarios were conducted to evaluate the effect of precipitation uncertainty on runoff simulation results. In Scenario I, model parameters were first calibrated using gauged rainfall data in the calibration period, and then the model was rerun using gauged rainfall data and the three SPPs in both the calibration and validation periods. Scenario I was mainly used to evaluate the streamflow simulation utility of the different SPPs using gauge-calibrated parameters [72]. In Scenario II, model parameters were recalibrated with individual satellite rainfall data during the calibration period; then, streamflow was simulated for both the calibration and validation periods using the three individual satellite-based parameters. Scenario II was adopted to determine whether the evaluated SPPs have the potential to be alternative data sources for hydrological simulations in data-poor or ungauged basins [9,10]. Table 3 shows calibrated parameter values in the GR4J model for different precipitation data inputs.

Table 3. Calibrated parameter values in the GR4J model for different precipitation data inputs.

Parameter	Rain Gauge	TRMM 3B42	CMORPH_CRT	PERSIANN_CDR
$x1$	1200	1737	1667	1998
$x2$	0.77	0.35	0.92	1.09
$x3$	25	37	23	35
$x4$	2.25	2.45	3.12	3.42

3. Results

3.1. Comparison of Rain Gauge and Satellite-Based Precipitation Data

3.1.1. Evaluation at the Grid and Basin Scales

Figures 2 and 3 present scatterplots of daily and monthly precipitation, respectively, from TRMM 3B42, CMORPH_CRT, and PERSIANN_CDR against rain gauge data at the grid scale. Intuitively, Figure 2 shows that the TRMM 3B42 product performed the best among the three SPPs for all metrics at the daily scale. Throughout the year, TRMM 3B42 had a much larger *CC*, but a smaller *RMSE, MAE,* and *Bias* than that of the other two products, with the *CC* = 0.62, *RMSE* = 7.7, *MAE* = 2.9, and *Bias* = 1.6%. Meanwhile, PERSIANN_CDR performed the worst, with the exception of the *Bias* (*Bias* = −5.4%), for which it was marginally better than that of CMORPH_CRT (*Bias* = −8.0%). The performances of the three products were also compared seasonally. According to the climate of the Red River Basin, May to October was considered to be the wet season, and November to April was considered to be the dry season [73]. For both the dry and wet seasons, TRMM 3B42 still showed a better performance than the other two products. All three SPPs achieved better estimations during the wet season than during the dry season.

On a monthly scale (Figure 3), the TRMM 3B42 product showed the best performance, regardless of season. Moreover, all three SPPs performed better for the wet season than for the dry season. In terms of *Bias*, TRMM 3B42, CMORPH_CRT, and PERSIANN_CDR products all showed underestimations during the dry season; such an underestimation in SPPs during the dry season has been reported in many studies [7,74,75]. In addition, the monthly results show much higher *CC* values with the observations than do the daily comparisons. This is because errors in the daily values are offset to some extent when added to the monthly values [4].

At the basin scale (Table 4), the daily and monthly *CC* values of the three SPPs improved, and the *MAE*s significantly decreased, respectively. Meanwhile, except for TRMM 3B42, the *Bias* values also showed a corresponding decreasing trend. From Table 4, the relative performances of the SPPs are similar to those at the grid scale. A distinct exception is that the TRMM 3B42 showed the largest overestimation in precipitation. Overall, all three SPPs had comparable and good performances for the monthly precipitation estimations.

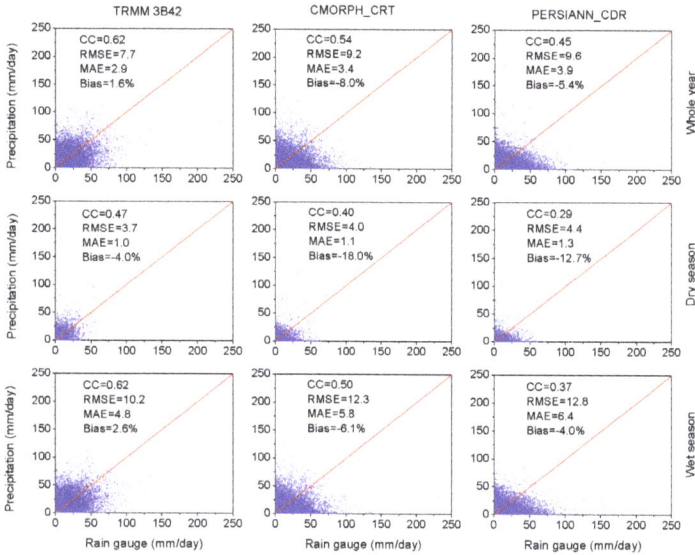

Figure 2. Scatterplots of daily precipitation from TRMM 3B42, CMORPH_CRT, and PERSIANN_CDR against ground observations at the grid scale: the three panels show the results from the whole year (upper panel), dry season (mid panel), and wet season (lower panel). The red line indicates a 1:1 correspondence.

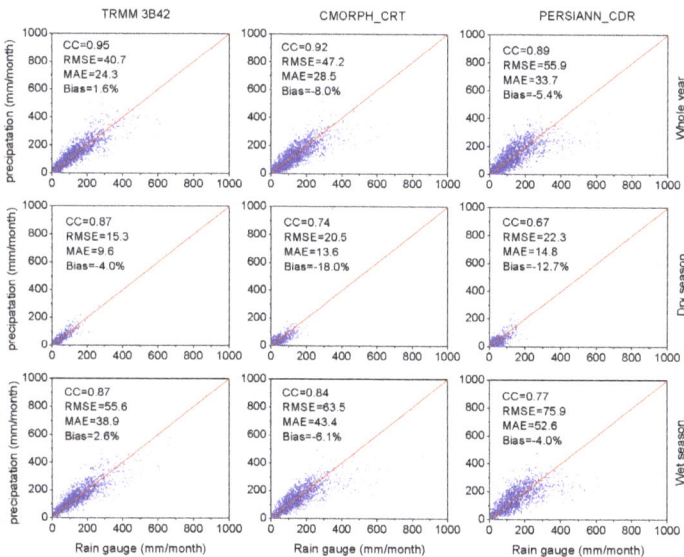

Figure 3. Scatterplots of monthly precipitation from TRMM 3B42, CMORPH_CRT, and PERSIANN_CDR against ground observations at the grid scale: the three panels show the results from the whole year (upper panel), dry season (mid panel), and wet season (lower panel). The red line indicates a 1:1 correspondence.

Table 4. Statistical indices between three satellite-based precipitation products (SPPs) and rain gauge data at the basin scale.

Basin Scale		Daily			Monthly			
		CC	RMSE	MAE	CC	RMSE	MAE	Bias
	TRMM 3B42	0.82	11.3	1.7	0.99	17.1	12.0	9.6%
Whole year	CMORPH_CRT	0.75	23.4	2.3	0.98	14.2	10.2	−0.9%
	PERSIANN_CDR	0.66	34.1	2.9	0.97	21.1	15.1	2.2%
	TRMM 3B42	0.57	2.7	0.7	0.96	6.2	4.6	4.8%
Dry season	CMORPH_CRT	0.49	4.7	0.8	0.90	10.3	7.0	−16.6%
	PERSIANN_CDR	0.42	6.8	1.1	0.83	10.6	8.0	−6.2%
	TRMM 3B42	0.81	19.8	2.7	0.96	23.4	19.3	10.4%
Wet season	CMORPH_CRT	0.66	41.8	3.8	0.97	17.3	13.4	2.0%
	PERSIANN_CDR	0.52	60.9	4.6	0.93	27.9	22.2	3.8%

3.1.2. Evaluation of Contingency

Figure 4 shows box plots of rainfall-detecting skill scores, including *POD, FOH, FAR, CSI,* and *HSS,* which were used to measure the contingency of the three SPPs. Among the three SPPs, TRMM 3B42 performed the best, having the highest *FOH, CSI,* and *HSS* values, and the lowest *FAR* value; PERSIANN_CDR performed the worst in terms of all five rainfall-detecting skill scores. These results illustrate that TRMM 3B42 yields the highest frequency of successful hits when rainfall really occurs, and the lowest erroneous detection rate when there is actually no rainfall. For the *POD,* TRMM 3B42 had a median value of 0.65, while that of CMORPH_CRT was 0.72; this implies that CMORPH_CRT is more likely to detect a larger fraction of precipitation events among all of the actual precipitation events. Note that the *POD, FOH, FAR, CSI,* and *HSS* of CMORPH_CRT tended to have slightly larger variations than they did for the other two products, indicating the considerable data instability of CMORPH_CRT. This instability can be partially attributed to the morphing processes in the CMORPH_RAW data, which determines the precipitation values as a weighted mean of PMW estimates from multiple sensors [74].

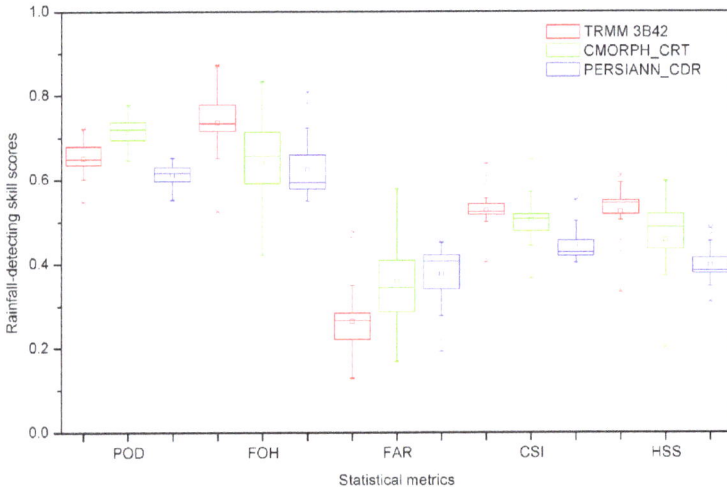

Figure 4. Box plots of rainfall-detecting skill scores for 25 rain gauges. The upper and lower edges of the large box mark the upper and lower quartiles (75% and 25%, respectively), the small box and the solid line within mark the mean and median value, and the upper and lower horizontal lines out of the large box mark the maximum and minimum, respectively.

3.1.3. Evaluation of Rainfall Intensity Distribution

Figure 5 displays the occurrence frequency distribution of daily precipitation for different rain intensity classes and their relative contributions to the amount of total rainfall. Non-rainy days have the largest occurrence frequency, accounting for 58–67% of the total days. TRMM 3B42 identified significantly more non-rainy days (67%) than did the gauged data (63%); this is because the TRMM data have less skill in correctly specifying moderate and light rain rates on short time intervals [48,50,76] identified less non-rainy days (58%), but more light rain (0–1 mm) days. PERSIANN_CDR identified a similar number of no-rain events (64%) as the observation; however, it deviated the most for the light rain event. In addition, all three SPPs overestimated the moderate rain intensity class (1–25 mm) and underestimated heavy rainfall (25–50 mm) and hard rainfall events (>50 mm), with the exception of the TRMM 3B42 product, which slightly overestimated the heavy rainfall event.

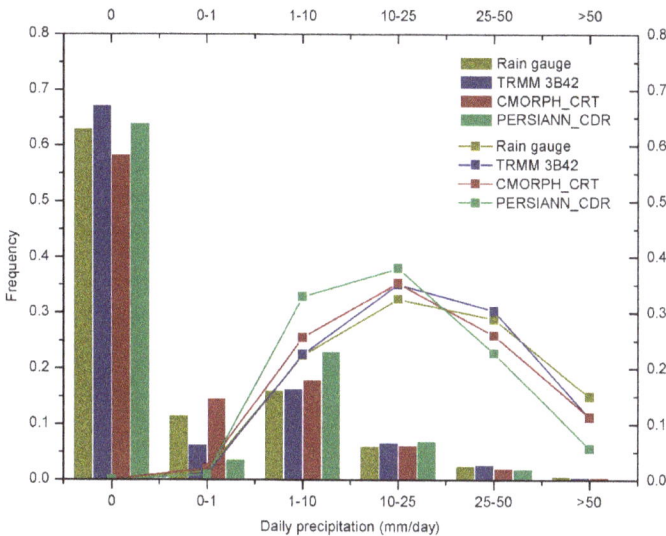

Figure 5. Occurrence frequency distribution (bars) of daily precipitation for different rain intensity classes and their relative contributions (lines) to total rainfall amount during 1998–2010.

For the distribution of relative contribution, PERSIANN_CDR showed the largest discrepancy, while the other two SPPs were very similar to the gauged data. It was found that although the frequency of light rainfall event was 4–15% of the total days, its contribution to the total rainfall amount was only 1.3% on average. On the contrary, although high rainfall occurred on only a small percentage of the total days, they had a significant contribution to the total rainfall amount. For example, the hard rainfall event had a frequency of just about 0.5%, but its contribution to the total rainfall amount was 10.7% on average. For the rain intensity of 10–25 mm, there was a similarly small percentage (6.5%) of total days for all four datasets, but this class contributed the most (35.2%) to the total rainfall amount. The different volume contribution performances greatly impacted the hydrologic simulations, since most of the hydrological processes within the models are sensitive to the total precipitation amount and rainfall intensity distribution [18,77].

3.2. Hydrologic Validation Using GR Models

3.2.1. Daily and Monthly Streamflow Simulations under Scenario I

Table 5 shows the model performance results simulated using different precipitation inputs with gauge-calibrated parameters under Scenario I. Generally, the streamflow simulated by rain gauge data agreed well with the observed hydrographs on both daily (*NSE* = 0.82, *CC* = 0.92, *Bias* = 0.6%) and monthly scales (*NSE* = 0.87, *CC* = 0.96, *Bias* = 1.8%). The GR4J and GR2M models were found to be robust and provided a sound basis for testing the applicability of the three SPPs.

Table 5. Comparison of daily and monthly observed and simulated streamflow under Scenario I.

Precipitation Product	Daily			Monthly		
	NSE	*CC*	*Bias* (%)	*NSE*	*CC*	*Bias* (%)
Rain gauge	0.82	0.92	0.6	0.87	0.96	1.8
TRMM 3B42	0.62	0.93	24.2	0.72	0.97	24.2
CMORPH_CRT	0.73	0.92	−7.5	0.82	0.96	−0.9
PERSIANN_CDR	0.53	0.88	−2.9	0.76	0.94	5.5

On a daily scale (Table 5 and Figure 6), CMORPH_CRT showed the best performance for streamflow simulations over the entire period, with relatively desirable *NSE* (0.73) and *CC* (0.92), and a high but acceptable *Bias* of −7.5%; PERSIANN_CDR had the lowest *NSE* (0.53) and *CC* (0.88), and a *Bias* of −2.9%. The TRMM 3B42 had better *NSE* (0.62) and *CC* (0.93) than did PERSIANN_CDR; however, a significant overestimation (24.2%) was found for the TRMM forced simulation. It is possible that the *Bias* (9.6%) of the daily TRMM 3B42 precipitation was magnified by the hydrological model, causing this large overestimation [17]. Moreover, all the three SPP-based streamflow simulations underestimated some high peak flows (e.g., 2001 and 2007) of the rainy season. This phenomenon can mainly be attributed to the precipitation estimate uncertainty of SPPs at the daily scale during heavy and hard rainfall events [72].

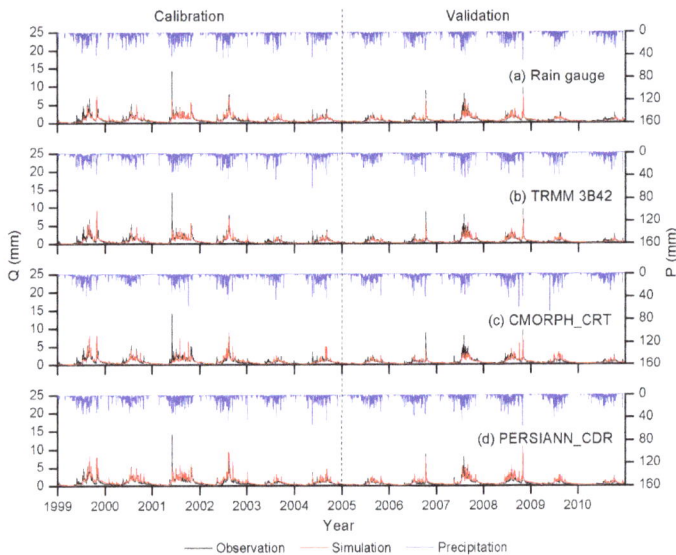

Figure 6. Simulated (red) and observed (black) daily streamflow under Scenario I from: (**a**) rain gauge data; (**b**) TRMM 3B42; (**c**) CMORPH_CRT; and (**d**) PERSIANN_CDR. Blue lines show precipitation data.

Figure 7 illustrates flow duration curves (FDCs) for the four simulations on the daily scale from 1999 to 2010 under Scenario I. The rain gauge, CMORPH_CRT, and PERSIANN_CDR simulations were all consistent with the observations. However, the FDC produced by the TRMM 3B42 simulation was apparently higher than that of the observations, which is consistent with the large *Bias* (24.2%; Table 5).

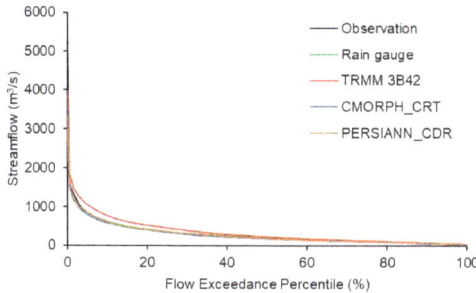

Figure 7. Daily discharge flow duration curves (FDCs) for observations, rain gauge simulation, TRMM 3B42 simulation, CMORPH_CRT simulation, and PERSIANN_CDR simulation under Scenario I.

The monthly streamflow simulations were generally found to be more accurate than the daily simulations in terms of *NSE* and *CC* values (Table 5 and Figure 8). Similar to the daily results, the CMORPH_CRT forced simulation had the best performance, with the highest *NSE* (0.82) and the lowest *Bias* (−0.9%). PERSIANN_CDR performed the second best (*NSE* = 0.76, *Bias* = 5.5%), while TRMM 3B42 performed the worst (*NSE* = 0.72, *Bias* = 24.2%) on the monthly scale. All the three SPPs' forced simulations had good agreements with the observed data, with *CC* values above 0.94. As shown in Figure 8, the monthly streamflow hydrographs from the three SPPs reasonably well matched the observed streamflow, predicting peak flow and low flow conditions perfectly. These results indicated that all three SPPs are suitable for monthly streamflow simulation purposes in the study area. In general, considering the results for the daily and monthly scales, the CMORPH_CRT product is most suitable for streamflow simulations using gauge-calibrated parameters over the URRB.

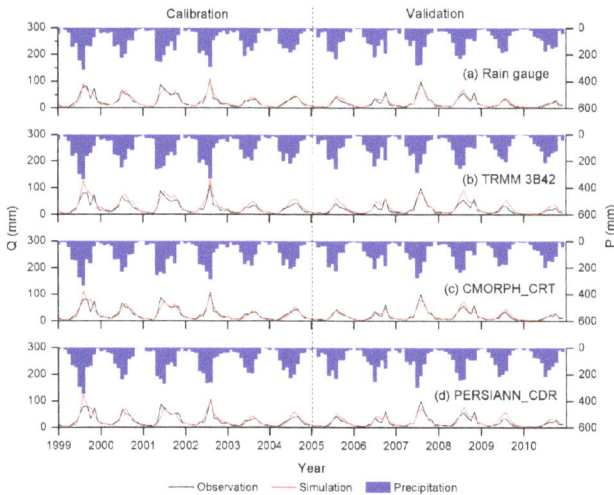

Figure 8. Simulated (red) and observed (black) monthly streamflow under Scenario I from: (**a**) rain gauge data; (**b**) TRMM 3B42; (**c**) CMORPH_CRT); and (**d**) PERSIANN_CDR. Blue bars show precipitation data.

3.2.2. Daily and Monthly Streamflow Simulations under Scenario II

Table 6 shows the model performance results for the three SPPs when using their own optimal parameters calibrated under Scenario II. The simulation results under Scenario II were clearly improved when compared with those under Scenario I, which is consistent with former studies [5,10,18,78]. For the daily scale (Table 6), the *NSE* values of streamflow simulated by the TRMM 3B42, CMORPH_CRT, and PERSIANN_CDR products increased to 0.76, 0.77, 0.63, and the *CC* kept its high values of 0.92, 0.92, and 0.88, respectively. Furthermore, the *Bias* values were also significantly reduced, except for the PERSIANN_CDR forced simulation. Even so, the underestimation problem for peak flow (e.g., 2001 and 2007) remained after the recalibration (Figure 9). Figure 10 shows the FDCs for the three SPP-driven simulations on a daily scale from 1999 to 2010 under Scenario II. The three individual simulations all agreed well with the FDC produced by observations.

Table 6. Comparison of daily and monthly observed and simulated streamflow under Scenario II.

Precipitation Products	Daily			Monthly		
	NSE	*CC*	*Bias* (%)	*NSE*	*CC*	*Bias* (%)
TRMM 3B42	0.76	0.92	−0.8	0.86	0.97	0.8
CMORPH_CRT	0.77	0.92	3.1	0.83	0.96	3.6
PERSIANN_CDR	0.63	0.88	3.4	0.79	0.94	5.5

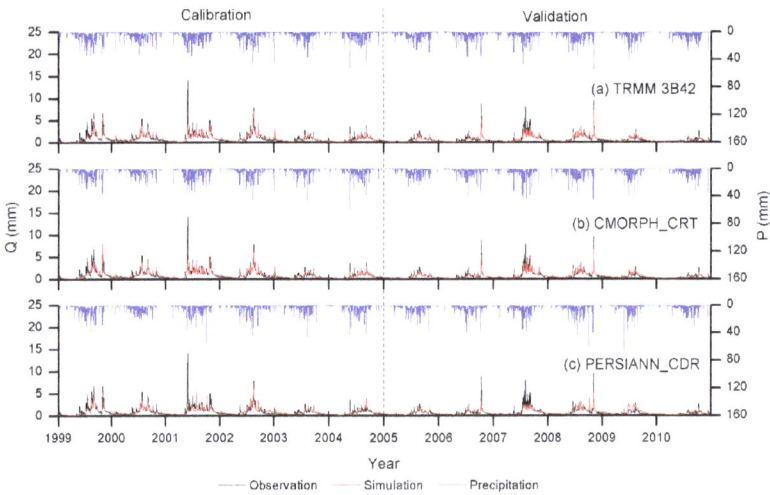

Figure 9. Simulated (red) and observed (black) daily streamflow under Scenario II from: (**a**) TRMM 3B42; (**b**) CMORPH_CRT; and (**c**) PERSIANN_CDR. Blue lines show precipitation data.

For the monthly scale (Table 6 and Figure 11), the *NSE* values for the three SPP-forced simulations also increased (0.86 for TRMM 3B42, 0.83 for CMORPH_CRT, and 0.79 for PERSIANN_CDR). Furthermore, the high *CC* values (>0.94) also indicated the good agreements between the three simulations and the observed data. However, only the TRMM 3B42 forced simulation saw a significant reduction in terms of *Bias* (from 24.2% to 0.8%). Additionally, it was found that the TRMM 3B42 forced simulation had the greatest improvements in *NSE* and *Bias* values from Scenario I to Scenario II for both the daily and monthly scales. Overall, the three SPP forced daily and monthly simulations all exhibited good performance, matching well with observations in this study area. The better agreement was achieved using TRMM 3B42 and CMORPH_CRT products, making them more suitable for performing hydrologic simulations with inadequate surface precipitation observations (e.g., in ungauged catchments).

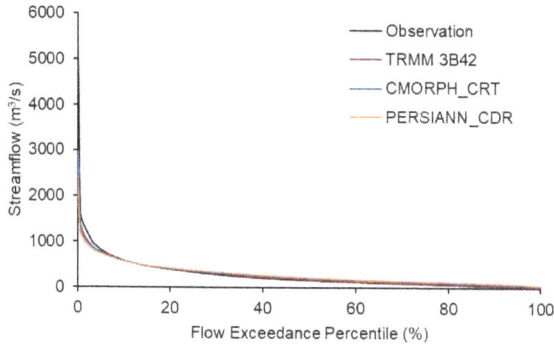

Figure 10. Daily discharge flow duration curves (FDCs) for observations, TRMM 3B42 simulation, CMORPH_CRT simulation, and PERSIANN_CDR simulation under Scenario II.

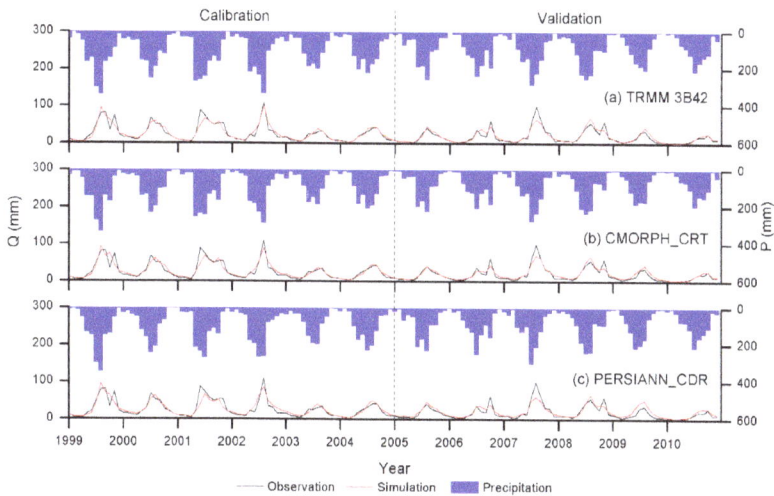

Figure 11. Simulated (red) and observed (black) monthly streamflow under Scenario II from: (**a**) TRMM 3B42; (**b**) CMORPH_CRT; and (**c**) PERSIANN_CDR. Blue bars show precipitation data.

3.2.3. Capability of Simulating Extreme Events

To further investigate the capability of the three SPPs to simulate extreme events, we defined the observed daily streamflow, exceeding its 90% quantile as high flow, and that less than its 50% quantile as low flow [79]. The observed high-flow and low-flow data were compared with the corresponding simulated flow with various precipitation inputs. In this section, the three simulations under Scenario II, plus the rain gauge simulation under Scenario I, were evaluated. The model performance results of high-flow and low-flow simulations were presented in Table 7. For the high-flow simulations, the gauged-based simulation generally had the best performance with desirable *NSE* (0.50), *CC* (0.67), and *Bias* (−11.6%) values. TRMM 3B42 and CMORPH_CRT products exhibited comparable performance of similar *NSE* (0.31 and 0.36), *CC* (0.65 and 0.68), and *Bias* values (−19.1% and −18.1%), whereas PERSIANN_CDR performed poorly with the *NSE* <0 and the largest underestimation (28.0%). For the low-flow simulations, the model performances of the four precipitation inputs were unacceptable, with the *NSE* values all below 0, indicating their poor value for low-flow simulation in this region. Additionally, there was an underestimation of the four precipitation inputs for the high-flow simulations, and an overestimation for the low-flow simulations, which is consistent with many other studies [5,10,78]. This is largely due to the underestimation or overestimation of different precipitation products during extreme rainfall events. As discussed, all three SPPs have poor modeling capabilities for extreme hydrologic events over the URRB.

Table 7. Comparison of daily high-flow and low-flow simulations using four precipitation inputs against observations.

Precipitation Product	High flow			Low flow		
	NSE	*CC*	*Bias* (%)	*NSE*	*CC*	*Bias* (%)
Rain gauge	0.50	0.67	−11.6	−0.06	0.79	15.9
TRMM 3B42	0.31	0.65	−19.1	−0.17	0.79	18.9
CMORPH_CRT	0.36	0.68	−18.1	−0.91	0.77	35.3
PERSIANN_CDR	−0.06	0.52	−28.0	−2.15	0.69	49.4

4. Discussion

SPPs have difficulty representing precipitation in mountainous regions, where precipitation is controlled by the orography and characterized by high spatiotemporal variability [8,80]. PMW data measured from low-orbit satellites and IR data from geostationary satellites are the main data sources for SPPs [17,81]. Specifically, PMW provides direct and accurate precipitation estimates at the cost of coarse temporal resolution, while IR provides useful information mainly about storm clouds based on the low temperatures of the top of these clouds: these data has fine temporal resolutions, but less accuracy [10,60]. In the mountainous regions, IR retrievals generally fail to capture light precipitation events and underestimate orographic rains, whereas PMW retrievals face challenges detecting orographic precipitation, especially in the cold season [82]. Therefore, all SPPs suffer from systematical overestimations or underestimations over the URRB, which is explained by the negative and positive values of *Bias* (Figure 2, Figure 3, and Table 4). In addition, SPPs perform better in the wet season than in the dry season.

According to our results of accuracy analyses, PERSIANN_CDR performs unsatisfactorily compared with the other two products (Figures 2 and 3), which is probably because PERSIANN_CDR is mainly based on IR data, while TRMM 3B42 and CMORPH_CRT combine PMW and IR data. The main difference between TRMM 3B42 and CMORPH_CRT is the gauge adjustment algorithm that is adopted. To develop CMORPH, PDF matching against the CPC unified daily gauge analysis was used to adjust the biases. Monthly GPCC rain gauge analyses and inverse-error variance weighting were used for the TRMM to adjust the biases [60]. However, the accuracy of TRMM generally outperforms that of CMORPH (Figures 2 and 3), which is probably due to the monthly gauge adjustment algorithm that is used in TRMM being superior to the PDF matching adopted in CMOPRH over this region. A similar finding was obtained in the previous studies over the Huaihe River basin in China [18]. At the basin scale, as expected, the *CC* and *MAE* values for SPPs improved, and the relative performances of the SPPs are similar to those at the grid scale (Table 4). Some distinct exceptions are that the TRMM 3B42 showed the largest overestimation in precipitation, while the PERSIANN_CDR showed underestimation in precipitation. This result indicated that the performance of these SPPs are not uniform with an increasing spatial scale, as a consequence of topographical variations [83]. Mei et al. [84] also stated that the size of catchments influences the satellite rainfall errors. Another possible reason is that there may not be enough rain gauges in the area to provide accurate estimates of rainfall for comparisons with satellite estimates at the basin scale.

Two parameters scenarios were designed to evaluate the effect of precipitation uncertainty on streamflow simulations. Under Scenario I (Table 5), CMORPH_CRT showed the best performance among the SPPs, which is mostly due to the relatively low *Bias* in precipitation input. In contrast, the TRMM 3B42-driven simulations overestimate the streamflow by 24.2% (Table 5), which is attributed to the large overestimation in the TRMM 3B42 precipitation input. Another possible explanation is that CMORPH_CRT used daily rain gauge analysis, and all of the bias correction algorithms were conducted at the daily scale directly, while the other two products conducted bias correction with rain gauge analysis at the monthly scale. Therefore, CMORPH_CRT performs better in streamflow simulation using the gauge-calibrated parameters. Previous studies have reported that the errors of SPPs are propagated into hydrological simulations [10,17,79]. An overestimation/underestimation of precipitation estimates can be transformed into a larger overestimation/underestimation in the simulated streamflow. In this study, the basin-scale *Bias* of TRMM 3B42 (9.6%), CMORPH_CRT (−0.9%), and PERSIANN_CDR (2.2%) resulted in corresponding *Bias* values for streamflow simulations of 24.2%, −7.5%, and −2.9% at the daily scale, and 24.2%, −0.9%, and 5.5% at the monthly scale, respectively. This comparison between the input and output *Bias* values in GR models indicates that there is a non-linear error propagation pattern [16].

Under Scenario II, a recalibrated hydrological model using the SPPs can greatly improve streamflow simulation performance, as the different parameter settings can compensate for errors in the satellite rainfall data (Table 6). However, a few studies have indicated that recalibration can sometimes cause parameter values to exceed their reasonable ranges [16,79]. The calibrated model parameters for rain gauge are all within the 80% confidence interval (Tables 2 and 3). However, some model parameters (i.e., x1 and x4) for SPP-forced calibration greatly exceed the 80% confidence interval, which may be attributed to the GR models being sensitive to the precipitation data and the size of catchments. However, for a specific catchment, since the underlying surface condition remains unchanged, the hydrological model parameters largely depend on the input data. If the forcing data change, the sensitive parameters will change accordingly in order to match the streamflow (Table 3). In spite of the influence of cancellation between parameter differences and precipitation bias on streamflow simulation, the three SPPs are able to produce a reasonably good streamflow under scenario II. For instance, TRMM 3B42-driven results presented a satisfactory model efficiency (*NSE* = 0.86) and smaller *Bias* (0.8% relative to the observations) than that of the rain gauge data at the monthly scale. However, the uncertainty in satellite-based precipitation data, together with parameter uncertainty and the structural uncertainty of hydrologic models, will result in uncertainty in streamflow predictions [85]. Therefore, a better understanding of parameter uncertainties and a comparison of different hydrological models will be the focus of future research.

Many studies have explored the applicability of different SPPs using hydrological models over basins of different scales. When the studied SPPs included TRMM 3B42, CMORPH_CRT, and PERSIANN_CDR, similar results were obtained as this study: the streamflow simulations of TRMM 3B42 and CMORPH_CRT performed better than that of PERSIANN_CDR. For example, Su et al. [27] found that TRMM 3B42 and CMORPH_CRT products showed acceptable performance in four SPPs, while PERSIANN_CDR showed little potential for streamflow simulations over the upper Yellow River Basin in China. For the Xixian Basin (upstream of the Huai River Basin), streamflow simulations using the Xinanjiang model found that TRMM 3B42 forced simulation fitted best with the observed streamflow series among three post-real-time research products; this was followed by the CMORPH_CRT-based simulation, and then the PERSIANN_CDR-based simulation [72]. Alazzy et al. [46] also drew similar conclusions according to the results of hydrologic simulation, testing four SPPs, including the three used in our study. According to Moriasi et al. [86], the models can be considered satisfactory if the *NSE* >0.5 and the absolute *Bias* <25%. In this study, the daily *NSE* and absolute *Bias* of PERSIANN_CDR were 0.53 and 2.9% under Scenario I, and 0.63 and 3.4% under Scenario II, indicating that the PERSIANN_CDR product also has the capability to produce acceptable streamflow simulation results by using the GR hydrological model in the URRB.

Inevitably, a few limitations were present in this study. First, compared to the dense rain gauge networks of previous studies [10,18], the rain gauge stations in our study region are relatively sparse and unevenly distributed, which may cause uncertainties in the rain gauge comparison and streamflow simulations. Moreover, it is difficult to ensure that the observed streamflow data for the Manhao station of the URRB is not affected by human activities and regional economic development, although the results of the hydrologic simulation for the study basin were reliable. For instance, the human consumption of water can lead to much lower observed discharge than actual natural discharge, particularly during droughts.

5. Conclusions

This study provides a comprehensive assessment of the three latest SPPs (TRMM 3B42 V7, CMORPH_CRT, and PERSIANN_CDR) based on rain gauge observations over the URRB for the period 1998–2010. The primary conclusions can be summarized as follows.

(1) On the grid scale, TRMM 3B42 performs the best, while PERSIANN_CDR performs the worst. Moreover, monthly SPP data have a much better correlation with gauge rainfall data than daily SPP data. Similar results are obtained at the basin scale, but with a high *Bias* for TRMM 3B42 (9.6%) and a much-improved *Bias* for CMORPH_CRT (−0.9%) and PERSIANN_CDR (2.2%).

(2) For the detection capability of precipitation events, TRMM 3B42 performs the best, while PERSIANN_CDR exhibits the worst performance. By comparison, CMORPH_CRT shows relatively better capability, but with larger fluctuation among different rain gauge stations.

(3) To different degrees, all three SPPs overestimate or underestimate no-rain (0 mm) and light rainfall (0–1 mm) events. Additionally, there is an overestimation of moderate rainfall events (1–25 mm) and an underestimation of heavy and hard rainfall events (>25 mm), indicating their poor ability to reflect extreme precipitation. For the distribution of relative contribution, the PERSIANN_CDR product deviates the most from gauge data.

(4) During hydrologic validation under Scenario I (gauge-calibrated parameters), the CMORPH_CRT product had the best consistency with observed streamflow series at both daily ($NSE = 0.73$) and monthly scales ($NSE = 0.82$), while TRMM 3B42 showed obvious overestimation (24.2%) for both daily and monthly streamflow simulations. Under Scenario II (individual-calibrated parameters), the performance of the recalibrated models significantly improved ($NSE >0.63$ for daily, $NSE >0.79$ for monthly); TRMM 3B42 and CMORPH_CRT performed better than PERSIANN_CDR. All three SPP-forced simulations showed underestimation for high-flow (18.1%–28%) and overestimation for low-flow (18.9–49.4%).

These findings clearly show the great potential for TRMM 3B42 and CMORPH_CRT products in hydrological applications over poorly gauged and inaccessible transboundary river basins in Southwest China, particularly for monthly time intervals, which are suitable for water resource management. However, all three SPPs underestimate and overestimate the occurrence frequency of daily precipitation for some rain intensity classes. Therefore, the local calibration of satellite-derived rainfall estimates and the merging of satellite estimates with rain gauge observations could be employed to alleviate these problems [87,88]. Future work will focus on the validation of higher-resolution SPPs (i.e., GPM), error corrections, spatial downscaling techniques, and their application in distributed hydrological modeling [9,89].

Supplementary Materials: The following are available online at http://www.mdpi.com/2072-4292/10/12/1881/s1, Table S1: precipitation statistics of the upper catchment of the Red River Basin during 1998–2010, Table S2: discharge statistics of the upper catchment of the Red River Basin during 1998–2010.

Author Contributions: Y.L. conceived and designed the study. Y.Z. performed the calculations and assessments. X.J., X.L., and X.L. participated in data processing. Y.Z. wrote the manuscript. All authors contributed to editing and reviewing the manuscript.

Funding: This work was funded by the Applied Basic Research Programs of Yunnan Province (Grant Number 2017FB071) and the National Key Research and Development Program of China (Grant Number 2016YFA0601601).

Acknowledgments: The authors would like to acknowledge the TRMM, CMORPH, and PERSIANN research communities for making the data available to international users. Also, the authors acknowledge the GR model developers for making the suite of lumped GR hydrological models available in an R package. The authors would also like to extend their acknowledgements to the editors and anonymous reviewers for their valuable comments and advice.

Conflicts of Interest: The authors declare no conflict of interest.

References

1. Hu, Q.; Yang, D.; Wang, Y.; Yang, H. Accuracy and spatio-temporal variation of high resolution satellite rainfall estimate over the Ganjiang River Basin. *Sci. China Technol. Sci.* **2013**, *56*, 853–865. [CrossRef]
2. Bai, P.; Liu, X. Evaluation of Five Satellite-Based Precipitation Products in Two Gauge-Scarce Basins on the Tibetan Plateau. *Remote Sens.* **2018**, *10*, 1316. [CrossRef]
3. Hou, A.Y.; Kakar, R.K.; Neeck, S.; Azarbarzin, A.A.; Kummerow, C.D.; Kojima, M.; Oki, R.; Nakamura, K.; Iguchi, T. The Global Precipitation Measurement Mission. *Bull. Am. Meteorol. Soc.* **2014**, *95*, 701–722. [CrossRef]
4. Ren, P.; Li, J.; Feng, P.; Guo, Y.; Ma, Q. Evaluation of Multiple Satellite Precipitation Products and Their Use in Hydrological Modelling over the Luanhe River Basin, China. *Water* **2018**, *10*, 677. [CrossRef]
5. Wang, Z.; Zhong, R.; Lai, C. Evaluation and hydrologic validation of TMPA satellite precipitation product downstream of the Pearl River Basin, China. *Hydrol. Process.* **2017**, *31*, 4169–4182. [CrossRef]
6. Boegh, E.; Thorsen, M.; Butts, M.B.; Hansen, S.; Christiansen, J.S.; Abrahamsen, P.; Hasager, C.B.; Jensen, N.O.; Keur, P.V.D.; Refsgaard, J.C. Incorporating remote sensing data in physically based distributed agro-hydrological modelling. *J. Hydrol.* **2004**, *287*, 279–299. [CrossRef]
7. Wang, W.; Lu, H.; Zhao, T.; Jiang, L.; Shi, J. Evaluation and Comparison of Daily Rainfall From Latest GPM and TRMM Products over the Mekong River Basin. *IEEE J. Sel. Top. Appli. Earth Obs. Remote Sens.* **2017**, *PP*, 1–10. [CrossRef]
8. Sun, Q.H.; Miao, C.Y.; Duan, Q.Y.; Ashouri, H.; Sorooshian, S.; Hsu, K.L. A Review of Global Precipitation Data Sets: Data Sources, Estimation, and Intercomparisons. *Rev. Geophys.* **2018**, *56*, 79–107. [CrossRef]
9. Wang, Z.; Zhong, R.; Lai, C.; Chen, J. Evaluation of the GPM IMERG satellite-based precipitation products and the hydrological utility. *Atmos. Res.* **2017**, *196*, 151–163. [CrossRef]
10. Wu, Z.; Xu, Z.; Wang, F.; He, H.; Zhou, J.; Wu, X.; Liu, Z. Hydrologic Evaluation of Multi-Source Satellite Precipitation Products for the Upper Huaihe River Basin, China. *Remote Sens.* **2018**, *10*, 840. [CrossRef]
11. Huffman, G.J.; Adler, R.F.; Arkin, P.; Chang, A.; Ferraro, R.; Gruber, A.; Janowiak, J.; Mcnab, A.; Rudolf, B.; Schneider, U. The Global Precipitation Climatology Project (GPCP) Combined Precipitation Dataset. *Bull. Am. Meteorol. Soc.* **1997**, *78*, 5–20. [CrossRef]
12. Ashouri, H.; Hsu, K.-L.; Sorooshian, S.; Braithwaite, D.K.; Knapp, K.R.; Cecil, L.D.; Nelson, B.R.; Prat, O.P. PERSIANN-CDR: Daily Precipitation Climate Data Record from Multisatellite Observations for Hydrological and Climate Studies. *Bull. Am. Meteorol. Soc.* **2015**, *96*, 69–83. [CrossRef]
13. Joyce, R.J.; Janowiak, J.E.; Arkin, P.A.; Xie, P. CMORPH: A Method That Produces Global Precipitation Estimates From Passive Microwave and Infrared Data at High Spatial and Temporal Resolution. *J. Hydrometeorol.* **2004**, *5*, 287–296. [CrossRef]
14. Huffman, G.J.; Adler, R.F.; Bolvin, D.T.; Gu, G.; Nelkin, E.J.; Bowman, K.P.; Hong, Y.; Stocker, E.F.; Wolff, D.B. The TRMM Multisatellite Precipitation Analysis (TMPA): Quasi-Global, Multiyear, Combined-Sensor Precipitation Estimates at Fine Scales. *J. Hydrometeorol.* **2007**, *8*, 38–55. [CrossRef]
15. Beck, H.E.; van Dijk, A.I.J. M.; Levizzani, V.; Schellekens, J.; Miralles, D.G.; Martens, B.; de Roo, A. MSWEP: 3-hourly 0.25° global gridded precipitation (1979–2015) by merging gauge, satellite, and reanalysis data. *Hydrol. Earth Syst. Sci.* **2017**, *21*, 589–615. [CrossRef]
16. Yong, B.; Ren, L.-L.; Hong, Y.; Wang, J.-H.; Gourley, J.J.; Jiang, S.-H.; Chen, X.; Wang, W. Hydrologic evaluation of Multisatellite Precipitation Analysis standard precipitation products in basins beyond its inclined latitude band: A case study in Laohahe basin, China. *Water Resour. Res.* **2010**, *46*. [CrossRef]
17. Tong, K.; Su, F.; Yang, D.; Hao, Z. Evaluation of satellite precipitation retrievals and their potential utilities in hydrologic modeling over the Tibetan Plateau. *J. Hydrol.* **2014**, *519*, 423–437. [CrossRef]
18. Sun, R.; Yuan, H.; Liu, X.; Jiang, X. Evaluation of the latest satellite–gauge precipitation products and their hydrologic applications over the Huaihe River basin. *J. Hydrol.* **2016**, *536*, 302–319. [CrossRef]

19. Gao, Z.; Long, D.; Tang, G.; Zeng, C.; Huang, J.; Hong, Y. Assessing the potential of satellite-based precipitation estimates for flood frequency analysis in ungauged or poorly gauged tributaries of China's Yangtze River basin. *J. Hydrol.* **2017**, *550*, 478–496. [CrossRef]

20. Bayissa, Y.; Tadesse, T.; Demisse, G.; Shiferaw, A. Evaluation of Satellite-Based Rainfall Estimates and Application to Monitor Meteorological Drought for the Upper Blue Nile Basin, Ethiopia. *Remote Sens.* **2017**, *9*, 669. [CrossRef]

21. Toté, C.; Patricio, D.; Boogaard, H.; van der Wijngaart, R.; Tarnavsky, E.; Funk, C. Evaluation of Satellite Rainfall Estimates for Drought and Flood Monitoring in Mozambique. *Remote Sens.* **2015**, *7*, 1758–1776. [CrossRef]

22. Yang, N.; Zhang, K.; Hong, Y.; Zhao, Q.; Huang, Q.; Xu, Y.; Xue, X.; Chen, S. Evaluation of the TRMM multisatellite precipitation analysis and its applicability in supporting reservoir operation and water resources management in Hanjiang basin, China. *J. Hydrol.* **2017**, *549*, 313–325. [CrossRef]

23. Awange, J.L.; Gebremichael, M.; Forootan, E.; Wakbulcho, G.; Anyah, R.; Ferreira, V.G.; Alemayehu, T. Characterization of Ethiopian mega hydrogeological regimes using GRACE, TRMM and GLDAS datasets. *Adv. Water Resour.* **2014**, *74*, 64–78. [CrossRef]

24. Dinku, T.; Ruiz, F.; Connor, S.J.; Ceccato, P. Validation and Intercomparison of Satellite Rainfall Estimates over Colombia. *J. Appl. Meteorol. Climatol.* **2010**, *49*, 1004–1014. [CrossRef]

25. Gebremichael, M.; Bitew, M.M.; Hirpa, F.A.; Tesfay, G.N. Accuracy of satellite rainfall estimates in the Blue Nile Basin: Lowland plain versus highland mountain. *Water Resour. Res.* **2014**, *50*, 8775–8790. [CrossRef]

26. Sorooshian, S.; AghaKouchak, A.; Arkin, P.; Eylander, J.; Foufoula-Georgiou, E.; Harmon, R.; Hendrickx, J.M.H.; Imam, B.; Kuligowski, R.; Skahill, B.; et al. Advanced Concepts on Remote Sensing of Precipitation at Multiple Scales. *Bull. Am. Meteorol. Soc.* **2011**, *92*, 1353–1357. [CrossRef]

27. Su, J.; Lü, H.; Wang, J.; Sadeghi, A.; Zhu, Y. Evaluating the Applicability of Four Latest Satellite–Gauge Combined Precipitation Estimates for Extreme Precipitation and Streamflow Predictions over the Upper Yellow River Basins in China. *Remote Sens.* **2017**, *9*, 1176. [CrossRef]

28. Behrangi, A.; Khakbaz, B.; Jaw, T.C.; Aghakouchak, A.; Hsu, K.; Sorooshian, S. Hydrologic evaluation of satellite precipitation products over a mid-size basin. *J. Hydrol.* **2011**, *397*, 225–237. [CrossRef]

29. Miao, C.; Ashouri, H.; Hsu, K.-L.; Sorooshian, S.; Duan, Q. Evaluation of the PERSIANN-CDR Daily Rainfall Estimates in Capturing the Behavior of Extreme Precipitation Events over China. *J. Hydrometeorol.* **2015**, *16*, 1387–1396. [CrossRef]

30. Beck, H.E.; Vergopolan, N.; Pan, M.; Levizzani, V.; van Dijk, A.I.J. M.; Weedon, G.P.; Brocca, L.; Pappenberger, F.; Huffman, G.J.; Wood, E.F. Global-scale evaluation of 22 precipitation datasets using gauge observations and hydrological modeling. *Hydrol. Earth Syst. Sci.* **2017**, *21*, 6201–6217. [CrossRef]

31. Poméon, T.; Jackisch, D.; Diekkrüger, B. Evaluating the Performance of Remotely Sensed and Reanalysed Precipitation Data over West Africa using HBV light. *J. Hydrol.* **2017**, *547*, 222–235. [CrossRef]

32. Kumar, D.; Pandey, A.; Sharma, N.; Flugel, W.A. Evaluation of TRMM-Precipitation with Raingauge Observation Using Hydrological Model J2000. *J. Hydrol. Eng.* **2015**, *115*, E5015007.

33. Su, F.; Hong, Y.; Lettenmaier, D.P. Evaluation of TRMM Multisatellite Precipitation Analysis (TMPA) and Its Utility in Hydrologic Prediction in the La Plata Basin. *J. Hydrometeorol.* **2008**, *9*, 622–640. [CrossRef]

34. Mantas, V.M.; Liu, Z.; Caro, C.; Pereira, A.J.S. C. Validation of TRMM multi-satellite precipitation analysis (TMPA) products in the Peruvian Andes. *Atmos. Res.* **2015**, *163*, 132–145. [CrossRef]

35. Simons, G.; Bastiaanssen, W.; Ngô, L.; Hain, C.; Anderson, M.; Senay, G. Integrating Global Satellite-Derived Data Products as a Pre-Analysis for Hydrological Modelling Studies: A Case Study for the Red River Basin. *Remote Sens.* **2016**, *8*, 279. [CrossRef]

36. Wang, W.; Lu, H.; Yang, D.; Sothea, K.; Jiao, Y.; Gao, B.; Peng, X.; Pang, Z. Modelling Hydrologic Processes in the Mekong River Basin Using a Distributed Model Driven by Satellite Precipitation and Rain Gauge Observations. *PLoS ONE* **2016**, *11*, e0152229. [CrossRef] [PubMed]

37. Beck, H.E.; Pan, M.; Roy, T.; Weedon, G.P.; Pappenberger, F.; van Dijk, A.I.J. M.; Huffman, G.J.; Adler, R.F.; Wood, E.F. Daily evaluation of 26 precipitation datasets using Stage-IV gauge-radar data for the CONUS. *Hydrol. Earth Syst. Sci. Discuss.* **2018**, 1–23. [CrossRef]

38. Rozante, J.; Vila, D.; Barboza Chiquetto, J.; Fernandes, A.; Souza Alvim, D. Evaluation of TRMM/GPM Blended Daily Products over Brazil. *Remote Sens.* **2018**, *10*, 882. [CrossRef]

39. Dezfuli, A.K.; Ichoku, C.M.; Huffman, G.J.; Mohr, K.I.; Selker, J.S.; Nick, V.D.G.; Hochreutener, R.; Annor, F.O. Validation of IMERG Precipitation in Africa. *J. Hydrometeorol.* **2017**, *18*. [CrossRef]

40. Sharifi, E.; Steinacker, R.; Saghafian, B. Assessment of GPM-IMERG and Other Precipitation Products against Gauge Data under Different Topographic and Climatic Conditions in Iran: Preliminary Results. *Remote Sens.* **2016**, *8*, 135. [CrossRef]

41. Prakash, S.; Mitra, A.K.; Aghakouchak, A.; Liu, Z.; Norouzi, H.; Pai, D.S. A preliminary assessment of GPM-based multi-satellite precipitation estimates over a monsoon dominated region. *J. Hydrol.* **2016**, *556*, 865–876. [CrossRef]

42. Anjum, M.N.; Ding, Y.; Shangguan, D.; Ahmad, I.; Ijaz, M.W.; Farid, H.U.; Yagoub, Y.E.; Zaman, M.; Adnan, M. Performance evaluation of latest integrated multi-satellite retrievals for Global Precipitation Measurement (IMERG) over the northern highlands of Pakistan. *Atmos. Res.* **2018**, *205*, 134–146. [CrossRef]

43. Tan, M.L.; Santo, H. Comparison of GPM IMERG, TMPA 3B42 and PERSIANN-CDR satellite precipitation products over Malaysia. *Atmos. Res.* **2018**, *202*, 63–76. [CrossRef]

44. Tan, M.; Duan, Z. Assessment of GPM and TRMM Precipitation Products over Singapore. *Remote Sens.* **2017**, *9*, 720. [CrossRef]

45. Tang, G.; Ma, Y.; Long, D.; Zhong, L.; Hong, Y. Evaluation of GPM Day-1 IMERG and TMPA Version-7 legacy products over Mainland China at multiple spatiotemporal scales. *J. Hydrol.* **2016**, *533*, 152–167. [CrossRef]

46. Alazzy, A.A.; Lü, H.; Chen, R.; Ali, A.B.; Zhu, Y.; Su, J. Evaluation of Satellite Precipitation Products and Their Potential Influence on Hydrological Modeling over the Ganzi River Basin of the Tibetan Plateau. *Adv. Meteorol.* **2017**, *2017*, 1–23. [CrossRef]

47. Liu, X.; Yang, T.; Hsu, K.; Liu, C.; Sorooshian, S. Evaluating the streamflow simulation capability of PERSIANN-CDR daily rainfall products in two river basins on the Tibetan Plateau. *Hydrol. Earth Syst. Sci.* **2017**, *21*, 169–181. [CrossRef]

48. Li, D.; Christakos, G.; Ding, X.; Wu, J. Adequacy of TRMM satellite rainfall data in driving the SWAT modeling of Tiaoxi catchment (Taihu lake basin, China). *J. Hydrol.* **2018**, *556*, 1139–1152. [CrossRef]

49. He, Z.; Yang, L.; Tian, F.; Ni, G.; Hou, A.; Lu, H. Intercomparisons of Rainfall Estimates from TRMM and GPM Multisatellite Products over the Upper Mekong River Basin. *J. Hydrometeorol.* **2017**, *18*. [CrossRef]

50. Zhu, H.; Li, Y.; Huang, Y.; Li, Y.; Hou, C.; Shi, X. Evaluation and hydrological application of satellite-based precipitation datasets in driving hydrological models over the Huifa river basin in Northeast China. *Atmos. Res.* **2018**, *207*, 28–41. [CrossRef]

51. Li, Y.; He, D.; Li, X.; Zhang, Y.; Yang, L. Contributions of Climate Variability and Human Activities to Runoff Changes in the Upper Catchment of the Red River Basin, China. *Water* **2016**, *8*, 414. [CrossRef]

52. Le, T.P.Q.; Garnier, J.; Gilles, B.; Sylvain, T.; Minh, C.V. The changing flow regime and sediment load of the Red River, Viet Nam. *J. Hydrol.* **2007**, *334*, 199–214. [CrossRef]

53. Le, H.; Sutton, J.; Bui, D.; Bolten, J.; Lakshmi, V. Comparison and Bias Correction of TMPA Precipitation Products over the Lower Part of Red–Thai Binh River Basin of Vietnam. *Remote Sens.* **2018**, *10*, 1582. [CrossRef]

54. Dang, T.H.; Coynel, A.; Orange, D.; Blanc, G.; Etcheber, H.; Le, L.A. Long-term monitoring (1960-2008) of the river-sediment transport in the Red River Watershed (Vietnam): Temporal variability and dam-reservoir impact. *Sci. Total Environ.* **2010**, *408*, 4654–4664. [CrossRef] [PubMed]

55. Li, Z.; Saito, Y.; Matsumoto, E.; Wang, Y.; Tanabe, S.; Vu, Q.L. Climate change and human impact on the Song Hong (Red River) Delta, Vietnam, during the Holocene. *Quat. Int.* **2006**, *144*, 4–28. [CrossRef]

56. Le, T.P.Q.; Seidler, C.; Kändler, M.; Tran, T.B.N. Proposed methods for potential evapotranspiration calculation of the Red River basin (North Vietnam). *Hydrol. Process.* **2012**, *26*, 2782–2790. [CrossRef]

57. Allen, R.G.; Pereira, L.S.; Raes, D. *Crop Evapotranspiration: Guidelines for Computing Crop Water Requirements*; FAO Irrigation and Drainage Paper No. 56; FAO: Rome, Italy, 1998.

58. Thiessen, A.H. Precipitation averages for large areas. *Mon. Weather Rev.* **1911**, *39*, 1082–1084. [CrossRef]

59. Huffman, G.J.; Bolvin, D.T. TRMM and Other Data Precipitation Data Set Documentation. Available online: https://pmm.nasa.gov/sites/default/files/document_files/3B42_3B43_doc_V7_180426.pdf (accessed on 10 October 2018).

60. Duan, Z.; Liu, J.; Tuo, Y.; Chiogna, G.; Disse, M. Evaluation of eight high spatial resolution gridded precipitation products in Adige Basin (Italy) at multiple temporal and spatial scales. *Sci. Total Environ.* **2016**, *573*, 1536–1553. [CrossRef] [PubMed]

61. Knapp, K.R. Scientific data stewardship of international satellite cloud climatology project B1 global geostationary observations. *J. Appl. Remote Sens.* **2008**, *2*, 023548. [CrossRef]

62. Gautheir, T.D. Detecting Trends Using Spearman's Rank Correlation Coefficient. *Environ. Forensics* **2001**, *2*, 359–362. [CrossRef]

63. Xu, R.; Tian, F.; Yang, L.; Hu, H.; Lu, H.; Hou, A. Ground validation of GPM IMERG and TRMM 3B42V7 rainfall products over southern Tibetan Plateau based on a high-density rain gauge network. *J. Geophys. Res. Atmos.* **2017**, *122*, 910–924. [CrossRef]

64. Nash, J.E.; Sutcliffe, J.V. River flow forecasting through conceptual models part I—A discussion of principles. *J. Hydrol.* **1970**, *10*, 282–290. [CrossRef]

65. Perrin, C.; Michel, C.; Andréassian, V. *A Set of Hydrological Models*; John Wiley & Sons, Inc.: Hoboken, NJ, USA, 2013; pp. 493–509.

66. Coron, L.; Thirel, G.; Delaigue, O.; Perrin, C.; Andréassian, V. The suite of lumped GR hydrological models in an R package. *Environ. Modell. Softw.* **2017**, *94*, 166–171. [CrossRef]

67. Perrin, C.; Michel, C.; Andréassian, V. Does a large number of parameters enhance model performance? Comparative assessment of common catchment model structures on 429 catchments. *J. Hydrol.* **2001**, *242*, 275–301. [CrossRef]

68. Perrin, C.; Michel, C.; Andréassian, V. Improvement of a parsimonious model for streamflow simulation. *J. Hydrol.* **2003**, *279*, 275–289. [CrossRef]

69. Mouelhi, S.; Michel, C.; Perrin, C.; Andréassian, V. Stepwise development of a two-parameter monthly water balance model. *J. Hydrol.* **2006**, *318*, 200–214. [CrossRef]

70. Okkan, U.; Fistikoglu, O. Evaluating climate change effects on runoff by statistical downscaling and hydrological model GR2M. *Theor. Appl. Climatol.* **2013**, *117*, 343–361. [CrossRef]

71. Mouelhi, S.; Madani, K.; Lebdi, F. A Structural Overview through GR(s) Models Characteristics for Better Yearly Runoff Simulation. *Open J. Mod. Hydrol.* **2013**, *03*, 179–187. [CrossRef]

72. Jiang, S.; Liu, S.; Ren, L.; Yong, B.; Zhang, L.; Wang, M.; Lu, Y.; He, Y. Hydrologic Evaluation of Six High Resolution Satellite Precipitation Products in Capturing Extreme Precipitation and Streamflow over a Medium-Sized Basin in China. *Water* **2017**, *10*, 25. [CrossRef]

73. Li, Y.; He, D.; Ye, C. Spatial and temporal variation of runoff of Red River Basin in Yunnan. *J. Geogr. Sci.* **2008**, *18*, 308–318. [CrossRef]

74. Shen, Y.; Xiong, A.; Wang, Y.; Xie, P. Performance of high-resolution satellite precipitation products over China. *J. Geophys. Res. Atmos.* **2010**, *115*. [CrossRef]

75. Peng, B.; Shi, J.; Ni-Meister, W.; Zhao, T.; Ji, D. Evaluation of TRMM Multisatellite Precipitation Analysis (TMPA) Products and Their Potential Hydrological Application at an Arid and Semiarid Basin in China. *IEEE J. Sel. Top. Appl. Earth Obs. Remote Sens.* **2014**, *7*, 3915–3930. [CrossRef]

76. Michaelides, S.; Levizzani, V.; Anagnostou, E.; Bauer, P.; Kasparis, T.; Lane, J.E. Precipitation: Measurement, remote sensing, climatology and modeling. *Atmos. Res.* **2009**, *94*, 512–533. [CrossRef]

77. Li, Z.; Yang, D.; Hong, Y. Multi-scale evaluation of high-resolution multi-sensor blended global precipitation products over the Yangtze River. *J. Hydrol.* **2013**, *500*, 157–169. [CrossRef]

78. Jiang, S.; Ren, L.; Hong, Y.; Yong, B.; Yang, X.; Yuan, F.; Ma, M. Comprehensive evaluation of multi-satellite precipitation products with a dense rain gauge network and optimally merging their simulated hydrological flows using the Bayesian model averaging method. *J. Hydrol.* **2012**, *452–453*, 213–225. [CrossRef]

79. Yuan, F.; Zhang, L.; Win, K.; Ren, L.; Zhao, C.; Zhu, Y.; Jiang, S.; Liu, Y. Assessment of GPM and TRMM Multi-Satellite Precipitation Products in Streamflow Simulations in a Data-Sparse Mountainous Watershed in Myanmar. *Remote Sens.* **2017**, *9*, 302. [CrossRef]

80. Maggioni, V.; Massari, C. On the performance of satellite precipitation products in riverine flood modeling: A review. *J. Hydrol.* **2018**, *558*, 214–224. [CrossRef]

81. Bitew, M.M.; Gebremichael, M. Assessment of satellite rainfall products for streamflow simulation in medium watersheds of the Ethiopian highlands. *Hydrol. Earth Syst. Sci.* **2011**, *15*, 1147–1155. [CrossRef]

82. Derin, Y.; Yilmaz, K.K. Evaluation of Multiple Satellite-Based Precipitation Products over Complex Topography. *J. Hydrometeorol.* **2014**, *15*, 1498–1516. [CrossRef]

83. Gebremicael, T.G.; Mohamed, Y.A.; van der Zaag, P.; Berhe, A.G.; Haile, G.G.; Hagos, E.Y.; Hagos, M.K. Comparison and validation of eight satellite rainfall products over the rugged topography of Tekeze-Atbara Basin at different spatial and temporal scales. *Hydrol. Earth Syst. Sci. Discuss.* **2017**, 1–31. [CrossRef]

84. Mei, Y.; Anagnostou, E.N.; Nikolopoulos, E.I.; Borga, M. Error Analysis of Satellite Precipitation Products in Mountainous Basins. *J. Hydrometeorol.* **2014**, *15*, 1778–1793. [CrossRef]

85. Sun, R.; Yuan, H.; Yang, Y. Using multiple satellite-gauge merged precipitation products ensemble for hydrologic uncertainty analysis over the Huaihe River basin. *J. Hydrol.* **2018**, *566*, 406–420. [CrossRef]

86. Moriasi, D.N.; Arnold, J.G.; Liew, M.W.V.; Bingner, R.L.; Harmel, R.D.; Veith, T.L. Model Evaluation Guidelines for Systematic Quantification of Accuracy in Watershed Simulations. *Trans. Asabe* **2007**, *50*, 885–900. [CrossRef]

87. Cheema, M.J.M.; Bastiaanssen, W.G.M. Local calibration of remotely sensed rainfall from the TRMM satellite for different periods and spatial scales in the Indus Basin. *Int. J. Remote Sens.* **2012**, *33*, 2603–2627. [CrossRef]

88. Hashemi, H.; Nordin, M.; Lakshmi, V.; Huffman, G.J.; Knight, R. Bias correction of long-term satellite monthly precipitation product (TRMM 3B43) over the conterminous United States. *J. Hydrometeorol.* **2017**, *18*. [CrossRef]

89. Zhang, Y.; Li, Y.; Ji, X.; Luo, X.; Li, X. Fine-Resolution Precipitation Mapping in a Mountainous Watershed: Geostatistical Downscaling of TRMM Products Based on Environmental Variables. *Remote Sens.* **2018**, *10*, 119. [CrossRef]

MDPI

St. Alban-Anlage 66

4052 Basel

Switzerland

Tel. +41 61 683 77 34

Fax +41 61 302 89 18

www.mdpi.com

Remote Sensing Editorial Office

E-mail: remotesensing@mdpi.com

www.mdpi.com/journal/remotesensing

www.ingramcontent.com/pod-product-compliance
Lightning Source LLC
Chambersburg PA
CBHW051715210326
41597CB00032B/5490